Methods in Enzymology

Volume 94

POLYAMINES

METHODS IN ENZYMOLOGY

EDITORS-IN-CHIEF

Sidney P. Colowick   Nathan O. Kaplan

Methods in Enzymology

Volume 94

Polyamines

EDITED BY

Herbert Tabor

Celia White Tabor

NATIONAL INSTITUTE OF ARTHRITIS, DIABETES, AND DIGESTIVE AND KIDNEY DISEASES
NATIONAL INSTITUTES OF HEALTH
BETHESDA, MARYLAND

1983

ACADEMIC PRESS

A Subsidiary of Harcourt Brace Jovanovich, Publishers

New York London
Paris San Diego San Francisco São Paulo Sydney Tokyo Toronto

ACADEMIC PRESS, INC.
111 Fifth Avenue, New York, New York 10003

United Kingdom Edition published by
ACADEMIC PRESS, INC. (LONDON) LTD.
24/28 Oval Road, London NW1 7DX

Library of Congress Cataloging in Publication Data
Main entry under title:

Polyamines.

(Methods in enzymology ; v. 94)
Bibliography: p.
Includes indexes.
1. Polyamines. I. Tabor, Herbert, Date.
II. Tabor, Celia White, Date. III. Series.
QP601.M49 vol. 94 574.19'25s 82-24318
[QP801.P638] [574.19'24]
ISBN 0-12-181994-9

PRINTED IN THE UNITED STATES OF AMERICA

83 84 85 86 9 8 7 6 5 4 3 2 1

Table of Contents

Section III. Genetic Techniques

A. Mutants in Biosynthetic Pathways

B. Preparation of Plasmids for Overproduction of Enzymes in Biosynthetic Pathway (*Escherichia coli*)

Section IV. Ornithine Decarboxylase, Arginine Decarboxylase, Lysine Decarboxylase

A. Enzyme Assays and Preparations

B. Enzyme Inhibitors

Section V. Adenosylmethionine Synthetase (Methionine Adenosyltransferase) and Adenosylmethionine Decarboxylase

A. Enzyme Assays and Preparations

B. Enzyme Inhibitors

Section VI. Putrescine Aminopropyltransferase (Spermidine Synthase) and Spermidine Aminopropyltransferase (Spermine Synthase)

A. Enzyme Assays and Preparations

B. Enzyme Inhibitors

Section VII. Amine Oxidases and Dehydrogenases

Section VIII. Spermidine Acetylation and Deacetylation

Section IX. Other Enzymes Involved in Polyamine Synthesis and Metabolism

Section X. Metabolism of 5′-Methylthioadenosine and 5-Methylthioribose

Section XI. Methods for the Study of Polyamines in Lymphocytes and Mammary Gland

Section XII. Analogs and Derivatives

Contributors to Volume 94

Article numbers are in parentheses following the names of contributors.
Affiliations listed are current.

P. R. ADIGA (58, 74), *Department of Biochemistry, Indian Institute of Science, Bangalore 560012, India*

L. ALHONEN-HONGISTO (41), *Department of Biochemistry, University of Helsinki, Unioninkatu 35, SF-00170 Helsinki 17, Finland*

GARY L. ANDERSON (48), *Department of Medicinal Chemistry, University of Iowa, Iowa City, Iowa 52242*

VALERIE J. ATMAR (21), *Department of Chemistry, New Mexico State University, Las Cruces, New Mexico 88003*

CYRUS J. BACCHI (33), *Haskins Laboratories, Pace University, New York, New York 10038*

O. BEFANI (54), *Institute of Applied Biochemistry, University of Rome–La Sapienza, 00185 Rome, Italy*

R. A. BENNETT (10), *Department of Physiology, The Milton S. Hershey Medical Center, Pennsylvania State University, Hershey, Pennsylvania 17033*

PHILIPPE BEY (31), *Centre de Recherche Merrell International, 67084 Strasbourg Cedex, France*

COLIN R. BIRD (60), *Botany School, University of Cambridge, Cambridge CB2 3EA, United Kingdom*

ELIZABETH A. BOEKER (18, 28), *Department of Chemistry and Biochemistry, Utah State University, Logan, Utah 84322*

STEPHEN M. BOYLE (17), *Faculty of Medicine, Memorial University of Newfoundland, St. John's, Newfoundland, Canada A1B 3V6*

GIOVANNA CACCIAPUOTI (8), *Department of Biochemistry, First Medical School, University of Naples, Via Constantinopoli 16, 80138 Naples, Italy*

EVANGELOS S. CANELLAKIS (30), *Department of Pharmacology, Yale University School of Medicine, New Haven, Connecticut 06511*

GIULIO L. CANTONI (64), *Laboratory of General and Comparative Biochemistry, National Institute of Mental Health, Bethesda, Maryland 20205*

MARIA CARTENÌ-FARINA (8, 62), *Department of Biochemistry, First Medical School, University of Naples, Via Constantinopoli 16, 80138 Naples, Italy*

PATRICK CASARA (31), *Centre de Recherche Merrell International, 67084 Strasbourg Cedex, France*

KAN CHANTRAPROMMA (70), *Department of Chemistry, Baker Laboratory, Cornell University, Ithaca, New York 14853*

PETER K. CHIANG (64), *Division of Biochemistry, Walter Reed Army Institute of Research, Washington, D.C. 20307*

SEYMOUR S. COHEN (47), *Department of Pharmacological Sciences, State University of New York at Stony Brook, Stony Brook, New York 11794*

MURRAY S. COHN (14, 38), *Consumer Product Safety Commission, Room 712B Westwood Towers, 5401 Westbard Avenue, Bethesda, Maryland 20207*

HERBERT L. COOPER (80), *National Cancer Institute, National Institutes of Health, Bethesda, Maryland 20205*

JAMES K. COWARD (48), *Department of Chemistry, Rensselaer Polytechnic Institute, Troy, New York 12181*

GARY R. DANIELS (21), *Department of Chemistry, New Mexico State University, Las Cruces, New Mexico 88003*

G. DOYLE DAVES, JR. (7), *Department of Chemistry, Lehigh University, Bethlehem, Pennsylvania 18015*

ROWLAND H. DAVIS (5, 16), *Department of Molecular Biology and Biochemistry,*

University of California, Irvine, Irvine, California 92717

FULVIO DELLA RAGIONE (8, 55), *Department of Biochemistry, First Medical School, University of Naples, Via Constantinopoli 16, 80138 Naples, Italy*

ACHILLES A. DEMETRIOU (67), *Department of Surgery, Albert Einstein College of Medicine, Bronx, New York 10461*

MARIO DE ROSA (62), *Institute for the Chemistry of Molecules of Biological Interest, Consiglio Nazionale delle Ricerche, Arco Felice, Naples, Italy*

TERHO ELORANTA (42, 45, 46), *Department of Biochemistry, University of Kuopio, 70101 Kuopio 10, Finland*

YASUO ENDO (6), *Department of Pharmacology, School of Dentistry, Tohoku University, 4-1 Seiryo-machi, Sendai 980, Japan*

ADOLPH J. FERRO (63), *Department of Microbiology, Oregon State University, Corvallis, Oregon 97331*

MARY LYNN FINK (61), *National Institute of Dental Research, National Institutes of Health, Bethesda, Maryland 20205*

EDMOND H. FISCHER (28), *Department of Biochemistry, University of Washington, Seattle, Washington 98195*

J. E. FOLK (61, 79, 80), *National Institute of Dental Research, National Institutes of Health, Bethesda, Maryland 20205*

JOHN R. FOZARD (34), *Department of Pharmacology, Centre de Recherche Merrell International, 16, rue d'Ankara, 67084 Strasbourg Cedex, France*

KAZUNOBU FUJITA (29), *Department of Nutrition, Jikei University School of Medicine, Tokyo 105, Japan*

PATRIZIA GALLETTI (8, 11), *Department of Biochemistry, First Medical School, University of Naples, Via Constantinopoli 16, 80138 Naples, Italy*

AGATA GAMBACORTA (62), *Institute for the Chemistry of Molecules of Biological Interest, Consiglio Nazionale delle Ricerche, Arco Felice, Naples, Italy*

BRUCE GANEM (70), *Department of Chemistry, Baker Laboratory, Cornell University, Ithaca, New York 14853*

JOHN G. GEORGATSOS (24), *Laboratory of Biochemistry, School of Science, Aristotelian University of Thessaloniki, Thessaloniki, Greece*

ANDRZEJ B. GURANOWSKI (64), *Instytut Biochemii, Akademia Rolnicza, PL-60-637 Poznan, Poland*

EDMUND W. HAFNER (12, 13, 17, 35), *UOP, Inc. Corporate Research Center, Ten UOP Plaza, Algonquin and Mt. Prospect, Des Plaines, Illinois 60016*

WILLIAM L. HANSON (33), *Department of Parasitology, College of Veterinary Medicine, University of Georgia, Athens, Georgia 30602*

SHIN-ICHI HAYASHI (22, 29), *Department of Nutrition, Jikei University School of Medicine, Tokyo 105, Japan*

J. HEANEY-KIERAS (78), *The Rockefeller University, New York, New York 10021*

JOHN S. HELLER (30), *Department of Pharmacology, Yale University School of Medicine, New Haven, Connecticut 06511*

JERALD L. HOFFMAN (36), *Department of Biochemistry, Schools of Dentistry and Medicine, University of Louisville, Louisville, Kentucky 40292*

ERKKI HÖLTTÄ (52), *Department of Biochemistry, University of Helsinki, Unioninkatu 35, SF-00170 Helsinki 17, Finland*

SEYMOUR H. HUTNER (33), *Haskins Laboratories, Pace University, New York, New York 10038*

KAZUTOMO IMAHORI (50, 51), *Tokyo Metropolitan Institute of Gerontology, 35-2 Sakaecho, Itabashi-ku, Tokyo 173, Japan*

MERVYN ISRAEL (69), *Division of Pharmacology, Dana-Farber Cancer Institute, Boston, Massachusetts 02115*

SHOSUKE ITO (81), *Institute for Comprehensive Medical Science, School of Medicine, Fujita-Gakuen University, Toyoake, Aichi 470-11, Japan*

J. JÄNNE (41), *Department of Biochemistry, University of Helsinki, Unioninkatu 35, SF-00170 Helsinki 17, Finland*

TAKAAKI KAMEJI (22), *Department of Nutrition, Jikei University School of Medicine, Tokyo 105, Japan*

K. KÄPYAHO (41), *Department of Biochemistry, University of Helsinki, Unioninkatu 35, SF-00170 Helsinki 17, Finland*

SEIICHI KAWASHIMA (50, 51), *Department of Biochemistry, Tokyo Metropolitan Institute of Gerontology, 35-2 Sakaecho, Itabashi-ku, Tokyo 173, Japan*

A. M. B. KROPINSKI (73), *Department of Microbiology and Immunology, Queen's University, Kingston, Ontario, Canada K7L 3N6*

GLENN D. KUEHN (21), *Department of Chemistry, New Mexico State University, Las Cruces, New Mexico 88003*

Z. KURYLO-BOROWSKA (78), *The Rockefeller University, New York, New York 10021*

DIMITRIOS A. KYRIAKIDIS (24, 30), *Laboratory of Biochemistry, School of Science, Aristotelian University of Thessaloniki, Thessaloniki, Greece*

PAUL R. LIBBY (56, 57), *Department of Experimental Therapeutics, Roswell Park Memorial Institute, Buffalo, New York 14263*

K. L. MALTMAN (73), *Department of Microbiology, University of British Columbia, Vancouver, British Columbia, Canada V6T 1W5*

KEVIN S. MARCHITTO (63), *Department of Microbiology, University of Texas Health Science Center at Dallas, Dallas, Texas 75235*

GEORGE D. MARKHAM (17, 35, 37), *Institute for Cancer Research, Fox Chase Cancer Center, Philadelphia, Pennsylvania 19111*

PETER P. MCCANN (33), *Merrell Research Center, Merrell Dow Pharmaceuticals, Inc., Cincinnati, Ohio 45215*

BRIAN METCALF (31), *Merrell Research Center, Merrell Dow Pharmaceuticals, Inc., Cincinnati, Ohio 45215*

JOHN L. A. MITCHELL (20), *Department of Biological Sciences, Northern Illinois University, DeKalb, Illinois 60115*

EDWARD J. MODEST (69), *Department of Biochemistry, Bowman Gray School of Medicine, Winston-Salem, North Carolina 27103*

B. MONDOVÌ (54), *Institute of Applied Biochemistry, University of Rome–La Sapienza, 00185 Rome, Italy*

DAVID R. MORRIS (18, 65), *Department of Biochemistry, University of Washington, Seattle, Washington 98195*

HENRY C. NATHAN (33), *Haskins Laboratories, Pace University, New York, New York 10038*

J. B. NEILANDS (77), *Department of Biochemistry, University of California, Berkeley, California 94720*

TAKAMI OKA (66), *Laboratory of Biochemistry and Metabolism, National Institute of Arthritis, Diabetes, and Digestive and Kidney Diseases, National Institutes of Health, Bethesda, Maryland 20205*

MASATO OKADA (50, 51), *Fujisawa Research Laboratory, Tokuyama Soda Co., Ltd., 2051 Endo, Fujisawa, Kanagawa 252, Japan*

ADRIANA OLIVA (11), *Department of Biochemistry, First Medical School, University of Naples, Via Constantinopoli 16, 80138 Naples, Italy*

TAIRO OSHIMA (68), *Mitsubishi-Kasei Institute of Life Sciences, Machida, Tokyo 194, Japan*

RAIJA-LEENA PAJULA (42, 46), *Department of Biochemistry, University of Kuopio, 70101 Kuopio 10, Finland*

CHRISTOS A. PANAGIOTIDIS (24), *Laboratory of Biochemistry, School of Science, Aristotelian University of Thessaloniki, Thessaloniki, Greece*

MYUNG HEE PARK (80), *National Institute of Dental Research, National Institutes of Health, Bethesda, Maryland 20205*

MARIE-LOUISE PART (34), *Department of Pharmacology, Centre de Recherche*

Merrell International, 16, rue d'Ankara, 67084 Strasbourg Cedex, France

THOMAS J. PAULUS (5, 16), *Martin Marietta Laboratories, 1450 South Rolling Road, Baltimore, Maryland 21227*

ANTHONY E. PEGG (10, 23, 26, 32, 39, 40, 43, 49, 55), *Department of Physiology, The Milton S. Hershey Medical Center, College of Medicine, Pennsylvania State University, Hershey, Pennsylvania 17033*

JOHN W. PERRY (66), *Laboratory of Biochemistry and Metabolism, National Institute of Arthritis, Diabetes, and Digestive and Kidney Diseases, National Institutes of Health, Bethesda, Maryland 20205*

LO PERSSON (25), *Department of Physiology and Biophysics, University of Lund, Sölvegatan 19, S-223 62 Lund, Sweden*

MARINA PORCELLI (11), *Department of Biochemistry, First Medical School, University of Naples, Via Constantinopoli 16, 80138 Naples, Italy*

HANNU PÖSÖ (32, 39), *Research Laboratories, State Alcohol Monopoly, Alko, Box 350, SF-00101 Helsinki 10, Finland*

AARNE RAINA (42, 45, 46), *Department of Biochemistry, University of Kuopio, 70101 Kuopio 10, Finland*

GIOVANNA ROMEO (62), *Department of Biochemistry, First Medical School, University of Naples, Via Constantinopoli 16, 80138 Naples, Italy*

ELSA ROSENGREN (25), *Department of Physiology and Biophysics, University of Lund, Sölvegatan 19, S-223 62 Lund, Sweden*

S. SABATINI (54), *C.N.R. Center for Molecular Biology, 00185 Rome, Italy*

KEIJIRO SAMEJIMA (45), *Tokyo Biochemical Research Institute, 3-41-8 Takada, Toshima-ku, Tokyo 171, Japan*

I. E. SCHEFFLER (15), *Department of Biology, University of California, San Diego, La Jolla, California 92093*

JAMES E. SEELY (23, 26, 32), *Department of Physiology, The Milton S. Hershey Medical Center, College of Medicine, Pennsylvania State University, Hershey, Pennsylvania 17033*

NIKOLAUS SEILER (1, 2, 3), *Centre de Recherche Merrell International, Filiale de The Dow Chemical Company, 67084 Strasbourg Cedex, France*

P. SEPPÄNEN (41), *Department of Biochemistry, University of Helsinki, Unioninkatu 35, SF-00170 Helsinki 17, Finland*

CHRISTINE E. SEYFRIED (65), *Department of Biochemistry, University of Washington, Seattle, Washington 98195*

RAM K. SINDHU (47), *Department of Pharmacological Sciences, State University of New York at Stony Brook, Stony Brook, New York 11794*

ALBERT SJOERDSMA (33), *Merrell Research Center, Merrell Dow Pharmaceuticals, Inc., Cincinnati, Ohio 45215*

RONALD G. SMITH (7), *Department of Developmental Therapeutics, The University of Texas M.D., Anderson Hospital and Tumor Institute, Houston, Texas 77030*

TERENCE A. SMITH (27, 53, 60), *Long Ashton Research Station, University of Bristol, Bristol BS18 9AF, United Kingdom*

K. S. SRIVENUGOPAL (58, 74), *Department of Biochemistry, University of Washington, Seattle, Washington 98195*

VICTOR STALON (59), *Department of Microbiology, Faculty of Sciences, Université Libre de Bruxelles, Avenue Emile Gryson, B-1070 Brussels, Belgium*

C. S. STEGLICH (15), *Department of Biology, University of California, San Diego, La Jolla, California 92093*

FRANK SUNDLER (25), *Department of Histology, University of Lund, Biskopsgatan 5, S-223 62 Lund, Sweden*

CELIA WHITE TABOR (4, 12, 13, 14, 17, 19, 35, 37, 38, 44, 67, 71, 72, 76), *National Institute of Arthritis, Diabetes, and Di-*

gestive and Kidney Diseases, National Institutes of Health, Bethesda, Maryland 20205

HERBERT TABOR (4, 12, 13, 14, 17, 19, 35, 37, 38, 44, 67, 71, 72, 76), *National Institute of Arthritis, Diabetes, and Digestive and Kidney Diseases, National Institutes of Health, Bethesda, Maryland 20205*

KUO-CHANG TANG (48), *Olin Chemical Group, R/D Center, New Haven, Connecticut 06511*

T. T. TCHEN (75), *Department of Chemistry, Wayne State University, Detroit, Michigan 48202*

JIRO TOBARI (75), *Department of Chemistry, St. Paul's University, Nishi-Ikebukuro, Toshima-ku, Tokyo 171, Japan*

P. TURINI (54), *Institute of Applied Biochemistry, University of Rome–La Sapienza, 00185 Rome, Italy*

ANIL K. TYAGI (19), *National Institute of Arthritis, Diabetes, and Digestive and Kidney Diseases, National Institutes of Health, Bethesda, Maryland 20205*

ROLF UDDMAN (25), *Department of Histology, University of Lund, Biskopsgatan 5, S-223 62 Lund, Sweden*

CORNELIUS A. VALKENBURG (7), *Department of Chemistry, University of Montana, Bozeman, Montana 59715*

JEAN-PAUL VEVERT (31), *Centre de Recherche Merrell International, 67084 Strasbourg Cedex, France*

R. A. J. WARREN (73), *Department of Microbiology, University of British Columbia, Vancouver, British Columbia, Canada V6T 1W5*

BANRI YAMANOHA (45), *Tokyo Biochemical Research Institute, 3-41-8 Takada, Toshima-ku, Tokyo 171, Japan*

PETER H. YU (9), *Psychiatric Research Division, Cancer and Medical Research Building, University of Saskatchewan, Saskatoon, Saskatchewan, Canada S7N 0W0*

IAN S. ZAGON (26), *Department of Anatomy, The Milton S. Hershey Medical Center, College of Medicine, Pennsylvania State University, Hershey, Pennsylvania 17033*

VINCENZO ZAPPIA (8, 11, 62), *Department of Biochemistry, First Medical School, University of Naples, Via Constantinopoli 16, 80138 Naples, Italy*

Preface

We are dedicating this volume to Dr. Sanford M. Rosenthal, who recognized the importance of polyamines at a very early date. In 1950, as chief of our laboratory, he initiated a program for the study of these compounds. Dr. Rosenthal was particularly skillful in combining the techniques of biochemistry, pharmacology, and physiology, and the program that he developed utilized all of these approaches. Before that time only very few laboratories had published in this field, and Dr. Rosenthal's familiarity with these earlier studies plus his prescient insight into the importance of the area led to the decision to study these compounds.

Since the publication in 1971 of Volumes XVII, Part A and XVII, Part B (Metabolism of Amino Acids and Amines), there has been a very large increase in the number of publications concerned with polyamines. This widespread interest indicated the need for a new volume covering the most recent advances in methodology. Unfortunately, we have not been able to include all of the published methods, but we have tried to include those procedures that have been most useful and are most frequently cited by others in this field.

HERBERT TABOR
CELIA WHITE TABOR

METHODS IN ENZYMOLOGY

EDITED BY

Sidney P. Colowick and Nathan O. Kaplan

VANDERBILT UNIVERSITY
SCHOOL OF MEDICINE
NASHVILLE, TENNESSEE

DEPARTMENT OF CHEMISTRY
UNIVERSITY OF CALIFORNIA
AT SAN DIEGO
LA JOLLA, CALIFORNIA

I. Preparation and Assay of Enzymes
II. Preparation and Assay of Enzymes
III. Preparation and Assay of Substrates
IV. Special Techniques for the Enzymologist
V. Preparation and Assay of Enzymes
VI. Preparation and Assay of Enzymes (*Continued*)
Preparation and Assay of Substrates
Special Techniques
VII. Cumulative Subject Index

METHODS IN ENZYMOLOGY

EDITORS-IN-CHIEF

Sidney P. Colowick Nathan O. Kaplan

VOLUME VIII. Complex Carbohydrates
Edited by ELIZABETH F. NEUFELD AND VICTOR GINSBURG

VOLUME IX. Carbohydrate Metabolism
Edited by WILLIS A. WOOD

VOLUME X. Oxidation and Phosphorylation
Edited by RONALD W. ESTABROOK AND MAYNARD E. PULLMAN

VOLUME XI. Enzyme Structure
Edited by C. H. W. HIRS

VOLUME XII. Nucleic Acids (Parts A and B)
Edited by LAWRENCE GROSSMAN AND KIVIE MOLDAVE

VOLUME XIII. Citric Acid Cycle
Edited by J. M. LOWENSTEIN

VOLUME XIV. Lipids
Edited by J. M. LOWENSTEIN

VOLUME XV. Steroids and Terpenoids
Edited by RAYMOND B. CLAYTON

VOLUME XVI. Fast Reactions
Edited by KENNETH KUSTIN

VOLUME XVII. Metabolism of Amino Acids and Amines (Parts A and B)
Edited by HERBERT TABOR AND CELIA WHITE TABOR

VOLUME XVIII. Vitamins and Coenzymes (Parts A, B, and C)
Edited by DONALD B. McCORMICK AND LEMUEL D. WRIGHT

VOLUME XXXIII. Cumulative Subject Index Volumes I–XXX
Edited by MARTHA G. DENNIS AND EDWARD A. DENNIS

VOLUME XXXIV. Affinity Techniques (Enzyme Purification: Part B)
Edited by WILLIAM B. JAKOBY AND MEIR WILCHEK

VOLUME XXXV. Lipids (Part B)
Edited by JOHN M. LOWENSTEIN

VOLUME XXXVI. Hormone Action (Part A: Steroid Hormones)
Edited by BERT W. O'MALLEY AND JOEL G. HARDMAN

VOLUME XXXVII. Hormone Action (Part B: Peptide Hormones)
Edited by BERT W. O'MALLEY AND JOEL G. HARDMAN

VOLUME XXXVIII. Hormone Action (Part C: Cyclic Nucleotides)
Edited by JOEL G. HARDMAN AND BERT W. O'MALLEY

VOLUME XXXIX. Hormone Action (Part D: Isolated Cells, Tissues, and Organ Systems)
Edited by JOEL G. HARDMAN AND BERT W. O'MALLEY

VOLUME XL. Hormone Action (Part E: Nuclear Structure and Function)
Edited by BERT W. O'MALLEY AND JOEL G. HARDMAN

VOLUME XLI. Carbohydrate Metabolism (Part B)
Edited by W. A. WOOD

VOLUME XLII. Carbohydrate Metabolism (Part C)
Edited by W. A. WOOD

VOLUME XLIII. Antibiotics
Edited by JOHN H. HASH

VOLUME XLIV. Immobilized Enzymes
Edited by KLAUS MOSBACH

VOLUME XLV. Proteolytic Enzymes (Part B)
Edited by LASZLO LORAND

VOLUME XLVI. Affinity Labeling
Edited by WILLIAM B. JAKOBY AND MEIR WILCHEK

VOLUME XLVII. Enzyme Structure (Part E)
Edited by C. H. W. HIRS AND SERGE N. TIMASHEFF

VOLUME XLVIII. Enzyme Structure (Part F)
Edited by C. H. W. HIRS AND SERGE N. TIMASHEFF

VOLUME XLIX. Enzyme Structure (Part G)
Edited by C. H. W. HIRS AND SERGE N. TIMASHEFF

VOLUME L. Complex Carbohydrates (Part C)
Edited by VICTOR GINSBURG

VOLUME LI. Purine and Pyrimidine Nucleotide Metabolism
Edited by PATRICIA A. HOFFEE AND MARY ELLEN JONES

VOLUME LII. Biomembranes (Part C: Biological Oxidations)
Edited by SIDNEY FLEISCHER AND LESTER PACKER

VOLUME LIII. Biomembranes (Part D: Biological Oxidations)
Edited by SIDNEY FLEISCHER AND LESTER PACKER

VOLUME LIV. Biomembranes (Part E: Biological Oxidations)
Edited by SIDNEY FLEISCHER AND LESTER PACKER

VOLUME LV. Biomembranes (Part F: Bioenergetics)
Edited by SIDNEY FLEISCHER AND LESTER PACKER

VOLUME LVI. Biomembranes (Part G: Bioenergetics)
Edited by SIDNEY FLEISCHER AND LESTER PACKER

VOLUME LVII. Bioluminescence and Chemiluminescence
Edited by MARLENE A. DELUCA

VOLUME LVIII. Cell Culture
Edited by WILLIAM B. JAKOBY AND IRA H. PASTAN

VOLUME LIX. Nucleic Acids and Protein Synthesis (Part G)
Edited by KIVIE MOLDAVE AND LAWRENCE GROSSMAN

VOLUME LX. Nucleic Acids and Protein Synthesis (Part H)
Edited by KIVIE MOLDAVE AND LAWRENCE GROSSMAN

VOLUME 61. Enzyme Structure (Part H)
Edited by C. H. W. HIRS AND SERGE N. TIMASHEFF

VOLUME 62. Vitamins and Coenzymes (Part D)
Edited by DONALD B. MCCORMICK AND LEMUEL D. WRIGHT

VOLUME 63. Enzyme Kinetics and Mechanism (Part A: Initial Rate and Inhibitor Methods)
Edited by DANIEL L. PURICH

VOLUME 64. Enzyme Kinetics and Mechanism (Part B: Isotopic Probes and Complex Enzyme Systems)
Edited by DANIEL L. PURICH

VOLUME 65. Nucleic Acids (Part I)
Edited by LAWRENCE GROSSMAN AND KIVIE MOLDAVE

VOLUME 66. Vitamins and Coenzymes (Part E)
Edited by DONALD B. MCCORMICK AND LEMUEL D. WRIGHT

VOLUME 67. Vitamins and Coenzymes (Part F)
Edited by DONALD B. MCCORMICK AND LEMUEL D. WRIGHT

VOLUME 68. Recombinant DNA
Edited by RAY WU

VOLUME 69. Photosynthesis and Nitrogen Fixation (Part C)
Edited by ANTHONY SAN PIETRO

VOLUME 70. Immunochemical Techniques (Part A)
Edited by HELEN VAN VUNAKIS AND JOHN J. LANGONE

VOLUME 71. Lipids (Part C)
Edited by JOHN M. LOWENSTEIN

VOLUME 72. Lipids (Part D)
Edited by JOHN M. LOWENSTEIN

VOLUME 73. Immunochemical Techniques (Part B)
Edited by JOHN J. LANGONE AND HELEN VAN VUNAKIS

VOLUME 86. Prostaglandins and Arachidonate Metabolites
Edited by WILLIAM E. M. LANDS AND WILLIAM L. SMITH

VOLUME 87. Enzyme Kinetics and Mechanism (Part C: Intermediates, Stereochemistry, and Rate Studies)
Edited by DANIEL L. PURICH

VOLUME 88. Biomembranes (Part I: Visual Pigments and Purple Membranes, II)
Edited by LESTER PACKER

VOLUME 89. Carbohydrate Metabolism (Part D)
Edited by WILLIS A. WOOD

VOLUME 90. Carbohydrate Metabolism (Part E)
Edited by WILLIS A. WOOD

VOLUME 91. Enzyme Structure (Part I)
Edited by C. H. W. HIRS AND SERGE N. TIMASHEFF

VOLUME 92. Immunochemical Techniques (Part E: Monoclonal Antibodies and General Immunoassay Methods)
Edited by JOHN J. LANGONE AND HELEN VAN VUNAKIS

VOLUME 93. Immunochemical Techniques (Part F: Conventional Antibodies, Fc Receptors, and Cytotoxicity)
Edited by JOHN J. LANGONE AND HELEN VAN VUNAKIS

VOLUME 94. Polyamines
Edited by HERBERT TABOR AND CELIA WHITE TABOR

VOLUME 95. Cumulative Subject Index Volumes 61–74, 76–80 (in preparation)
Edited by MARTHA G. DENNIS AND EDWARD A. DENNIS

VOLUME 96. Biomembranes [Part J: Membrane Biogenesis: Assembly and Targeting (General Methods, Eukaryotes)] (in preparation)
Edited by SIDNEY FLEISCHER AND BECCA FLEISCHER

VOLUME 97. Biomembranes [Part K: Membrane Biogenesis: Assembly and Targeting (Prokaryotes, Mitochondria, and Chloroplasts)] (in preparation)
Edited by SIDNEY FLEISCHER AND BECCA FLEISCHER

Methods in Enzymology

Volume 94

POLYAMINES

Section I

Analytical Methods for Amines

[1] Thin-Layer Chromatography and Thin-Layer Electrophoresis of Polyamines and Their Derivatives

By NIKOLAUS SEILER

Until a decade ago an impressive amount of work had been performed in polyamine biochemistry that was based on surface-chromatographic separations. In 1973 Bachrach[1] quoted 30 different solvents that had been suggested for paper chromatography, 11 solvents for thin-layer chromatography, and 12 different buffers for paper electrophoretic separations of the nonderivatized polyamines.

Most of the suggested methods were used for specific purposes and never gained widespread attention, but certain versions of paper electrophoresis[2,3] found wide application in the quantitative assay of urinary polyamines until recently, when it was recognized that spermine concentrations, as measured after paper electrophoretic separation, were higher than those found with other procedures owing to an interfering compound.

Little progress has been made since 1973 with regard to polyamine separations by surface chromatography. Not even the introduction of high-performance thin-layer silica gel plates by several companies has had an effect in this regard, because more sensitive and automated methods have been developed (see this volume [2,3]).

The main difficulty in the application to polyamine assay of paper- and thin-layer chromatography is the high polarity of the compounds, which makes their isolation from salt-containing aqueous solutions difficult, and amines and amino acids move together, so that extensive separations are required that are not easily obtained by unidimensional development of thin-layer chromatograms.

Thin-layer electrophoresis is slightly more advantageous in this respect. If the pH for the separations is appropriate, neutral amino acids stay near the origin, and only basic amino acids and peptides move toward the cathode together with the amines.

Thin-Layer Chromatography

The most complete study of thin-layer chromatographic determination of polyamines and their acetyl derivatives was reported by Hammond and

[1] U. Bachrach, "Function of Naturally Occurring Polyamines." Academic Press, New York, 1973.

[2] A. Raina, *Acta Physiol. Scand., Suppl.* **218,** 1 (1963).

[3] D. H. Russell, C. C. Levy, S. C. Schimpf, and I. A. Hawk, *Cancer Res.* **31,** 1555 (1971).

ISBN 0-12-181994-9

Herbst.[4] Based on experience with paper chromatography, they chose two solvents and studied the chromatographic behavior of polyamines and basic amino acids, i.e., of those compounds that are eluted together in the same fraction of an ion-exchange chromatographic cleanup procedure.[4]

Using Whatman CC-41 cellulose thin-layer plates and diethylene glycol monoethyl ether–propionic acid–water (14 : 3 : 3) saturated with NaCl, the following R_f values were obtained: spermine, 0.13; spermidine, 0.24; histidine, 0.29; ornithine, 0.34; histamine, 0.35; 1,3-diaminopropane, 0.36; putrescine, 0.40; arginine, 0.42; lysine, 0.43; cadaverine, 0.47; N^1-acetylspermidine, 0.57; N^8-acetylspermidine, 0.58; ethanolamine, 0.65; monoacetylspermine, 0.72; diacetylspermidine, 0.81; N^1,N^{12}-diacetylspermine, 0.83.

Regardless of the fact that it takes 5 hr to develop the chromatogram, the limitations of the method are obvious: it is not possible to determine putrescine in the presence of lysine and arginine. Unfortunately it is difficult to eliminate these basic amino acids completely by a cleanup procedure. Even extraction of the polyamines with butanol[2,5] may leave traces in the polyamine fraction, and thus prohibit the determination of low concentrations of putrescine.

The system is suitable, however, for separating polyamines from acetylpolyamines, and it may therefore be considered in metabolic studies with labeled precursors if the study does not necessitate the determination of both N^1- and N^8-acetylspermidine at the same time.

A better separation of these two biologically important spermidine conjugates[6,7] has been achieved by developing silica gel 60 G-24 plates (Brinkmann Instruments, Westbury, New York) using chloroform–methanol–ammonium hydroxide (2 : 2 : 1),[8] but no data exist for this system for other polyamine derivatives.

There is little doubt that the application of high-performance thin-layer plates can improve polyamine separations and, especially, reduce the separation time. Moreover, staining with ninhydrin[4] can be substituted by a convenient and more sensitive staining method with fluorescamine (see below). This method has the additional advantage that the fluorescent spots can be extracted with organic solvents and the extracts used for

[4] J. E. Hammond and E. Herbst, *Anal. Biochem.* **22,** 474 (1968).

[5] S. M. Rosenthal and C. W. Tabor, *J. Pharmacol. Exp. Ther.* **116,** 131 (1956).

[6] N. Seiler, J. Koch-Weser, B. Knödgen, W. Richards, C. Tardif, F. N. Bolkenius, P. J. Schechter, G. Tell, P. S. Mamont, J. R. Fozard, U. Bachrach, and E. Grosshans, *Adv. Polyamine Res.* **3,** 197 (1981).

[7] F. N. Bolkenius and N. Seiler, *Int. J. Biochem.* **13,** 287 (1981).

[8] J. Blankenship and T. Walle, *Adv. Polyamine Res.* **2,** 97 (1978).

determination of radioactivity. Nevertheless the limitations for free polyamine separations are severe, as is clear even from these short considerations, so that applications will be restricted to specific analytical problems.

Thin-Layer Electrophoresis

The Separation System

Using the buffer of the first reported paper electrophoretic procedure for the determination of di- and polyamines,[9] rapid separations of the polyamines can be achieved on silica gel and cellulose thin-layer plates, usually with very regular spots.[10] The method was therefore used by us to separate polyamines and their metabolites, mostly for the demonstration of the formation of radiolabeled polyamines from precursors.[11,12]

Thin-layer electrophoresis is carried out with a cooled plate apparatus (Camag, Muttenz, Switzerland). For the determination of the polyamines and related compounds (i.e., for compounds that are positively charged at pH 4.8), the samples are applied on the plates as streaks or small spots (diameter 3 mm) at a distance of 3 cm from one edge. Then the plates are evenly sprayed with pyridine–acetate buffer pH 4.80[9] (100 ml of pyridine, 75 ml of glacial acetic acid, and 30 g of citric acid monohydrate in 2375 ml of deionized water), and portions of 20 ml of the same buffer are poured into the troughs of the apparatus before each run. Plates and troughs are connected by buffer-soaked filter-paper strips.

Separations are usually carried out at 600 V (1–1.5 V/cm). If the apparatus is cooled to 0°, a current of 90–100 mA is usually observed with 20 × 20 cm plates of 250 mm layer thickness.

Figure 1 shows separations of polyamines, acetylpolyamines, and basic amino acids on (A) silica gel 1500 (Schleicher & Schuell, Dassell, Federal Republic of Germany) and (B) silica gel 60 (Merck, Darmstadt, Federal Republic of Germany) thin-layer plates.

Under these electrophoretic conditions, neutral amino acids remain near the origin and acidic amino acids move toward the anode. They do not interfere with polyamine separations.

The relative mobilities of the various compounds are similar on silica gel 1500, silica gel G (plaster of Paris as binder), and cellulose layers. However, the comparison of Fig. 1A with 1B shows that the mobility

[9] F. G. Fischer and H. Bohn, *Hoppe-Seyler's Z. Physiol. Chem.* **308,** 108 (1957).
[10] N. Seiler and B. Knödgen, *J. Chromatogr.* **131,** 109 (1977).
[11] H. A. Fischer, H. Korr, N. Seiler, and G. Werner, *Brain Res.* **39,** 197 (1972).
[12] N. Seiler and M. J. Al-Therib, *Biochem. J.* **144,** 29 (1974).

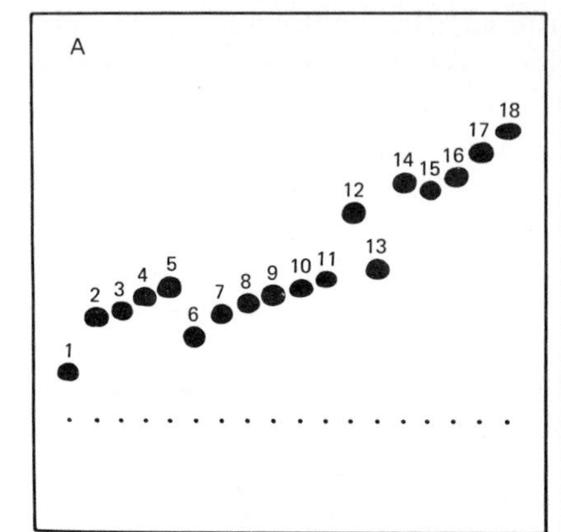

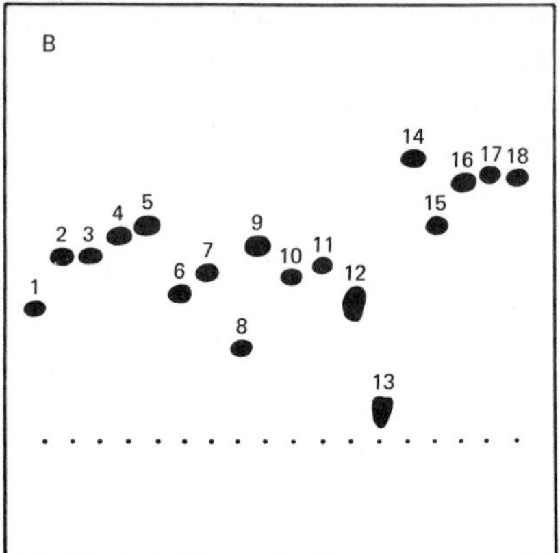

FIG. 1. Thin-layer electrophoretograms of polyamines, acetylpolyamines, and basic amino acids. Electrophoretic conditions: cooled plate apparatus (Camag, Muttenz, Switzerland), 600 V, 50 min; 0°; pyridine–acetic acid buffer, pH 4.8.[9] (A) Silica gel 1500 plate (Schleicher & Schuell, Dassell, Federal Republic of Germany). (B) Silica gel 60 plate (Merck, Darmstadt, Federal Republic of Germany) (both 20 × 20 cm). 1 = 4-aminobutyric acid (GABA); 2 = arginine; 3 = histidine; 4 = lysine; 5 = ornithine; 6 = homocarnosine; 7 = putreanine; 8 = N^1-acetylspermine; 9 = monoacetylputrescine; 10 = N^8-acetylspermidine; 11 = N^1-acetylspermidine; 12 = spermidine; 13 = spermine; 14 = ethanolamine; 15 = histamine; 16 = putrescine; 17 = cadaverine; 18 = 1,3-diaminopropane. The figures were copied from original electrophoretograms. The distance from plate edge to origin is 40 mm; the spots (approximately 60 nmol) were visualized by reaction with ninhydrin.

pattern is totally different on Merck plates. The irregular shapes and low mobilities of the spermidine and spermine spots indicate a strong interaction with the thin-layer matrix, which obviously increases with the number of positive charges and the length of the aliphatic chain. Spermine and monoacetylspermine interact especially strongly. Since the same observations were made using high-performance thin-layer plates of the same company,[10] it is assumed that the polyamines interact with the polyvinylalcohol binder in these plates.

4-Aminobutyrate, the basic amino acids, and the monoamines (monoacetylputrescine, ethanolamine) move faster on the Merck than on the Schleicher & Schuell plates. Neutral amino acids also tend to move somewhat toward the cathode on the Merck plates. This might indicate a lower effective pH than is achieved on the Schleicher & Schuell silica gel plates (or cellulose layers) under the electrophoretic conditions.

Practical experience with electrophoretic separations of polyamines on thin layers is presently restricted to a few types of silica gel and cellulose layers. Other types of commercial plates may interact with the polyamines to a different degree. Since the interactions with the layer matrix are well reproducible, it is possible to use them for achieving certain separations. For instance, the separation of putrescine and ethanolamine is more complete on Merck plates than on the other thin layers, and spermine interferes with N^1-acetylspermidine only on Schleicher & Schuell plates. 1,3-Diaminopropane and cadaverine, on the other hand, are favorably separated on thin-layer plates of this type (Fig. 1).

It is self-evident that prolongation of the separation time will improve the separation of the slowly moving compounds, whereby putrescine and the other diamines may move into the paper strip and be lost. It is possible to decrease the separation time if only fast-moving components of a mixture (putrescine, spermidine, spermine) are to be separated.

The above separation system was chosen because of the regular spots obtainable and its applicability to polyamine assay and because of the determination of certain amino acids (4-aminobutyrate, glutamic acid, aspartic acid, glutathione).[13] However, other buffers (with different pH and composition) may also turn out to be useful for the thin-layer electrophoresis of polyamines. It may at least be worthwhile to test those buffer solutions that have previously been used with advantage in paper electrophoretic separations.[1]

Staining with Ninhydrin and Fluorescamine

Staining of thin-layer chromatograms with ninhydrin, extraction of the colored spots, and colorimetric estimation have previously been used successfully.[4] The same method is applicable to thin-layer electrophoretograms.

The plates are dried in an oven for 30 min at 90°. They are then sprayed uniformly with ninhydrin reagent (1% ninhydrin and 1% 2,4,6-trimethylpyridine in absolute ethanol) and heated for 15 min at 60°. The colored spots are scraped off, and silica gel is collected in small centrifuge tubes. (A method for scraping out spots with a simple homemade device is described by Seiler[14].) The colored reaction product is eluted with 1 ml of 70% ethanol by mixing. After centrifugation, absorbance of the supernatants is measured at 575 nm. With this method, 5–50 nmol of the amines can be measured (SD ± 1% for the staining method and colorimetry).[4] Direct scanning of the plates is also feasible but has not been used for polyamine determinations.

More recently, spraying with fluorescamine or an *o*-phthalaldehyde reagent has been suggested for the detection and determination of amino acids and peptides, drugs, etc., on thin-layer chromatograms. These tech-

[13] S. Sarhan, N. Seiler, J. Grove, and G. Bink, *J. Chromatogr.* **162,** 561 (1979).
[14] N. Seiler, *Methods Biochem. Anal.* **18,** 259 (1970).

niques were compared with the ninhydrin method,[15] but were not found to be advantageous, owing to high background fluorescence and low yields in fluorescent derivatives. However, commercial thin-layer plates can be submitted to a dipping technique with fluorescamine that gives low homogeneous background fluorescence and reproducible yields of fluorescent derivatives. This method was originally designed for the estimation of certain amino acids after thin-layer electrophoretic separations,[13] but was also found to be suitable for the determination of polyamines and their derivatives on electrophoretograms.[7]

The procedure is as follows: The dry thin-layer plates are heated to 50° before staining. Then they are dipped rapidly into a tank containing an alkaline solution (to ensure optimal reaction condition). For the preparation of this solution, 100 ml of saturated NaOH in methanol are mixed with 100 ml of 1-butanol, and the mixture is gradually diluted with 600 ml of toluene. The plate is removed from the tank after 30 sec and dried in a horizontal position for a few minutes at room temperature and then completely for 10 min at 110°.

The staining solution contains 10 mg of fluorescamine in 100 ml of a mixture of acetone–1-butanol (1 : 1). A stainless steel or glass tank (22 × 20 × 0.7 cm) containing approximately 200 ml of the fluorescamine solution is advantageous for dipping. The plates are cooled to room temperature, and then they are dipped twice into the fluorescamine solution with drying periods of 1–2 min. About 15 ml of the fluorescamine solution are needed per 20 × 20 cm plate. The solution should be prepared daily.

The fluorescamine derivatives are formed gradually. Before quantitative evaluation the plates are therefore stored for at least 2 hr at room temperature. If they are protected from dust and light, quantitative evaluation is, however, possible even 48 hr after staining.

Direct scanning of fluorescence is the most advantageous technique for quantitative evaluation. Fluorescence is activated at 320 nm, and total emitted light is measured (using a 420 nm cutoff filter in the path of the emitted light). It is also possible to extract the fluorescamine derivatives with methanol and to measure fluorescence intensity conventionally. Methanol extracts can also be used for the determination of radioactivity. In contrast to the colored reaction products with ninhydrin, the fluorescamine derivatives do not decrease counting efficiency by fluorescence quenching.

The relationship using *in situ* scanning of fluorescence is linear between recorded curve areas and the amount of compound in the range

[15] E. Schilz, K. D. Schnackerz, and R. W. Gracy, *Anal. Biochem.* **79,** 33 (1977).

from 200 to 2000 pmol per spot. In this range the reproducibility of the procedure (including tissue extraction and sample application) is ±8–10%.

Conclusion

The methods described in this chapter cannot compete in sensitivity, specificity, and precision with other existing methods of polyamine assay[16,17] (see also this volume [2,4]). However, they are simple and rapid in the sense that they allow one to perform many separations in parallel and require only relatively unsophisticated, versatile equipment.

Surface chromatographic separations of nonderivatized polyamines should be especially useful for the identification of radiolabeled precursors of the polyamines and of the polyamines themselves. Although this possibility has been utilized in the past,[11,12,18] it has not been fully exploited in polyamine metabolic studies.

Both methods, thin-layer chromatography and thin-layer electrophoresis, separate not only the amines and their conjugates, but in the same systems certain amino acids and metabolic intermediates. Thus, it is possible to resolve, for instance, tissue or cell extracts to a considerable extent, especially if the two methods are used in two directions on the same thin-layer plate. Autoradiographs of the plates then give a good picture of the metabolic reactions that took place.

The specificity of some enzymatic assays can be considerably increased if these simple but powerful separation methods are employed. Examples are the demonstration of N^8-acetylspermidine formation by nuclear acetyl-CoA:spermidine acetyltransferase[8] and the degradation of N^1-acetylspermidine to putrescine, in contrast with the deacetylation of N^8-acetylspermidine by liver homogenates.[7,18]

It is therefore concluded that these more recent versions of the older thin-layer chromatography and electrophoretic methods for polyamine assays may still remain a valuable tool in future work.

[16] N. Seiler, *Clin. Chem.* (*Winston-Salem, N.C.*) **23,** 1519 (1977).

[17] N. Seiler, *in* "Polyamines in Biomedical Research" (J. M. Gaugas, ed.), p. 435. Wiley, New York, 1980.

[18] N. Seiler, *in* "Polyamines in Biology and Medicine" (D. R. Morris and L. J. Marton, eds.), p. 127. Dekker, New York, 1981.

[2] Liquid Chromatographic Methods for Assaying Polyamines Using Prechromatographic Derivatization

By NIKOLAUS SEILER

The structural features of the polyamines are unfavorable for sensitive detection by optical or electrochemical methods. Their reaction with reagents forming derivatives with high molar extinction coefficients or intense fluorescence has been successfully applied in the past. With the availability of simple and sensitive electrochemical detectors,[1] the attachment of groups suitable for electrochemical detection became an attractive possibility for the measurement of polyamines in column eluates.

Among the many suitable reagents,[2,3] the following acid chlorides have been suggested for prechromatographic derivatization of the polyamines: benzoyl chloride,[4] *p*-toluenesulfonyl chloride,[5] quinoline-8-sulfonyl chloride,[6] and 5-dimethylaminonaphthalene-1-sulfonyl chloride (Dns-Cl).[7] 5-Di-*n*-butylaminonaphthalene-1-sulfonyl chloride[8] has advantages over Dns-Cl, especially in the mass spectrometric identification and determination of amines, but has not been used for polyamine analyses in published work.

In addition to the acid chlorides, fluorescamine has been suggested for prechromatographic derivatization.[9] Reaction with *o*-phthalaldehyde and 2-mercaptoethanol provides a further possibility[10] which might be utilized in the future. The latter two reagents react only with primary amino groups, in contrast with the acid chlorides which form derivatives not only with primary and secondary amino groups, but also with imidazole nitrogen and phenolic hydroxyls and even with some alcohols. Fluores-

[1] P. T. Kissinger, *Anal. Chem.* **49,** 447 (1977).

[2] J. F. Lawrence and R. W. Frei, "Chemical Derivatization in Liquid Chromatography." Elsevier, Amsterdam, 1976.

[3] N. Seiler and R. Demisch, *in* "Handbook of Derivatives for Chromatography" (K. Blau and G. S. King, eds.), p. 346. Heyden, London, 1977.

[4] J. W. Redmond and A. Tseng, *J. Chromatogr.* **170,** 479 (1979).

[5] T. Sugiura, T. Hayashi, S. Kawai, and T. Ohno, *J. Chromatogr.* **110,** 385 (1975).

[6] E. Roeder, I. Pigulla, and J. Troschuetz, *Fresenius' Z. Anal. Chem.* **288,** 56 (1977).

[7] N. Seiler and M. Wiechmann, *Hoppe-Seyler's Z. Physiol. Chem.* **348,** 1285 (1967).

[8] N. Seiler, T. Schmidt-Glenewinkel, and H. H. Schneider, *J. Chromatogr.* **84,** 95 (1973).

[9] K. Samejima, *J. Chromatogr.* **96,** 250 (1974).

[10] T. P. Davis, C. W. Gehrke, C. W. Gehrke, Jr., K. O. Gerhardt, T. D. Cunningham, C. Kuo, H. D. Johnson, and C. H. Williams, *Clin. Chem.* (*Winston-Salem, N.C.*) **24,** 1317 (1978).

ISBN 0-12-181994-9

camine and *o*-phthalaldehyde–2-mercaptoethanol are therefore more selective reagents than the acid chlorides. Nevertheless, the acid chlorides, more specifically Dns-Cl, have been utilized much more for prechromatographic derivatization of the polyamines than has fluorescamine.

Since the acid chlorides do not differ too greatly in their reactivity, the same principles of derivative formation are valid for all these reagents. The experience available with Dns-Cl[3,11–13] can therefore be utilized for other acid chlorides.

Reaction is normally achieved in alkaline-buffered (pH 8–10) solutions containing an organic solvent in order to ensure a homogeneous reaction medium. Similar conditions are also used for the derivatization with fluorescamine.[14] Reagents must be applied in excess in order to obtain quantitative reaction with all amino groups of the polyamines.

One of the advantages of the prechromatographic derivatization method is the solubility of the derivatives in organic solvents. This allows their extraction from aqueous phases, and thus their convenient accumulation even from relatively large volumes of tissue extracts and body fluids.

Another advantage of forming chemically defined, stable derivatives is the possibility of their chromatographic isolation. If labeled precursors have been used in metabolic studies, the determination of both radioactivity and concentration is possible from the same sample; i.e., the specific radioactivity can be determined with great precision. Moreover, the unambiguous identification of the isolated derivatives is possible by mass spectrometry. Thin-layer chromatographic procedures are in this regard especially practical because they allow one to perform several consecutive separations in one-dimensional or two-dimensional chromatograms with visual control of each separation step. Thin-layer chromatographic procedures are normally faster than column chromatographic methods because they allow the handling of many samples in parallel. Their disadvantage is that they demand a lot of work in contrast to the fully automated column chromatographic procedures. Moreover, the separations that can be achieved with modern high-pressure columns are superior to those obtained on thin-layer plates.

[11] N. Seiler, *Methods Biochem. Anal.* **18,** 259 (1970).

[12] N. Seiler and M. Wiechmann, *in* "Progress in Thin-Layer Chromatography and Related Methods" (A. Niederwieser and G. Pataki, eds.), Vol. 1, p. 95. Ann Arbor Sci. Publ., Ann Arbor, Michigan, 1970.

[13] N. Seiler, *Res. Methods Neurochem.* **3,** 409 (1975).

[14] K. Samejima, M. Kawase, S. Sakamoto, M. Okada, and Y. Endo, *Anal. Biochem.* **76,** 392 (1976).

The Dansylation Method

Reaction of polyamines with 5-dimethylaminonaphthalene-1-sulfonyl chloride, thin-layer chromatographic separation of the Dns derivatives, and subsequent quantitative evaluation of the thin-layer plates either by *in situ* scanning of fluorescence or after extraction, by conventional fluorescence measurement[7,11-13] was the first sensitive method for the polyamines[7] and permitted for the first time the precise measurement of putrescine in tissues.[15] The method has been used in many laboratories, frequently in modifications of the original procedure. For the reasons mentioned above, the dansylation method is still a powerful tool, although fully automated column chromatographic procedures have widely replaced the method in the routine analysis of tissues and body fluids.

Purification of Dns-Cl

Impure samples of Dns-Cl are a frequent cause for unsatisfactory analytical results. The following procedure is suitable for the purification even of outdated commercial samples: 10 g of the product are dissolved in 20 ml of distilled toluene. A column (12 × 100 mm) of silica gel 60 (70–230 mesh) (Merck, Darmstadt, Federal Republic of Germany) is prepared and the solution is passed through the column. Dns-Cl is then eluted from the column with toluene. The impurities remain adsorbed to the silica gel. Evaporation to dryness *in vacuo* yields an oil that crystallizes to form deep orange crystals (mp 69°).

Reaction with Dns-Cl

In the presence of excess reagent, putrescine, spermidine, and spermine, as well as their monoacetyl derivatives, react quantitatively; i.e., all primary and secondary amino groups are labeled. Unfavorable reaction conditions (incorrect pH or reaction time, insufficient reagent) can lead to incompleteness of the reaction, especially with spermine, owing to partial reaction with this polyfunctional molecule.

Tissue Extracts. Tissue samples are homogenized with 10 volumes of 0.2–0.4 *M* perchloric acid and centrifuged; samples of the supernatant varying from several microliters to several milliliters are used for assay. Routinely, volumes of 100–500 μl (corresponding to 10–50 mg of tissue) are preferred. Samples with internal standards are prepared by addition to the supernatant of known amounts of the polyamines, corresponding approximately to the amounts of the amines in the samples. External standards are carried through the procedure in similar amounts. In addition

[15] N. Seiler and A. Askar, *J. Chromatogr.* **62,** 121 (1971).

two blanks (containing the appropriate amount of perchloric acid) are prepared.

Three times the sample volume of a solution of Dns-Cl in acetone or dioxane (containing 10 mg/ml) are added to each sample, and the mixture is saturated with $Na_2CO_3 \cdot 10\ H_2O$. The stoppered glass tubes are mixed and then stored overnight at room temperature. An increase in reaction velocity is obtained by sonifying the samples in ultrasonic equipment designed for cleaning purposes. The reaction is complete under these conditions within 2–3 hr at room temperature. Enhancement of reaction velocity by heating to 60° is also feasible; however, phase separation can occur.

Since Dns derivatives are light sensitive, unnecessary exposure to intensive light sources should be avoided. Dim light in the laboratory is satisfactory, however.

Excess reagent (recognizable by its yellow color) is removed by addition of a solution of 5 mg of proline dissolved in 20 μl of water. After sonification for 3–5 min or storage for 30 min the Dns amides can be extracted with 3–5 volumes of toluene; i.e., in a routine determination with 5–10 ml. The aqueous phase is separated from the toluene phase by centrifugation.

If the concentration of the amines to be determined is high, as is usually the case for spermidine and spermine, small aliquots of the toluene extract can be directly submitted to separation. In other cases (for instance, for putrescine determinations in tissues) the toluene extract is evaporated to dryness and the residue is redissolved in a small volume of toluene.

Urine. In urine, total polyamines are usually determined after hydrolysis for 16 hr at 110° with 6 *N* HCl in sealed glass tubes, but free and conjugated polyamines can be determined in nonhydrolyzed samples. In both cases the reaction with Dns-Cl is carried out in the same way as with tissue extracts, but it is advisable to increase the reagent concentration to 30 mg of acetone per milliliter. If an adequate separation method is subsequently used, total polyamines can be determined in urine hydrolyzates without any cleanup step.[16–20]

In order to avoid excessive amounts of ammonia in the hydrolyzed urine samples, Dreyfuss *et al.*[17] recommended treatment with urease be-

[16] O. Heby and G. Andersson, *J. Chromatogr.* **145,** 73 (1978).
[17] G. Dreyfuss, R. Dvir, A. Harell, and R. Chayen, *Clin. Chim. Acta* **49,** 65 (1973).
[18] N. D. Brown, R. B. Sweet, J. A. Kintzios, H. D. Cox, and B. P. Doctor, *J. Chromatogr.* **164,** 35 (1979).
[19] J. H. Fleisher and D. H. Russell, *J. Chromatogr.* **110,** 335 (1975).
[20] F. L. Vandemark, G. J. Schmidt, and W. Slavin, *J. Chromatogr. Sci.* **16,** 465 (1978).

fore hydrolysis. Others remove ammonia by evaporation of $NaHCO_3$- or Na_2CO_3-saturated samples of hydrolyzed urine before allowing it to react with Dns-Cl.

Blood. Whole blood (0.25 ml), or the same volume of washed erythrocytes, is mixed with 0.75 ml of water in order to achieve hemolysis. Proteins are precipitated by mixing with 1 ml of 10% perchloric acid, and the supernatant is allowed to react with Dns-Cl, as was described for tissue extracts.[21]

Cleanup Procedures

PREDERIVATIZATION CLEANUP

These methods were suggested for the concentration of polyamines from tissues and body fluids and for the improvement of subsequent chromatographic separations. They are generally applicable and are not restricted to prechromatographic derivatization methods. Among the many known procedures,[22–28] two are described in detail because they have been used for the determination of both free, conjugated, and total polyamines in urine.

Procedure 1. Preseparation on Dowex 50W-X8 Columns.[27] Purification of the ion-exchange resin (Dowex 50W-X8, 200–400 mesh; Serva, Heidelberg, Federal Republic of Germany) is achieved by washing it with large volumes of 1 *N* NaOH, water, and 6 *N* HCl. The resin is stored at 3°, suspended in 6 *N* HCl. Columns are prepared from polypropylene pipettor tips supported on a cotton wool plug inserted into the constricted end of the tip. For 2-ml urine samples (hydrolyzed or nonhydrolyzed), a total of 5 ml of resin is placed into the column. The resin columns are connected by polyethylene tubes with a peristaltic pump with silicon pumping tubes and washed to neutrality with water that had been distilled over phosphoric acid (deionized water is not suitable). Two-milliliter samples of urine (or the aqueous solution of the dried residue of hydrolyzed urine) are mixed with 2 ml of ethanol and allowed to stand for 1 hr at 3°, after which the precipitate is removed by centrifugation. The clear supernatants are applied to the columns by pumping at a rate of 0.8–1 ml/min.

[21] Y. Saeki, N. Uehara, and S. Shirakawa, *J. Chromatogr.* **145,** 221 (1978).
[22] H. Shimizu, J. Kakimoto, and I. Sano, *J. Pharmacol. Exp. Ther.* **143,** 199 (1964).
[23] H. Inoue and A. Mizutani, *Anal. Biochem.* **56,** 408 (1973).
[24] K. Fujita, T. Nagatsu, K. Maruta, M. Ho, H. Senka, and K. Miki, *Cancer Res.* **36,** 1320 (1976).
[25] T. Hayashi, T. Sugiura, S. Kawai, and T. Ohno, *J. Chromatogr.* **145,** 141 (1978).
[26] M. Kai, T. Ogata, K. Haraguchi, and Y. Ohkura, *J. Chromatogr.* **163,** 151 (1979).
[27] N. Seiler and B. Knödgen, *J. Chromatogr.* **164,** 155 (1979).
[28] M. M. Abdel-Monem and J. L. Merdink, *J. Chromatogr.* **222,** 363 (1981).

Subsequently the columns are washed at the same pumping rate with 20 ml of 2 *N* HCl. The polyamine fractions, containing both monoacetyl-polyamines and nonconjugated polyamines, are eluted with 25 ml of 6 *N* HCl. The samples are evaporated to dryness at a pressure of about 15 mm Hg. The residues are redissolved in 0.2 *N* perchloric acid, then the samples are derivatized as usual.

The same procedure can be applied to 0.2 *N* perchloric acid tissue extracts and blood samples.

Procedure 2. Precleaning on Silica Gel.[28] This procedure is a version of the method of Grettie *et al.*[29] Aliquots (6 ml) of urine are adjusted to pH 9.0 with 1 *N* NaOH using an automatic titrator. A 2-ml aliquot is taken up into a syringe and added to a silica gel cartridge (Sep-Pak, Waters Associates, Milford, Massachusetts). The cartridge is washed with water (5 ml), and the polyamines are then eluted with 10 ml of 0.1 *N* hydrochloric acid.

Postderivatization Cleanup

Alkaline Hydrolysis. It has been mentioned that Dns-Cl and other acid chlorides can react with a number of compounds such as phenols, phenolic acids, and alcohols, including sugars. Chromatograms of the Dns derivatives of polyamines, including their monoacetyl derivatives, can be considerably improved if these derivatives and those of the amino acids are removed prior to separation. The following procedure[30] has been shown to be suitable for polyamine determinations in urine[27] and tissue samples.

The residues of the toluene extracts containing the Dns derivatives are dissolved in 150 μl of 5 *M* KOH in methanol, and the samples are treated for 30 min at 50°. After the addition of 1.5 ml of aqueous solutions of 100 mg each of KH_2PO_4 and Na_2HPO_4, the Dns derivatives are extracted with 4 ml of toluene.

Chromatography on Silica Gel Columns. The following separation of the mixture of Dns derivatives on small silica gel columns removes Dns derivatives that are either less or more polar than the polyamine derivatives.[31]

Disposable 5-ml polypropylene pipettor tips are closed at the constricted end with a cotton wool plug and filled with 2 g of silica gel 60 (0.06–0.2 mm, Merck, Darmstadt, Federal Republic of Germany). The toluene solutions of the Dns derivatives are evaporated, and the residues are redissolved in 3 ml of toluene. These solutions are applied to the silica

[29] D. P. Grettie, D. Bartos, F. Bartos, R. G. Smith, and R. A. Campbell, *Adv. Polyamine Res.* **2,** 13 (1978).

[30] N. Seiler and K. Deckardt, *J. Chromatogr.* **107,** 227 (1975).

[31] N. Seiler, B. Knödgen, and F. Eisenbeiss, *J. Chromatogr.* **145,** 29 (1978).

gel columns, and the columns are first washed with two 3-ml portions of toluene and then with 5 ml of toluene–triethylamine (10 : 1). These eluates are discarded. The polyamine derivatives are eluted together with other Dns-amine derivatives with 4 ml of ethyl acetate. The ethyl acetate solutions are evaporated to dryness. The residues are submitted to chromatographic separation.

Separation and Quantitative Evaluation

Thin-Layer Chromatography

Separation. Usually 20 × 20 cm silica gel thin-layer plates are used with a layer thickness of 100 or 200 μm.[7,11–13,15,16,19,27,32] These can be silica gel G layers or commercial plates with a variety of binders. High-performance thin-layer plates (silica gel 60 F_{254} HPTLC, Merck, Darmstadt, Federal Republic of Germany,[33] or comparable products) allow one to produce very small spots, and the time for development is significantly reduced. Alumina[17,34] and kieselguhr plates[35] in conjunction with appropriate solvents have also been suggested for polyamine determinations.

The advantages of thin-layer chromatography are rapidity, the adaptability of the method to specific separation problems, the possibility of visual control of the separations, and the application of various methods of quantitative evaluation. Moreover, autoradiographs can be prepared for the identification of metabolites from labeled precursors.[36] The disadvantage is the amount of manipulation necessarily required.

Usually the plates are activated for 1 hr at 110° before use. For one-dimensional separations, the samples are applied in small spots to the thin-layer plate 3 cm from one edge. The distance between two neighboring spots is 2 cm. The formation of small spots is aided if solutions in toluene are applied, since this solvent does not move the Dns amides chromatographically.

Sample application is one of the critical points of quantitative thin-layer chromatography. In our experience any precise graduated pipette or microsyringe yields reproducible results, provided that the solution is applied without touching the plate with the pipette tip. Even small losses of silica gel from the origin cause serious losses if the diameter of the spot

[32] A. S. Dion and E. J. Herbst, *Ann. N. Y. Acad. Sci.* **171,** 723 (1970).

[33] N. Seiler and B. Knödgen, *J. Chromatogr.* **131,** 109 (1977).

[34] M. M. Abdel-Monem, K. Ohno, N. E. Newton, and C. E. Weeks, *Adv. Polyamine Res.* **2,** 37 (1978).

[35] K. Igarashi, I. Izumi, K. Hara, and S. Hirose, *Chem. Pharm. Bull.* **22,** 451 (1974).

[36] U. Bachrach and N. Seiler, *Cancer Res.* **41,** 1205 (1981).

at the origin is kept within 2–3 mm. The use of precision pipettes with Teflon plungers is especially convenient. A fully automated device is now commercially available (CAMAG, Muttenz, Switzerland).

In order to ensure reproducibility even under not very well defined environmental conditions, separations are usually carried out in filter paper-lined chromatographic tanks by ascending chromatography. But one can prepare several two-dimensional chromatograms on a single normal or high-performance thin-layer plate in horizontal tanks,[13,33] if extensive separations are required. It is possible to quantitatively evaluate the two-dimensional chromatograms,[33] but usually one-dimensional separations are preferred.

For the separation of the Dns derivatives both of polyamines and of monoacetylpolyamines, one run with chloroform–carbon tetrachloride–methanol (14 : 6 : 1) is suitable.[27,36] However, spermine is rather close to other spots and near the solvent front, and sometimes the (large) Dns-ammonia spot interferes with the Dns-monoacetylspermidine spots. It is therefore preferable to separate free and acetylpolyamines in separate runs.

Plates destined for monoacetylpolyamine determinations are run first 2–3 times with methyl or ethyl acetate in order to remove the nonpolar Dns derivatives, including Dns polyamines and Dns-ammonia, from the zone that is subsequently used for acetylpolyamine separations with the above solvent mixture.[27]

Two-dimensional development with methyl acetate in the first direction and the above solvent mixture in the second direction permits the determination of all acetylpolyamines, including monoacetylspermine and diacetylspermine.[37]

Thin-layer chromatographic separations of Dns-monoacetylputrescine, and monoacetylspermidines have also been reported by Abdel-Monem and co-workers.[34]

For the one-dimensional separation of the Dns derivatives of putrescine, spermidine, and spermine, a number of solvent mixtures are suitable.

For the determination of polyamines in tissue extracts and urine, silica gel thin-layer plates are developed first with cyclohexane–ethyl acetate (1 : 1) (two runs) followed by one run with cyclohexane–ethyl acetate (3 : 2) in the same direction.

In tissues rich in 4-aminobutyric acid, such as the brain, development with benzene–cyclohexane–methanol (85 : 15 : 1.5) followed by cyclohex-

[37] N. Seiler, F. N. Bolkenius, B. Knödgen, and K. Haegele, *Biochim. Biophys. Acta* **676,** 1 (1981).

ane–ethyl acetate (3 : 2) and cyclohexane–ethyl acetate (1 : 1) (one run each) is preferable.[38]

Extracts of cultured cells usually contain only small amounts of interfering compounds. It is therefore feasible to determine polyamines after development of the chromatograms with cyclohexane-diethyl ether (3 : 2) (two runs). Since putrescine is especially cleanly separated from ammonia, this solvent is also suitable for putrescine determinations in tissues.

Chloroform-triethylamine (10 : 1) was developed for the separation of putrescine from its homologs.[11,12] This or a 5 : 1 mixture is useful for polyamine determinations in hydrolyzed urine samples.[16,19]

Quantitative Evaluation. The principles of quantitative determination of Dns derivatives on thin-layer plates have been extensively described[39,40] and reviewed.[11–13] Since the Dns derivatives are subject to irreversible destruction on active surfaces by light, it is essential to avoid exposure of dry plates to light. Heptane does not cause the Dns polyamines to migrate chromatographically. If elution and conventional fluorometry is the method of quantitative evaluation, it is advisable to store the plates in an *n*-heptane containing tank, in which the solvent is allowed to ascend the thin layer, until the plates are evaluated.

The elution method for Dns derivatives is the same as for other absorbing or fluorescing compounds. After chromatography the fluorescent zones are quickly visualized with the aid of a 360-nm UV source. If fluorometry is to be performed in 3-ml cuvettes, the zones containing the Dns derivatives are scraped off with one of the commercially available zone extractors or with a simple homemade device[11] that collects the adsorbent in a 10-ml centrifuge tube. Shaking with 5 ml of ethyl acetate, dioxane, or methanol gives adequate extraction. After centrifugation, the fluorescence intensity of the supernatant is estimated using a conventional spectrofluorometer (activation of fluorescence at 360–365 nm, fluorescence intensity measured at 510 nm).

Techniques suitable for the extraction of normal-sized spots with 30–50 μl of solvent and quantitative evaluation by fluorescence measurement in microcuvettes or by mass spectrometry are described elsewhere.[13,33,41]

The fluorescence quantum yield of the Dns derivatives is severely quenched by oxygen. It is therefore not possible to compare fluorescence intensities of standard solutions prepared from pure crystallized derivatives with extracts from thin-layer plates, in order to determine, for in-

[38] N. Seiler and T. Schmidt-Glenewinkel, *J. Neurochem.* **24,** 791 (1975).

[39] N. Seiler and B. Knödgen, *Org. Mass. Spectrom.* **7,** 97 (1973).

[40] N. Seiler and M. Wiechmann, *Fresenius' Z. Anal. Chem.* **220,** 109 (1966).

[41] N. Seiler and M. Wiechmann, *Hoppe-Seyler's Z. Physiol. Chem.* **350,** 1493 (1969).

stance, recovery of the procedure. Either the standards should be carried through the chromatographic step (in which case the losses of the chromatographic procedure and extraction are neglected) or the solutions should be purged with nitrogen for at least 15 min in order to minimize quenching.

The sensitivity of the method is limited by unavoidable background fluorescence and by the sensitivity of the fluorometer. In sample volumes of 300 μl, 5 pmol of the polyamines can be determined.

The fluorescence measurement is not the step limiting the accuracy of the procedure. Reproducibility of the elution method and the direct scanning technique are the same, namely, ±2–5% SD, for the determination of 10–100 pmol of a polyamine in tissue. Recovery of the polyamines throughout the whole procedure is better than 90% for 10 pmol amounts.

The method of direct scanning is especially useful for the determination of several separated spots and has therefore been frequently used for polyamine determinations.

Previously it had been suggested that spraying the plates subsequent to chromatography with 20 ml of a solution of triethanolamine in 2-propanol (1 : 4)[40] stabilized the Dns derivatives and increased fluorescence quantum yield. Today mechanically stable commercial thin-layer plates are available and it is possible to substitute spraying by a technically less difficult dipping procedure[27]: the air-dried plates are rapidly inserted into a chromatography tank filled with a 10% solution in cyclohexane of paraffin oil or silicon oil (for instance Rhodorsil S1 710, Prolabo, Paris, France). After drying for 5 min in a hood, the dipping procedure is repeated. Fluorescence intensity is stable on oil-impregnated plates for at least a week if they are stored in the dark.

The air-dried oil-impregnated plates are equilibrated for 30 min with the atmosphere of the room in which scanning is performed. Direct scanning can be done with one of the many commercially available attachments to fluorometers. A mercury arc lamp in conjunction with a Woods filter (360–365 nm) is adequate for fluorescence activation. Usually the total emitted fluorescence light is recorded (using a 420 nm cutoff filter).

The spots obtained with the chromatographic procedures described above are normally symmetrical, so that the recorded curves approximate Gaussian shape. Quantitation is therefore possible by peak height measurements or by calculating the area under the curves from the height and width at half height. The use of electronic integrators accelerates and improves the quantitation.

The areas under the curve are directly proportional to spot concentrations over a very wide range (5 pmol to 10 nmol). Reproducibility of polyamine and acetylpolyamine determinations is ±5–10% in this range, if determined on tissue extracts.

High-Pressure Liquid Chromatography

The application of microparticle alumina or silica gel columns instead of thin-layer plates for the separation of Dns polyamines turned out not to be practical. Owing to changes in activity of the adsorbents, the separations were not sufficiently reproducible for routine determinations.[34] However, the lipophilic character of the Dns derivatives suggested that separation on reversed-phase columns with aliphatic chains chemically bonded to silica cores would be worth attempting. The first separations with this technique were reported by Kneifel.[42] Subsequently a number of groups worked out assay procedures that differ in the type of column used, the solvents, and the elution mode and time.[20,21,31,34,42,43] Brown *et al.*[18] add 1-heptanesulfonic acid to the water–acetonitrile gradient. It is not clear, however, to what extent the separations are influenced by ion-pairing since, as was mentioned, the Dns derivatives have already a high affinity to the bonded hydrophobic phase.

It is presently not possible to judge the advantages and disadvantages of the various procedures, but it is clear that all commercial columns with bonded C_{18}-aliphatic chains can give satisfactory separations if a suitable water–methanol or water–acetonitrile gradient is selected. Our procedure[31] was devised to measure small putrescine concentrations in tissues; i.e., in order to achieve a complete separation of the homologous diamine derivatives, an elution program of 40 min was not considered to be a handicap.

Before reaction with Dns-Cl, a known amount of 1,6-diaminohexane · 2 HCl is added to the tissue extracts as an internal standard. In order to avoid overloading of the column with amino acid derivatives and side products of the derivatization reaction (i.e., Dns-dimethylamine), the above-mentioned silica gel column chromatographic cleanup step is usually applied. The residue with the Dns polyamines is dissolved in methanol and immediately before separation the methanol solutions are diluted with water to give a methanol–water mixture of 3 : 1.

The gradient is prepared from a mixture of methanol–water (1 : 1) and methanol. After equilibration of the column [250 × 3 mm, Lichrosorb RP8 (7 μm particles); Merck, Darmstadt, Federal Republic of Germany] at a rate of 1 ml/min with water–methanol (57.5 : 42.5) for 10 min, 10- to 50-μl aliquots of the Dns-derivative solutions are applied and gradient elution is started with the same methanol concentration, which is then increased

[42] H. Kneifel, *Chem.-Ztg.* **101,** 165 (1977).

[43] K. Macek, Z. Deyl, J. Jiranek, and M. Smrz, *Sb. Lek.* **80,** 2 (1978).

linearly by an increment of 0.5% per minute. From 20 to 30 min the methanol increment is 1.5% per minute, and it is further increased to 3% per minute after 30 min. Elution with pure methanol is continued for 4 min to remove impurities. At 40 min the gradient is switched to the initial water–methanol mixture. Before the next run the column is equilibrated again with 57.5% methanol for 10 min.

For detection, an 8-μl flow cell is used in conjunction with a fluorescence spectrophotometer. To minimize the scattered light, a Woods filter is placed in the path of the activating light and a 420 nm cutoff filter in the emitted light (activation of fluorescence at 360 nm, fluorescence measurement at 510 nm). The signal is recorded with a two-channel recorder at two sensitivities in order to obtain both the less intense signal of putrescine and the more intense signals of spermidine and spermine.

The recorded peaks can be evaluated by measurement of peak height and width at half-height, because they are nearly of Gaussian shape. The mean standard deviation of measurements repeated on different days is less than $\pm 7\%$ for the range between 20 pmol and 2 nmol of the various amines, and the relationship between amount of substance and recorded peak areas is linear over this range.

Owing to their high molar extinction coefficients,[11] Dns derivatives can also be monitored in column eluates by UV absorption.[34,43]

The above and related separations of the Dns polyamine derivatives are limited by the fact that the isomeric monoacetyl spermidines cannot be separated. In order to enable their determination, Abdel-Monem and Merdink[28] suggested a two-step procedure: the mixture of Dns derivatives is first separated on a 250 × 2.5 mm Micropack CN-10 column (10 μm particle size, Varian, Palo Alto, California) with a solvent composed of *n*-hexane–2-propanol (100 : 3) (solvent A) and *n*-hexane–dichloromethane–2-propanol (10 : 5 : 1) (solvent B). The sample is eluted with a solvent gradient, the gradient changing from 100% of solvent A to 100% of solvent B within 15 min at a flow rate of 3 ml/min. The elution is then continued with solvent B for an additional 5 min. The fraction with the Dns monoacetylpolyamines is eluted after 15–20 min. It is collected and evaporated to dryness. The residue is dissolved in 100 μl of chloroform, and 25 μl of the resulting solution are applied to a silica gel column (Ultrasphere-Si, 150 × 4.4 mm; 5-μm silica gel particles; Altex, Berkeley, California; or, alternatively, a 250 × 4.6 mm column packed with 10 μm silica gel, Alltech Associates, Deerfield, Illinois). Elution is achieved with chloroform-2-propanol (100 : 6), the flow rate being changed from 1 ml/min to 2 ml/min in 10 min and then maintained at this rate for an additional 10 min. Under these conditions it is possible to determine monoacetylputrescine and N^1- and N^8-acetylspermidine in urine.

Derivatization with Acid Chlorides Other than Dns-Cl

Among the acid chlorides suggested for preseparation derivatization of the polyamines[4-6] only benzoyl chloride deserves consideration because of the limited experience with the other derivatives.

Although there are many advantages of Dns-Cl as a prechromatographic derivatization reagent, it has also one major disadvantage: it tends to produce a number of side products by reacting with itself. Thus one always observes the formation of Dns-dimethylamine and of a series of other products.[11,12] Benzoyl chloride is advantageous in this regard, and presumably it is the small number of side products of the benzoylation reaction that allows the application of short elution programs for the separation of cadaverine, putrescine, spermidine, and spermine.[4,44] The following procedure was suitable for polyamine determinations in brain.[44]

Perchloric acid extracts are prepared and adjusted to pH 7 by addition of KOH solution. After 60 min at 4° the perchlorate is removed by centrifugation. The supernatants are lyophilized, and the residues are dissolved in 1 ml of water. Before derivatization a solution of 1,6-diaminohexane is added as internal standard; 1 ml of 2 *N* NaOH (1 ml) is then added, followed by 5 μl of benzoyl chloride. After intense shaking, reaction is complete within 20 min at room temperature. Two milliliters of a saturated NaCl solution are added; extraction of the benzoyl derivatives is achieved with 2 ml of diethyl ether. The ether layer is dried *in vacuo* and the residue is dissolved in 200 μl of methanol.

The methanol solution (5–50 μl) is applied to a 300 $\times$ 3.9 mm μ-Bondapak C_{18} column (10 μm particle size, Waters Associates, Milford, Massachusetts). Elution is achieved in the isocratic mode with methanol–water (3 : 2) at a flow rate of 2 ml/min.

Spermine is eluted under these conditions (with considerable peak broadening) at about 9–10 min.

Quantitative determination is carried out by recording absorbance at 254 nm.

The following recoveries were obtained[44]: putrescine, 61.3 $\pm$ 2.4%; spermidine, 50.0 $\pm$ 4.0%; spermine, 71.3 $\pm$ 7.0%. Using maximum detector sensitivity, a full-scale peak was obtained with approximately 200 pmol of putrescine.[4]

Fluorescamine

Fluorescamine, 4-phenylspiro[furan-2(3*H*),1′-phthalan]-3,3′-dione, is a nonfluorescent compound that at pH 8–9.5 forms fluorescent derivatives with compounds containing primary amino groups, with an absorp-

[44] R. Porta, R. L. Doyle, S. B. Tatter, T. E. Wilens, R. A. Schatz, and O. Z. Sellinger, *J. Neurochem.* **37,** 723 (1981).

tion maximum at 390 nm and a maximal fluorescence emission at 475 nm.[3] Its main application has been in methods using postseparation derivatization.

In 1974 Samejima described the separation of preformed fluorescamine derivatives by reversed-phase column chromatography,[9] but the method[14] requires preseparation of the di- and polyamines on (carboxymethyl)cellulose before derivatization with fluorescamine, because of insufficient resolution of the fluorescamine derivatives. Similarly, the suggestion of derivatizing the polyamines at the origin of a thin-layer plate and subsequently separating the fluorescamine derivatives[45] did not find practical applications.

It was observed, however, that reaction of fluorescamine with a variety of biogenic amines other than the polyamines is inhibited by Ni^{2+}, and a method for the separation of the fluorescamine derivatives of the polyamines could be worked out in such a way that free and total polyamines could be determined in human serum.[26] The procedure is as follows: to 0.5 ml of serum, 0.4 nmol of 1,6-diaminohexane is added as an internal standard. Proteins are precipitated with 0.1 ml of 3 *M* perchloric acid. Neutralization of the supernatant with 1.5 *M* KOH and centrifugation removes potassium perchlorate.

For the determination of free polyamines, 1 ml of 0.1 *M* sodium phosphate buffer, pH 7.0, is added to the supernatant. This solution is mixed with 0.3 ml of chloroform and 0.3 ml of methanol. The aqueous layer is derivatized with fluorescamine.

For total polyamine determinations the above supernatant is diluted with 1 ml of concentrated HCl and heated for 12 hr at 115°. The dry residue is dissolved in 1 ml of water.

Cellex P (H^+ form) columns (30 × 15 mm) are prepared essentially as described by Kremzner and Wilson[46] and are equilibrated with 0.01 *M* sodium phosphate buffer, pH 6.0. The sample solutions are applied to the columns, and the columns are washed with 2 ml of 0.01 *M* sodium phosphate buffer, pH 6.0, 1 ml of water, and 1.5 ml of 0.05 *M* sodium chloride solution. The polyamines are eluted with 2 ml of 3 *M* sodium chloride. For reaction with fluorescamine, 0.5 ml of 0.4 *M* borate buffer (pH 9) and 0.2 ml of 20 m*M* nickel(II) sulfate are added to the polyamine fractions, and then 0.5 ml of a 1 m*M* solution of fluorescamine in acetone is added with vigorous mixing.

The fluorescamine derivatives are extracted with 0.4 ml of ethyl acetate after addition of 1 ml of 0.3 *M* succinic acid and 1 g of sodium chloride.

The organic layer is removed and diluted with 3 ml of cyclohexane. The fluorescamine derivatives are now extracted from the organic phase

[45] F. Abe and K. Samejima, *Anal. Biochem.* **67,** 298 (1975).

[46] L. T. Kremzner and I. B. Wilson, *J. Biol. Chem.* **238,** 1714 (1963).

with 0.2 ml of 0.4 *M* borate buffer, pH 10.0. Reagent blanks are carried through the procedure.

Aliquots (100 μl) of the borate buffer solutions containing the fluorescamine derivatives are submitted to separation on a 150 × 4 mm Lichrosorb RP-18 (5 μm particle size; Merck, Darmstadt, Federal Republic of Germany) column using a linear water–methanol gradient. Initial proportion, 45% methanol; final mixture, 80% methanol (after 25 min); column temperature, 30°; flow rate, 1 ml/min. With a 20-μl flow cell, fluorescence is activated at 390 nm, and fluorescence emission is recorded at 490 nm. The lower limit of detection is reported to be 5–10 pmol for spermine. Linear relationships were observed between the ratios of the peak heights of the polyamines to those of the internal standard in the range 10–200 pmol. The recovery of the polyamines was 97 ± 3% for putrescine and cadaverine, and 75 ± 5% for spermidine and spermine.

Conclusion

Among the many methods now available for polyamine assay[47,48] (see also this volume [3,4]), the preseparation derivatization procedures deserve special attention. Some of the advantages have been briefly discussed above. In addition, the following aspects are worth considering.

1. Thin-layer chromatographic versions of this type of method are especially suitable for rapid semiquantitative screening purposes. Literally hundreds of samples can be separated in parallel, without the necessity of expensive equipment, and visually examined. This type of application may turn out to be useful in the isolation of polyamine-deficient mutants and in related applications.

2. Prechromatographic derivatization procedures are most probably methods that can be improved significantly, again without the need for sophisticated apparatus. There are two obvious ways to achieve this. First, the fluorescent ligand can be improved. This includes not only an increase in fluorescence quantum yield, but, more important, the improvement of the selectivity of the reaction and especially the reduction in the formation of side-reaction products. This would significantly decrease the noise level of the chromatograms and thus allow the utilization of the full sensitivity of existing detection systems. It would furthermore allow the use of microcolumns.

[47] N. Seiler, *Clin. Chem.* (*Winston-Salem, N.C.*) **23,** 1519 (1977).

[48] N. Seiler, *in* "Polyamines in Biomedical Research" (J. M. Gaugas, ed.), p. 435. Wiley, New York, 1980.

Second, the application of radiolabeled ligands is feasible. These would allow automated quantitative evaluation by the convenient methods of β-scintillation counting. Even more convenient are double-isotope derivative assays. This method has been suggested among others for the sensitive assay of 4-aminobutyric acid[49] and a number of biogenic amines.[50,51] A known amount of [^{14}C]4-aminobutyric acid (or of a biogenic amine) is added to the tissue sample, and derivatization is achieved with the commercially available [^{3}H]Dns-Cl. After extensive chromatographic purification of the labeled Dns derivative, the amount of amino acid or amine can be conveniently estimated from the ^{14}C : ^{3}H-ratio. An analogous procedure could readily be applied to polyamine assays.

In general, sensitive prechromatographic derivatization methods are feasible with relatively low-cost, versatile instrumentation. Their application is, therefore, especially applicable in those laboratories that have no need for daily routine polyamine assays.

[49] S. R. Snodgrass and L. L. Iversen, *Nature (London)* **241,** 154 (1973).
[50] M. Recasens, J. Zwiller, G. Mack, J. P. Zanetta, and P. Mandel, *Anal. Biochem.* **82,** 8 (1977).
[51] T. J. Paulus and R. H. Davis, this volume [5].

[3] Ion-Pair Partition Chromatographic Separation of Polyamines and Their Monoacetyl Derivatives

By NIKOLAUS SEILER

The formation of ion pairs between readily ionizable organic compounds as a basis for their isolation by extraction, or their separation and determination by liquid–liquid chromatography, has provoked considerable interest. The principles of this technique and some practical considerations have been the subject of several reviews.[1,2]

The natural polyamines have the prerequisites to form ion pairs with organic acids. Their separation in this form was therefore considered to be a worthwhile attempt.

The assay method requires single ion species of each component. For the generation of single ion species of spermidine, and spermine, respec-

[1] J. H. Knox, J. N. Done, A. F. Fell, M. T. Gilbert, H. Pryde, and R. A. Wall, "High-performance Liquid Chromatography." Edinburgh Univ. Press, Edinburgh, 1979.
[2] G. Schill, R. Modin, K. O. Borg, and B. A. Persson, *in* "Handbook of Derivatives for Chromatography" (K. Blau and G. King, eds.), p. 500. Heyden, London, 1977.

ISBN 0-12-181994-9

tively, relatively high concentrations of the counterion are needed. 1-Octanesulfonic acid proved to be suitable for the ion-pair formation with the polyamines. It did not interfere with the postcolumn derivatization with *o*-phthalaldehyde–2-mercaptoethanol,[3] a method well established in automated polyamine determinations using ion-exchange column chromatography for separation.

Owing to the excellent separatory quality of the reversed-phase column, it was possible to separate not only the polyamines, but also their monoacetyl derivatives within a time comparable to that normally needed for polyamine separations using the most advanced ion-exchange column chromatographic methods. The method described in the following text is indeed the first sensitive, routine, automatable method for the assay of acetylpolyamines, polyamines, and histamine.

Method[4]

For the separations a 3.9 × 300 mm μ-Bondapak C_{18} column (10 μm particles; Waters, Paris, France) is used. It is guarded by a precolumn (100 × 3 mm) filled with pellicular silica material with C_{18} brushes (CO:PELL ODS, 30–38 μm; Whatman Inc., Clifton, New Jersey). The elution system consists of gradients that are prepared by continuous mixing of two buffer solutions: (A) 0.1 *M* sodium acetate adjusted to pH 4.50 with acetic acid and containing 10 m*M* octane sulfonate; (B) 0.2 *M* sodium acetate (adjusted to pH 4.50 with acetic acid) plus acetonitrile (10 : 3) containing 10 m*M* octane sulfonate. For the preparation of the buffer solutions, tap water is distilled over phosphoric acid, acetic acid over sulfuric acid, and acetonitrile over phosphorus pentoxide.

The *o*-phthalaldehyde reagent is prepared by dissolving 50 g of boric acid, 44 g of KOH, and 3 ml of Brij 35 solution (30% solution in water, Merck, Darmstadt, Federal Republic of Germany) per liter of distilled water. To this solution 3 ml of 2-mercaptoethanol and 400 mg of *o*-phthalaldehyde dissolved in 5 ml of distilled methanol are added. The reagent is usually stored overnight before use and is prepared daily. Specific precautions (such as storage under nitrogen) do not seem necessary.

The thermostatted (35 ± 0.5°) column is equilibrated for 6 min with buffer A at a flow rate of 1.5 ml/min. Then, the sample is applied (usually a loop injector valve is used with a 250-μl loop) and gradient elution is started. A linear gradient with an increment of 4% per minute of buffer B separates completely putrescine, cadaverine, histamine, N^1-acetylspermi-

[3] M. Roth and A. Hampai, *J. Chromatogr.* **83,** 353 (1973).
[4] N. Seiler and B. Knödgen, *J. Chromatogr.* **221,** 227 (1980).

dine, N^8-acetylspermidine, 1,7-diaminoheptane (internal standard), spermidine, N^1-acetylspermine, and spermine within 27 min.[5] Shorter elution programs may be chosen if one does not need to determine acetylpolyamines. In principle, this is feasible by increasing initially buffer B concentration and by increasing the steepness of the linear gradient. An effort in this respect, however, was not made, because the method was mainly used for both polyamine and acetylpolyamine determinations.

For the determination of both free and conjugated polyamines in urine, a somewhat longer elution program is advisable. In practice the following two-step linear gradient proved to be adequate[4] and gave results identical to those obtained with the previously published dansylation method.[6]

After equilibration with buffer A for 6 min, a linear gradient is prepared from buffer A and buffer B with an increment of 2% per minute of buffer B for 30 min. At this time the gradient contains 60% buffer B. For the remaining time the increment of buffer B is increased to 4% per minute, and elution is completed. Resetting to buffer A is usually done at 45 min after commencing the run, if no unusual impurities were eluted from the column. Accordingly, one separation requires 51 min.

Column eluent and the *o*-phthalaldehyde reagent are mixed in a 1 : 1 ratio. As is usual for postcolumn derivatization with *o*-phthalaldehyde–2-mercaptoethanol, fluorescence is activated (using a 150 W xenon arc lamp) at 345 nm, and fluorescence intensity is measured at 455 nm and recorded at two sensitivities, which usually differ by a factor of 10 in the case of urine samples and by a factor of 20 in the case of tissue samples.

Sample preparation is not different from that used in ion-exchange column chromatographic procedures: 0.2 *N* perchloric acid tissue or cell extracts can be applied on the column directly after appropriate dilution with buffer A and filtration through a 0.2 μm filter. Acidified urines or urine hydrolyzates are diluted with buffer A and separated after filtration. A considerable improvement in separations and especially a prolongation of the lifetime of precolumn and column can be achieved by using a cleanup step[6,7] (see this volume [2]).

For quantitation, standard mixtures are run at regular intervals, and the peaks of the individual amines are related to that of 1,7-diaminoheptane, the internal standard. Since the separations are very reproducible, and the peak shapes are nearly Gaussian, even peak height measurements give satisfactory results. In the concentration range between 5 μM and

[5] N. Seiler, F. N. Bolkenius, B. Knödgen, and K. Haegele, *Biochim. Biophys. Acta* **676,** 1 (1981).

[6] N. Seiler and B. Knödgen, *J. Chromatogr.* **164,** 155 (1979).

[7] M. M. Abdel-Monem and J. L. Merdink, *J. Chromatogr.* **222,** 363 (1981).

5 nM of the amines, standard deviations were ±8.1% for putrescine, ±3.9% for spermidine, and ±6.4% for spermine. Less than 50 pmol of any of the above-mentioned amines can be measured under completely routine assay conditions; the method is not different in this regard from advanced ion-exchange chromatographic procedures.

Discussion

There is no doubt that similarly effective procedures can be worked out using other commercially available reversed-phase columns, other sufficiently strong organic acids of hydrophobic nature for ion pairing, and other buffers than were used in this first ion-pair partition chromatographic method for polyamine and acetylpolyamine determination. Indeed, Wagner *et al.*[8] have already suggested a version of the method. They use the same column type but form a gradient of the following two buffers: (A) 0.1 M NaH_2PO_4–acetonitrile (49 : 1) containing 8 mM octane sulfonate and 0.1 mM EDTA (pH adjusted to 2.55 with 3 M H_3PO_4); (B) 0.2 M NaH_2PO_4–acetonitrile (7 : 3) containing 8 mM octane sulfonate and pH adjusted to 3.10 with 3 M H_3PO_4. Elution is achieved at a flow rate of 1.5 ml/min by a linear gradient starting with 15% buffer B and leading within 30 min to 85% buffer B. At this time the system is reset to the initial conditions.

If the eluent is run through a UV detector (254 nm) before it is mixed with the *o*-phthalaldehyde reagent, it is possible to determine quantitatively a number of *S*-adenosyl derivatives (*S*-adenosyl-L-methionine, *S*-adenosyl-L-homocysteine, decarboxylated *S*-adenosylmethionine, 5′-deoxy-5′-methylthioadenosine, among others), and some aromatic amino acids by absorptimetry, and subsequently the polyamines by fluorometry in the form of their isoindole derivatives. The above *S*-adenosyl derivatives give no derivatives, or only weakly fluorescing ones, with *o*-phthalaldehyde–2-mercaptoethanol and do not interfere, therefore, with the polyamine assay. Another version of ion-pair reversed-phase separation of polyamines was mentioned in an abstract.[9]

The method is without doubt superior, with respect to completeness and rapidity of separations, to any existing column chromatographic assay procedure for the polyamines. Ion-exchange columns have usually higher capacities, but this is not a problem for the assay of polyamines in tissues and body fluids by reversed-phase ion-pair liquid chromatography. However, the method has other limitations which are most probably due

[8] J. Wagner, C. Danzin, and P. S. Mamont, *J. Chromatogr.* **227,** 349 (1982).

[9] M. D. Johnson, S. Swaminathan, and G. T. Bryan, *Fed. Proc., Fed. Am. Soc. Exp. Biol.* **40,** 1868 (1981).

to the presence of hydrophobic peptides in tissue extracts and body fluids. The following observations support this assumption.

Although the separation procedure has been designed to separate all amino acids cleanly from the basic dipeptides carnosine and homocarnosine, and from monoacetylputrescine,[4] one nevertheless observes a number of unidentified compounds interfering with the determination of the histidine peptides and with monoacetylputrescine, respectively. Some unknown compounds elute even near the region where spermine appears. In an attempt to estimate acetylpolyamines in (nonhydrolyzed) cerebrospinal fluid, a very large number of peaks appeared on the chromatograms which prevented the determination even of putrescine and spermidine. After hydrolysis, however, the polyamine determinations could be carried out with no difficulty.

By passing solutions in the absence of octane sulfonate through cartridges that contain the same material with C_{18} brushes that is used in the precolumn, some of the hydrophobic compounds can be removed, but a systematic study of the practical usefulness of this cleanup step is lacking.

Another limitation of the method is that it does not separate 1,3-diaminopropane and putrescine. Large amounts of dopamine in the samples also interfere with putrescine determinations.

Despite these limitations, the method of reversed-phase ion-pair liquid chromatography, even in its present version, opens new possibilities in the routine determination not only of the classical polyamines, but also of their biologically and pharmacologically important analogs. Owing to its sensitivity and rapidity, the method may also be used in the future for the assay of the activity of a variety of enzymes involved in polyamine metabolism and may thus substitute for some of the currently preferred methods that employ radiolabeled precursors.

[4] Quantitative Determination of Naturally Occurring Aliphatic Diamines and Polyamines by an Automated Liquid Chromatography Procedure

By CELIA WHITE TABOR and HERBERT TABOR

The naturally occurring aliphatic polyamines and diamines can be separated with the following automated procedures on a sulfonated-type ion-exchange column. Quantitative determinations can be carried out on 2–40

ISBN 0-12-181994-9

nmol of each amine.[1] In addition to 1,3-diaminopropane, putrescine, cadaverine, spermidine, and spermine, the procedures separate a number of other naturally occurring derivatives, such as monoacetylputrescine, monoacetylspermidine, carbamoylputrescine, hydroxyputrescine, and glutathionylspermidine.

The method can be used for a variety of biological materials, including urine, animal tissues, bacterial extracts, and yeast extracts. These materials can be assayed without any requirement for preliminary purification of the samples by chromatography on a separate cation exchange column, as required by many of the other methods used for amine analyses.

Method I is the procedure that we have used most frequently. It is a single buffer system and requires $2\frac{1}{2}$ hr for a full analysis of diamines, triamines, and tetraamines. Adenosylmethionine and decarboxylated adenosylmethionine can also be determined.

Method II gives better separation of the monoamines, the basic amino acids, and adenosylmethionine than Method I, but requires 8 hr. It involves elution by a single buffer followed by a gradient elution.

Method III is a system for the accelerated elution of the polyamines and requires 75 min.

Method IV is particularly useful for the separation of the two isomers of monoacetylspermidine and requires 2 hr.

General Procedures

We have used a Beckman amino acid analyzer, Model 121, equipped with programming and automated loading accessories, and a recorder modified so that a full-scale deflection is obtained with 40 nmol of putrescine. Ninhydrin detection is used but the procedure can be modified so that fluorescent detection systems (such as those based on *o*-phthalaldehyde) can be used.

Column. We have used the cation-exchange resin Durrum DC-6A, packed in a 0.9 × 10 cm Beckman glass column. The resin is regenerated automatically after each run by a 15-min wash period with 0.2 *N* NaOH–0.1% EDTA followed by a 40-min equilibration with the buffer used for the elution.

Ninhydrin Reagent. Ninhydrin solution is prepared from the ninhydrin reagent kit of Pierce Co. (catalog No. 21010) omitting the titanous chloride solution. After N_2 is bubbled through this mixture for 30 min, 2.4 g of cyclohexylenedinitrilotetracetic acid (CYDTA) and 0.28 g of ascorbic

[1] H. Tabor, C. W. Tabor, and F. Irreverre, *Anal. Biochem.* **55,** 457 (1973). See also H. Tabor and C. W. Tabor, *J. Biol. Chem.* **250,** 2648 (1975); C. W. Tabor, H. Tabor, and E. W. Hafner, *ibid.* **253,** 3671 (1978).

acid are added. The mixture is then poured into the Beckman ninhydrin container supplied with the amino acid analyzer. N_2 is bubbled through this container with the ninhydrin solution for 30 min before sealing the container. (We have found that the above formulation, which is recommended in the manual for the Hitachi amino acid analyzer, eliminates the problems of clogging of the reaction coil due to metallic deposits.)

Running Conditions. The column temperature is 54°; the pressure is 125–150 psi. The flow rate of the buffer is 70 ml/hr; the flow rate of the ninhydrin solution is 35 ml/hr. The ninhydrin reaction coil is immersed in a 100° water bath.

Preparation of Samples

Escherichia coli B, grown to a density of 3.3×10^8 cells/ml in 50 ml of a phosphate-based minimal medium,[2] are harvested and extracted for 2 hr with 5 ml of 10% trichloroacetic acid at 0°. The extract is centrifuged, and the supernatant fluid is extracted three times with three volumes of diethyl ether. The residual ether is removed *in vacuo*. Analysis is carried out on 0.30 ml of this extract (equivalent to 10^9 cells) in a final volume of 1 ml. In a control analysis carried through this procedure none of the amines in the standard mixture used for Fig. 3 are found in the ether extract. This same procedure can be used for the preparation of extracts from other bacteria, viruses, fungi, or cultured eukaryotic cells.

Animal or plant tissues are rapidly removed and frozen in liquid nitrogen. The frozen sample (360 mg) is ground at 4° with 5 ml of 10% trichloroacetic acid and 400 mg of sand. After 2 hr at 4°, the precipitate is removed by centrifugation at 0°. The supernatant solution is extracted three times with three volumes of ether and the ether is removed *in vacuo*. Analysis is carried out on 0.3 ml of the extract (equivalent to 22 mg of tissue) in a final volume of 1 ml.

Urine samples that contain no protein are filtered through a Millipore filter 47 mm in diameter, 0.45 μm pore size. Urine samples that contain protein are carried through the trichloroacetic acid precipitation prior to filtration. Hydrolyzed samples of urine are prepared by refluxing the urine with an equal volume of concentrated HCl for 24 hr. The HCl is removed *in vacuo*. The residue is dissolved in 0.001 *N* HCl and filtered through a Millipore filter.

The samples are either used directly or diluted with sufficient water to obtain final concentrations of each amine below 40 nmol/ml. Prior to injection, each sample is adjusted to pH 3.25 by the addition of one-tenth

[2] H. J. Vogel and D. M. Bonner, *J. Biol. Chem.* **218,** 97 (1956).

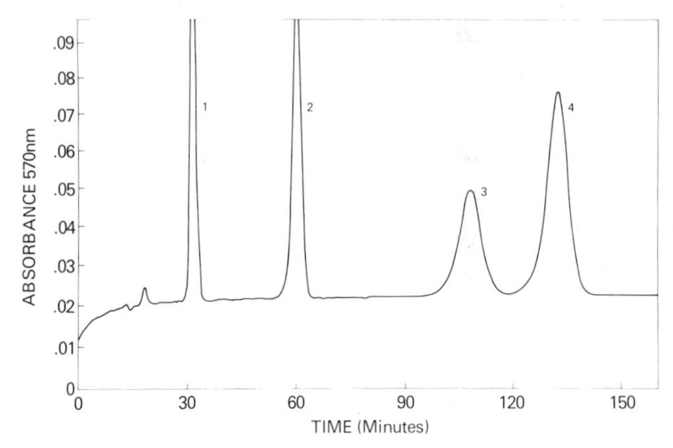

FIG. 1. Single-buffer system for elution of polyamines (Method 1). A mixture containing 20 nmol of each amine is applied to the column. The peak numbers represent the following amines: (1) putrescine; (2) spermidine; (3) agmatine; and (4) spermine. We have also found that adenosylmethionine is eluted at 27 min and that decarboxylated adenosylmethionine is eluted at 90 min.

the volume of the concentrated Beckman sodium citrate buffer (2.0 *M* in Na^+, pH 3.25).

Method I. Single-Buffer (High-Salt) Elution System

Reagents

Buffer I: sodium citrate, dihydrate, 406.5 g; sodium chloride, 1074 g; potassium chloride, 420.6 g; water (deionized or freshly distilled), 8 liters. Dissolve the salts with magnetic stirring. Add 45 ml of 4 *N* HCl, then add deionized water to exactly 12 liters. The buffer is filtered (with pressure or vacuum) through a 0.45 μm pore size HA Millipore filter (catalog No. SSWP 142-50) with a suitable filter holder (Millipore catalog No. YY2214230) and is stored at 4° in tightly closed screw-capped 1-liter sterile bottles containing 0.2 ml of caprylic acid per liter.

Procedure. The column is equilibrated with buffer I for 15 min; then the sample (0.5 ml) is loaded automatically onto the column. The elution is with buffer I for $2\frac{1}{2}$ hr, followed by a regeneration cycle. The elution pattern is shown in Fig. 1.

Comments. Buffer I elutes a large amount of ninhydrin-positive material in the first 20 min after loading a physiological sample, leading to excessively high base lines. To avoid this problem when using this method we have programmed the analyzer to divert the column eluate to drain for the first 17 min. This prevents material being eluted in this period from reacting with ninhydrin.

Method II. Two-Step System, with a Low-Salt Elution, Followed by a Gradient

Reagents. All buffers are filtered through a Millipore filter (142 mm in diameter, 0.45 μm pore size) to remove any particulate material.

Buffer IIA: A concentrated stock solution of sodium citrate (3.5 *M* Na^+) is prepared by dissolving 1373 g of sodium citrate · $2H_2O$ in distilled water, to a final volume of 4 liters. Buffer IIA (sodium citrate, 0.35 *M* Na^+) is prepared just before use by mixing 100 ml of this concentrate, 3.75 ml of 4 *N* HCl, 0.2 ml of caprylic acid, and sufficient water to adjust the final volume to 1 liter. The final pH is 6.32.

Buffer IIB. A sodium citrate (0.35 *M* Na^+)–sodium chloride (1 *M* Na^+) mixture is prepared by dissolving 137 g of sodium citrate · 2 H_2O and 234 g of NaCl in distilled water to a final volume of 4 liters; 15 ml of 4 *N* HCl and 0.8 ml of caprylic acid are added. The final Na^+ concentration is 1.35 *M* Na^+ and the final pH is 5.81.

Procedure. The column is equilibrated at 30° with buffer IIA [sodium citrate (0.35 *M* Na^+); pH 6.32] for 30 min. The sample is then loaded onto the column and elution with buffer IIA is carried out for $1\frac{1}{2}$–2 hr at 30°. The elution is then changed to a gradient (at 54°) between buffer IIB and buffer I and programmed over a 6-hr period by the pattern in Fig. 2 with

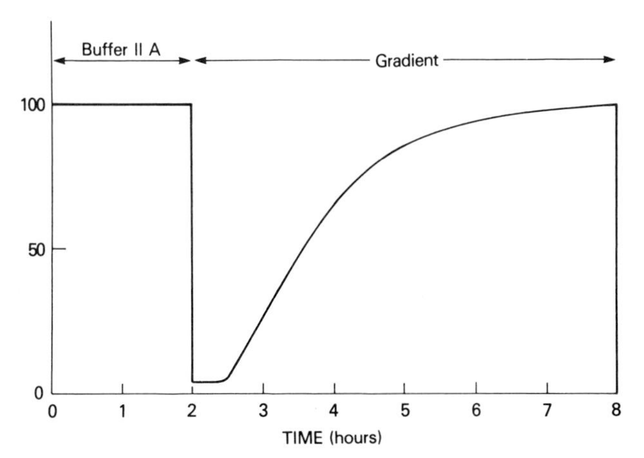

FIG. 2. Pattern used with Ultrograd for Method II. A single buffer is used for 2 hr (Buffer IIA), followed by a 6-hr gradient, as described in the text.

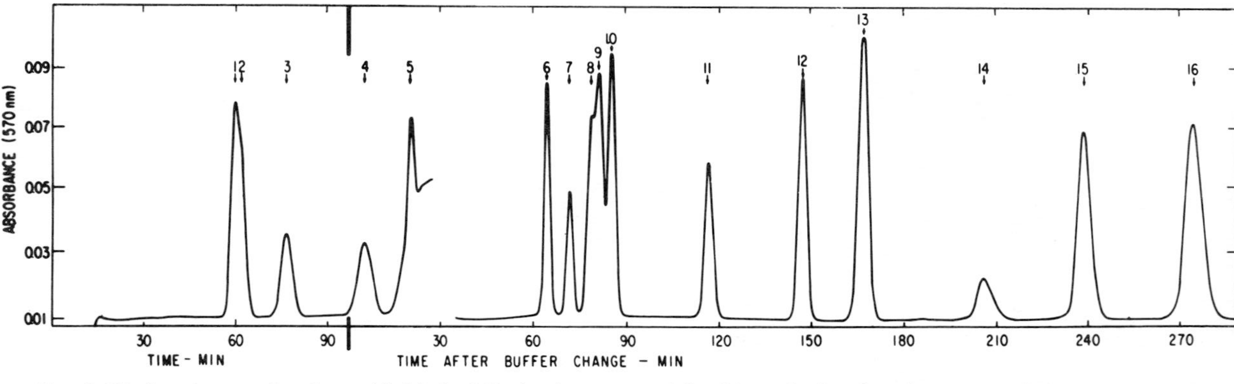

FIG. 3. Elution pattern of amines with Method II. A mixture containing 20 nmol of each amine was applied to the column. The vertical bar on the abscissa at 95 min indicates when the buffer and temperature were changed. (The appearance of the buffer change at the colorimeter occurs 22 min after the buffer change and causes a variable increase in the reading which returns to the base line again 13 min later. In this figure this artifactual peak has been deleted.)

The numbers above the peaks represent the amines listed below: (1) ammonia; (2) monoacetyl-1,3-diaminopropane; (3) monoacetyl-1,4-diaminobutane (monoacetylputrescine); (4) monocarbamyl-1,4-diaminobutane (monocarbamylputrescine); (5) arginine; (6) 1,4-diamino-2-hydroxybutane (hydroxyputrescine); (7) 1,3-diaminopropane; (8) N^1-monoacetylspermidine; (9) N^8-monoacetylspermidine; (10) 1,4-diaminobutane (putrescine); (11) 1,5-diaminopentane (cadaverine); (12) *N*-(3-aminopropyl)-1,3-diaminopropane; (13) spermidine; (14) agmatine; (15) *N*,*N*′-bis(3-aminopropyl)-1,3-diaminopropane; and (16) spermine. For the source of these compounds, see reference 1. [Reprinted from H. Tabor, C. W. Tabor, and F. Irreverre, *Anal. Biochem.* **55,** 457 (1973).]

the use of a gradient maker (LKB Ultrograd[3]). Alternatively, a three-vessel gradient can be made with the buffers described by Tabor *et al.*[1] to achieve the same gradient pattern. The elution pattern of a variety of amines obtained with the latter gradient is shown in Fig. 3.

When this method is used to determine only the more strongly basic substances (i.e., diamines, triamines, or tetraamines), the 2-hr elution with buffer IIA can be omitted. A 30-min preliminary elution with buffer IIB (with the eluate directed to the drain) is used before starting the gradient; the temperature is 54°.

Method III. Rapid (Single-Buffer) Elution

Reagents

Buffer III: A potassium citrate (0.35 M K^+–2 M KCl) mixture is prepared by dissolving 151 g of potassium citrate · H_2O and 596 g of KCl in distilled water to a final volume of 4 liters; 15 ml of 4 N HCl are then added. The final concentration of K^+ is 2.35 M.

Procedure. Buffer III is used at 54°. The column eluate is diverted to drain for 10 min after the elution is started.

Comments. A very much more rapid elution than that of Method I is obtained with this procedure. The elution times (in minutes) are as follows: putrescine, 24; cadaverine, 35; spermidine, 37.5; spermine, 63.5; agmatine, 74.

Method IV. A Single-Buffer System for the Separation of N^1-Monoacetylspermidine and N^8-Monoacetylspermidine

The two isomers of monoacetylspermidine are not separated well by Methods I–III. Method IV has been developed to achieve this separation. However, the method cannot be used if 1,3-diaminopropane is present, since the separation of this diamine from N^8-monoacetylspermidine is poor.

Reagents

Buffer IV: NaCl, 234 g; concentrated Beckman sodium citrate (2 M Na^+, pH 3.25), 400 ml; H_2O to 4 liters; the final pH is 2.95; the final Na^+ concentration is 1.2 M.

Comment. The elution times (in minutes) are N^1-monoacetylspermidine, 82; N^8-monoacetylspermidine, 91.5; hydroxyputrescine, 78; 1,3-diaminopropane, 92.5; putrescine, 107.

[3] The small mixing chamber that is usually used with the Ultrograd should not be used, since it introduces many air bubbles into the system.

Other Methods for Automated Cation-Exchange Chromatography of Polyamines

The methods described above give satisfactory separation of a large number of diamines, polyamines, and various derivatives of these amines, which have been described in microorganisms, in animal tissues, or in plant tissues.

Other methods have also been published that offer greater sensitivity by the use of fluorometric assays and shorter elution times; however, these methods have usually been limited to the separation of only putrescine, spermidine, and spermine.[4,5] A more sensitive and rapid method has been reported by Villaneuva and Adlakha[6] for the separation of a large number of phenolic amines, indoleamines, and basic amino acids, as well as monoamines, diamines, spermidine, and spermine.

Several other more rapid methods for putrescine, spermidine, and spermine have also been described, but there is little information in the descriptions on the separation of a variety of other basic compounds found in natural products from these three amines.[7]

[4] L. J. Marton and P. L. Y. Lee, *Clin. Chem.* (*Winston-Salem, N.C.*) **21,** 1721 (1975).

[5] G. Milano, M. Schneider, P. Cambon, J. L. Boublil, J. Barbe, N. Renee, C. M. Lalanne, *J. Clin. Chem. Clin. Biochem.* **18,** 157 (1980).

[6] V. R. Villaneuva and R. C. Adlakha, *Anal. Biochem.* **91,** 264 (1978).

[7] Short-column nonautomated methods have been described for the determination of putrescine, spermidine, and spermine by Y. Endo, *J. Chromatogr.* **205,** 155 (1981); H. Inoue and A. Mizutani, *Anal. Biochem.* **56,** 408 (1973). See also this series, Vol. 6 [90].

[5] A Double-Isotope Derivative Assay for Polyamines

By THOMAS J. PAULUS and ROWLAND H. DAVIS

The reaction of dansyl-Cl with polyamines is widely used for polyamine determinations.[1-3] Dansyl polyamines can be quantitated by direct fluorometric scanning after isolation by thin-layer chromatography on silica gel plates.[2,3] A simple modification of the basic dansyl-Cl assay

[1] Polyamines: putrescine (1,4-diaminobutane), spermidine, spermine, cadaverine (1,5-diaminopentane). Dansyl-Cl: 5-dimethylaminonaphthalene-1-sulfonyl chloride.

[2] N. Seiler, *Res. Methods Neurochem.* **3,** 409 (1975).

[3] S. S. Cohen, "Introduction to the Polyamines." Prentice-Hall, Englewood Cliffs, New Jersey, 1971.

ISBN 0-12-181994-9

procedure is to use [^{3}H]dansyl-Cl to measure the specific radioactivity of a known amount of [^{14}C]polyamine before and after addition to a sample containing an unknown amount of polyamine. The decrease in specific radioactivity is a measure of the polyamine content of the sample. A double-isotope derivative assay using dansyl-Cl was first used by Snodgrass and Iverson[4] for the determination of γ-aminobutyric acid and certain other amino acids from brain tissue. More recently, Airhart *et al.*[5] have expanded this technique to measure the specific radioactivities of amino acids labeled *in vivo* as well as amino acid levels in cultured cells. Here we describe the application of the double-isotope derivative assay to measure polyamine levels in acid extracts of *Neurospora crassa*.[6] We have also used the technique to measure the specific radioactivities of putrescine and spermidine labeled *in vivo*.[7]

Assay Method

Principle. The method depends upon the dilution of the specific radioactivity of a known quantity of [^{14}C]polyamine by a sample solution. Samples to which [^{14}C]polyamine is added are allowed to react with [^{3}H]dansyl-Cl. The dansyl polyamines are isolated by thin-layer chromatography, eluted into a scintillation vial, and counted. The ratio of ^{14}C cpm to ^{3}H cpm is directly related to the specific radioactivity of the [^{14}C]polyamine through Eq. (1).

$$\mathrm{SR}_{\text{polyamine}} = (^{14}\mathrm{C\ cpm}/^{3}\mathrm{H\ cpm}) \times \mathrm{SR}_{\text{dansyl-Cl}} \times N \quad (1)$$

where SR is specific radioactivity in counts per minute per nanomole, and N is the nanomoles of dansyl-Cl per nanomole of polyamine (i.e., 2 for putrescine and cadaverine, 3 for spermidine, and 4 for spermine). Actual specific radioactivities of samples and standards need not be calculated if the specific radioactivity of dansyl-Cl is constant. The ratio of ^{14}C cpm to ^{3}H cpm in the derivatized mixture is compared to that of the undiluted, derivatized ^{14}C standard, and the amount of polyamine in the test sample can be calculated by Eq. (2)[5]

$$\mathrm{nmol}_{\text{sample}} = \mathrm{nmol}_{\text{standard}} \left[\frac{(^{14}\mathrm{C\ cpm}/^{3}\mathrm{H\ cpm})_{\text{standard}}}{(^{14}\mathrm{C\ cpm}/^{3}\mathrm{H\ cpm})_{\text{mix}}} - 1 \right] \quad (2)$$

[4] S. R. Snodgrass and L. L. Iverson, *Nature (London) New Biol.* **241,** 154 (1973).

[5] J. Airhart, J. Kelley, J. E. Brayden, R. B. Low, and W. S. Stirewalt, *Anal. Biochem.* **96,** 45 (1979).

[6] T. J. Paulus and R. H. Davis, *J. Bacteriol.* **145,** 14 (1981).

[7] T. J. Paulus and R. H. Davis, *Biochem. Biophys. Res. Commun.* **104,** 228 (1982).

Materials

Putrescine, spermidine, spermine, cadaverine (hydrochlorides, Sigma Chemical Co.)
Dansyl-Cl (Sigma Chemical Co.)
Perchloric acid, 0.4 *M*
Na_2CO_3, anhydrous
Reagent-grade acetone, benzene, 1-butanol, chloroform, cyclohexane, dioxane, ethyl acetate, 2-propanol, triethanolamine.
Silica gel thin-layer chromatograms (Polygram sil G, 20 × 20 cm, 0.25 mm on plastic backing, Brinkmann Instruments Inc.).
[1,4-^{14}C]Putrescine dihydrochloride (90.2 mCi/mmol), [1,4-^{14}C]spermidine trihydrochloride (98.7 mCi/mmol), [1,4-^{14}C]spermine tetrahydrochloride (74.0 mCi/mmol), [1,5-^{14}C]cadaverine dihydrochloride (107.7 mCi/mmol), [*methyl*-^{3}H]dansyl chloride (36.7 Ci/mmol). All isotopes may be purchased from New England Nuclear.

Derivatization. Mycelia of *N. crassa* are cultured and harvested as previously described.[6,8,9] Mycelial pads (15–30 mg dry weight) are extracted three times with 1.0 ml of cold 0.4 *M* PCA. Supernatants, after low-speed centrifugation, are combined and saved for assay. A known amount of [^{14}C]polyamine is added to 0.1 ml of the PCA extract in a glass conical centrifuge tube. For perchloric acid extracts of *N. crassa,* a 0.5 nmol addition of [^{14}C]putrescine, [^{14}C]cadaverine, or [^{14}C]spermine (20 dpm/pmol) and a 10.0 nmol addition of [^{14}C]spermidine (2.0 dpm/pmol) are used routinely. However, the amount of standard may be varied to be approximately equal to the estimated polyamine content of the sample. A 0.1-ml aliquot of [^{3}H]dansyl-Cl (2.0–20.0 dpm/pmol, 1.5 mg per milliliter of acetone) is added. The reaction mixture is vortexed vigorously, made alkaline by the addition of 20–30 mg of Na_2CO_3, and then vortexed again. The tube is sealed with a rubber stopper and incubated in the dark at 37° for 1 hr. The reaction mixture is extracted with 0.5–1.0 ml of benzene. The benzene layer is removed and evaporated to dryness in 1.5-ml plastic conical tubes (Eppendorf microfuge tubes). This is most conveniently performed by placing the tubes in a fume hood overnight; however, the process can be accelerated to less than 1 hr by warming the samples to 45° and applying a gentle stream of air. Dried samples can be stored (in the dark) several days at room temperature without affecting the results. The yield under these reaction conditions is approximately 50%. The reaction can be incubated overnight at room temperature with slightly better yields. Complete recovery of polyamines is unnecessary, however. Therefore, the concentration of [^{3}H]dansyl-Cl was purposely kept low to

[8] R. H. Davis and F. J. de Serres, this series, Vol. 17A, p. 79.
[9] R. H. Davis and T. J. Paulus, this volume [16].

conserve isotope and to minimize extraneous spots. Other conditions (e.g., benzene extraction and evaporation) were picked for convenience. Partially dansylated polyamines remain at the origin during chromatography and do not interfere with the assay.

Thin-Layer Chromatography. Samples are redissolved in 50 μl of benzene, and 5–25 μl are spotted on a silica gel thin-layer chromatogram. The chromatograms are developed in the dark in either one or two dimensions. For one-dimensional development, the solvent is ethyl acetate–cyclohexane (2 : 3, v/v). Seven to nine samples may be spotted 2 cm apart and 2 cm from the bottom of a 20 × 20 cm plate. This is convenient for routine measurements of putrescine and spermidine, and occasionally of spermine. Low levels of spermine, and the presence of other dansylated products migrating near dansylspermine in this solvent system[10,11] make two-dimensional chromatography preferable for spermine measurements. The two-dimensional solvent system uses ethyl acetate–cyclohexane (2:3, v/v) in the first dimension, and chloroform–1-butanol–dioxane (48 : 1 : 1, v/v/v)[12] in the second dimension. Two-dimensional development is essential for complete separation of dansylcadaverine and lowers blank values significantly. The latter observation is most significant when low polyamine levels are expected and [^{3}H]dansyl-Cl of high specific radioactivity is required.

After development, the chromatogram is air-dried and the dansyl polyamines are eluted from the silica gel into scintillation vials by either of two methods.

METHOD 1. The chromatogram is sprayed heavily with triethanolamine–2-propanol (1 : 4, v/v) and air-dried again. The fluorescent spots are located under a low-energy UV light and marked with a pencil. The entire spot is cut from the chromatogram with scissors and placed in a scintillation vial with 5 ml of scintillation fluid (toluene with 5 g of 2,5-diphenyloxazole per liter). After allowing at least 4 hr for elution, the chromatogram slice can be removed from the vial, and the vial counted. Spraying the chromatogram with the triethanolamine reagent is necessary for efficient elution of the derivatives from the plate by the toluene-based scintillation fluid. The spray also stabilizes and enhances the fluorescence.[2] Vials are counted in a Beckman LS230 scintillation counter. The ^{14}C cpm to ^{3}H cpm ratio is determined after subtracting the proportion of ^{14}C cpm in the ^{3}H channel (determined using nonradioactive dansyl-Cl and [^{14}C]polyamines). The counting efficiency was about 20% for ^{3}H and >90% for ^{14}C with approximately 20% of the ^{14}C cpm in the ^{3}H channel.

[10] K. W. Nickerson, L. D. Dunkle, and J. L. Van Etten, *J. Bacteriol.* **129,** 173 (1971).
[11] T. J. Paulus, P. T. Kiyono, and R. H. Davis, *J. Bacteriol.* **152,** 291 (1982).
[12] L. Alhonen-Hongisto and J. Jänne, *Biochem. Biophys. Res. Commun.* **93,** 1005 (1980).

METHOD 2. After locating the spots under a low-energy UV light, the silica gel is removed from the plastic backing by "vacuum scraping" with the tip of a 5-in. disposable pipette fitted with a plug of glass wool halfway down the barrel; the silica gel is trapped against the glass wool by vacuum. The pipette is then placed tip down in a scintillation vial. Dansylpolyamines are eluted from the trapped silica gel into the vial with two 1-ml portions of acetone added to the large end of the pipette. A small amount of silica gel is occasionally washed into the vial but does not affect the results. The acetone is evaporated by warming the samples gently on a hot plate, and 5 ml of scintillation fluid is added. The samples are counted and analyzed as described for Method 1. Method 2 was found to be very convenient for two-dimensional chromatograms, whereas Method 1 was most useful for large numbers of chromatograms developed only in one dimension.

Blank Values. The critical blank value is the amount of ^3H background. This value is determined by constructing several samples in which [^3H]dansyl-Cl is added to tubes containing only perchloric acid. These perchloric acid blanks are prepared and processed with the normal contingent of samples. The dansyl polyamine areas of the blank chromatograms are cut and processed. The ^3H cpm background determined in this manner is from 0 to 400 cpm (after subtraction of machine background) and is dependent upon the specific activity of the [^3H]dansyl-Cl. Machine background is used routinely for the ^{14}C cpm blank. The cause of the relatively high ^3H background is not known; however, degradation or side products of the dansylation reaction during the development of the chromatogram are possibilities. The addition of L-proline (5 mg) toward the end of the dansylation reaction to remove unreacted dansyl-Cl had no effect on the ^3H background. Cleaning all glassware with chromic acid prior to use did not alter the blank values or the results of the assay. Activation of the silica gel plate (100–110° for 1 hr) also had no effect on the ^3H background.

Isotope Separation. A partial fractionation of tritiated and nontritiated dansyl derivatives is effected by chromatography on silica gel. An enrichment (up to fourfold) of ^3H radioactivity is seen in the lower half of a visibly symmetrical fluorescent spot; ^{14}C radioactivity from the polyamine moiety distributes evenly. This pattern is observed for [^3H]dansyl[^{14}C]putrescine, [^3H]dansyl[^{14}C]spermidine, and [^3H]dansyl-NH_2 (spermine and cadaverine have not been tested). A mixture of unreacted [^3H]- and [^{14}C]dansyl-Cl also exhibited an enrichment of ^3H in the lower portion of the spot when chromatographed on silica gel in a pentane–dichloromethane solvent (1 : 4, v/v). This means that when [^3H]dansyl-Cl is used for the double-isotope derivative assay the entire spot must be eluted from the chromatogram. Summing the radioactivity in both portions of the spot yields the same results as eluting the entire spot. The use of [^{14}C]dansyl-Cl with [^{3}H]polyamines would avoid this problem completely. However, [^{14}C]dansyl-Cl is more expensive than [^{3}H]dansyl-Cl and would add significantly to the cost of the assay.

Application

Pool Measurements. We have used the double-isotope derivative assay to measure polyamine pools from wild-type and mutant strains of *N. crassa* with a wide range of polyamine content.[6] The results are shown in this volume.[9]

We have nominally measured less than 5 pmol of putrescine using this assay; however, the true limit of detection is directly related to the variability of the standard (undiluted) ratios and the amount of polyamine added. The range of coefficients of variation for putrescine standard ratios determined in several separate experiments is 3.3 to 10.3% ($N \geq 4$). Thus, for the addition of 500 pmol to a sample solution, the valid detection limit is 17–57 pmol. Therefore, the limit of the assay for *N. crassa* is presently ±0.1 nmol per milligram dry weight. It is possible that the sensitivity of the assay can be improved by further optimization of the chromatographic procedures, e.g., the addition of carrier dansyl derivatives to the benzene extract prior to spotting.[4] We have not tested this extensively.

Specific Radioactivities of Polyamines Labeled in Vivo. We have used the double-isotope derivative assay procedure to measure the specific radioactivities of putrescine and spermidine following the administration of L-[U-^{14}C]ornithine to an exponentially growing culture of *N. crassa*.[7] Polyamines can be partially purified and concentrated from perchloric acid extracts of harvested mycelia using the ion-exchange column procedure of Inoue and Mizutani.[13] This procedure recovers polyamines in good yields (>90%) and removes a large portion of the amino acids and NH_3 that might compete for [^{3}H]dansyl-Cl. Polyamines are eluted in 6 *N* HCl, which is removed by evaporation at 55° under vacuum. The dried sample is redissolved in 0.4 *M* perchloric acid and allowed to react with [^{3}H]dansyl-Cl of known specific activity. Radioactive polyamine standards are not added in this case. Chromatograms (20 × 20 cm) are developed in two dimensions and Method 2 is used for elution of dansyl derivatives. Specific radioactivities are calculated from Eq. (1). The concentration of the [^{3}H]dansyl-Cl solution is measured by absorbance at 369 nm (ε = 3690)[14]; specific radioactivity of [^{3}H]dansyl-Cl solution is

[13] H. Inoue and A. Mizutani, *Anal. Biochem.* **56,** 408 (1973).

[14] W. Gray, this series, Vol. 25, p. 121.

measured by counting an aliquot of the solution (acetone is evaporated prior to addition of scintillation fluid).

Discussion

The double-isotope derivative assay has been useful for the measurement of polyamine pools from *N. crassa*. The technique has the advantages of being convenient, specific, and reasonably sensitive (17–57 pmol). While the traditional dansyl assay with direct fluorometric scanning of thin-layer plates has equal sensitivity and avoids the extensive use of radioactivity,[2,3] fluorometric scanning devices are not always available. Scintillation counters, however, are standard equipment in many laboratories. Therefore, the double-isotope derivative assay will be of use to those laboratories that anticipate a limited number of polyamine assays and in which a scanning fluorometer is not available. A major advantage of the double-isotope derivative assay is that it can be readily adapted to measure specific radioactivities of polyamines labeled *in vivo*. This may be very useful in investigations of polyamine metabolism.

Acknowledgments

This research was supported by grants from National Institutes of Health (AM-20083) and the American Cancer Society (BC-366).

[6] A Simple Method for the Determination of Polyamines and Histamine and Its Application to the Assay of Ornithine and Histidine Decarboxylase Activities

By YASUO ENDO

Various automated methods using ion-exchanger columns have been developed for the determination of putrescine and polyamines, but a simple manual method will be described here. The method is based on the separation of each amine on a small phosphorylated cellulose (P-cellulose) column (0.6 × 3 cm) and on its determination by spectrofluorometry or spectrocolorimetry.[1] With this method many samples could be analyzed concurrently by using many columns. Further, the method could be

[1] Y. Endo, *J. Chromatogr.* **205,** 155 (1981).

ISBN 0-12-181994-9

applied to simultaneous assay of ornithine and histidine decarboxylase activities.[2]

Procedures

Preparation of P-Cellulose Columns. The capacity and/or particle size of P-cellulose differs depending on the manufacturer. In this study, P-cellulose obtained from Brown Co., Berlin, New Hampshire (capacity 0.99 mEq/g) was used. The exchanger was swollen in water and the fine particles were removed by decantation and discarded. Then the exchanger was washed with 0.1 *M* NaOH, water, 0.1 *M* HCl, and water on a glass filter. After the exchanger was packed into columns (3×0.6 cm),[3] the columns were washed with 0.1 *M* NaOH (5 ml) and water (5 ml) and equilibrated with 0.03 *M* sodium phosphate buffer (pH 6.2) by passing 5 ml of the buffer through the column.

Preparation of Tissue Extracts. Animal tissues were homogenized in more than 10 volumes of 0.4 *M* $HClO_4$. The homogenates were centrifuged (5000 *g*, 5 min), and the supernatants were adjusted to pH 4–6 with 2 *M* KOH in an ice bath with magnetic stirring. The KOH solution was added through a fine tube with monitoring with a pH meter. Then the precipitate was removed by centrifugation (5000 *g*, 5 min), and the supernatants were used for the chromatography.

Separation Procedure. The neutralized supernatants (less than 5 ml) were applied to the P-cellulose columns. The separation procedure is summarized in Table I. Borate buffer (0.2 *M*, pH 8.5) was prepared from 0.2 *M* H_3BO_3 and 0.2 *M* Na_2CO_3. After washing the column with buffers 1, 2, and 3, histamine, putrescine, spermidine, spermine, and histone H1 were eluted step by step from the column, in this order, with borate buffers containing different amounts of NaCl.

Determination of the Amines

The eluate from the column containing each amine was subjected to the following reactions. Histamine was allowed to react with *o*-phthalaldehyde (OPA),[4] and putrescine and polyamines with fluorescamine (FA).[5] For the determination of large quantities of putrescine and poly-

[2] Y. Endo, *Biochem. Pharmacol.* **31,** 1643 (1982).

[3] The packed columns were cycled several times during the treatment with 0.1 *M* NaOH, water, and 0.03 *M* sodium phosphate buffer as described above. Columns with a 10-ml reservoir were used.

[4] P. A. Shore, A. Burkhalter, and V. H. Cohn, Jr., *J. Pharmacol. Exp. Ther.* **127,** 182 (1959).

[5] S. Udenfriend, S. Stein, P. Böhlen, W. Dairman, W. Leimgruber, and M. Weigele, *Science* **178,** 871 (1972).

TABLE I
SEPARATION PROCEDURE IN PHOSPHORYLATED CELLULOSE COLUMN CHROMATOGRAPHY[a]

Elution step and buffers (NaCl concentration in the buffers)	Elution volume (ml)	Separation
(1) 0.03 *M* phosphate buffer, pH 6.2	3.0	NH_3, many primary amines, amino acids, and peptides (discarded fraction)[b]
(2) 0.06 *M* phosphate buffer, pH 6.2	5.0	
(3) 0.1 *M* borate buffer, pH 8.5 (0.025 *M* NaCl)	1.5	
	3.0	Histamine, 1-methylhistamine for blank
	3.0	
(4) 0.2 *M* borate buffer, pH 8.5 (0.2 *M* NaCl)	3.0	Putrescine, cadaverine, agmatine
(5) 0.2 *M* borate buffer, pH 8.5 (0.4 *M* NaCl)	8.0	Spermidine
(6) 0.2 *M* borate buffer, pH 8.5 (0.6 *M* NaCl)	8.0	Spermine
(7) 0.2 *M* borate buffer, pH 8.5 (0.8 *M* NaCl)	5.0	Histone H1

[a] From Endo.[1]

[b] The following substances were eluted completely with buffers 1 and 2: aminoguanidine, tryptophan, 5-methoxytryptophan, tyrosine, phenylalanine, dopa, histidine, lysine, arginine, ornithine, epinephrine, norepinephrine, dopamine, metanephrine, normetanephrine, tyramine, serotonin, tryptamine, 5-methoxytryptamine, melatonin, glutathione, dithiothreitol, cysteine, histidylglycine, and carnosine.

amines (more than 20 nmol in the sample), these amines could be determined spectrocolorimetrically by the reaction with 2,4,6-trinitrobenzene sulfonate (TNBS). These reagents were prepared daily.

OPA Reaction. The sample (3 ml) was mixed with 2 *M* NaOH (0.4 ml), then OPA reagent (0.1% in methanol) (0.2 ml) was added and mixed by shaking. After the reaction mixture had been allowed to stand for 5 min at room temperature, 3.5 *M* H_3PO_4 (0.4 ml) was added to the mixture. The fluorescence was measured at 440 nm with excitation at 360 nm.

FA Reaction. While the sample solution (3 ml) was being vigorously shaken on a vortex-type mixer, FA reagent (15 mg/100 ml of dioxane) (1.0 ml) was added rapidly, by means of a syringe, at room temperature. The reaction was carried out for about 5 sec. The fluorescence was measured at 475 nm with excitation at 390 nm.

TNBS Reaction. TNBS reagent (100 mg of dimethyl sulfoxide per 150 ml) (1.0 ml) was added to the sample (3 ml). The reaction was carried out at 50° for 10 min and terminated by cooling the reaction mixture in water. Then the absorbance at 420 nm was measured.

Table II shows the sensitivity, linearity, and stability of these methods. Examples of the separation patterns of the amines are shown in Fig. 1. The recoveries of these amines were almost complete (more than 90%).

TABLE II
SENSITIVITY, LINEARITY, AND STABILITY[a]

Amines	Reactions	Sensitivity[b] (nmol)	Linearity[c] (nmol)	Stability[d] (hr)
Histamine	OPA[e]	0.15	0–100	1
Putrescine	FA[e]	0.9	0–50	3
Spermidine	FA	0.7	0–50	3
Spermine	FA	0.6	0–50	3
Histone H1	FA	0.025 (0.5 μg)	0–1	1
Putrescine	TNBS[e]	22	0–150	1
Spermidine	TNBS	20	0–150	1
Spermine	TNBS	21	0–150	1
Histone H1	TNBS	3.3 (70 μg)	0–30	1

[a] From Endo.[1]
[b] Amount of amines that gave a fluorescence intensity or absorbance five times higher than that of the reagent blank.
[c] Range of concentration of amines that gave a linear relationship between fluorescence intensity or absorbance and the concentration of amines.
[d] Minimum time during which the fluorescence intensity or absorbance remained constant.
[e] OPA, *o*-phthalaldehyde; FA, fluorescamine; TNBS, 2,4,6-trinitrobenzene sulfonate.

Application to the Simultaneous Assay of Ornithine and Histidine Decarboxylase Activities

Since putrescine and histamine could be separated completely from large quantities of ornithine and histidine by the P-cellulose column chromatography, the present method could be applied to the assay of the two enzymes.[2]

Preparation of Enzyme Solutions. Tissues were homogenized with 5–30 volumes of ice-cold 0.02 *M* sodium phosphate buffer (pH 6.2) containing pyridoxal 5′-phosphate (20 μ*M*) and dithiothreitol (0.2 m*M*). Homogenates were centrifuged at 20,000 *g* for 20 min. To the supernatants, P-cellulose powder[6] was added (20 mg/ml) and suspended by shaking. Then the suspension was centrifuged (5000 *g*, 5 min), and the resulting supernatant was used as the enzyme solution. This P-cellulose treatment could remove putrescine and histamine included in the enzyme solution

[6] P-Cellulose powder was prepared as follows. P-Cellulose equilibrated with 0.03 *M* sodium phosphate buffer (pH 6.2) was washed with water and acetone and dried on a glass filter with a drier.

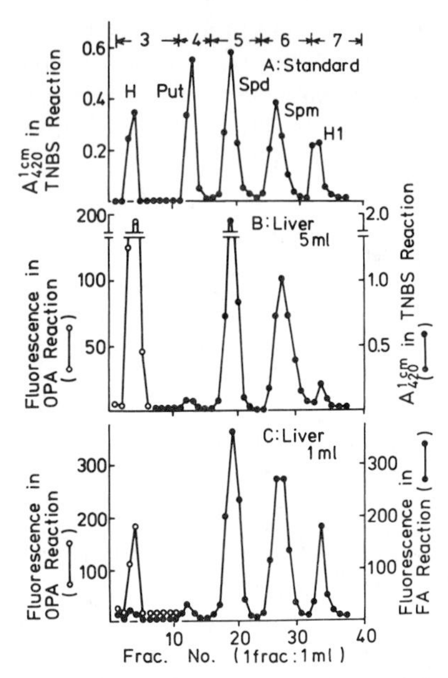

FIG. 1. Separation of histamine (H), putrescine (Put), spermidine (Spd), spermine (Spm), and histone H1 (H1) on a phosphorylated cellulose column.[1] Each substance was eluted step by step with buffers 1–7 (Table I). Fractions (1 ml) were collected, diluted to 3 ml with each elution buffer and subjected to the *o*-phthalaldehyde (OPA), 2,4,6-trinitrobenzene sulfonate (TNBS), or fluorescamine (FA) reaction. (A) 0.4 *M* $HClO_4$ (4 ml) containing 100 nmol of each amine and 10 nmol of calf thymus histone H1 was neutralized with KOH, and the supernatant was applied to the column. (B) Neutralized extract from rat liver (5 ml) corresponding to about 0.5 g of the tissue was applied. In the OPA reaction, the control fluorescence intensity was 100 for 1 nmol of histamine. (C) The same sample as B (1 ml) corresponding to about 0.1 g of the liver was applied. Control fluorescence intensities in the OPA and FA reactions were 100 for 0.2 nmol of histamine and 100 for 10 nmol of putrescine, respectively.

without any loss of the enzyme activities and markedly lowered the blank value in the enzymatic reactions.

Enzymatic Reaction. The enzymatic reactions of the two enzymes were carried out in a single reaction mixture. The reaction mixture contained 0.2 ml of 0.2 *M* sodium phosphate buffer (pH 6.7), pyridoxal 5′-phosphate (50 nmol), dithiothreitol (500 nmol), aminoguanidine sulfate (50 nmol),[7] L-histidine · HCl (1 μmol), L-ornithine · HCl (1 μmol), and the

[7] An inhibitor of histaminase or diamine oxidase [M. A. Beaven and R. E. Shaff, *Biochem. Pharmacol.* **24,** 979 (1975)]. I found that aminoguanidine inhibits ornithine decarboxylase (about 50% at 0.5 m*M*). Therefore, in the absence of significant amounts of diamine oxidase (amine oxidase), this agent should be omitted.

enzyme solution in a final volume of 1 ml. The reaction was carried out at 37° for 1–3 hr and terminated by the addition of 0.4 *M* $HClO_4$ (2.5 ml). Another reaction mixture was usually used as the blank for the reaction without incubation. Putrescine and histamine in the reaction mixture were determined as described above.

Remarks

1-Methylhistamine was eluted from the column together with histamine. This amine, however, showed no fluorescence in the OPA reaction. Other known interfering substances in the OPA reaction were removed from the histamine fraction during the chromatography.[1] Cadaverine and agmatine, which were not separated from putrescine, are not present in detectable amounts in animal tissues.[8–10]

The capacities of cellulose exchangers are variable with different lots or the particle size. It is necessary, therefore, to adjust the size of the column or to design the volume of elution buffers after preliminary examination.

With respect to the manual methods using small exchanger columns for the separation of polyamines, there are a few methods comparable to the present one. Small exchanger columns are currently used for the preparation of salt-free samples or for partial purification. Inoue and Mizutani used a small Dowex 50 column for the preparation of samples for paper electrophoresis.[11] I reported previously a method using a CM-cellulose column (0.6 × 10 cm) for the separation of histamine, putrescine, and polyamines.[12] The present method has some advantages over the previous one: (*a*) each amine is eluted from the column with a smaller volume of buffer; therefore, the separation can be carried out without the use of a fraction collector and it is possible to analyze many samples simultaneously; (*b*) higher fluorescence intensities of polyamines in the FA reaction (2–3 times) were obtained by the use of elution buffers containing larger amounts of NaCl; (*c*) the method can be applied to the simultaneous assay of ornithine and histidine decarboxylase activities; (*d*) histone H1 can be determined in the same way.[1]

[8] H. Tabor, C. W. Tabor, and F. Irreverre, *Anal. Biochem.* **55,** 457 (1973).
[9] L. J. Marton, O. Heby, C. B. Wilson, and P. L. Y. Lee, *FEBS Lett.* **41,** 99 (1974).
[10] S. I. Harik, G. W. Pasternak, and S. H. Snyder, *Biochim. Biophys. Acta* **304,** 753 (1973).
[11] H. Inoue and A. Mizutani, *Anal. Biochem.* **56,** 408 (1973).
[12] Y. Endo, *Anal. Biochem.* **89,** 235 (1978).

[7] Gas Chromatographic–Mass Spectrometric Analysis of Polyamines and Polyamine Conjugates

By G. DOYLE DAVES, JR., RONALD G. SMITH, and CORNELIUS A. VALKENBURG

Selected ion monitoring (SIM), an analytical method in which a mass spectrometer (usually in combination with a gas chromatograph) continuously monitors the ion current at preselected mass-to-charge (m/z) ratios for the detection and quantitative determination of ions, has been developed as an assay of polyamines.[1,2] When used with deuterium-labeled internal standards for each analyte, this assay combines high selectivity with sensitivity to the picomole level. These advantages are best utilized when the accurate quantitation of specific polyamines at low concentrations is desired. For this assay, derivatization of polyamines by fluoroacetylation is usual; however, trimethylsilylation has also been used.

The discovery of polyamines as monoacetylated conjugates in urine[3] and 1-*N*-acetylspermidine in human serum[4] has created the need for a quantitative method for the analysis of these compounds. Deuterium-labeled analogs of these conjugates have been synthesized, allowing for their determination by SIM. Schiff base conjugates of polyamines with pyridoxal or pyridoxal phosphate have been identified in urine[5]; their quantitative analysis should be possible by the SIM method following the synthesis of appropriate deuterium-labeled analogs.

Synthesis of Deuterium-Labeled Internal Standards

The individual polyamines differ with respect to their extraction efficiencies, adsorption to glass[6] and other materials, chromatographic behavior, and susceptibility to decomposition due to trace metals, enzymes, or bacteria. These differences preclude the use of a single internal stan-

[1] R. G. Smith and G. D. Daves, Jr., *Biomed. Mass Spectrom.* **4,** 146 (1977).
[2] R. G. Smith, G. D. Daves, Jr., and D. P. Grettie, *Adv. Polyamine Res.* **2,** 23 (1978).
[3] T. Walle, *in* "Polyamines in Normal and Neoplastic Growth" (D. H. Russell, ed.), p. 355. Raven, New York, 1973.
[4] R. G. Smith, D. Bartos, F. Bartos, D. P. Grettie, W. Frick, R. A. Campbell, and G. D. Daves, Jr., *Biomed. Mass Spectrom.* **5,** 515 (1978).
[5] R. Aigner-Held, R. A. Campbell, and G. D. Daves, Jr., *Proc. Natl. Acad. Sci. U.S.A.* **76,** 6652 (1979).
[6] D. P. Grettie, D. Bartos, F. Bartos, R. G. Smith, and R. A. Campbell, *Ad. Polyamine Res.* **2,** 13 (1978).

ISBN 0-12-181994-9

$H_2NCD_2CH_2CH_2CD_2NH_2$
Putrescine-d_4

$H_2NCD_2CH_2CH_2CH_2CD_2NH_2$
Cadaverine-d_4

$H_2NCD_2CH_2CH_2NHCD_2CH_2CH_2CD_2NH_2$
Spermidine-d_6

$H_2NCD_2CH_2CH_2NHCD_2CH_2CH_2CD_2NHCH_2CH_2CD_2NH_2$
Spermine-d_8

O
||
$CH_3CNHCD_2CH_2CH_2CD_2NH_2$
Acetylputrescine-d_4

O
||
$CH_3CNHCD_2CH_2CH_2NHCD_2CH_2CH_2CD_2NH_2$
N^1-Acetylspermidine-d_6

O
||
$H_2HCD_2CH_2CH_2NHCD_2CH_2CH_2CD_2NHCCH_3$
N^8-Acetylspermidine-d_6

FIG. 1. Synthetic deuterium-labeled analogs.

dard for quantitative analysis of multiple polyamines. Use of homologous internal standards for the analysis of compounds having hydrogen bonding or other internal interactions must be approached with caution. An isotopically labeled analog is an ideal internal standard that has chemical and physical properties essentially identical with those of the analyte and assures negligible differences in extraction, derivatization, and chromatographic properties.

Deuterium-labeled polyamines were prepared by modifications of known procedures[7–10] and resulted in products of high isotopic purity that are labeled exclusively at the positions shown in Fig. 1.

Putrescine-d_4. A mixture of succinonitrile (0.80 g, 10 mmol) and platinum oxide (100 mg) in 20 ml of *O*-deuterioethanol (95% solution in deuterium oxide, 99% isotopically pure) and 3 ml of deuterium chloride (37% solution in deuterium oxide, 99% isotopically pure) was shaken under 2 atmospheres of deuterium (99.5% isotopically pure) until gas uptake ceased. Crude product, obtained by removal of catalyst and evaporation of the solvents *in vacuo,* was recrystallized from 100% ethanol to give

[7] H. A. Fischer, *J. Labelled Compd.* **11,** 141 (1975).
[8] H. A. Fischer, N. Seiler, and G. Werner, *J. Labelled Compd.* **7,** 175 (1971).
[9] D. T. Dubin and S. M. Rosenthal, *J. Biol. Chem.* **235,** 776 (1960).
[10] H. Tabor, S. M. Rosenthal, and C. W. Tabor, *J. Biol. Chem.* **233,** 907 (1958).

0.66 g (41%) of putrescine-d_4 · 2HCl, mp 310° (lit. mp 315°[11] for unlabeled putrescine · 2HCl).

Cadaverine-d_4. Similar reduction of glutaronitrile yielded cadaverine-d_4 · 2HCl, mp 258–260° (lit. mp 275°[11] for the undeuterated compound).

Spermidine-d_6 and Spermine-d_8. To a suspension of 300 mg of putrescine-d_4 · 2HCl in 40 ml of ethanol was added 0.4 ml of 4.8 *N* aqueous sodium hydroxide. After solution was effected, 120 μl of acrylonitrile was added slowly. The mixture was stirred at room temperature for 24 hr, then 4.0 ml of 6 *N* HCl was added and the solvents were removed under reduced pressure. To the solid residue in 30 ml of deuterium oxide and 2.0 ml of 37% deuterium chloride (in deuterium oxide) was added 50 mg of platinum oxide, and the mixture was shaken under 2 atmospheres of deuterium for 36 hr. The catalyst was removed by filtration, the solvents were evaporated under reduced pressure, and the solid residue was dissolved in water and placed on a column (2 × 20 cm) of ion exchange (Bio-Rad AG 50W-X2, prepared by washing with 0.1 *N* HCl).[12] The column was eluted using a gradient achieved by placing 300 ml of water in a mixing vessel and adding 700 ml of 2.5 *N* HCl at a rate equal to the column elution rate. Fractions (15 ml) were collected, and elution of polyamines was monitored using ninhydrin. Fractions 28–37 were pooled and evaporated to yield 32 mg of spermidine-d_6 · 3HCl, and fractions 52–67 yielded 117 mg of spermine-d_8 · 4HCl (yields following recrystallization from ethanol).

1-N-Acetylspermidine-d_6. Using the method of Tabor *et al.*,[13] 0.3 g of spermidine-d_6 · 3HCl in 1.2 ml of water and 6 ml of pyridine was stirred while 0.5 ml of acetic anhydride was added during 30 min. After 24 hr the reaction mixture was evaporated *in vacuo* and the residue was dissolved in 0.75 *N* NH_4OH and placed onto a column (3 × 20 cm) of Amberlite CG-50 type II, NH_4^+ ion form.[14] The column was eluted with 100 ml of 0.75 *N* NH_4OH and then with 2.8 *N* NH_4OH. Fractions were monitored with ninhydrin. Solvent was removed from ninhydrin-positive fractions after pooling and the residue was twice dissolved in methanol followed by evaporation. The residue was then dissolved in 2-propanol (chilled in an ice bath) and a cold solution of 1 : 1 2-propanol–conc. HCl was added dropwise until acid to litmus. Crystals formed overnight to yield 16 mg of

[11] G. Klein and D. Boser, *Arch. Pharm. Ber. Dtsch. Pharm. Ges.* **270,** 374 (1932).
[12] C. W. Tabor and S. M. Rosenthal, this series, Vol. 6, p. 615.
[13] H. Tabor, C. W. Tabor, and L. de Meis, this series, Vol. 17B, p. 829.
[14] A stepwise gradient modification of M. Tsuji, T. Nakajima, and I. Sano, *Clin. Chim. Acta* **52,** 161 (1975).

1-*N*-acetylspermidine-d_6 · 2HCl, mp 171–176° (lit.[13] mp 173–178° for undeuterated analog).

8-N-Acetylspermidine-d_6. Crude acetylputrescine-d_4 prepared as described[13] was purified by ion-exchange chromatography on Amberlite CG-50 (see above) using 1.7 *N* NH_4OH to elute acetylputrescine-d_4. Acetylputrescine-d_4 (0.84 g) and 0.36 g of acrylonitrile in 1.1 ml of 4.8 *N* sodium hydroxide and 6.3 ml of ethanol were stirred at room temperature for 18 hr and then heated on a steam bath for 1 hr. The solution was cooled, the pH was adjusted to 3 with hydrochloric acid, and the solution was evaporated *in vacuo*. The residue was dissolved in 20 ml of *O*-deuterioethanol and 3 ml of 37% aqueous deuterium chloride; 100 mg of platinum oxide was added and the mixture was shaken under 2 atmospheres of deuterium for 36 hr. The catalyst was removed, the solvent was evaporated, and the residue was dissolved in 0.75 *N* NH_4OH and subjected to ion-exchange chromatography on Amberlite CG-50 (see above) using 1.7 *N* NH_4OH for elution of 8-*N*-acetylspermidine-d_6 (following elution of some unchanged acetylputrescine-d_4). Recrystallization from ethanol yielded 52 mg of 8-*N*-acetylspermidine-d_6, mp 201–206° (lit.[13] mp 204–205.5°).

Biological Sample Preparation and Storage

Isotopically labeled internal standards were added to biological samples before processing to ensure that processing losses are accounted for (i.e., ratios of endogenous polyamines to added standards are unaffected by sample processing). Biological samples must undergo a separation step before derivatization and analysis by SIM, since large sample residues due to salts interfere with manipulation of the small volumes of solvents and reagents used during derivatization. In addition, residual proteins interfere with trifluoroacetylation. Serum samples were prepared by chromatography on specially prepared silica gel[6] using 0.01–0.03 *N* hydrochloric acid for elution of polyamines. This eluate, when dried, gave a residue of polyamine salts that is stable and suitable for storage.

Derivatization

Derivatization for the gas chromatographic separation of polyamines has usually been achieved by trifluoroacetylation; occasionally trimethylsilylation has been used. The trifluoroacetylation reaction generates trifluoroacetic acid (TFA) which catalyzes the removal of conjugating groups by transacylation. Therefore, quantitative determination of acetylated polyamines has required a modification to maintain the integrity of these conjugates.

Trifluoroacetylation of Polyamines. Samples are derivatized in 0.3-ml screw-cap conical vials by the action of 100 μl of trifluoroacetic anhydride for 10–15 min in a sonic bath (or 3 min in a steam bath). The excess reagent was removed under a stream of nitrogen and the sample was reconstituted in a small volume (10–50 μl) of dichloromethane. This procedure adds one trifluoroacetyl group to each primary and secondary amine. It also converts acetylated polyamines to the same derivatives obtained from free polyamines and is therefore used for assays of total polyamines. Longer heating periods (2–3 hr at 80–100°) result in overderivatization.

Trifluoroacetylation of Acetylpolyamines. To the dried sample in a 1-dram vial was added 30–40 mg of anhydrous sodium carbonate followed by 0.2–0.8 ml of trifluoroacetic anhydride. The vial was sealed with a Teflon-lined screw cap and heated on a steam bath for 15–30 min. After cooling to room temperature the vial was carefully opened (evolution of CO_2) and 0.5 ml of dichloromethane was added. A saturated solution of aqueous sodium carbonate was then added slowly until the evolution of carbon dioxide ceased. The organic phase was separated and washed twice with water to remove remaining carbonate and then transferred to a dry vial and evaporated to dryness. The residue was reconstituted in 25 μl of dichloromethane.

Trimethylsilylation. Dried samples in 0.3-ml screw-cap vials are treated with a 2:2:1 mixture of pyridine, *N,O*-bis(trimethylsilyl)acetamide, and trimethylchlorosilane and placed in a sonic bath or a steam bath for 15 min. In this procedure all N hydrogens are replaced by trimethylsilyl (TMS) groups.

Gas Chromatography and Mass Spectrometry

Gas Chromatography. The TFA polyamine derivatives were chromatographed on a 2 m × 2 mm (i.d.) glass column packed with 3% OV-17 coated on either 100/200 Gas Chrom Q or 80/100 Chromosorb W-HP. Before installing a new column it was flushed overnight at 280° with a slow flow of oxygen-free helium (or nitrogen). Further improvement, as determined by chromatographic peak shape, is sometimes possible using a silylating column conditioner. The glass interface and injection port were kept above 240° to prevent peak broadening.

The wide range in volatility of the polyamine derivatives (TFA and TMS) necessitates temperature programming, usually 120–300° at 15°/min. The corresponding derivatives of acetylated putrescine and spermi-

REFERENCES FOR MASS SPECTRA AND SPECTRAL DATA

Compound	TFA[a]			TMS[a]
	EI	Positive CI	Negative CI	EI
Putrescine	3	16[b]	16[b]	1,[b] 15[b]
Cadaverine	3	—	—	15[b]
Spermidine	1,[b] 3	16[b]	16[b]	15[b]
Spermine	2,[b] 3	16[b]	16[b]	15[b]
1-*N*-Acetylspermidine	4	—	—	—
8-*N*-Acetylspermidine	4	—	—	—
1-*N*-Spermidine–pyridoxal Schiff base	5	—	—	—
8-*N*-Spermidine–pyridoxal Schiff base	5	—	—	—

[a] TFA, Trifluoroacetyl derivatives; TMS, trimethylsilyl derivatives; CI, chemical ionization; EI, electron ionization.
[b] Deuterium-labeled analog included.

dine also elute under these conditions. The TFA derivatives of the polyamine–pyridoxal Schiff bases require temperature programming to 400° on a 3% Dexil 300 column.[5]

Mass Spectra. Mass spectra and spectral data for a number of derivatized polyamines and polyamine conjugates have been published (see the table). The spectra of the TFA polyamines are typified by numerous fragmentations and rearrangements. A characteristic rearrangement of 1-*N*-acetylspermidine was used to identify this isomer in human serum.[4] Spectra of polyamine TMS derivatives exhibit a number of novel rearrangements.[15] Positive and negative chemical ionization (methane) spectra of the TFA derivatives exhibit relatively few ions and are dominated by $[M + H]^+$ and $[M - HF]^-$ ions.[16] Monitoring of high-abundance negative ions represents a further increase in SIM sensitivity.

Selected Ion Monitoring

This method has been used for quantitative analysis of polyamines in urine and serum and to demonstrate the presence of polyamine–pyridoxal Schiff base conjugates in urine. Quantitative analysis of a polyamine or acetylpolyamine by SIM is accomplished by monitoring simultaneously an ion (e.g., $[M - CF_3]^+$) characteristic of the polyamine TFA derivative

[15] R. G. Smith and G. D. Daves, Jr., *J. Org. Chem.* **43,** 2178 (1978).
[16] J. R. Shipe, Jr., D. F. Hunt, and J. Savory, *Clin. Chem.* (*Winston-Salem, N.C.*) **25,** 1564 (1979).

and the corresponding ion for the added deuterium-labeled analog. The ratios of the resulting ion abundances and the known level of added internal standard establish the level of the endogenous analyte. Large excesses of the internal standards are possible because of the high isotopic enrichment factor (EF) achieved during synthesis of the isotopically labeled analogs. The EF is a measure of the contribution by the labeled ion to the observed intensity at the *m/z* value for the corresponding ion of the unlabeled compound, i.e.,

$$EF = [I_L/(I_U + I_L)] \times 100 \text{ atom } \%$$

where I_U and I_L are the ion intensities of an unlabeled ion and its fully labeled analog. The enrichment factors for the synthetic deuterated polyamines (as determined from $[M - CF_3]^+$ ions in spectra of the TFA derivatives), are 99.83–99.95%. Ions were monitored by a four-channel SIM accessory to a DuPont 21-491B mass spectrometer. Monitoring the four polyamines was possible by manually switching the channels to different *m/z* values between elution of the various polyamine derivatives.

The standard curves of this assay provide a linear response over the range of 1.0–100 pmol for each polyamine when 100 pmol of the internal standards are used. The precision of measurement for samples of this size is very good with coefficients of variation less than 8%. Polyamine concentrations in pooled normal serum have been determined with equal precision and the values are comparable to those obtained from high-pressure cation-exchange chromatography and radioimmunoassay.[17]

[17] F. Bartos, D. Bartos, D. P. Grettie, R. A. Campbell, L. J. Marton, R. G. Smith, and G. D. Daves, Jr., *Biochem. Biophys Res. Commun.* **75,** 915 (1977).

Section II

Analytical and Preparative Methods for Adenosylmethionine, Decarboxylated Adenosylmethionine, and Related Compounds

[8] High-Performance Liquid Chromatographic Analysis of Adenosyl-Sulfur Compounds Related to Polyamine Biosynthesis

By VINCENZO ZAPPIA, MARIA CARTENÌ-FARINA, PATRIZIA GALLETTI, FULVIO DELLA RAGIONE, and GIOVANNA CACCIAPUOTI

The adenosyl-5′-sulfur nucleosides represent key intermediates in the biosynthesis of aliphatic polyamines, as well as in the transmethylation pathway.[1] A variety of methods for estimation of *S*-adenosylmethionine and its metabolic products have been proposed in the literature.[2-4] However, they are generally quite laborious and permit the estimation of only a single molecular species. On the other hand, the study of the metabolic pathway involving *S*-adenosylmethionine (AdoMet) and its related adenosyl-sulfur compounds [*S*-adenosyl-(5′)-3-methylthiopropylamine (decarboxylated AdoMet), *S*-adenosylhomocysteine (AdoHcy), and 5′-methylthiopropylamine (MTA) (see scheme in Fig. 1)] can be greatly enhanced by the possibility of a quantitative estimation of these molecules.

This chapter describes a rapid high-pressure liquid chromatography method for the simultaneous estimation of several AdoMet metabolites.[5,6] This procedure represents an excellent tool for the study of the overall metabolism of AdoMet in cell extracts, and it is also suitable for checking the chemical purity of adenosyl-sulfur compounds used in biochemical assays.

Reference Compounds

Since commercial *S*-adenosylmethionine is contaminated by numerous impurities, the compound is routinely prepared from cultures of *Saccharomyces cerevisiae* and isolated by ion-exchange chromatography.[7]

[1] F. Salvatore, E. Borek, V. Zappia, H. G. Williams-Ashman, and F. Schlenk, "The Biochemistry of Adenosylmethionine." Columbia Univ. Press, New York, 1977.

[2] F. Salvatore, R. Utili, V. Zappia, and S. K. Shapiro, *Anal. Biochem.* **41,** 16 (1971).

[3] T. O. Eloranta, E. O. Kayander, and A. M. Raina, *Biochem. J.* **160,** 287 (1976).

[4] J. Seidenfeld, J. Wilson, and H. G. Williams-Ashman, *Biochem. Biophys. Res. Commun.* **95,** 1861 (1980).

[5] V. Zappia, P. Galletti, M. Porcelli, C. Manna, and F. Della Ragione, *J. Chromatogr.* **189,** 399 (1980).

[6] F. Della Ragione, M. Cartenì-Farina, M. Porcelli, G. Cacciapuoti, and V. Zappia, *J. Chromatogr.* **226,** 243 (1981).

[7] S. K. Shapiro and D. J. Ehninger, *Anal. Biochem.* **15,** 323 (1966).

ISBN 0-12-181994-9

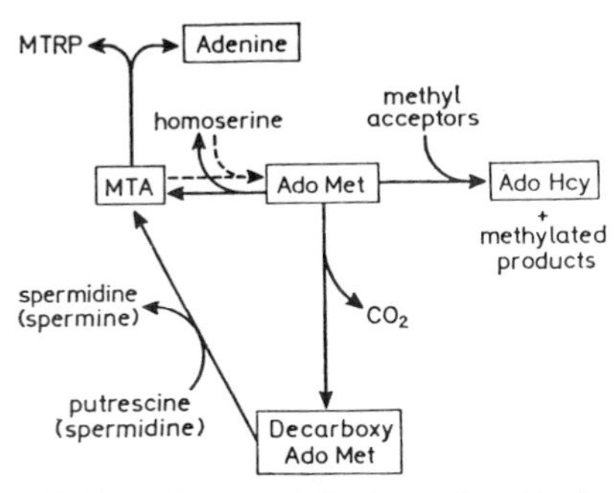

Fig. 1. The main metabolic pathways of *S*-adenosylmethionine in mammalian tissues. The molecules in the boxes can be separated and analyzed by the aid of the reported high-performance liquid chromatographic systems. AdoMet = *S*-adenosyl-L-methionine; AdoHcy = *S*-adenosyl-L-homocysteine; decarboxy-AdoMet = *S*-adenosyl-(5′)-3-methylthiopropylamine; MTA = 5′-methylthioadenosine; MTRP = 5-methylthioribose 1-phosphate. From Zappia *et al.*[5]

S-Adenosyl-L-[*methyl*-^{14}C]methionine is supplied by the Radiochemical Centre, Amersham United Kingdom. Decarboxylated AdoMet is obtained by enzymatic decarboxylation of AdoMet.[8] The decarboxylated sulfonium compound is purified according to the procedure of Zappia *et al.*[9] described in this volume [11]. 5′-Methylthioadenosine and 5′-[*methyl*-^{14}C]methylthioadenosine are prepared by acid hydrolysis of adenosylmethionine (pH 4.5, 100° for 30 min). The reaction is monitored spectrophotometrically using adenosine deaminase from *Aspergillus oryzae*.[10] *S*-Adenosyl-(5′)-3-methylthiopropanol (internal standard) is prepared by partial deamination of decarboxylated AdoMet.[11] Adenine, adenosine, and *S*-adenosylhomocysteine are obtained from Sigma Chemical Co., St. Louis, Missouri.

All standard solutions are prepared in water at a concentration of 2 m*M*.

High-Pressure Liquid Chromatography Equipment

A model LC-65T Perkin-Elmer liquid chromatograph equipped with a model LC-15 high-performance recording UV detector, operating at fixed wavelength of 254 nm, is used. The columns (25 cm × 4.6 mm) are prepacked with Partisil 10 SCX and Partisil 10 ODS RP-18, both from

[8] R. B. Wickner, C. W. Tabor, and H. Tabor, *J. Biol. Chem.* **245,** 2132 (1970).

[9] V. Zappia, P. Galletti, A. Oliva, and A. De Santis, *Anal. Biochem.* **79,** 535 (1977).

[10] T. K. Sharpless and R. Wolfenden, this series, Vol. 12, p. 126.

[11] V. Zappia, G. Cacciapuoti, G. Pontoni, and A. Oliva, *J. Biol. Chem.* **255,** 7276 (1980).

Whatman. Integration is performed electronically using a Spectra-Physic Minigrator.

Injection of sample was done via Model 70-10 sample injector valve and Model 70-10 loop filler port (Rheodyne). A Hamilton syringe (25–50 μl) is used to inject the sample volumes.

HPLC Analysis of Adenosyl-Sulfur Compounds

Operating Conditions. The separation of a standard mixture of AdoMet and its metabolites by cation-exchange HPLC (Partisil 10 SCX) is reported in Fig. 2. AdoHcy, adenine, MTA, AdoMet, *S*-adenosyl-(5′)-3-methylthiopropanol (internal standard), and decarboxylated AdoMet are resolved at room temperature, with 0.5 *M* ammonium formate buffer pH 4 as eluent. The flow rate, 3 ml/min, produces a pressure drop of 3000 psi.

Sensitivity, Reproducibility, and Linearity. The concentration of adenosyl-sulfur compounds is determined from calibration curves constructed by plotting the ratio of peak area measurement of the compounds to internal standard (10 nmol), over the concentration range. *S*-Adenosyl-(5′)-3-methylthiopropanol, a synthetic analog of decarboxylated AdoMet, is used as internal standard.

The lower limit of detection for the tested compounds in our chromatographic system depends on the detector-sensitive noise level and column efficiency. In the system used, 50 pmol of the above-mentioned compounds are easily detectable and the peak shape remains constant up to 50 nmol.

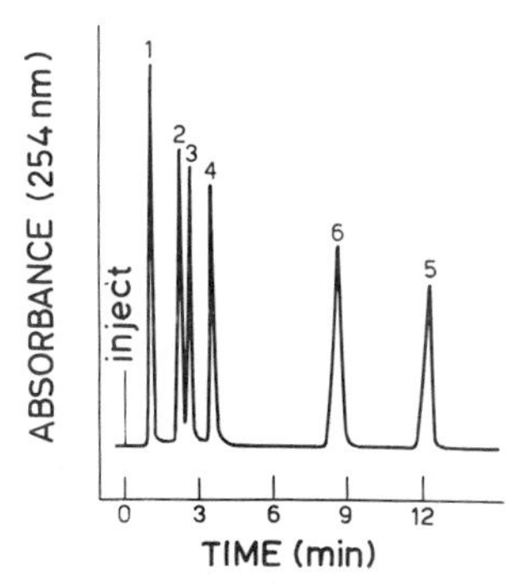

FIG. 2. High-performance liquid chromatographic separation of adenine and adenosyl-sulfur compounds. 1 = AdoHcy; 2 = adenine; 3 = MTA; 4 = AdoMet; 5 = decarboxy-AdoMet; 6 = *S*-adenosyl-(5′)-3-methylthiopropanol (internal standard). The compounds were added in the amount of 10 nmol each. Column, Partisil 10 SCX; temperature, ambient; sensitivity, 2.048 absorbance units full scale; eluent, ammonium formate, 0.5 *M*, pH 4.0; flow rate, 3.0 ml/min. From Zappia *et al.*[5]

RETENTION TIMES OF NATURAL ADENOSYL-SULFUR COMPOUNDS[a,b]

Compounds	Retention time (sec)	Variation coefficient (%)
Adenine	142	3
S-Adenosyl-L-homocysteine	92	2
S-Adenosyl-L-methionine	225	2
5′-Methylthioadenosine	170	3
5′-Methylthioinosine	82	2
S-Adenosyl-(5′)-3-methylthiopropylamine	750	3.5
S-Adenosyl-(5′)-3-methylthiopropanol (internal standard)	530	3.5

[a] From Zappia *et al.*[5]
[b] Operating conditions as in Fig. 2.

Excellent reproducibility of retention times and peak areas is always obtained. The coefficients of variation of the retention times for five consecutive injections of standard solutions are reported in the table.

Response curves are linear for all five tested compounds in amounts ranging from 5 to 50 nmol. The wide linear range for the different sulfur-containing compounds is more than adequate for their analysis in biological samples.

"Enzymatic Peak Shift" for Detection of the Thioethers. The "enzymatic peak shift" is used to verify peak identities. This technique utilizes the specificity of adenosine deaminase (adenosine aminohydrolase, EC 3.5.4.4) from *Aspergillus oryzae*.[10] Previous data indicate that this enzyme is very effective in deaminating adenosyl-5′-thioethers,[12] whereas it is unable to convert the corresponding adenosyl-5′-sulfonium compounds.[13] Figure 3 shows the effect of deamination on the retention times of AdoMet, adenine, and MTA. After the enzymatic reaction the peak of MTA is replaced by a new peak, with a lower retention time, corresponding to methylthioinosine (MTI). The peaks of AdoMet and adenine are not altered by the reaction. The same procedure can also be applied in order to identify AdoHcy, which can be quantitatively deaminated to *S*-inosyl-L-homocysteine, with a lower retention time. The "enzymatic peak shift" represents a useful tool for the positive identification of the investigated thioethers.

[12] F. Schlenk, C. R. Zydek-Cwick, and N. K. Hutson, *Arch. Biochem. Biophys.* **142,** 144 (1971).
[13] V. Zappia, P. Galletti, M. Cartenì-Farina, and L. Servillo, *Anal. Biochem.* **58,** 130 (1974).

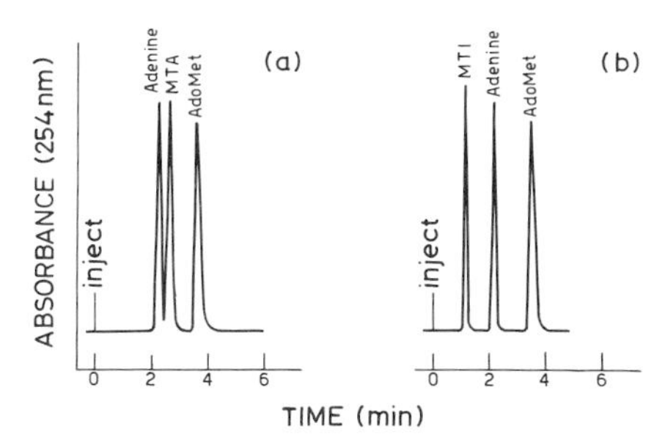

FIG. 3. Peak shift caused by reaction of nonspecific adenosine deaminase on a standard solution of adenine, MTA, and AdoMet. (a) Separation of a standard solution containing 10 nmol each of adenine, MTA, and AdoMet. (b) The same solution after reaction with 20 μg of nonspecific adenosine deaminase for 15 min at room temperature. 5′-Methylthioadenosine is quantitatively converted into 5′-methylthioinosine (MTI). From Zappia *et al.*[5]

Separation of Adenosyl-Sulfur Compounds from Biological Samples

Preliminary Purification of Biological Samples. About 1 g of freshly excised tissue is homogenized with 1.5 *M* perchloric acid (1 : 4, w/v). After centrifugation, the deproteinized supernatant is chromatographed through a Dowex 50 (H^+ form) column (100–200 mesh, resin bed 2 × 0.2 cm) previously equilibrated with 0.1 *M* HCl. Elution is carried out stepwise with 50 ml of 0.1 *M* HCl and 20 ml of 1.8 *M* HCl to remove contaminating nucleotides and nucleosides, and the desired metabolites are collected with 6 *M* HCl. The 6 *M* acid eluate, containing AdoMet, AdoHcy, decarboxylated AdoMet, and residual amounts of adenine and MTA, is concentrated under reduced pressure to 1 ml, and the pH is adjusted to 4.0. The recovery of AdoMet, AdoHcy, and decarboxylated AdoMet is quantitative, whereas MTA and adenine are partly eluted with 1.8 *M* HCl. The internal standard, 0.5 μmol of *S*-adenosyl-(5′)-3-methylthiopropanol, is added to the tissue during the homogenization with chloric acid.

HPLC Analysis. Samples amounting to 20 mg of tissue are injected for each HPLC analysis, performed as above described. Figure 4 shows a chromatogram from such an analysis of rat liver extract.

Analysis of 5′-Methylthioadenosine in Biological Samples

An accurate measurement of MTA concentration in the various tissues is a condition for a proper evaluation of the physiological roles of the

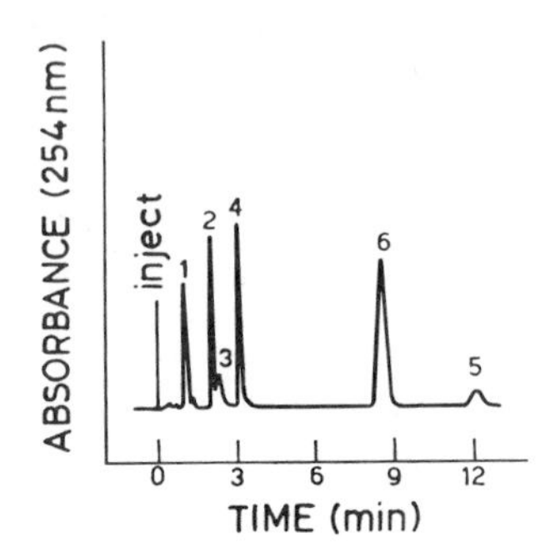

Fig. 4. Separation of adenosyl-sulfur compounds in rat liver extract. Sample of 20 μl equivalent to 20 mg of fresh tissue was injected into the column. The sample cleanup procedure is described in the text. Operating conditions were as described for Fig. 2. From Zappia *et al.*[5]

molecule.[14] On the other hand, only few data are reported in the literature[4,13] on the cellular concentration of the thioether.

The HPLC separation of adenosyl-sulfur compounds described above is not adequate for an actual evaluation of MTA in biological samples. In fact the MTA recovery in the cleanup procedure is not quantitative; furthermore, when adenine is present in large amounts it may hide the presence of small amounts of MTA, which has a retention time close to that of adenine (Fig. 4). Therefore for estimation of MTA in biological materials two alternative HPLC procedures, combined with an isotopic dilution technique, have been developed in the authors' laboratory.[6]

The first procedure involves a Dowex 50 chromatography followed by HPLC separation on a reversed-phase column. The second requires two chromatographic steps before the cation-exchange HPLC analysis.

Preliminary Purification of Sample. Figure 5 summarizes the multistep sequence of the two proposed analytical procedures. Freshly excised tissue (0.5 g) is homogenized with ice-cold 1.5 *M* perchloric acid (1 : 4 w/v). ^{14}C-Methyl-labeled MTA (2 nmol/g, 140,000 cpm/nmol) is added directly to the tissue during the homogenization step. After centrifugation the deproteinized supernatant is neutralized with 1 *M* KOH, centrifuged, and chromatographed through a Dowex 50 (H^+ form) column (100–200 mesh, resin bed 2 × 0.2 cm) previously equilibrated with 1.5 *M* perchloric acid to collect MTA. The acid eluate is then analyzed, as reported in Fig. 5, either according to procedure A or procedure B.

Procedure A: The eluate is neutralized with KOH; after centrifugation, the supernatant is concentrated to 0.5 ml under reduced pressure. A

[14] H. G. Williams-Ashman, J. Seidenfeld, and P. Galletti, *Biochem. Pharmacol.* **31**, 277 (1982).

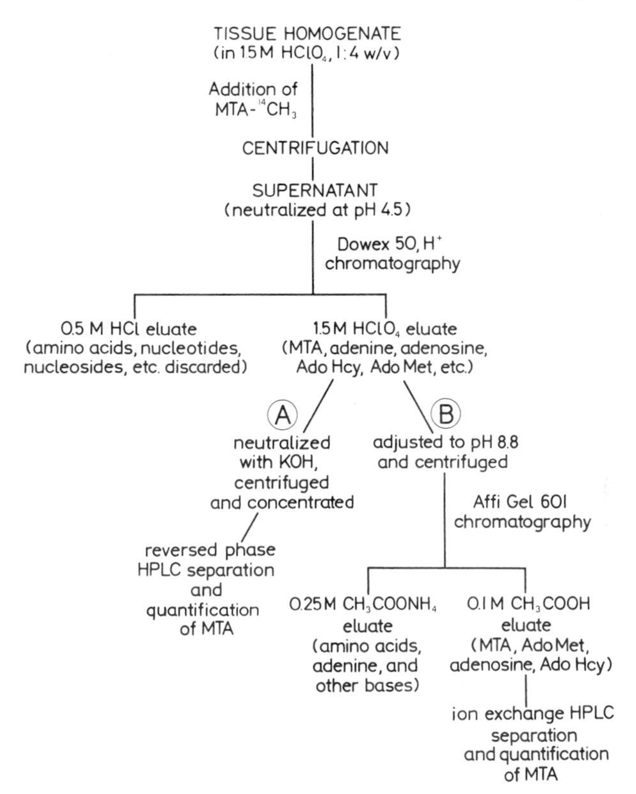

FIG. 5. Procedures for estimation of 5′-methylthioadenosine (MTA) in tissues. For abbreviations, see text. From Della Ragione *et al.*[6]

100-μl aliquot of this sample is analyzed by reversed-phase (Partisil 10 ODS) HPLC.

Procedure B: The perchloric acid eluate from the Dowex 50 column, adjusted to pH 8.8, is applied to a 0.5 × 2 cm column of phenylboronate resin (Affi-Gel 601) equilibrated with 0.25 *M* ammonium acetate (pH 8.8). The elution is carried out with 20 ml of the same buffer in order to remove all the non-*cis*-diol compounds; MTA is then eluted with 10 ml of 0.1 *M* acetic acid and concentrated to 500 μl under reduced pressure. A 100-μl aliquot of this sample is analyzed by ion-exchange (Partisil 10 SCX) HPLC.

Operating Conditions. The elution from the reversed-phase column (Partisil 10 ODS) is carried out with a 7:93 (v/v) mixture of anhydrous methanol and 7 m*M* acetic acid at a flow rate of 1.5 ml/min. Ammonium formate buffer, 0.25 *M*; pH 4.0, is used as eluent for the Partisil 10 SCX column, with a flow rate of 0.6 ml/min.

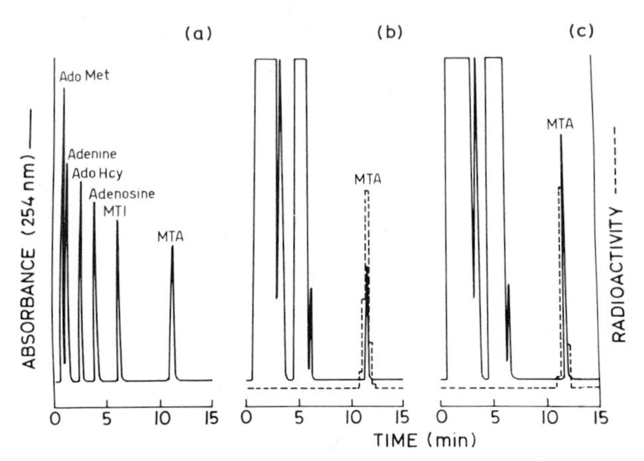

FIG. 6. (a) Separation of MTA and related reference compounds detected at 254 nm. The compounds were added in the amount of 0.2 nmol each. Column, Partisil 10 ODS; temperature, ambient; sensitivity, 0.5 absorbance unit full scale; eluent, 7:93 (v/v) mixture of anhydrous methanol and 7 m*M* acetic acid; flow rate, 1.5 ml/min. (b) Chromatogram of a rat liver extract after Dowex-50 chromatography. (c) Chromatogram of the sample shown in Fig. 2b coinjected with MTA. Modified from Della Ragione *et al.*[6]

A representative chromatogram of a standard mixture of MTA and related compounds (adenosine, adenine, AdoHcy, MTI, AdoMet) is shown in Fig. 6a. The chromatographic separation is carried out by reversed-phase column as described above. All the molecules are eluted within 20 min from their injection. Figure 6b reports a typical chromatogram relative to an analysis of rat liver extract. The thioether is extracted from 500 mg of wet tissue according to procedure A. Samples amounting to 100 mg of tissue are injected for HPLC analysis. Coinjection with a nonlabeled MTA standard results in an increase of the peak area coincident with the radioactive peak (Fig. 6c). The purity of the identified thioether is checked by the "enzymatic peak shift" method.

The alternative procedure proposed involves a strong cation-exchange HPLC. Figure 7a shows the elution profile of a standard mixture of MTA and related compounds processed with Partisil 10 SCX, and Fig. 7b shows the chromatogram of rat liver extracts cleaned-up according to procedure B.

MTA Quantification. The concentration of MTA can be determined by comparison of absorbance-integrated peak areas (or peak height) to standard curves obtained for solutions of pure MTA over the concentration range 0.1–5 nmol.

In order to calculate the recovery, methyl-labeled MTA is added directly to the homogenate. After HPLC analysis, 0.5-ml fractions of

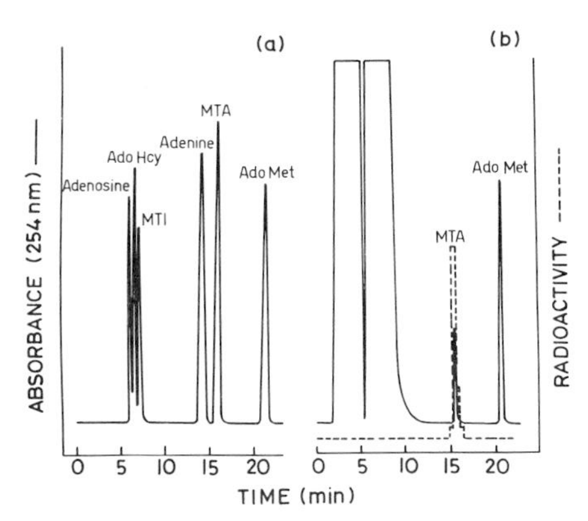

FIG. 7. (a) Separation of reference compounds detected at 254 nm. The compounds were added in the amounts of 0.2 nmol each. Column, Partisil 10 SCX; temperature, ambient; sensitivity, 0.5 absorbance unit full scale; eluent, ammonium formate buffer (0.25 *M*, pH 4.0); flow rate, 0.6 ml/min. (b) Chromatogram of a rat liver extract after Dowex 50 plus Affi-Gel 601 treatment. For abbreviations, see text. Modified from Della Ragione *et al.*[6]

eluates are collected and mixed up in the scintillation vials with 4 ml of Instagel (Packard). The radioactivity is measured in a TriCarb liquid scintillation spectrometer (Packard Model 3380).

For the quantification of MTA, an isotope dilution technique is used owing to the small amounts of the thioether present in the tissues. The average recovery of ^{14}C-labeled MTA added to the rat tissues is 82–85% for procedure A and 75–80% for procedure B. The isotope dilution technique appears to be necessary because of this variability in the yield.

The MTA concentration was determined five times for each rat tissue by both methods. The average for both methods agreed within 15%, demonstrating the accuracy and reproducibility of the two analyses. The levels of the thioether in the examined tissues ranged from 0.6 to 3 nmol/g. The sensitivity of the method, which permits the determination of MTA concentration as low as 200 pmol/g, is sufficient for its application in biological samples.

Comments

The described methods offer several advantages over previous ones. A rapid and excellent separation of the compounds can be achieved at room temperature with high sensitivity, selectivity, and efficiency. The results are quantitative, and minimal sample preparation is required. Since an isocratic elution is used, no reequilibration time is required between the analyses. The procedure is also employed in our laboratories for the assay of enzymatic reactions where AdoMet and its related compounds are involved as substrates or products, i.e., MTA phosphorylase (EC 2.4.2.1, purine-nucleoside phosphorylase) and AdoMet decarboxylase (EC 4.1.1.50).

Acknowledgments

This work was supported by grants from the Consiglio Nazionale delle Ricerche, Rome, Italy: Project "Control of Neoplastic Growth" 81.01487.96 and Research Group "Structure and Function of Biological Macromolecules" 81.00384.04.

[9] Determination of Tissue *S*-Adenosylmethionine by Radioenzymatic Assay

By PETER H. YU

Principle

The method is based on the enzymatic measurement of the conversion of *S*-adenosylmethionine and tritium-labeled dopamine to tritiated 3-methoxytyramine in the presence of catechol *O*-methyltransferase (EC 2.1.1.6, COMT). The enzyme is quite specific. When [G-^{3}H]dopamine is used in the assay, only one radioactive product, 3-methoxy[^{3}H]tyramine is formed. The linearity of the quantitative estimation of *S*-adenosylmethionine (AdoMet) is greatly improved by including S-[methyl-^{14}C]-adenosylmethionine to act as an internal standard in each assay. The relationship of the amount of AdoMet to the change of radioactivity in this double-isotope technique is shown in the following reactions.[1]

$$n(a + x)\ [^3\text{H}]\text{dopamine} + \begin{matrix} nx\ \text{AdoMet} \\ na\ \text{AdoMet-}^{14}\text{CH}_3 \end{matrix} \xrightarrow[\text{Mg}^{2+}]{\text{COMT}} \begin{matrix} nx\ \text{3-MT-}^3\text{H} \\ na\ \text{3-MT-}^3\text{H-}^{14}\text{CH}_3 \end{matrix}$$

If $x = 0$

$$\frac{\text{3-MT-}^3\text{H}}{\text{3-MT-}^{14}\text{C}} = R_a = \frac{na\ (\text{specific activity-}^3\text{H})}{na\ (\text{specific activity-}^{14}\text{C})_a} \tag{1}$$

If $x > 0$

$$\frac{\text{3-MT-}^3\text{H}}{\text{3-MT-}^{14}\text{C}} = R_{a+x} = \frac{n(a + x)\ (\text{specific activity-}^3\text{H})}{na\ (\text{specific activity-}^{14}\text{C})_{a+x}} \tag{2}$$

$$(\text{Specific activity-}^{14}\text{C})_{a+x} = a/(a + x)\ (\text{specific activity-}^{14}\text{C})_a \tag{3}$$

[1] P. H. Yu, *Anal. Biochem.* **86,** 498 (1978).

ISBN 0-12-181994-9

Equation (2) can be rewritten as:

$$R_{a+x} = \frac{n(a + x)\ (\text{specific activity-}^3\text{H})}{na[a/(a + x)]\ (\text{specific activity-}^{14}\text{C})_a} \tag{4}$$

Then Eq. (4) divided by Eq. (1) yields

$$\frac{R_{a+x}}{R_a} = \frac{(a + x)^2}{a^2} \tag{5}$$

Therefore,

$$x = [(R_{a+x}/R_a)^{1/2} - 1]a \tag{6}$$

The symbols a and x refer to quantities of reagents (weight or molar weight), and n represents the fraction of *S*-adenosylmethionine reacting. ^{3}H and ^{14}C denote radioactivity (dpm). Specific activity is expressed as curies per mole. X is the amount of *S*-adenosylmethione to be determined, and a is the slope of x vs $[(R_{a+x}/R_a)^{1/2} - 1]$.

Reagents

S-[methyl-^{14}C]-adenosylmethionine, 53 *M* (Ci/mmol)

3,4-[G-^{3}H]Dihydroxyphenylethylamine, 6.5 Ci/mmol

Catechol *O*-methyltransferase is partially purified by a modification of the procedure of Nikodejevic *et al.*[2] The rat livers (10 g) are homogenized in 10 volumes of KCl (1.15%) and centrifuged at 17,000 *g* for 30 min. An enzyme fraction is obtained from the supernatant acid precipitation at pH 5.2 and ammonium sulfate precipitation (30–50% saturation). The enzyme is then further purified by column chromatography on Sephadex G-100. The specific activity of the enzyme is 3 units per milligram of protein; 1 unit is defined as 1 μmol of 3-methoxytyramine formed per 60-min incubation.

Procedure

Step 1. Preparation of Tissue Extracts. Tissues are homogenized in 10 volume of ice cold 2% perchloric acid. *S*-[methyl-^{14}C]-adenosylmethionine internal standard (4 nCi or 76 pmol/ml) is added. The insoluble precipitates are removed by centrifugation at 12,000 *g* for 10 min at 4°, washed with distilled water, and recentrifuged. The pH of the combined supernatant is adjusted to neutral with 0.5 *N* NaOH.

Step 2. Incubation. Aliquots (20–100 μl) of the extract are added to a catechol *O*-methyltransferase reaction mixture (100 μl) containing 50

[2] B. Nikodejevic, S. Senoh, J. W. Daly, and C. R. Creveling, *J. Pharmacol. Exp. Ther.* **174,** 83 (1970).

μmol of Tris buffer (pH 8.6), 3 μmol of dithiothreitol, 1.25 μmol of $MgCl_2$, 0.25 μmol (1 μCi) of [G-^{3}H]dopamine, adenosine deaminase (0.05 unit), and catechol *O*-methyltransferase (about 0.5 unit). The mixture is incubated for 40 min at 37°.

Step 3. Extraction and Measurement of 3-Methoxytyramine. The reaction is terminated by adding 600 μl of 0.5 *M* borate buffer (pH 10.0). The labeled end product, 3-methoxytyramine, is extracted from the incubation mixture by shaking with 1 ml of toluene–isoamyl alcohol (3 : 2, v/v). After centrifugation, 800 μl of the organic phase is transferred to a clean tube containing 600 μl of 0.4 *N* HCl, shaken on a vortex mixer, and then centrifuged. The aqueous HCl phase (500 μl) is transferred to a scintillation counting vial, mixed with 10 ml of ACS scintillation fluid, and counted by a ^{3}H/^{14}C double-counting program in a liquid scintillation counter (Beckman LS 7500) using the H-number method of quench monitoring. The counting efficiency is 33% and 59% with respect to ^{3}H and ^{14}C, respectively.

Authentic *S*-adenosylmethione as well as *S*-adenosylmethionine added to tissue homogenates before addition of perchloric acid are included for standard determination. More than 95% of the *S*-adenosylmethionine added to tissues is recovered. The mean difference between duplicate determinations is 2.5 ± 0.3% of the values.

Remarks

The radioenzymatic method described here permits the measurement of as little as 1 ng of *S*-adenosylmethionine, equivalent to that in 0.1 mg of brain, or 0.02 mg of liver, or 50 μl of human plasma. This method also has the advantage that, since the reaction is very specific, purification of the samples before assay is not necessary. Furthermore, the sensitivity of this catechol *O*-methyltransferase system is approximately 100-fold greater than that of the hydroxyindole *O*-methyltransferase (HIOMT) system developed by Baldessarini and Kopin.[3] It has been reported that the K_m values of catechol *O*-methyltransferase and hydroxyindole *O*-methyltransferase for *S*-adenosylmethionine are 3.2×10^{-6} *M*[4] and 2.5×10^{-4} *M*,[5] respectively. The much higher affinity of *S*-adenosylmethionine for catechol *O*-methyltransferase than hydroxyindole *O*-methyltransferase clearly explain why the former enzyme is much more suitable for a microassay of *S*-adenosylmethionine. Histamine *N*-methyltransferase, al-

[3] R. J. Baldessarini and I. J. Kopin, *J. Neurochem.* **13,** 769 (1966).

[4] L. Flohe and K. P. Schwabe, *Biochim. Biophys. Acta* **220,** 469 (1970).

[5] R. L. Lin, N. Narasimhachari, and H. E. Himwich, *Biochem. Biophys. Res. Commun.* **54,** 751 (1973).

though it exhibits a high affinity for histamine, and can be used for the determination of histamine,[6] does not have a high affinity toward S-adenosylmethionine ($K_m = 3.8 \times 10^{-4}\ M$).[7] This limits its application for the microassay of S-adenosylmethionine.

The purity of radiolabeled dopamine or S-adenosylmethionine should be checked routinely. We have observed that both [^{3}H]dopamine and S-[methyl-^{14}C]adenosylmethionine deteriorate in 3 months at −20°, and this significantly affects the assays owing to high background. Repeated freezing and thawing should be avoided.

[6] D. J. Salberg, L. B. Hough, D. E. Kaplan, and E. F. Domino, *Life Sci.* **21,** 1439 (1977).
[7] D. D. Brown, R. Tomchick, and J. Axelrod, *J. Biol. Chem.* **234,** 2948 (1959).

[10] Determination of Cellular Decarboxylated S-Adenosylmethionine Content

By Anthony E. Pegg and R. A. Bennett

Decarboxylated S-adenosylmethionine is the source of the aminopropyl groups of the polyamines spermidine and spermine.[1] Under normal physiological conditions, its concentration is much lower than that of S-adenosylmethionine itself, necessitating sensitive methods for its determination.[2] The method described here is based on isocratic high-performance liquid chromatography (HPLC) separation using a cation-exchange medium and quantitation by ultraviolet (UV) absorbance.[3] Other methods are available, including an enzymatic assay using spermidine synthase[2] or reversed-phase ion-pair chromatography using a gradient for elution.[4]

Reagents

Trichloroacetic acid, 5% (w/v)

Dowex 50W-X4. The resin is washed four times each with 10 volumes of 6 *N* HCl until no more yellow color is observed in the wash. The acid is then removed by washing with distilled water and small columns (bed volume 1.5 ml) prepared in Pasteur pipettes plugged

[1] H. Tabor and C. W. Tabor, *Adv. Enzymol. Relat. Areas Mol. Biol.* **36,** 203 (1972).
[2] H. Hibasami, J. L. Hoffman, and A. E. Pegg, *J. Biol. Chem.* **255,** 6675 (1980).
[3] A. E. Pegg, H. Pösö, K. Shuttleworth, and R. A. Bennett, *Biochem. J.* **202,** 519 (1982). See also this volume [11].
[4] J. Wagner, C. Danzin, and P. S. Mamont, *J. Chromatogr.* **227,** 349 (1982).

ISBN 0-12-181994-9

with glass wool or in disposable polypropylene chromatography tubes (Kontes, Vineland, New Jersey) to a bed volume of 3 ml.

Boronate-affinity columns. Affi-Gel 601 boronate affinity resin (Bio-Rad Laboratories, Richmond, California) is swollen for at least 30 min in 0.25 *M* ammonium acetate, pH 8.8, and then packed in 1-ml disposable pipette tips (Fisher Scientific, Pittsburgh, Pennsylvania) fitted with glass wool plugs so that each tip contains about 0.1 ml of resin. The columns are stored at 4° and used within 5 days.

Ammonium acetate, 0.25 *M* and 0.5 *M*, pH 8.8

Ammonium formate, 0.5 *M* (pH 4.0). This solution is used as the mobile phase for HPLC and is prepared using deionized water purified with a water purification system from Hydro Service and Supplies (Durham, North Carolina) and filtered through a 0.45 μm 47 mm Mitex filter (Millipore Corp., Bedford, Massachusetts) using a Millipore stainless-steel screen, glass analytical filter holder.

Methylglyoxal bis(guanylhydrazone), 12.5 μM

Preparation of Sample Extracts

Extracts from cell cultures are prepared by addition at 0–2° of 2 ml of 5% (w/v) trichloroacetic acid to the cell pellets followed by sonication in a sonicating water bath to ensure full extraction of the precipitate. After 15 min at 0–2°, the samples are centrifuged at 10,000 *g* for 15 min, and the supernatants are applied to columns (1.5 ml bed volume) of Dowex 50W-X4 (H^+ form). The columns are washed with 5 ml of 0.5 *N* HCl and then eluted with 5 ml of 6 *N* HCl. The 6 *N* HCl eluate is evaporated to dryness under reduced pressure at 40°, reconstituted in 0.25 ml of water, and stored frozen at −20° until analysis.

Tissue extracts are prepared by homogenization of samples of about 2 g wet weight (removed as rapidly as possible) in 12 ml of 5% (w/v) trichloroacetic acid at 0–2° using a Polytron homogenizer (Brinkmann Instruments, Westbury, New York). The extracts are centrifuged at 10,000 *g* for 20 min, and the supernatant is applied to columns (3 ml bed volume) of Dowex 50W-X4 (H^+ form). The columns are washed with 20 ml of 1.5 *N* HCl and then eluted with 6 ml of 6 *N* HCl. The 6 *N* HCl eluate is then evaporated and reconstituted as described above.

In order to remove impurities that increase the absorbance background and decrease the life of the HPLC columns, the samples are first purified on boronate affinity columns. An equal volume of 0.5 *M* ammonium acetate (pH 8.8) is added to the samples dissolved in water, and the pH is adjusted to between 8 and 9 by addition of 10 μl of concentrated NH_4OH. The sample is then applied immediately to columns (0.1 ml bed volume) of Affi-Gel 601. The column is washed with 0.5 ml of ammonium acetate, pH 8.8, and then eluted with 1 ml of 0.1 *M* formic acid. This eluate is collected in a weighed tube, which is reweighed to obtain an accurate indication of the volume, and an aliquot (0.2–1.0 ml) of the solution is applied to the HPLC column.

HPLC Analysis

Columns are prepared as follows: Partisil-10 SCX (1.1–1.2 g) (purchased from Whatman Inc., Clifton, New Jersey) is stirred gently in 0.5 *M* ammonium formate, pH 4.0, at reduced pressure for 15 min prior to loading into a Micromeritics Instrument Corp. (Norcross, Georgia) Model 705 stirred-slurry column packer. The packer is then attached to the solvent delivery system and a 0.46 × 10 cm stainless-steel column, which is packed at a flow rate of 4 ml/min until full.

High-performance liquid chromatography is performed using a Micromeritics Instrument Corp. Model 750 solvent delivery system, Model 730 universal injector, and Model 731 column oven. Sample components are detected by monitoring the column effluent at 254 nm with a Waters Associates (Milford, Massachusetts) Model 440 UV absorbance monitor. The recorder output is recorded on a Laboratory Data Control (Riviera Beach, Florida) Model 3401 recorder.

Analysis is carried out at a flow rate of 1 ml/min and a temperature of 50°. Decarboxylated *S*-adenosylmethionine is quantitated by comparison of peak heights to those of known quantities of the authentic compound. The response is directly proportional to peak height up to at least 3 nmol of decarboxylated *S*-adenosylmethionine, and the minimum amount for accurate quantitation is 50 pmol. The elution times vary slightly from column to column, but on average decarboxylated *S*-adenosylmethionine is eluted at 11 min, well resolved from all other UV-absorbing compounds that are eluted within the first 5 min. An internal standard can be added to the samples in order to compensate for any losses during sample injection. Methylglyoxal bis(guanylhydrazone) (Aldrich, Milwaukee, Wisconsin) (MGBG) is suitable, since it elutes at 7.5 min and absorbs strongly at 254 nm. Therefore, 0.02 ml of 12.5 μM MGBG is added per milliliter of eluate from the boronate affinity column.

Tissue and Cell Levels

The content of decarboxylated *S*-adenosylmethionine in normal rat tissues ranged from 0.9 nmol/g in heart and intestine to 2.5 nmol/g in ventral prostate.[3] This represents 3–5% of the content of *S*-adenosylmethionine itself. Chronic treatment with inhibitors of ornithine decar-

Content of Decarboxylated S-Adenosylmethionine, S-Adenosylmethionine, and Polyamines in SV3T3 Cells[a]

Treatment	Decarboxylated S-adenosylmethionine (pmol/million cells)	Putrescine (nmol/million cells)	Spermidine (nmol/million cells)
Control	0.8 ± 0.4	0.28 ± 0.09	3.18 ± 0.22
5 mM α-difluoromethyl-ornithine	426 ± 46	<0.02	0.10 ± 0.05
5 mM α-difluoromethyl-ornithine + 50 μM putrescine	0.5 ± 0.3	0.61 ± 0.12	3.36 ± 0.26

[a] Cells were grown for 3 days as described.[3]

boxylase depletes the tissue concentrations of putrescine and spermidine. In the absence of these acceptors of the aminopropyl group, decarboxylated *S*-adenosylmethionine levels increase manyfold and can exceed those of *S*-adenosylmethionine itself.[3–6] Cultured mouse fibroblasts contain very low levels of decarboxylated *S*-adenosylmethionine under normal growth conditions (see the table). When exposed to the potent inhibitor of ornithine decarboxylase α-difluoromethylornithine, putrescine and spermidine levels are reduced to less than 3% of normal, but decarboxylated *S*-adenosylmethionine increases 500-fold. This increase is totally abolished when exogenous putrescine is provided (see the table). The change in decarboxylated *S*-adenosylmethionine levels from less than 1% of the *S*-adenosylmethionine content (which is about 90 pmol/million cells) to more than 400% could lead to derangement of cellular methylation processes and may contribute to the pharmacological effects of α-difluoromethylornithine.[3,5] This increase is not unique to mouse fibroblasts, but has been observed also in several other cell types including rat prostate,[6] HTC cells,[6] 9L rat brain tumor cells,[7] and trypanosomes.[8]

[5] A. E. Pegg, H. Pösö, and R. A. Bennett, *in* "Biochemistry of *S*-adenosylmethionine and Related Compounds" (E. Usdin, R. T. Borchardt, and C. R. Creveling, eds.), p. 547. Macmillan, New York.

[6] P. S. Mamont, C. Danzin, J. Wagner, M. Siat, A.-M. Joder-Ohlenbusch, and N. Claverie, *Eur. J. Biochem.* **123,** 499 (1982).

[7] A. E. Pegg and L. J. Marton, unpublished observations.

[8] C. J. Bacchi, J. Garofalo, D. Mockenhaupt, P. P. McCann, K. Diekema, A. E. Pegg, H. C. Nathan, L. Chunosoff, A. Sjoerdsma, and S. H. Hutner, *Mol. Biochem. Parasitol.* (in press).

[11] Methods for the Preparation and Assay of S-Adenosyl-(5')-3-methylthiopropylamine (Decarboxylated Adenosylmethionine)

By VINCENZO ZAPPIA, PATRIZIA GALLETTI, ADRIANA OLIVA, and MARINA PORCELLI

S-Adenosyl-(5')-3-methylthiopropylamine (decarboxylated AdoMet) represents a key intermediate in the polyamine biosynthesis pathway.[1-3] The sulfonium compound is formed by enzymatic decarboxylation of *S*-adenosyl-L-methionine (AdoMet) and, in turn, is the donor of a propylamine moiety to putrescine or spermidine, yielding spermidine and spermine, respectively.[4]

The availability of pure decarboxylated AdoMet free of contaminating *S*-adenosylmethionine (AdoMet) is essential for a correct evaluation of the actual activities of spermidine EC 2.5.1.16 and spermine synthases EC 2.5.1.-, as well as for the determination of their kinetic parameters.[4]

The chemical synthesis of the decarboxylated AdoMet is quite complicated and yields a racemic compound.[5] On the other hand, the enzymatic synthesis by a specific AdoMet decarboxylase[6] has to be followed by a separation of the decarboxylated product from the substrate.

In this chapter two different chromatographic procedures are reported for quantitative separation of the two sulfonium compounds yielding pure decarboxylated AdoMet.[7]

Chemicals and Enzymes

S-Adenosyl-L-[*carboxyl*-^{14}C]methionine is supplied by the Radiochemical Centre, Amersham, United Kingdom; in some experiments, the compound is prepared from L-[*carboxyl*-^{14}C]methionine by biosynthesis

[1] H. Tabor and C. W. Tabor, *Annu. Rev. Biochem.* **45,** 285 (1976).

[2] V. Zappia, M. Cartenì-Farina, and P. Galletti, *in* "The Biochemistry of Adenosylmethionine" (F. Salvatore, E. Borek, V. Zappia, H. G. Williams-Ashman, and F. Schlenk, eds.), p. 473. Columbia Univ. Press, New York, 1977.

[3] V. Zappia, M. Cartenì-Farina, and M. Porcelli, *in* "Transmethylation" (E. Usdin, R. T. Borchardt, and C. R. Creveling, eds.), p. 95. Elsevier/North-Holland, Amsterdam, 1979.

[4] V. Zappia, G. Cacciapuoti, G. Pontoni, and A. Oliva, *J. Biol. Chem.* **255,** 7276 (1980).

[5] G. A. Jamieson, *J. Org. Chem.* **28,** 2397 (1963).

[6] R. B. Wickner, C. W. Tabor, and H. Tabor, *J. Biol. Chem.* **245,** 2132 (1970).

[7] V. Zappia, P. Galletti, A. Oliva, and A. De Santis, *Anal. Biochem.* **79,** 535 (1977). See also this volume [43] and W. H. Bowman, C. W. Tabor, and H. Tabor, *J. Biol. Chem.* **248,** 2480 (1973) for other methods for the preparation of decarboxylated AdoMet.

METHODS IN ENZYMOLOGY, VOL. 94

ISBN 0-12-181994-9

with *Saccharomyces cerevisiae* and isolated by ion-exchange chromatography.[8,9] Since the commercial nonradioactive AdoMet is contaminated by numerous impurities, it is always synthesized by the aforementioned procedure with minor modifications.[10] To minimize decomposition, the compound is stored at −20° in 0.1 *M* HCl. All other chemicals are obtained from commercial conventional sources.

Enzymatic decarboxylation of AdoMet is performed using the enzyme partially purified from *Escherichia coli*, according to Wickner *et al.*,[6,9a] except for the DEAE-cellulose chromatographic step.

Nonspecific adenosine deaminase (adenosine aminohydrolase, EC 3.5.4.4) is purified from *Aspergillus oryzae* powder (Sanzyme, Calbiochem) according to Kaplan[11] and Sharpless and Wolfenden.[12]

Chromatographic Procedures

Purification of Decarboxylated AdoMet by Dowex-1 Chromatography

Enzymatic Decarboxylation of AdoMet. AdoMet is enzymatically decarboxylated using adenosylmethionine decarboxylase purified from *E. coli*.[6] The reaction is followed by measuring the $^{14}CO_2$ release from carboxyl-labeled AdoMet. The enzymatic mixture contains 0.8 mmol of triethanolamine buffer, pH 7.4, 0.4 mmol of $MgSO_4$, 70 μmol of *S*-adenosyl-L-[*carboxyl*-^{14}C]methionine (109 × 10^3 cpm/μmol), and 50 mg of decarboxylase in a final volume of 5 ml. After a 180-min incubation, the mixture is deproteinized with 2.5 ml of 1.5 *M* perchloric acid and is centrifuged. With these conditions the enzymatic decarboxylation of AdoMet, even after longer incubation times, never exceeds 30–40% completion. Therefore, a separation of the decarboxylated product from the substrate is necessary.

Dowex 50 H^+ Chromatography. In order to separate the two sulfonium compounds from the contaminants of the incubation mixture, the supernatant is processed through a Dowex-50 H^+ column (100–200 mesh, resin bed 2 × 10 cm). The resin is washed with 100 ml of 1 *M* HCl and 200

[8] S. K. Shapiro and D. J. Ehninger, *Anal. Biochem.* **15,** 323 (1966).

[9] V. Zappia, F. Salvatore, C. R. Zydek-Cwick, and F. Schlenk, *J. Labeled Compd.* **4,** 230 (1968).

[9a] See also this volume [37] for a more recent preparation of this enzyme.

[10] V. Zappia, C. R. Zydek-Cwick, and F. Schlenk, *J. Biol. Chem.* **244,** 4499 (1969).

[11] N. O. Kaplan, this series, Vol. 2, p. 475.

[12] T. K. Sharpless and R. Wolfenden, this series, Vol. 12A, p. 126.

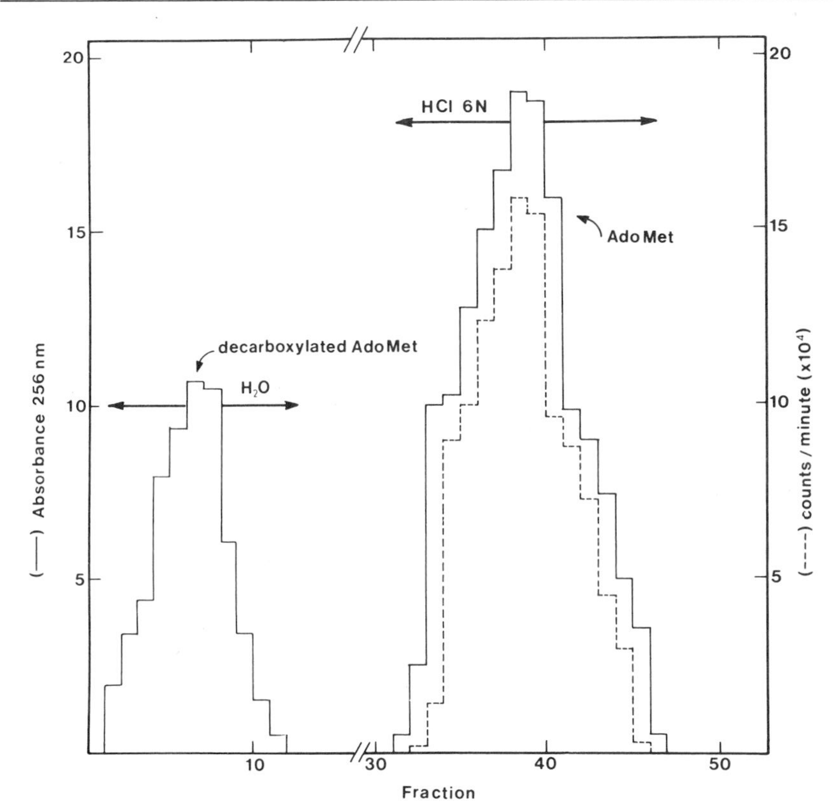

FIG. 1. Elution pattern of a mixture of *S*-adenosyl[*carboxyl*-^{14}C]methionine and *S*-adenosyl-(5′)-3-methylthiopropylamine chromatographed on Dowex 1 (OH^- form). *S*-Adenosylmethionine (AdoMet, 50.9 μmol; 109×10^3 cpm/μmol) and *S*-adenosyl-(5′)-3-methylthiopropylamine (decarboxylated AdoMet, 19.1 μmol) are added to a column 1.5 cm in diameter. Fractions of 5 ml are collected. From Zappia *et al.*[7]

of 2 *M* HCl, then the two sulfonium compounds are eluted simultaneously with 6 *M* HCl. The eluate is evaporated to dryness under reduced pressure, and the residue is dissolved in 50 ml of water.

Dowex-1 OH^- Chromatography. The solution is then applied to a Dowex-1 OH^- column (100–200 mesh, resin bed 1.5×10 cm) equilibrated with water. The elution pattern is shown in Fig. 1. The decarboxylated sulfonium compound does not bind to the positive groups of the anionic resin and is eluted with water, whereas the AdoMet retained by the column can be eluted with HCl. The separation, as well as the recovery, is quantitative.

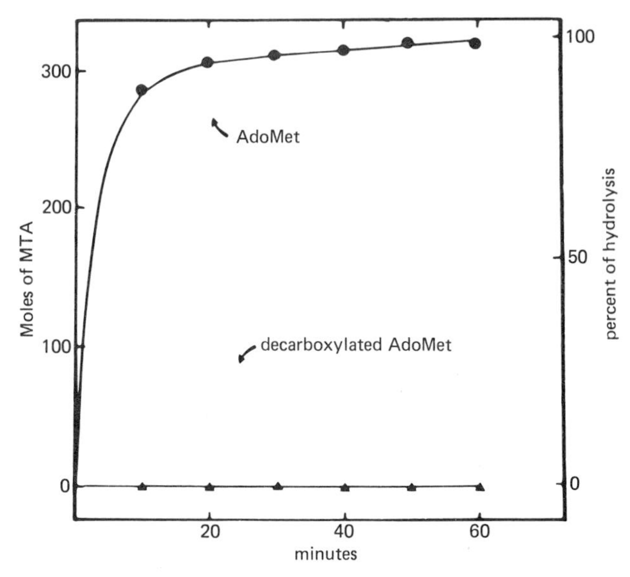

FIG. 2. Time course of the acid hydrolysis of *S*-adenosylmethionine (AdoMet) (●) and decarboxylated AdoMet (▲). The hydrolysis is performed as indicated in the text. 5-methylthioadenosine (MTA) formed at the indicated times is measured spectrophotometrically at 265 nm after deamination by adenosine deaminase from *Aspergillus oryzae*. Decarboxylated AdoMet and AdoMet are not deaminated by the enzyme. From Zappia *et al.*[7]

Separation of Decarboxylated AdoMet by a Single-Step Chromatography after Acid Hydrolysis of AdoMet

Decarboxylated AdoMet, as its parent sulfonium compound, shows an extreme lability toward alkaline hydrolysis. Hydrolysis in 0.1 *M* NaOH for a few minutes at 25° releases adenine, as can be observed by the shift of the maximum of absorbance from 260 to 268 nm and by the decrease of the molar absorbance.[10] Conversely, the chemical stabilities in acid of the two sulfonium compounds are quite different. After hydrolysis at 100° and pH 4.5, AdoMet is fragmented into 5′-methylthioadenosine (MTA) and homoserine, whereas the decarboxylated sulfonium compound is not hydrolyzed. The kinetics of acid hydrolysis can be followed spectrophotometrically by measuring the MTA formed in the reaction by its deamination with a "nonspecific" adenosine deaminase.[13-15] As shown in Fig. 2,

[13] F. Schlenk and C. R. Zydek-Cwick, *Biochem. Biophys. Res. Commun.* **31,** 437 (1968).

[14] F. Schlenk, C. R. Zydek-Cwick, and N. Y. Hutson, *Arch. Biochem. Biophys.* **142,** 144 (1971).

[15] V. Zappia, P. Galletti, M. Cartenì-Farina, and L. Servillo, *Anal. Biochem.* **58,** 130 (1974).

after a 20-min incubation at 100° and pH 4.5, AdoMet is almost completely converted into MTA and homoserine, whereas decarboxylated AdoMet remains unmodified, even after prolonged incubation under the same conditions. This results suggest that the carboxyl group of AdoMet is responsible for an intramolecular nucleophilic attack on carbon atom 4 of the amino acid side chain, the hydrolytic products being homoserine and MTA.

By taking advantage of the different behavior of the two molecules in acid, a second method of purification, which involves only a single chromatographic step, can be employed.

Enzymatic Decarboxylation of AdoMet and Acid Hydrolysis of the Products. The composition of the enzymatic mixture is the same as reported above, except that the concentration of *S*-adenosyl-L-[*carboxyl*-^{14}C]methionine (282×10^3 cpm/μmol) is 8.5 μmol/ml. After a 180-min incubation, about 30% AdoMet decarboxylation is achieved. The mixture is deproteinized with 1.5 *M* perchloric acid and centrifuged; the acidity of the supernatant is reduced to pH 4.5 with 3 *M* $KHCO_3$, and potassium perchlorate is removed by centrifugation at 9000 *g* for 15 min. In order to hydrolyze quantitatively the residual amounts of AdoMet, the supernatant is then placed in a boiling water bath for 60 min.

Dowex 50 Na^+ Chromatography. The hydrolyzed mixture is applied to a Dowex 50, Na^+ column (100–200 mesh, resin bed 1.5×9 cm) previously equilibrated with 0.1 *M* NaCl. As indicated in Fig. 3, the hydrolytic products of AdoMet are quantitatively eluted with 0.1 *M* NaCl (homoserine) and 2 *M* HCl (MTA), while decarboxylated AdoMet is collected with 6 *M* HCl. Even with this alternative procedure, the recovery of pure decarboxylated AdoMet is quantitative.

Purity of S-Adenosyl-(5′)-3-methylthiopropylamine: Analytical Tests

The conventional analytical procedure of paper and thin-layer chromatography in various solvent systems[10] are unable to separate AdoMet from its decarboxylated product. Conversely, high-voltage paper electrophoretic methods,[9,16] which permit a good separation between the two molecules, can be profitably used for purity analysis. A good separation can also be achieved by ion-exchange paper chromatography. The R_f values of the two sulfonium compounds and their hydrolytic products, separated by ion-exchange paper chromatography as well as the electrophoretic mobilities, are reported in the table.

[16] H. Pösö, P. Hannonen, and J. Jänne, *Acta Chem. Scand.* **30,** 807 (1976).

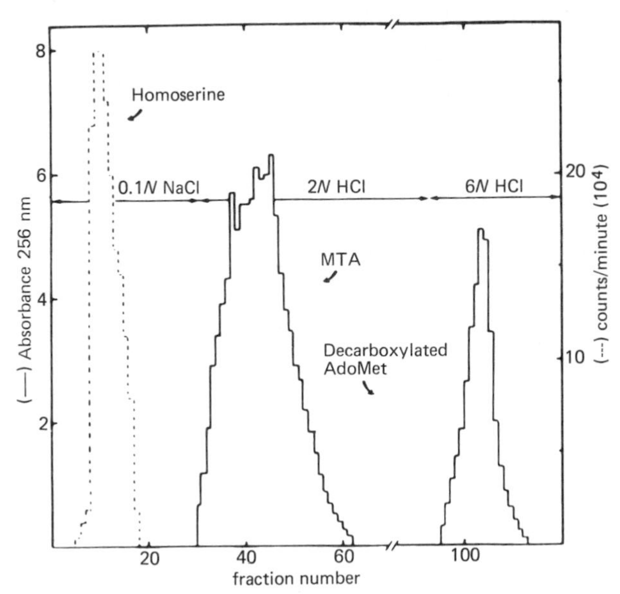

FIG. 3. Elution pattern of a mixture of [*carboxyl*-^{14}C]homoserine, 5′-methylthioadenosine, and decarboxylated AdoMet chromatographed on Dowex 50 (Na^+ form). Homoserine (31.2 μmol; 282 × 10^3 cpm/μmol), 5′-methylthioadenosine (MTA, 31.2 μmol), and *S*-adenosyl-(5′)-3-methylthiopropylamine (decarboxylated AdoMet, 11.25 μmol) are added to a column 1.5 cm in diameter. Fractions of 5 ml are collected. From Zappia *et al.*[3]

HIGH-VOLTAGE ELECTROPHORESIS AND ION-EXCHANGE CHROMATOGRAPHY OF *S*-ADENOSYL-(5′)-3-METHYLTHIOPROPYLAMINE AND RELATED COMPOUNDS ON CM- AND DEAE-CELLULOSE[a]

Compound	Paper chromatography (R_f values × 100)		High-voltage electrophoresis[d] (cm V^{-1} hr^{-1} × 10^3)
	DEAE-cellulose[b]	CM-cellulose[c]	
Decarboxylated AdoMet	78	6.3	17
AdoMet	72	41	8
Methylthioadenosine	53	40	2.7
Homoserine	70	87	1.7

[a] From Zappia *et al.*[3]
[b] Solvent system: 100 m*M* acetate buffer/2-propanol (97 : 3, v/v), pH 6.5.
[c] Solvent system: 100 m*M* acetate buffer/2-propanol (97 : 3, v/v), pH 5.5.
[d] Solvent system: 50 m*M* acetate buffer, pH 4.0.

A satisfactory separation of the two sulfonium compounds can also be carried out by high-pressure liquid chromatography,[17] as described in this volume [8].

All of these analytical tests have shown that decarboxylated AdoMet, prepared according to the described procedures, is pure and completely free of AdoMet.

Comments

The separation methods reported here are relatively simple, not time-consuming, and permit the isolation of decarboxylated AdoMet on a large scale. The only limit for obtaining large amounts of *S*-adenosyl-(5′)-3-methylthiopropylamine is the low extent of enzymatic decarboxylation of AdoMet which, using the enzyme from *Escherichia coli* through the $(NH_4)_2SO_4$ fractionation step, never exceeds 50% yield. However, the yield can be significantly enhanced using the same enzyme, purified through the DEAE-cellulose column step.[16]

The first preparative procedure requires two chromatographic steps; the separation of AdoMet and decarboxylated AdoMet on Dowex 1 depends on the differences in binding of the two compounds. The negatively charged carboxyl group of AdoMet is firmly bound to the resin, whereas the decarboxylated AdoMet is not retained. To minimize decomposition of decarboxylated AdoMet, which occurs at neutral pH values, chromatography has to be carried out rapidly and the samples must be acidified soon after elution.

The second method, requiring only a single chromatographic step after acid hydrolysis, is more rapid. The yields are quantitative with both methods and the compound obtained is pure and completely free of AdoMet.

An alternative procedure for preparative isolation of enzymatically obtained decarboxylated AdoMet is described by Pösö *et al.*[16] The method involves chromatography of the enzymatic mixture on Dowex 50, H^+ column, followed by preparative paper electrophoresis and rechromatography on Dowex 50 H^+. The obvious limitation of this procedure is the electrophoretic step, in that the amount of the product that can be loaded on a Whatman No. 1 paper sheet (20 × 30 cm) is less than 0.5 μmol.

Hibasami *et al.*[18] have described different procedures for the determination of decarboxylated AdoMet content in mammalian tissues. Three

[17] V. Zappia, P. Galletti, M. Porcelli, C. Manna, and F. Della Ragione, *J. Chromatogr.* **189,** 399 (1980).

[18] H. Hibasami, J. L. Hoffman, and A. E. Pegg, *J. Biol. Chem.* **255,** 6675 (1980). See also this volume [10].

methods were used: (*a*) high pressure liquid chromatographic separation followed by quantitation using UV absorbance; (*b*) intravenous injection of labeled methionine to rats and separation by high-voltage electrophoresis with subsequent determination of radioactivity of tissue soluble extracts; (*c*) separation by high-voltage electrophoresis followed by an isotope dilution assay with spermidine synthase. The three procedures gave comparable results, and the content of decarboxylated AdoMet in several rat organs was found to range between 0.9 and 2.5 nmol/g wet weight of tissue.

Acknowledgments

This work was supported by grants from the Consiglio Nazionale delle Ricerche, Rome, Italy: Project "Control of Neoplastic Growth" 81.01487.96 and Research Group, "Structure and Function of Biological Macromolecules" 81.00384.04.

Section III

Genetic Techniques

A. Mutants in Biosynthetic Pathways
Articles 12 through 16

B. Preparation of Plasmids for Overproduction of Enzymes in Biosynthetic Pathway (*Escherichia coli*)
Article 17

[12] Mass Screening for Mutants in the Biosynthetic Pathway for Polyamines in *Escherichia coli:* A General Method for Mutants in Enzymatic Reactions Producing CO_2

By CELIA WHITE TABOR, HERBERT TABOR, and EDMUND H. HAFNER

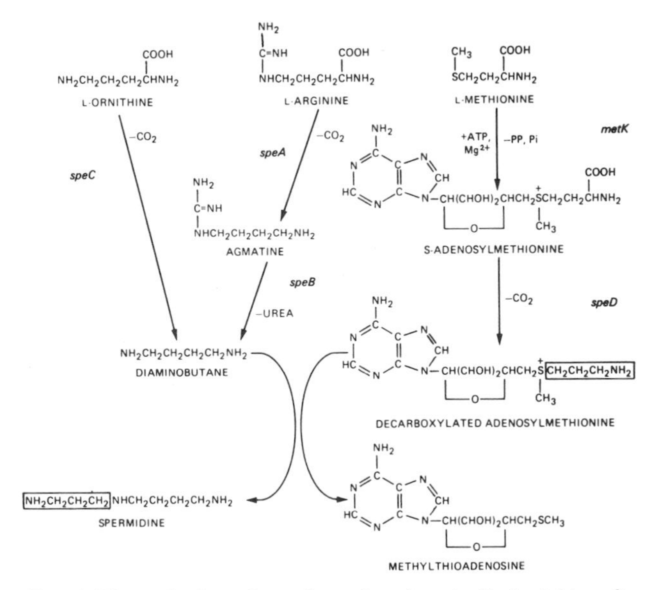

FIG. 1. Biosynthetic pathway for polyamines in *Escherichia coli.*

Selection of mutants in the biosynthetic pathways for putrescine and spermidine in *Escherichia coli* is difficult because mutations in single steps result in little effect on growth rate. Thus, there is no good selective medium for screening for mutants, and measurement of each enzymatic reaction in individual mutagenized clones is necessary.

The following general method was developed[1] for the rapid screening of many individual clones of *E. coli* for mutations in any pathway producing $^{14}CO_2$ from a labeled substrate. The $^{14}CO_2$ formed from individual clones is trapped as insoluble $BaCO_3$ (immediately above each colony) on paper saturated with barium hydroxide.

This method has been useful also for determining the phenotype of mutants in mapping experiments (e.g., bacteriophage P1 transductions or Hfr crosses) and for detecting the presence of cloned genes on plasmids

[1] H. Tabor, C. W. Tabor, and E. W. Hafner, *J. Bacteriol.* **128,** 485 (1976).

ISBN 0-12-181994-9

during the construction of enzyme-overproducing strains with recombinant DNA techniques. The methods can also be used to screen other bacteria, yeast, and mammalian cells for mutants in the biosynthetic pathway for amines.[2]

I. Mutants in Adenosylmethionine Decarboxylase (*speD*): *In Vivo* Assay[3]

We have used ^{14}COOH-labeled methionine as the substrate for this assay because the formation of $^{14}CO_2$ from this substrate in *E. coli* K12 *in vivo* results largely (>95%) from the combined action of adenosylmethionine synthetase and adenosylmethionine decarboxylase.[3]

Equipment and Materials

A semiautomatic dispensing device. An inexpensive device, such as the Accudrop made by Dynatech Laboratories, Inc., Alexandria, Virginia, is very useful for the rapid dispensing under sterile conditions of 25–150 μl into each well of a standard 96-well Microtiter plate.

Disposable Microtiter plates with 96 wells (Falcon catalog No. 3040 or equivalent) and covers (Falcon, catalog No. 3042)

A set of 48-prong stampers (described by Weiss and Milcarek[4]) is useful to transfer bacteria from a template of clones on an agar plate to the Microtiter plate, or from one Microtiter plate to another.

Barium hydroxide-impregnated paper: Whatman 3 MM paper (11.7 by 7.5 cm; i.e., to fit the Microtiter plate) is soaked in saturated barium hydroxide, blotted between filter papers, and stored in a desiccator over KOH solution until use. Care must be taken to avoid prolonged contact of the paper with atmospheric CO_2. (Note that NaOH or KOH solutions cannot be used to wet the filter paper, as soluble carbonates are formed that diffuse.)

L-[^{14}COOH]Methionine

[2] All the procedures presented here depend on the conversion of the labeled substrate to $^{14}CO_2$ by the reactions indicated rather than indirectly by various undefined side reactions in the cell or in a crude extract. This is true for the strain (*E. coli* K12) used in this study, as shown by the marked decrease in the amount of $^{14}CO_2$ produced when the assays are carried out with mutants lacking the enzyme being studied. This assumption may not necessarily be true, however, when other microorganisms or tissues are used, or under other cultural conditions. In general, the *in vitro* assays described here are more specific in this regard than the *in vivo* procedures.

[3] C. W. Tabor, H. Tabor, and E. W. Hafner, *J. Biol. Chem.* **253,** 3671 (1978).

[4] B. Weiss and C. Milcarek, this series, Vol. 29, p. 180.

Culture media

LB agar: 10 g of tryptone, 5 g of yeast extract, 10 g of NaCl, and 15 g of agar per liter of water

Vogel-Bonner minimal medium: Prepare as described in this series,[5] except for the omission of glucose. Any additional nutrients required by the strain are added.

Procedure

Step 1. Cells are mutagenized with *N*-methyl-*N'*-nitro-*N*-nitrosoguanidine.[6] The mutagenized culture is then plated for single colonies on LB plates, and the plates are incubated overnight at 37°. Individual colonies are transferred with sterile toothpicks or the "pinwheel" device of Weiss and Milcarek[4] to another LB plate in a pattern of the same size as the 48-prong stampers[7] and the plates are incubated overnight at 37°.

Step 2. Preparation of Microtiter plate I. To each well of a sterile 96-well Microtiter plate is added 100 μl of minimal medium containing the nutrients required by the strain and 0.02–0.04% glucose.[8]

Step 3. Microtiter plate I is inoculated from the template on the LB plate (step 1) with the help of the 48-prong stampers. The Microtiter plate is then covered with a plastic lid, wrapped loosely in aluminum foil to decrease evaporation, and incubated overnight at 37°. At the end of the incubation period the Microtiter plate is checked to be certain that there is equal growth in each well; the absence of growth in any well is recorded.[9]

Step 4. A second Microtiter plate (plate II) is filled with 100 μl of the same medium, containing [^{14}COOH]methionine (5–10 μCi/μmol, approximately 10^6 cpm/ml), and 0.2% glucose. This plate is then inoculated from the wells of plate I using the 48-prong stamper. The plate is immediately covered with Whatman 3 MM paper that has been impregnated with

[5] This series, Vol. 17A, p. 5.

[6] See this series, Vol. 17A, p. 9.

[7] A paper template with dots corresponding to the 48-prong stamper is placed under the agar plate to facilitate the correct placement of the colonies.

[8] This low glucose concentration limits the yield of cells and permits all the cultures to reach the same cell density before the assays for enzyme activity are carried out. Since the organisms have been heavily mutagenized, some of those isolated grow very slowly.

[9] For recording the data, we have found it convenient to prepare copies of a pattern consisting of a rectangle containing 96 squares arranged in a similar fashion to the Microtiter plates; i.e., 12 vertical rows with 8 squares per row, numbered to correspond to the numbering of the 96-well Microtiter plate.

barium hydroxide.[10] A plastic lid is then firmly secured on the dish with rubber bands. The plate is wrapped in aluminum foil and incubated overnight at 37°.[11,12]

Step 4a (Alternative Step). Alternatively at the end of step 3 (i.e., after the overnight incubation of the clones with limiting glucose), instead of transferring the cultures to a new Microtiter plate, one can add directly to each well of plate I 25 μl of a mix containing minimal medium, 2% glucose, and 25,000 cpm (25 nmol) of L-[^{14}COOH]methionine. The plate is then covered with barium hydroxide paper as described above.

Step 5. Remove the filter paper and wash it in a small pan containing acetone. Remove the acetone from the filter paper by air drying, and then heat for 5 min at 110°.

Step 6. Place the paper on Kodak XAR5 film for at least a day.

A typical autoradiogram has black spots above the wells that contain colonies that have active adenosylmethionine decarboxylase, and very light gray spots or no spots above the cells that lack adenosylmethionine decarboxylase.[13] Figure 2 shows an autoradiogram that was obtained from

[10] It is desirable to have the barium hydroxide-impregnated paper fit tightly against the top of the Microtiter plate. Therefore we usually press the damp paper against the top of the plate with a round glass rod or tube. We also place a 11.7 × 7.5 cm piece of plastic-covered paper (cut from a sheet of Benchkote) under the lid, so that the barium hydroxide paper is pressed tightly against the Microtiter dish during the incubation.

[11] We have found that a 3–4 hr incubation time is equally satisfactory.

[12] It is likely that temperature-sensitive mutants could be selected by using a 30° incubation temperature in steps 1 and 3, followed by a 43° temperature and a 1-hr incubation time in step 4a.

[13] When the autoradiogram indicates a mutant lacking adenosylmethionine decarboxylase, the appropriate clone on the original template is grown in liquid culture and a cell extract is assayed as described in this volume [37]. After the absence of the enzyme is confirmed, the gene is mapped by standard procedures (see this series, Vol. 17A [1]; J. H. Miller, "Experiments in Molecular Genetics." Cold Spring Harbor Lab., Cold Spring Harbor, New York, 1972). The mutated gene is then usually transferred by P1 transduction to another strain that has not been mutagenized.

In order to transfer the *speD* gene into other strains, we have found it convenient first to introduce a Tn*10* transposon close to the *speD* gene. Since *speD* is located at 2.7 min on the *E. coli* chromosome, we accomplished this by transduction of the *speD* strain with P1 from a strain that has Tn*10* at about 3.2 min on the chromosome (*zad*::Tn*10*; strain SJ2 of J. Jakowski, obtained from Dr. Barbara Bachmann of the *E. coli* Genetics Center of Yale University). *speD* can then be transferred to any other strain by another P1 transduction, selecting for tetracycline-resistant transductants. *speD* and *zad*::Tn*10* are sufficiently close to be cotransduced at a high frequency (ca. 40%).

The various mutants obtained as described in this chapter and in this volume [13] can be combined into a single strain by either P1 transduction or Hfr recombination to produce strains that lack all the enzymes that can make polyamines [E. W. Hafner, C. W. Tabor, and H. Tabor, *J. Biol. Chem.* **254,** 12419 (1979); H. Tabor, E. W. Hafner, and C. W. Tabor, *J. Bacteriol.* **144,** 952 (1980)].

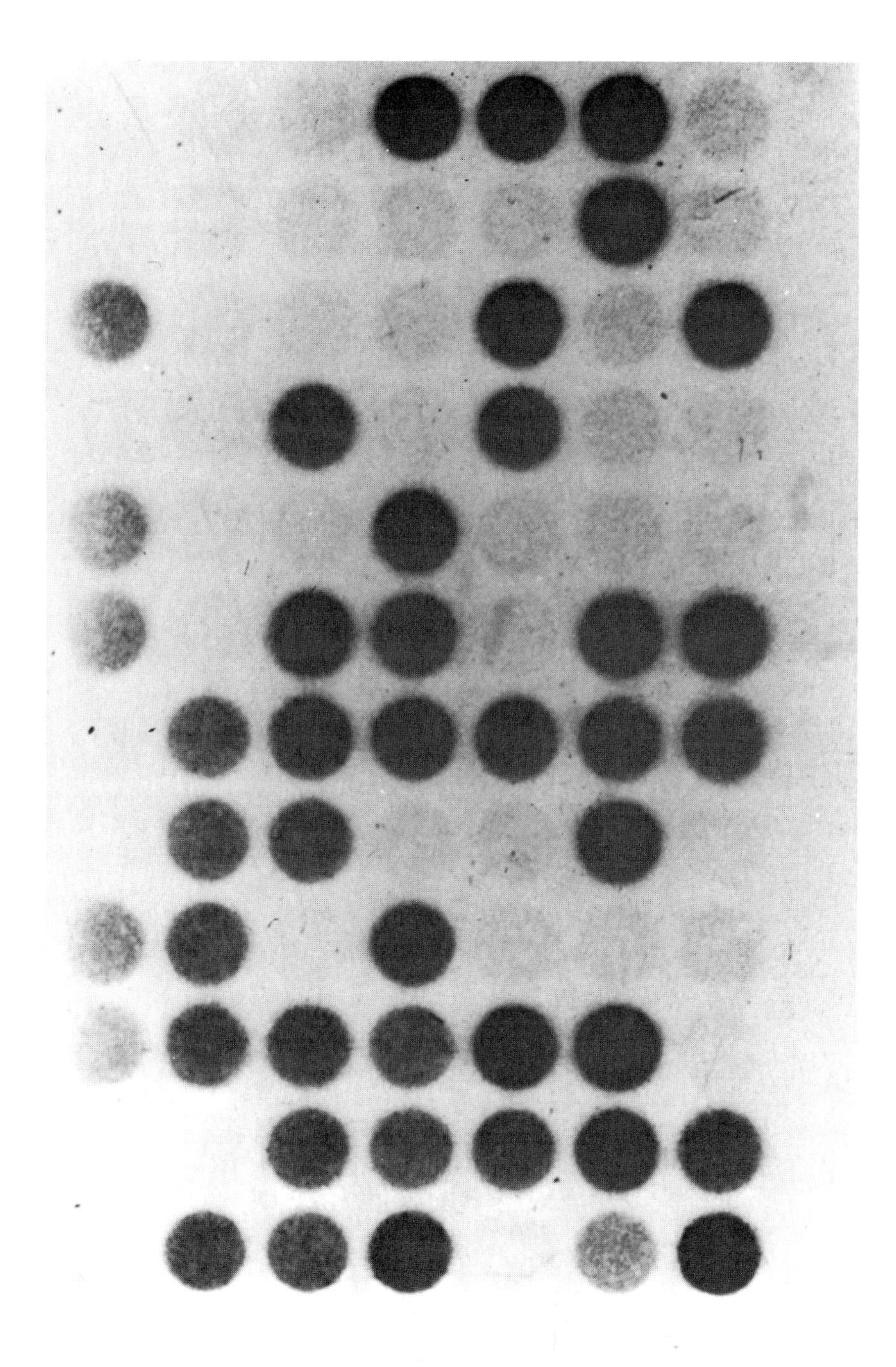

FIG. 2. Autoradiogram showing dark spots formed by the $^{14}CO_2$ released by cultures with adenosylmethionine decarboxylase and faint spots (or no spots) above cultures lacking the enzyme. (Reprinted from reference 1.)

colonies with and without adenosylmethionine decarboxylase; these colonies were wild-type and mutant recombinants in a genetic mapping experiment.

II. Mutants in Ornithine Decarboxylase, Arginine Decarboxylase, and Lysine Decarboxylase

Modifications of the *in vivo* method (Section I) have been used for the detection and assay of mutants in these three decarboxylases and are described below. However, in general we have found better results if the assays are carried out with cell lysates *in vitro* instead of with intact cells. (The *in vitro* assays resulted in much lighter spots when enzyme-negative strains were assayed, presumably because they are less likely to reflect multistep side reactions unrelated to the enzyme being studied).

Ornithine Decarboxylase (*speC*)

In Vivo[14]

The assay is carried out as described in Section I for adenosylmethionine decarboxylase, except that 3 × 10^6 cpm of L-[^{14}COOH] ornithine[17] (20–30 μCi/μmol) are used per milliliter of the mix instead of the ^{14}C-labeled methionine. Best results are obtained when the cells are at low cell density; i.e., by limiting the initial glucose concentration to 0.01–0.02% in steps 2 and 3 above and by carrying out the final incubation with label and glucose in step 4a for only 2–3 hr.

In Vitro[17]

The initial steps are the same as described above (Section I, steps 1–3) except that the initial glucose concentration is 0.02%. After the overnight incubation the cultures receive 25 μl (per well) of fresh medium with 2% glucose and are incubated for 2–4 hr at 37° before the next steps.

Step 1. The plates are centrifuged for 20 min at room temperature in a Beckman Model TJ-6 centrifuge with a TH-4 rotor in supports No. 270-341980. The medium is removed by rapid inversion of the plate with one or two whiplike motions.

Step 2. To each well is then added 50 μl of a 1% suspension of toluene (prepared by a 30-sec sonic oscillation of a buffer containing 0.02 *M* Tris-

[14] Other techniques for selecting mutants in *speA*, *speB*, or *speC* have been used by Morris and Jorstad[15] and by Maas.[16]

[15] D. R. Morris and C. M. Jorstad, *J. Bacteriol.* **101,** 731 (1970).

[16] W. K. Maas, *Mol. Gen. Genet.* **119,** 1 (1972).

[17] E. W. Hafner, C. W. Tabor, and H. Tabor, *J. Bacteriol.* **132,** 832 (1977).

HCl buffer, pH 8.0, 0.004 *M* dithiothreitol, 0.001 *M* EDTA, and 1% toluene). The toluene suspension is most conveniently added with the semiautomatic dispenser described in Section I. The cell pellets are suspended by mixing with the 48-prong stamper; the plates are covered with a plastic lid and incubated at 37° for 30 min to permit the cell walls to become permeable.

Step 3. Fifty microliters of mix A (see below) are then added to each well with the aid of the semiautomatic dispenser. The plates are covered with barium hydroxide-impregnated paper, as described in Section I, and incubated at 37° for 2–3 hr. The remainder of the procedure is the same as in steps 5 and 6 of Section I.

Mix A contains 20 m*M* Tris-HCl, pH 8.1, 5 m*M* dithiothreitol, 50 μM pyridoxal phosphate, 100 μM L-[^{14}COOH]ornithine (0.3 μCi/ml), and 1 mg of bovine serum albumin per milliliter.

Arginine Decarboxylase (*speA*)

In Vivo

The procedure is identical to that used for ornithine decarboxylase except for the use of DL-[^{14}COOH]arginine (2×10^7 cpm/μmol, 0.3 μCi/ml).[18,19]

In Vitro

The assay is the same as that described for ornithine decarboxylase, except for the use of mix B instead of mix A. Mix B contains 0.1 *M* HEPES-HCl, pH 8.5, 5 m*M* dithiothreitol, 10 μM pyridoxal phosphate, 40 μM L-arginine, 20 m*M* $MgCl_2$, 0.35 μCi of DL-[^{14}COOH]arginine per milliliter (specific activity 20 μCi/μmol), and 1 mg of bovine serum albumin per milliliter.

Lysine Decarboxylase[20]

Wild-type *E. coli* K12 has so little lysine decarboxylase that it is difficult to assay for the disappearance of this enzyme. Therefore, we first mutagenize the culture, and select a mutant (*cadR*) that overproduces

[18] To lower the blank values, the ^{14}C-labeled substrates should be dissolved in 0.01 *N* HCl and evaporated to dryness *in vacuo* just before use.

[19] [^{14}COOH]Arginine and [^{14}COOH]lysine are no longer listed in the commercial catalogs and must be ordered specially. It is likely that uniformly labeled materials would be satisfactory, but we have not checked these materials in the assays.

[20] H. Tabor, E. W. Hafner, and C. W. Tabor, *J. Bacteriol.* **144,** 952 (1980).

lysine decarboxylase.[21] This *cadR* strain is then mutagenized again and mutants are selected that no longer are able to decarboxylate lysine.

For both of these steps lysine decarboxylase is measured essentially as described for adenosylmethionine decarboxylase (Section I). However, step 4a of that assay is modified, so that the final glucose concentration is 0.4%, and growth is permitted to occur for 4 hr at 37° before the addition of labeled lysine (1.4 nmol of DL-[^{14}COOH]lysine; 10,000 cpm; 25 μl volume per well).[18,19]

III. Agmatine Ureohydrolase (*speB*)[17]

This enzyme is assayed *in vivo* by a coupled reaction in which L-[*guanido*-^{14}C]arginine is converted to [^{14}C]urea by the sequential action of arginine decarboxylase and agmatine ureohydrolase. This urea is excreted into the medium and is converted to $^{14}CO_2$ by added urease, as shown by Morris and Koffron.[22]

The specific procedures are the same as described in Section I for adenosylmethionine decarboxylase except for the medium used for step 4. To the minimal medium containing the necessary supplements and 0.2% glucose is added crude jack bean urease (10 mg/ml). The suspension is centrifuged and the supernatant is sterilized by ultrafiltration. The wells are filled with 100 μl of this medium and inoculated as described in Section I (step 4). Then, to each well, 25 μl of 0.3 m*M* L-[*guanido*-^{14}C]arginine (1 μCi/ml) is added. The procedure is continued as described in Section I.

IV. Adenosylmethionine Synthetase (*metK*)[23]

The assay depends on coupling endogenous adenosylmethionine synthetase and endogenous adenosylmethionine decarboxylase to form $^{14}CO_2$

[21] The *cadR* strains can also be obtained by selecting clones that are resistant to aminoethylcysteine (thiosine). P. S. Popkin and W. K. Maas [*J. Bacteriol.* **141,** 485 (1980)] have shown that such strains (*lysP*) are derepressed for lysine decarboxylase as well as being defective in lysine transport. *lysP* mutants map in the same position as *cadR*, although the identity of *lysP* and *cadR* has not been established with certainty.

[22] D. R. Morris and K. L. Koffron, *J. Bacteriol.* **94,** 1516 (1967).

[23] Adenosylmethionine synthetase (*metK*) mutants have usually been isolated by selecting strains that are resistant to ethionine. This selection procedure is based on the derepression of methionine synthesis in *metK* mutants; the resulting increased methionine levels antagonize the ethionine toxicity. To select *metK* mutants by this procedure,[17] we have plated cultures onto minimal agar plates containing 42, 20, and 9 m*M* DL-ethionine (for growth at 30°, 37°, and 42°, respectively, since ethionine is more toxic at higher temperatures). To obtain a mutant with a temperature-sensitive adenosylmethionine synthetase, mutants are selected that are resistant to 9 m*M* ethionine at 42°, but sensitive to 42 m*M* ethionine at 30°.

from L-[^{14}COOH]methionine. As mentioned above,[2] essentially all the $^{14}CO_2$ from [^{14}COOH]methionine is formed by this coupled pathway. Thus, in organisms containing an active adenosylmethionine decarboxylase, we have been able to screen for mutants in adenosylmethionine synthetase by determining the formation of $^{14}CO_2$ from [^{14}COOH]methionine during a 1-hr incubation.

The cultures are grown overnight with limiting glucose as described in Section I. Then, 25 μl of fresh medium containing 2% glucose are added per well, and the incubation is continued at 37° for 3–4 hr. At this time, 25 μl of fresh medium containing 15 μg of chloramphenicol are added, and the incubation is continued at 37° for 40 min. Then, 25 μl of medium containing 5000–10,000 cpm of L-[^{14}COOH]methionine (60 μCi/μmol) are added per well. The Microtiter dish is covered with barium hydroxide-impregnated paper and the incubation is continued at 37° for 1 hr.[24] The remainder of the procedure is as described in Section I.

[24] Best results are obtained when the incubation time is limited to 1 hr. After long incubation times, $^{14}CO_2$ is formed via other pathways, as well as by the decarboxylation of adenosylmethionine.

[13] Localized Mutagenesis of Any Specific Region of the *Escherichia coli* Chromosome with Bacteriophage Mu

By HERBERT TABOR, EDMUND W. HAFNER, and CELIA WHITE TABOR

Deletion mutants are desirable in studies of polyamine biosynthesis since such organisms completely lack the specific biosynthetic steps and do not revert. One common method for obtaining such deletion mutants is the insertion and deletion of bacteriophage Mu. Bacteriophage Mu is a temperate phage that inserts into the *Escherichia coli* chromosome at apparently random loci during lysogenization.[1–7] Mu insertion effectively inactivates the gene into which insertion is made, and this mutation is

[1] A. I. Bukhari, *Annu. Rev. Genet.* **10,** 389 (1976).
[2] A. I. Bukhari and A. L. Taylor, *J. Bacteriol.* **105,** 844 (1971).
[3] A. I. Bukhari and D. Zipser, *Nature (London) New Biol.* **236,** 240 (1972).
[4] M. Couturier, *Cell* **7,** 155 (1976).
[5] E. Daniell, R. Roberts, and J. Abelson, *J. Mol. Biol.* **69,** 1 (1972).
[6] M. M. Howe and E. G. Bade, *Science* **190,** 624 (1975).
[7] A. L. Taylor, *Proc. Natl. Acad. Sci. U.S.A.* **50,** 1043 (1963).

METHODS IN ENZYMOLOGY, VOL. 94

ISBN 0-12-181994-9

usually nonreverting.[1,4,6,8] Moreover, Mu infection at low multiplicities generally results in only one mutation per chromosome, in contrast to other mutagens (for example, nitrosoguanidine) which often yield multiple mutations.[9,10] One limitation in its use is that the frequency of its insertion into any average-sized gene is expected to be low, i.e., approximately 1 in 4000, assuming 4000 genes per *E. coli* chromosome.[11] Therefore, a search for a particular Mu-induced mutation requires one to examine several thousand colonies in order to have a reasonable chance of success when no selection techniques are available.

We have developed two methods for improving the chances of finding any desired Mu insertion mutation.[12] Both methods involve the infection of an Hfr donor strain with phage Mu, followed by the transfer of a small segment of the donor chromosome to a suitable F^- recipient. This transfer is accomplished by interrupted mating and is followed by selection of those recombinants that have received a Mu prophage from the donor. The two methods differ in the manner of the selection of these lysogens. Both methods give rise to a culture in which each cell contains a Mu phage randomly inserted within that region of the host chromosome transferred early by the Hfr strain used. Consequently, when a 5-min segment of the *E. coli* chromosome (i.e., 1/20 of the total chromosomal length) is so mutagenized, the frequency of a Mu insertion mutation is expected to be 20-fold greater for any gene in this region than following a random infection with Mu. This 20-fold increase means that one may expect to find a specific Mu mutant with the probability of approximately 1 : 200.

1. General Methods

Phage and bacterial strains are listed in the table. The media were essentially those described by Howe[13] and by Miller.[14] Superbroth[13] contained 32 g of tryptone, 20 g of yeast extract, 5 g of NaCl, and 5 ml of 1 *N* NaOH per liter, supplemented with $CaCl_2$ (0.005 *M*) and $MgCl_2$ (0.003 *M*). Minimal medium was that of Vogel and Bonner,[15] supplemented at concentrations recommended by Miller[14] with the specific requirements for

[8] A. I. Bukhari, *J. Mol. Biol.* **96,** 87 (1975).

[9] N. Guerola, J. L. Ingraham, and E. Cerdá-Olmedo, *Nature* (*London*) *New Biol.* **230,** 122 (1971).

[10] M. P. Oeschger and M. K. B. Berlyn, *Mol. Gen. Genet.* **134,** 77 (1974).

[11] J.-S. Hong and B. N. Ames, *Proc. Natl. Acad. Sci. U.S.A.* **68,** 3158 (1971).

[12] H. Tabor, E. W. Hafner, and C. W. Tabor, *J. Bacteriol.* **132,** 359 (1977).

[13] M. M. Howe, *Virology* **54,** 93 (1973).

[14] J. H. Miller, "Experiments in Molecular Genetics." Cold Spring Harbor Lab., Cold Spring Harbor, New York, 1972; see also J. R. Roth, this series, Vol. 17a [1].

[15] H. J. Vogel and D. M. Bonner, *J. Biol. Chem.* **218,** 97 (1956); see also this series, Vol. 17a, p. 5.

BACTERIAL AND BACTERIOPHAGE STRAINS

A. Bacterial strains

Strain No.	Genotype[a]	Source
HT200[b]	HfrH	B. Bachmann[c]
PK18[d]	HfrPK18 *thi* Δ(*lac pro*)	P. Kahn
CA158	HfrH *lacZ supE pyrD thi*	F. Jacob (via J. Beckwith and E. B. Konrad)
CSH50	F^- *ara* Δ(*lac pro*) *rpsL thi*	Cold Spring Harbor collection[e]
CSH55	F^- Δ(*lac pro*) *supE nalA thi*	Cold Spring Harbor collection[e]
CSH75	F^- *ara leu lacY proC purE gal trp his argG malA strA xyl mtl ilv metA thi*	Cold Spring Harbor collection[e]
AT2699	F^- *argG6 metC69 hisG1 serA6 thyA3 malA1 gal-6 leuYl mtl-2 rpsL tsx-64 supE44*(?)	A. T. Taylor via B. Bachmann
EWH154 (PL8-31)	F^- *thr-1 leu-6 thi-1 proA2 hisC3 metG87 metK86 serA25* Δ(*speC glc*) *rpsL-25 lacYl galK2 mtl-1 xyl-5 ara-14 supE44*(?)	Hunter *et al.* via B. Bachmann

B. Bacteriophage strains

Mu strain	Phenotype	Source
Mu c^+	Wild type	M. Howe
Mu *c*25	Clear mutant	M. Howe
Mu *cts*62 *Kam*1010	Amber, heat-inducible mutant	M. Howe

[a] For map position and gene designations, see Bachmann and Low.[19]

[b] The origin of transfer of HfrH (Hfr Hayes) is at 97 min. The direction of transfer is clockwise.[16,17]

[c] Dr. Barbara Bachmann, Curator, Yale Genetics Stock Center, Yale University, New Haven, Connecticut 06510.

[d] The origin of transfer of HfrPK18 is between *metC* and *rpsL*. The direction of transfer is counterclockwise.[18]

[e] These strains may be obtained from the Cold Spring Harbor Laboratory.[14]

the strain (see the table). In experiments designed to select mutants defective in putrescine or spermidine biosynthesis, all media contained 10^{-4} *M* putrescine or spermidine.

[16] W. Boram and J. Abelson, *J. Mol. Biol.* **62,** 171 (1971).

[17] K. B. Low, *Bacteriol. Rev.* **36,** 587 (1972).

[18] P. L. Kahn, *J. Bacteriol.* **96,** 205 (1968).

[19] B. J. Bachmann and K. B. Low, *Microbiol. Rev.* **44,** 1 (1980).

Mating procedures[20,21] were those described by Miller.[14] The temperature of the mating was 30° instead of 37°, because some of the strains contained a temperature-inducible Mu prophage. At 30° mating is slower and 50-min mating times were used to obtain suitable numbers of *leu*$^+$ or *ara*$^+$ recombinants when HfrH was the donor strain.

2. Bacteriophage Mu Preparations

Mu stocks and lysogens were prepared following published procedures.[13,16,22] We have found that the preparation of phage stocks of high titer requires the addition of sodium citrate (final concentration of 0.05 *M*) just before lysis. In our hands, in the absence of this citrate, phage titers rapidly fell (over 99% after 1 day at 4°). Presumably, adsorption of the phage to bacterial debris occurs, as suggested by Howe,[13] and citrate may prevent this.

Mu cts62 Kam1010. This phage, obtained by us from Dr. M. Howe, is derived from the temperature-inducible Mu *cts*62 phage[13] and contains an amber mutation in the *K* gene. Although the use of Mu with an amber mutation is not required for the methods described here, M. Howe (personal communication) suggested that this strain be used to facilitate the conversion of Mu *cts* insertions into deletions in future work. The usual methods for this conversion involve selection of the survivors following induction of Mu *cts* lysogens by incubation at 44°; among these survivors are some organisms that no longer contain Mu.[23,24] These nonlysogenic survivors, growing on a culture plate containing an excess of phage released by the induced lysogens, would normally be subject to lysis due to reinfection by Mu; such lysis by reinfection does not occur with Mu *cts Kam,* since this amber prophage cannot form infective particles upon induction if a nonsuppressor bacterial host is used.

Mu *cts*62 *Kam*1010 was stored as a lysogen in the *supE* strain CA158. For the preparation of free phage, a fresh culture of the lysogen was grown at 30° in superbroth to a cell count of 2×10^8 cells/ml and then heated at 44° for 30 min with good aeration. Sodium citrate was added to yield a final concentration of 0.05 *M*, and the incubation was continued at 30°. Lysis occurred about 20 min later. Immediately after lysis, the cell debris was removed by centrifugation for 5–10 min at 5000 *g*, and the

[20] W. Hayes, *J. Gen. Microbiol.* **16,** 97 (1957).

[21] F. Jacob and E. L. Wollman, eds., "Sexuality and the Genetics of Bacteria." Academic Press, New York, 1961.

[22] A. I. Bukhari and M. Metlay, *Virology* **54,** 109 (1973).

[23] M. Bachhuber, W. J. Brill, and M. M. Howe, *J. Bacteriol.* **128,** 749 (1976).

[24] R. S. Buxton, *J. Bacteriol.* **121,** 475 (1975).

supernatant solution was sterilized either by ultrafiltration (Millipore Swinnex-25 filter, 0.45 μm pore size) or by the addition of chloroform. When the preparation was used to infect a culture, additional $CaCl_2$ was added to the mixture in order to complex the citrate.

Mu c^+ (Wild-Type) Phage. This phage was stored as a prophage in *E. coli* HT200; the supernatant of a fresh culture was used as a low-titer source of the phage. Phage stocks of higher titer were prepared by the confluent plate method,[16] using *E. coli* CSH55 as a host, or by a procedure similar to that used to prepare Mu *c*25 phage (see below).

Mu c25 (Clear-Plaquing Mutant). A culture of *E. coli* CSH50 at $\sim 5 \times 10^7$ cells/ml was infected with $\sim 5 \times 10^6$ pfu/ml of Mu *c*25 and incubated at 37°. As soon as lysis began, as indicated by a fall in the absorbance at 540 nm, sodium citrate was added to a final concentration of 0.05 *M*. When lysis was complete, the cell debris was removed by centrifugation and the supernatant solution was sterilized by ultrafiltration or by the addition of chloroform.

3. Preparation of Mu Lysogens

Escherichia coli HfrH Mutagenized with Mu c^+ [E. coli HT200::(Mu c^+ random)].[25] A fresh culture of the *E. coli* HfrH strain, HT200, was infected with Mu c^+ at a bacterial density of 5×10^7 cells/ml and a phage multiplicity of 0.1. The mixture was incubated at 37° with vigorous aeration until lysis was observed by a fall in the absorbance at 540 nm. On occasion, the cell density became too high (i.e., $>1 \times 10^9$ cells/ml) before lysis occurred. When this happened, the culture was diluted 10-fold and the incubation was continued until lysis occurred. Forty percent of the surviving organisms were lysogenic for Mu, as shown by phage release when tested[13] on a lawn of a Mu-sensitive strain (CSH55). These lysogenic organisms were also immune to Mu *c*25. The culture was stored for short periods of time in 40% glycerol at −20°.

Escherichia coli HfrH Mutagenized with Mu cts62 Kam [E. coli CA158::(Mu cts62 Kam random)]. This lysogen was prepared essentially as described in the preceding paragraph except that the incubation tem-

[25] The notation used to describe Mu lysogens is that utilized by Howe and others.[13,22,23] The symbol "::" indicates that the Mu prophage given in parentheses following the symbol is inserted into the host chromosome at the locus specified immediately before the symbol. We also use here the designation illustrated by *E. coli* CA158::(Mu *cts Kam,* random) to indicate that the culture contains a mixture of lysogens with Mu *cts Kam* at different, presumably random locations on the chromosome. Hfr = high frequency of recombination; pfu = plaque-forming unit; Mu *cts* = heat-inducible Mu bacteriophage (i.e., contains a heat-sensitive repressor).

perature was 30°. For the preparation of a comparable lysogen in nonsuppressor strains in which viable progeny of this amber phage are not formed, it would be necessary to add the infecting phage at a multiplicity of infection >1.

Escherichia coli CSH75 rha::(Mu cts62 Kam). Mu *cts*62 *Kam*1010 was prepared as described above and used to infect a culture of *E. coli* CSH75 (F^- ara^- leu^-) at a multiplicity of infection >1. After several hours at 30°, aliquots containing 10^3 bacteria were plated on McConkey agar plates containing rhamnose, and a rhamnose-negative (white) colony was isolated. This organism did not grow at either 30° or 44° on a minimal medium containing the necessary auxotrophic requirements and rhamnose as the carbon source. When the medium contained glucose in place of rhamnose, growth was observed at 30° but not at 44°, since at the latter temperature the lysogen was induced.

Escherichia coli CA158 leu::(Mu cts62 Kam) and E. coli CSH50 leu::(Mu cts62 Kam). The *E. coli* HfrH *supE* strain, CA158, was infected with Mu *cts*62 *Kam*1010 at 30°. About 1–2% of the survivors were auxotrophic as expected.[6,7] A leucine auxotroph was isolated among these, *E. coli* CA158 *leu*::(Mu *cts*62 *Kam*). This leu^- mutation, resulting from the Mu *cts Kam* insertion, was transferred in a mating experiment at 30° to *E. coli* CSH50 (F^- ara^-, leu^+, pro^-). Ara^+ recombinants were selected and scored for 44° sensitivity; the temperature-sensitive recombinants were also leucine auxotrophs. One of these [F^- ara^+ pro^+ leu^-::(Mu *cts*62 *Kam*)] was used as the recipient organism in method A (below).

Escherichia coli Hfr PK18::[Mu c (random)] and E. coli Hfr PK18::[Mu cts62 Kam (random)]. These lysogens were prepared essentially as described above except for the use of *E. coli* PK18 as the recipient strain.[18]

4. Two Methods for Insertion of Mu into a Desired Gene (Such as speD)[26]

Two different methods have been developed for increasing the frequency of Mu insertions into specific chromosomal regions over the frequency found in a randomly Mu-mutagenized culture. Both methods use

[26] Even though these methods were developed for the construction of deletion mutants in the genes concerned with polyamine biosynthesis, they are equally suitable for obtaining deletions in any other gene by using an Hfr strain with an origin close to the gene in question. In this section we have used HfrH, since the gene for adenosylmethionine decarboxylase (*speD*) is at 2.7 min on the *E. coli* chromosome map. In the method for obtaining deletion mutants in *speA, speB,* and *speC* (below), HfrPK18 is used in order to obtain Mu insertions that map at about 63.5 min on the *E. coli* chromosome map.

an Hfr cross to donate a small piece of DNA from a randomly Mu-mutagenized donor strain to a suitable recipient, but they differ in how the subsequent selection of the desired lysogenic recombinants is made. Presumably, the methods are suitable for use with any Hfr donor. The examples below are with HfrH.

Method A

This method involves the transfer of DNA from a Mu c^+-mutagenized HfrH donor into an F^- recipient that is lysogenic for Mu *cts*62 *Kam*1010. This temperature-inducible recipient prophage must reside at a locus distant from the chromosomal region transferred early by the donor (Fig. 1). The presence of this prophage in the recipient serves two functions.

1. It prevents zygotic induction[21] when Mu c^+ prophage is transferred to the recipient by the Hfr cross.
2. It allows for the selection of Mu c^+ recombinant lysogens among the recombinants. After the cross, incubation at 44° results in lysis of all F^- cells that did not receive a Mu c^+ prophage, since presumably only those bacteria with Mu c^+ have a heat-stable (dominant) repressor that prevents the induction of Mu *cts Kam*.

By interrupting the mating, one can restrict the amount of DNA that is transferred. Furthermore, by the use of a selective marker, such as arabinose in this case, one can eliminate recombinants that, because of the early separation of some mating pairs,[16] have received very little (<4%) of the donor chromosome. The combination of a limited time of transfer during the mating experiment, selection for an early marker, and selection of a heat-stable organism resulted in the isolation of those organisms that contain Mu c^+ randomly distributed within a selected small region of the chromosome. Details of the method are given in the following examples.

Step 1. A fresh culture of the HfrH strain, HT200::(Mu c^+ random), was prepared by diluting the stored culture (0.05 ml; 1×10^6 organisms)[27] to 50 ml with Superbroth and incubating the culture at 30° with aeration until the cell density was 1.5×10^8 bacteria per milliliter. This culture was then mixed with 50 ml of CSH75 *rha*::(Mu *cts*62 *Kam*) grown at 30° to a cell density of 2×10^8 bacteria per milliliter. After the mating had proceeded for 50 min in a 1-liter flask at 30° with gentle shaking, 20 mg of streptomycin were added to counterselect against the donor bacteria, and the mating was interrupted by vigorous vortexing (20-ml portions, 1-min

[27] It is important that the inoculum contain more than 10^4 organisms to permit a statistically random distribution of the different lysogens.

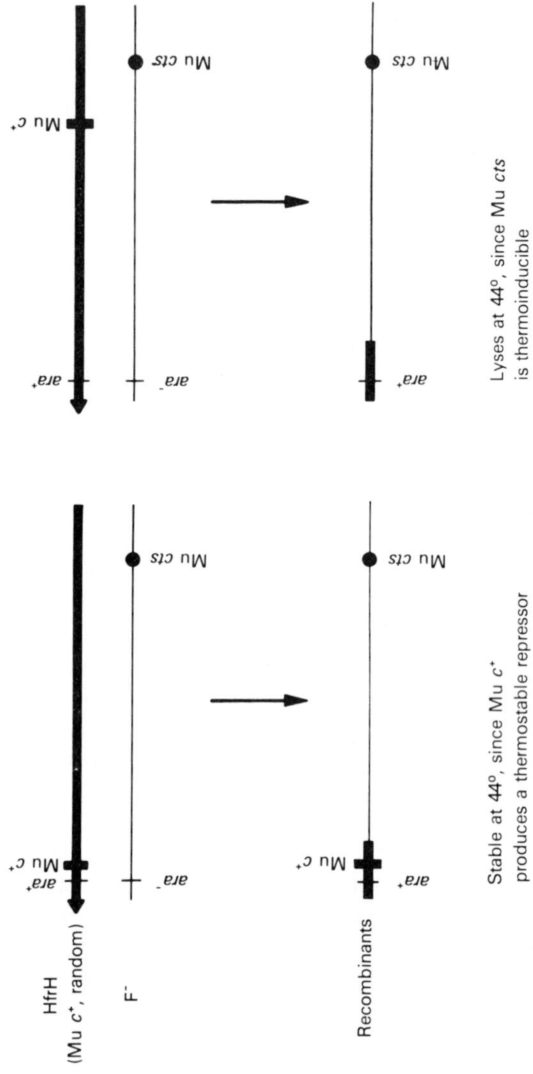

FIG. 1. Schematic representation of method A. Left-hand section indicates insertion of Mu c^+ into area of interest (near *ara*). Right-hand side represents a mating in which the Mu c^+ is present elsewhere in the Hfr donor.

periods). We estimate that most recombinants have received 4% to 8% of the donor chromosome.

Step 2. The mixture was diluted with 100 ml of superbroth and incubated at 30° for 2.5 hr to allow expression of recombinant DNA, then at 43° for 2–3 hr to induce heat-sensitive lysogens (i.e., recipients that did not receive wild-type Mu), and finally at 30° for 2 hr to complete lysis of induced lysogens.

Step 3. Portions (0.1 ml) of a 10^{-2} dilution were plated on minimal plates containing 1.3% arabinose as the carbon source and amino acids, purines, pyrimidines, and vitamins. These plates were incubated at 30° for 6 hr and then overnight at 44°. Approximately 100 ara^+ recombinants were found on each plate.

Step 4. The ara^+ recombinants were transferred to fresh arabinose-containing plates. To demonstrate the mutant enrichment obtained with this method, a search for auxotrophic mutants among the recombinants was made.[14] The rapid stamping techniques of Weiss and Milcarek[28] were used for replica platings.

Sixty-one auxotrophic mutants were found in the 1612 recombinants tested. Over 90% of these mutants were in genes comprising the portion of the chromosome donated early by HfrH. The assignment of the various mutations to genes in this area was confirmed by complementation tests with *E. coli* F′104.[17] The various recombinants were also tested for adenosylmethionine decarboxylase by a mass-screening method.[29] Four mutants were found that were deficient in this enzyme (*speD*).

Method B

In this method, an F^- leu^-: : (Mu *cts*62 *Kam*) recipient, auxotrophic for leucine because of its Mu insertion, is crossed by interrupted mating with an HfrH strain that has been randomly mutagenized with Mu *cts*62 *Kam* (Fig. 2). All leu^+ recombinants are expected to have lost the Mu *cts*62 *Kam* originally present in the leucine operon of the recipient strain. However, since the donor strain has been mutagenized randomly with Mu *cts*62 *Kam* before the mating, a small percentage of the leu^+ recombinants have a Mu *cts*62 *Kam* prophage, located at a different, but nearby, location. To select these leu^+ recombinants, the following procedure was developed which is based on the immunity of these lysogenic recombinants to infection by Mu *c*25.

[28] B. Weiss and C. Milcarek, this series, Vol. 29, p. 180.

[29] H. Tabor, C. W. Tabor, and E. W. Hafner, *J. Bacteriol.* **128,** 485 (1976). See also this volume [12].

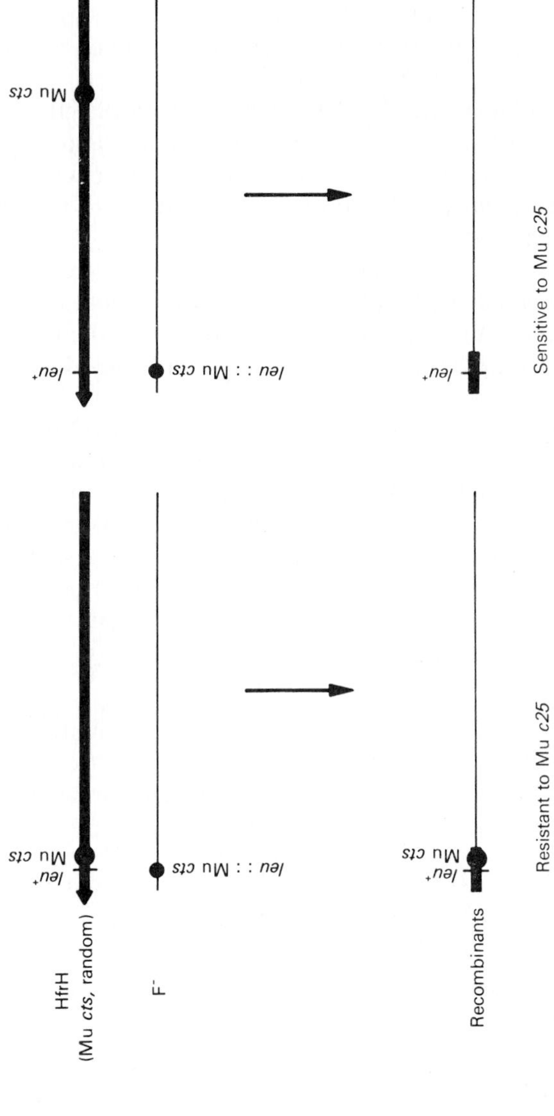

FIG. 2. Schematic representation of method B. Left-hand section indicates insertion of Mu *cts* into area of interest (near *leu*). Right-hand side represents a mating in which the Mu *cts* is present elsewhere in the Hfr donor.

Step 1. A cross between the Hfr strain, CA158::(Mu *cts*62 *Kam*, random), and a recipient F^- strain, CSH50 *leu*::(Mu *cts*62 *Kam*), was carried out in superbroth in a 1:2 ratio of donor to recipient ($\sim 2 \times 10^8$ cells/ml) at 30°. After 50 min, the mating mixture was interrupted by vigorous vortexing. The mixture was diluted 10-fold with Superbroth containing streptomycin (120 μg/ml) and incubated at 30° for 5–6 hr.

Step 2. To eliminate nonlysogenic recombinants, aliquots were treated with dilutions of Mu *c*25 (clear) to yield multiplicities of infection from 1 to 10. After incubation at 30° for 20 min, the mixtures were diluted 10-fold and incubated at 30° for an additional 6 hr. At the correct multiplicity of infection, this treatment resulted in the lysis of most of the nonlysogens and survival of the lysogens.[30]

Step 3. Aliquots of a 1:100 dilution were spread onto selective plates (i.e., minimal plates, containing amino acids, purines, pyrimidines, and vitamins, but no leucine) to select leu^+ recombinants.

Step 4. The individual leu^+ recombinants were then examined for newly acquired auxotrophic requirements or other mutations; to prevent growth of any contaminating leu^- recipients, media not containing leucine were used.

Twenty-three auxotrophic mutants were obtained in the 982 recombinants that were tested. In addition, 4 *speD* mutants were obtained.

5. Method for Obtaining Deletion Mutants in *speD*

Deletions from *speD* lysogenic strains containing Mu *cts Kam* were selected by a procedure similar to that of Bachhuber *et al.*[23] The lysogenic strain was grown in superbroth or LB broth at 30° to a bacterial count of 1×10^8 cells/ml. The culture was then incubated at 42–44° for 6 hr with

[30] High multiplicities of infection with Mu *c*25 must be avoided because multiple infections can kill lysogens. For this reason, several dilutions of Mu *c*25 were used and the flask with the dilution just causing detectable lysis (as indicated by $A_{540\ nm}$ readings) was selected for step 3. [The optimal amount of Mu *c*25 did not cause gross lysis since the nonrecombinant leu^-::(Mu *cts Kam*) recipients are the most numerous cell type and are immune to Mu *c*25.]

A better way of selecting the flask with the correct amount of Mu *c*25 was to carry each flask through step 3; i.e., to measure the number of leu^+ recombinants in each flask. In a culture that had received the optimum amount of Mu *c*25, the number of leu^+ recombinants found was about 3–5% of the total leu^+ recombinants found without Mu *c*25 treatment. The vast majority of those leu^+ recombinants that survived the Mu *c*25 treatment were lysogenic for Mu *cts Kam;* i.e., they contained a new Mu *cts Kam* from the Hfr donor. The presence of Mu *cts* in these leu^+ recombinants was easily confirmed by observing no growth when the isolates were tested at 44°.

shaking in order to induce lysis and to select heat-resistant survivors. Samples (0.1 ml) were plated on LB agar and incubated overnight at 44°.

To test which of the heat-resistant survivors no longer contained phage genes (i.e., at least the immunity *c* gene), the survivors were tested for sensitivity to Mu *c*25. Approximately 0.5% of the heat-resistant survivors were sensitive to Mu *c*25, although this value varied considerably in different experiments. Complementation tests with Mu *cts Cam* showed that some of these strains also lacked the *C* gene of the Mu phage genome.

6. Method for Obtaining Deletion Mutants in *speA*, *speB*, and *speC*

For this construction we chose Hfr PK18 as the donor strain; Hfr PK18 donates *metC* early in the *metC speC metK* (*speA speB*) *serA lysA* direction.

Step 1. Hfr PK18 : : Mu c^+, random was crossed with a *rha* : : Mu *cts Kam*, *thyA*$^+$*serA6* derivative of strain AT2699 (*metC*) (Method A in Section 4); 42–44° resistant *serA*$^+$ recombinants (i.e., double lysogens) were selected. A *metC*$^+$ *serA*$^+$ *lysA* : : Mu c^+ recombinant was then isolated by standard replica-plating tests, screening for a lysine requirement. This strain was converted to *rha*$^+$ by P1*vir* transduction in order to remove the Mu *cts Kam* prophage.

Step 2. Hfr PK18 : : Mu *cts Kam*, random, was then crossed by interrupted mating with this *metC*$^+$ *serA*$^+$ *lysA* : : Mu c^+ strain. The *lysA*$^+$ Mu *c*25 (clear mutant)-resistant recombinants were selected at 32° (method B, Section 4). These recombinants were assayed for the enzymatic activities corresponding to the *speA*, *speB*, and *speC* genes by the mass screening methods.[29] Among 1800 recombinants tested, 11 *speC* : : Mu *cts Kam*, 4 *speB* : : Mu *cts Kam*, and 6 *speA* : : Mu *cts Kam* were obtained. The *speA* : : Mu *cts Kam* were *speB*$^+$, and the *speB* : : Mu *cts Kam* mutants were *speA*$^+$. One of the *speA* : : Mu *cts Kam* mutants (designated P44) was used for step 3.

Step 3. A *sup*$^+$ derivative of strain P44 was constructed by crossing strain P44, which is *sup argG rpsL*, with an Hfr strain that was *sup*$^+$ *arg*$^+$ *rpsL*$^+$ (such as HT200) and selecting *sup*$^+$ *arg*$^+$ *rpsL* recombinants. (The *sup* marker was assayed by the presence (*sup*) or the absence (*sup*$^+$) of phage release upon 42° induction of the Mu *cts Kam* prophage in strain P44.)

Step 4. Deletions spanning *speA* : : Mu *cts Kam* were selected by the method outlined in Section 5. The *sup*$^+$ *speA* : : Mu *cts Kam* lysogen prepared in step 3 was heat induced by incubation at 42–44° on LB plates, and the Mu *c*25-sensitive survivors were isolated. One colony among these was found that also lacked agmatine ureohydrolase activity (*speB*). We presume that this strain contains a deletion mutation spanning *speA* and *speB* (i.e., Δ(*speA speB*)).

Step 5. The Δ(*speA speB*) mutation isolated in step 4, as well as the Δ*speD* mutation isolated as described in Section 5, were introduced into the Δ(*speC glc*) strain EWH154 in standard P1*vir* transductions, selecting for $serA^+$ and leu^+, respectively. Coincidentally with the transduction of Δ(*speA speB*), $metK^+$ was introduced.

Comments

The methods presented here permit the isolation of a collection of Mu lysogens of *E. coli,* each containing the phage Mu integrated into a limited, well-defined part of the *E. coli* chromosome. Since these procedures markedly increase the frequency of a desired mutant over that found in a randomly mutagenized culture, they facilitate the isolation of mutants whose detection requires difficult screening methods. These methods, however, require knowledge of the approximate map position of the locus one desires to mutate and also an Hfr strain that transfers this locus early during mating.

The two methods of insertion of Mu into a gene presented in this paper appear to be equally suitable for the selection of various mutations. Method A is simpler, as it does not require either the use of Mu *c*25 or the availability of a recipient strain with a prophage in any specific locus. Method B is preferable when lysogenization with a temperature-inducible prophage is desired, as, for example, for the eventual conversion of Mu *cts* insertions to deletions, as discussed earlier.

The advantage of mutagenesis with Mu is that the mutational event is relatively well defined[6] compared with those produced by many other mutagens (such as nitrosoguanidine, which results in multiple mutations within a single chromosome[9,10]). Furthermore, mutations that result from insertions of Mu are stable. Two limitations of Mu mutations are that one does not expect to obtain conditional-lethal mutations and that occasionally Mu insertions and DNA deletions may occur at the same time.[31]

The concept used here is similar to the idea of "localized mutagenesis," developed by Hong and Ames,[11] in which isolated bacteriophage P1 was treated with a mutagen before being used in transduction experiments. Although high frequencies of mutation have usually been reported with this method, in some instances much lower mutation frequencies were

[31] T. Cabezón, M. Faelen, M. De Wilde, A. Bollen, and R. Thomas, *Mol. Gen. Genet.* **137,** 125 (1975).

observed. A disadvantage of the method is the likelihood of obtaining multiple mutations in the heavily mutagenized segment of DNA involved, and the mutations obtained are not deletions. In favor of the Hong–Ames technique is that it allows the isolation of conditional lethal mutations.

[14] Mass Screening for Mutants in the Polyamine Biosynthetic Pathway in *Saccharomyces cerevisiae*

By HERBERT TABOR, CELIA WHITE TABOR, and M. S. COHN

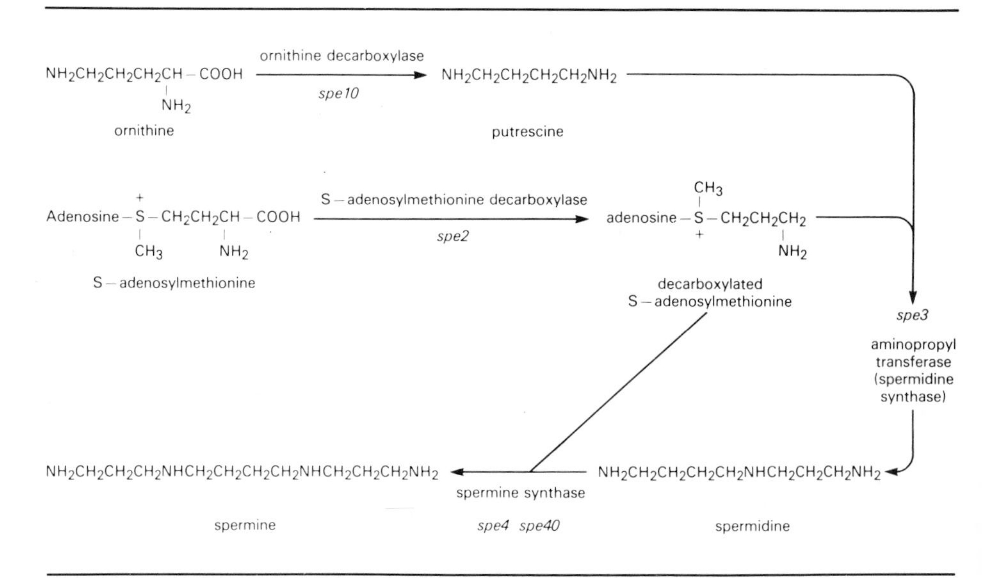

FIG. 1. Polyamine biosynthetic pathway in *Saccharomyces cerevisiae*. The genes involved in each step are indicated beneath the arrows.

Mass Screening for Mutants of Ornithine Decarboxylase (*spe10*)[1–3]

In searching for mutants in ornithine decarboxylase in *S. cerevisiae*, we use a modification of the rapid screening method presented in this

[1] M. S. Cohn, C. W. Tabor, and H. Tabor, *J. Bacteriol.* **134,** 208 (1978).

[2] M. S. Cohn, C. W. Tabor, and H. Tabor, *J. Bacteriol.* **142,** 791 (1980).

[3] We have used this direct screening method for detecting mutants of ornithine decarboxylase since we had thought that it would not be possible to select mutants by methods

ISBN 0-12-181994-9

volume [12]. In this procedure, after mutagenesis individual colonies are distributed into 96-well Microtiter plates and lysed, and the individual extracts are tested for their ability to produce $^{14}CO_2$ from carboxy-labeled ornithine.

Since *S. cerevisiae,* growing in rich medium, has a very low ornithine decarboxylase activity it is very difficult to select mutants that lack this enzyme. To avoid this problem we first used the procedure described below to select mutant strains that have unusually high ornithine decarboxylase activity. One of these strains was then further mutagenized, and individual colonies were tested for the loss of ornithine decarboxylase. After mutants lacking ornithine decarboxylase were obtained, the genetic defect was transferred to a nonmutagenized background by standard genetic techniques.

Procedure

Step 1. A culture of *S. cerevisiae* is mutagenized with 0.02% *N*-methyl-*N*′-nitro-*N*-nitrosoguanidine.[4] Cells are then spread on rich agar plates (YPAD medium),[1,5] and incubated at 25° until large colonies are obtained. The colonies should be well separated from each other.

Step 2. Each well of 96-well Microtiter plates is filled with 100 μl of mix A with a semiautomatic dispenser (see this volume [12]).

Mix A contains 2 m*M* dithiothreitol, 10 μ*M* potassium phosphate buffer, pH 7.5, 1 m*M* $MgCl_2$, 0.1 m*M* pyridoxal phosphate, 0.1 m*M* EDTA, and 0.5 mg of Zymolyase 5000 per milliliter (Miles Laboratories, Inc., code No. 32-090-1).

Step 3. Single colonies are picked at random from the YPAD plate (step 1) with the flat end of a 5-inch sterile wooden swab stick or the flat end of a sterile toothpick, and a portion of each colony is suspended in each well to obtain roughly the same cell density in all the wells.

Step 4. The plates are covered with a plastic lid and incubated at 37° for 1 hr to digest the cell walls.

---

based on a requirement for amines for growth, in view of the large amount of amines present intracellularly in yeast. However, such selection procedures have been used successfully by P. A. Whitney and D. R. Morris in their isolation of ornithine decarboxylase mutants of yeast [*J. Bacteriol.* **134,** 214 (1978)]. K. Hosaka and S. Yamashita [*Eur. J. Biochem.* **116,** 1 (1981)] have also developed a method for obtaining ornithine decarboxylase mutants based on selection for a requirement for amines for growth after treatment of the mutagenized culture with nystatin. The reader is referred to these references for the details of these alternative procedures.

[4] S. G. Kee and J. E. Haber, *Proc. Natl. Acad. Sci. U.S.A.* **72,** 1179 (1975).

[5] A rich medium is used to facilitate the growth of the heavily mutagenized organisms.

Step 5. After this incubation, 1.1 nmol of L-[^{14}COOH]ornithine (56 μCi/μmol) in 25 μl of 0.01 *M* potassium phosphate buffer, pH 7.5, is added to each well with the aid of the semiautomatic dispenser. The plate is then covered with barium hydroxide-impregnated filter paper and a plastic lid, as described in this volume [12]. The plates are then incubated at 37° for 4 hr. The papers are dried and subjected to autoradiography as described in this volume [12].

After development the X-ray film shows predominantly dark spots above the corresponding wells. However, in the position of a few wells, there are no spots (or sometimes faint gray spots). To confirm that these cells are indeed amine-requirers, the patches corresponding to these wells on the YPAD plate of step 1 are streaked on amine-free[6] SD plates[7]; after growth on the first SD plate, the partially depleted organisms are restreaked on a second and then a third SD plate to complete the depletion of internal amines. Isolates that have no ornithine decarboxylase do not form colonies on the third SD plate; if the mutation is leaky, only tiny colonies are found. Good growth occurs if putrescine, spermidine, or spermine is added to the plates (10^{-5} *M* final concentration).

The absence of ornithine decarboxylase in these mutants is further confirmed by growing them in liquid culture (SD medium), disrupting the cells by passage through a French press, and assaying the extracts for ornithine decarboxylase as described in this volume [12].

We have checked 10 mutants that we have isolated by the above procedure by complementation assays and have found that all mapped in

[6] Special precautions are required to eliminate all traces of amines from the medium in order to show clearly an absolute requirement of ornithine decarboxylase mutants for amines for growth. Unless such precautions are carried out, slight growth is seen on the SD plates. We have found that as little as 10^{-10} *M* spermine or spermidine will permit some growth of an amine-requiring mutant.

Water stored in conventional carboys is unsatisfactory; presumably a slight contamination by organisms in these carboys results in enough amines to permit some growth. The water used for media should be freshly distilled, and sterilized by Millipore filtration. We have found that autoclaving in a standard steam autoclave is unsatisfactory since amine additives are added to the steam to prevent boiler scale. This problem is especially significant if the medium is acid. An autoclave using distilled water as a local source of steam or a pressure cooker can be used instead of Millipore filtration.

The agar used for the plates should be checked by analysis or by absence of growth of a known amine-requiring mutant (such as *spe10-3*) to determine that it is amine-free, since we have found some differences in the growth of mutants lacking ornithine decarboxylase on different types of agar.

All glassware should be washed with dilute HCl. The acid should be removed by washing the glassware with freshly distilled water. The glassware is then sterilized in a special autoclave (see above) or is dry-sterilized at 150°.

[7] SD medium contains 0.67% yeast nitrogen base without amino acids (Difco), 2% dextrose, 2% agar, and any supplements needed to satisfy auxotrophic requirements (0.01%).

the same gene. The position of this gene is not known yet, nor whether the same gene is involved in the mutants isolated by Whitney and Morris and by Hosaka and Yamashita. Other work has indicated that the mutants that we have isolated are probably in a regulatory gene for ornithine decarboxylase rather than in the structural gene.

Mass Screening for Mutants of *S. cerevisiae* Deficient in *S*-Adenosylmethionine Decarboxylase (*spe2*)[1]

Cultures are mutagenized and placed in a Microtiter plate as described in the preceding section, except for the use of mix B. Mix B contains 1 m*M* dithiothreitol, 100 m*M* potassium phosphate (pH 8.0), 0.1 m*M* EDTA, 2.5 m*M* putrescine dihydrochloride, and 0.5 mg of Zymolyase 5000 per milliliter. After 1 hr at 37° to lyse the cells, 75 pmol (in 20 μl) of *S*-adenosyl-L-[^{14}COOH]methionine (54.6 μCi/μmol; New England Nuclear) are added to each well with the automatic dispenser. The remainder of the procedure is the same as described for ornithine decarboxylase.

Extracts of *spe2* isolates are tested for adenosylmethionine decarboxylase as described in this volume [12].

Spe2 mutants are closely linked to *arg1*. *arg1* has been shown by Hilger and Mortimer[8] to be on chromosome XV. *spe2* mutants have a long doubling time (about 12 hr) in the absence of added amines.

Selection of Mutants Lacking Spermidine Aminopropyltransferase (*spe4* and *spe40*)[2]

When *spe10* strains are almost depleted of amines by subculturing from rich medium onto two successive SD amine-free plates at 25° (see the preceding section), growth becomes very slow, and, if enough organisms are plated, a diffuse lawn is found. A few larger isolated colonies emerge after several days. These colonies no longer have an absolute requirement for amines for growth; they do not have any spermidine aminopropyltransferase activity by direct *in vitro* analysis (see below) and have no spermine (but they do have normal putrescine and spermidine) by automated amine analysis (see this volume [4]).

Assay for Spermidine Aminopropyltransferase

The assay mix contains 100 nmol of decarboxylated adenosylmethionine[9] and enzyme (dialyzed overnight against buffer A) in a final

[8] F. Hilger and R. K. Mortimer, *J. Bacteriol.* **141,** 270 (1980).

[9] For methods describing the preparation of decarboxylated adenosylmethionine, see this volume [11] and this series, Vol. 17B [190].

volume of 160 μl of buffer A [50 mM potassium phosphate (pH 8.2)–1 mM EDTA]. The reaction is started by adding 0.47 μCi of [^{14}C]spermidine trihydrochloride (specific activity 4.7 μCi/μmol) in 40 μl of H_2O. After incubation for 2 hr at 37°, 2 ml of 10% trichloroacetic acid is added. The precipitate is pelleted by centrifugation at 4°. The supernatant solution is extracted twice with 6 ml of diethyl ether, and residual ether is removed under vacuum.

To assay the amount of [^{14}C]spermine formed by the action of spermidine aminopropyltransferase, samples are placed on the chromatographic cation exchange column (as described in Method 1 of this volume [14]) and the eluate is diverted to a fraction collector (5 ml fractions). The formation of [^{14}C]spermine from [^{14}C]spermidine is detected by the elution of ^{14}C at the position of spermine. Alternatively, the labeled supernatant fraction can be chromatographed on paper with carrier spermine, as described in this volume [44].

Genetic Studies

We have found by genetic analysis that these mutants fall into two classes: one, *spe4,* may be in the structural gene for spermidine aminopropyltransferase; the other, *spe40,* appears to be a regulatory gene for this enzyme. The *spe40* mutation is tightly linked to the *spe10* mutation, but *spe4* is not linked to *spe10*. *spe10spe40* strains grow at almost the wild-type rate, while *spe10spe4* mutants have a 10-hr doubling time in the absence of amines.

[15] Selection of Ornithine Decarboxylase-Deficient Mutants of Chinese Hamster Ovary Cells

By C. STEGLICH and I. E. SCHEFFLER

We describe here a procedure for the selection of Chinese hamster ovary (CHO) cells lacking appreciable activity of the enzyme ornithine decarboxylase (ODC). Wild-type (ODC$^+$) cells that have incorporated high specific activity tritiated ornithine into their polyamine pools are killed efficiently during brief storage of the cells in the frozen state. During this storage time, radioactive decay leads to a lethal irradiation. Ornithine decarboxylase-deficient cells escape this radiation suicide. We take advantage of the fact that the parental cell line, CHO-K1, is auxo-

ISBN 0-12-181994-9

trophic for proline[1] owing to a double mutation that prevents the cells from making proline, either from glutamate or from ornithine. In particular, these cells have no ornithine aminotransferase (EC 2.6.1.13) activity.[2] Consequently, these cells use ornithine only as a precursor to polyamine biosynthesis. Normal *pro*$^+$ hamster cells (V79) convert about 15% of exogenous radioactively labeled ornithine to acid-precipitable counts (presumably protein), which could significantly affect the efficiency of the selection procedure described here.

Methods

Chinese hamster ovary cells are maintained in Dulbecco's modified Eagle's medium (DME) with 2% fetal calf serum, 4% newborn calf serum, nonessential amino acids, penicillin (100 units/ml), and streptomycin (100 μg/ml) in a 37° incubator with an atmosphere of 10% CO_2 in air. Putrescine is added at a final concentration of 0.5 m*M* from a 50 m*M* solution made in distilled water and sterilized by filtration through a 0.2 μm filter (Nalgene).

For mutagenesis, 2.4×10^6 cells are plated at 10^5 cells per 100 mm in diameter tissue culture plate (Lux). Twenty-four hours later, ethylmethane sulfonate (Sigma) is added at 300 μg/ml for 18 hr. At the end of the treatment, cells are washed with TD buffer[3] and fed growth medium containing putrescine. Survival after the mutagen treatment for our cells under these conditions is about 4%. Selection is begun after a 2- or 3-day recovery period. We have preliminary indications that a 3-day phenotypic lag yields a higher frequency of mutants. The optimum lag for mutant isolation has not yet been determined.

The enrichment procedure is outlined in Fig. 1. Mutagenized cells are plated at 10^5 cells per 100 mm in diameter plate in medium without putrescine containing 5 μCi/ml, L-[2,3-^{3}H]ornithine (15.4 Ci/mmol; New England Nuclear). After 2 days, the tritiated ornithine medium is removed and the plates are washed twice with TD buffer and once with cold freezing medium (DME with 20% fetal calf serum, 10% dimethyl sulfoxide) and placed at −80° for 2 weeks with about 2 ml of freezing medium in the plate. The cells are thawed by adding growth medium containing putrescine. Alternatively, the cells could be removed from the plate by trypsinization and frozen in a suspended state. This suicide cycle can be repeated if necessary.

[1] F. Kao and T. T. Puck, *Genetics* **55,** 513 (1967).

[2] R. J. Smith and J. H. Phang, *J. Cell. Physiol.* **98,** 475 (1979).

[3] To buffer: 0.025 *M* Tris, 0.137 *M* NaCl, 0.005 *M* KCl, 0.7 m*M* Na_2HPO_4, pH 7.4. I. E. Scheffler and G. Buttin, *J. Cell. Physiol.* **81,** 199 (1973).

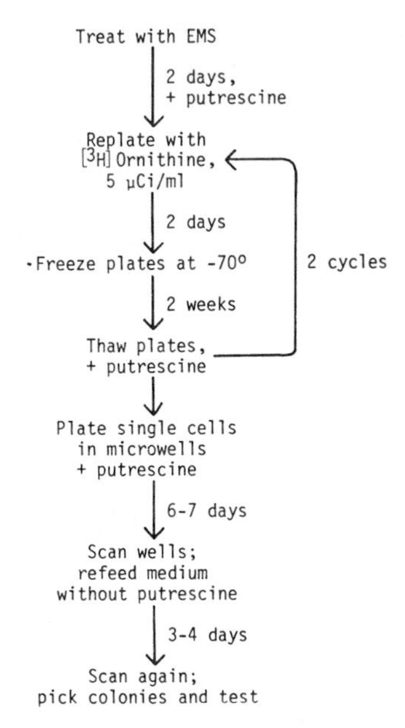

FIG. 1. Selection for putrescine auxotrophs.

If the number of survivors is low, individual colonies are isolated directly and tested for a putrescine requirement for growth. If the number of survivors is large, a sample of them is plated in microwells (Falcon Micro Test II) at one cell per well in growth medium containing 5×10^{-4} M putrescine. After 1 week the wells are scanned and those with colonies are noted. At this time the medium is changed to medium without putrescine; after an additional 3–4 days the individual colonies are observed again to note those whose growth or appearance is poor in the absence of putrescine. Such colonies are picked and retested for their ability to proliferate and form colonies in medium with or without putrescine.

Sample Results

We have used this procedure in two independent selections with mutagenized CHO cells to obtain several putrescine-requiring cell lines. In the pilot experiment two rounds of suicide selection followed by screening 1630 colonies in micro wells yielded 1 mutant clone. This mutant has a stringent requirement for putrescine for proliferation and has about 3% of the maximum inducible ornithine decarboxylase activity of the wild-type parental cell line. This mutant has been described in detail elsewhere.[4]

In this first experiment we allowed a 2-day phenotypic lag after mutagenesis, and survival after suicide selection was fairly high. A second experiment used a 3-day lag and a single round of selection with a very low survival. Roughly one-fourth of the surviving clones were putrescine auxotrophs. This second experiment more nearly approaches the efficiency of selection that we calculate to be possible based on a reconstruction experiment with a known mixture of mutant and wild-type cells (which could also be distinguished by additional genetic markers). Wild-type CHO cells incubated for 48 hr in 5 μCi of tritiated ornithine per milliliter and stored at $-80°$ had about a 10^{-4} chance of survival, whereas about 40% of the mutant cells survived this treatment. Thus, a single round of selection should yield a 4000-fold enrichment for mutant cells.

Discussion

One may ask about the chances of isolating such mutants from other mammalian cell lines. Two factors need to be considered. First, in other cell lines the conversion of tritiated ornithine to other amino acids is significant and the absence of ornithine decarboxylase activity alone may not protect against lethal irradiation by radioactive decay of accumulated isotope. In any case it would be advisable to determine for each individual circumstance the optimum cell density, concentration of tritiated ornithine, time allowed for incorporation, and storage time at $-80°$ for an efficient killing of wild-type cells. Second, the mutation in the ODC^- cells is recessive, and there are indications that it is in the structural gene for ornithine decarboxylase (unpublished observations). As long as this gene has not been mapped, the problem of ploidy and its relationship to the frequency at which mutants are generated in a population has to be considered. Chinese hamster ovary cells are pseudodiploid, and they have been shown to be hemizygous at a number of loci and chromosome segments.[5] By chance ODC^- mutants may be found more frequently in CHO cell populations because they may have only one functional gene. This may not be the case for other cell lines.

[4] C. Steglich and I. E. Scheffler, *J. Biol. Chem.* **257,** 4603 (1982).

[5] L. Siminovitch, *Cell* **7,** 1 (1976).

[16] Uses of Arginaseless Cells in the Study of Polyamine Metabolism (*Neurospora crassa*)

By ROWLAND H. DAVIS and THOMAS J. PAULUS

It is desirable in certain studies of polyamine metabolism to deprive cells of polyamines without resorting to mutations of polyamine-specific enzymes or to inhibitors of these enzymes. In *Neurospora crassa,*[1,2] yeast,[3] and bacteria,[4] this is done by creating a state of ornithine deprivation. In *N. crassa,* this leads eventually to depletion of polyamines, derepression of ornithine decarboxylase, and the appearance of cadaverine.[5] Most important, the control of polyamine synthesis in an intact pathway may be studied after restoration of ornithine.[2,6]

Ornithine deprivation can be brought about in *N. crassa* mutants that lack arginase.[1] These strains grow quite normally on a minimal medium. However, in these strains, ornithine cannot be formed from externally added arginine. External arginine, moreover, will feedback-inhibit *de novo* ornithine synthesis. The end result of arginine supplementation is starvation for polyamines, a state that is reversed by the addition of ornithine, putrescine, or spermidine.[1] Mutants of other cell types, such as yeast and mammalian cells, can in principle be ornithine-deprived in a similar way. This phenomenon is quite similar to the situation in *Escherichia coli,* described in 1970 by Morris and Jorstad.[4] In *E. coli,* which lacks arginase, arginine decarboxylase is the first step on one of two routes to putrescine. The other route is ornithine decarboxylase. Mutants lacking arginine decarboxylase, when grown in arginine, starve for ornithine and polyamines.

Methods

Strains. The wild-type strain of *N. crassa,* 74A, and strains carrying *aga* (allele UM-906) and *spe-1* (allele TP-138) mutations are available from the Fungal Genetics Stock Center, Humboldt State University Foundation, Arcata, California 95521. The *aga* mutant has no detectable argi-

[1] R. H. Davis, M. B. Lawless, and L. A. Port, *J. Bacteriol.* **102,** 299 (1970).
[2] T. J. Paulus and R. H. Davis, *J. Bacteriol.* **145,** 14 (1981).
[3] P. A. Whitney and D. R. Morris, *J. Bacteriol.* **134,** 214 (1978).
[4] D. R. Morris and C. M. Jorstad, *J. Bacteriol.* **101,** 731 (1970).
[5] T. J. Paulus, P. T. Kiyono, and R. H. Davis, *J. Bacteriol.* **152,** 291 (1982).
[6] R. H. Davis, unpublished observations.

ISBN 0-12-181994-9

nase[1]; the *spe-1* mutant has no detectable ornithine decarboxylase.[5] The wild-type and *aga* strains grow well in minimal medium; the *aga* strain acquires a polyamine requirement in the presence of arginine. The *spe-1* strain and the arginine-grown *aga* strain grow well when supplemented with 1 m*M* putrescine or 1 m*M* spermidine.

Growth. Stock maintenance and the growth of strains for mass conidial inocula are done by methods fully described in this series.[7] Briefly, it involves growth in flasks or tubes of solid Vogel's medium N, with sucrose as carbon source and supplements as necessary. Stationary cultures are grown at a constant temperature (25–35°) in Erlenmeyer flasks of liquid medium; 10 ml of medium in a 50-ml flask is convenient for determining the dependence of dry weight upon supplements to media.

Exponential cultures are started by adding washed, filtered conidia (5–10 days old) to 1000 ml of Vogel's medium contained in a 1000-ml boiling flask. A convenient inoculum is about 1×10^6 conidia per milliliter, final concentration (10 Klett units with the No. 54 filter). Cultures are aerated and mixed continuously by sparging with hydrated air through a glass tube thrust to the bottom of the flask. Dry weight is monitored by removing samples of known volume, collecting the cells on Whatman No. 540 filter circles (1-inch), and drying immediately with acetone as filtration continues. By 8 hr at 25°, dry weight will be 0.2 mg/ml, which will rise to 1 mg/ml by 15–17 hr in the case of wild type. All methods are given in detail in Vol. 17A of this series.[7]

Normal doubling time of *N. crassa* is about $2\frac{3}{4}$ hr at 25°. The *aga* strain, grown in 1 m*M* arginine, has a doubling time of 5–6 hr.[1]

Determination of Polyamines. The procedure used to assay polyamines is described in this volume.[8]

Assay of Ornithine Decarboxylase. Ornithine decarboxylase is assayed in 0.3-ml reaction mixtures containing 0.6 μmol of L-[1-^{14}C]ornithine (specific radioactivity ca 0.5×10^6 cpm/μmol), 15 nmol of pyridoxal phosphate, 30 μmol of K^+ phosphate, pH 7.1, 0.5 μmol of 2-mercaptoethanol, and 0.3 μmol of EDTA. The reactions are run in 13 × 100 mm tubes in the mouth of which an accordion-pleated fan of Whatman No. 1 filter paper (1 × 4 cm) with 10 μl of monoethanolamine–methyl Cellosolve (3 : 1, v/v) is inserted.[9] The last ingredient is added with a pipette tip past the fan. The tubes are stoppered, incubated at 37°, and stopped with 0.2 ml of 10% trichloroacetic acid after an appropriate interval. After 90 min of further incubation (stoppered) at 37°, the fans are put

[7] R. H. Davis and F. J. de Serres, this series, Vol. 17A, p. 79.

[8] T. J. Paulus and R. H. Davis, this volume [5].

[9] D. R. Morris and A. B. Pardee, *Biochem. Biophys. Res. Commun.* **20,** 697 (1965).

ORNITHINE-DEPRIVED *aga* CULTURES

Strain	Medium[a]	Doubling time (hr)	ODC[b] activity	Polyamine pools[b] (nmol/mg dry wt)				
				PUT	SPD	SPM	CAD	APC
74A	Minimal	2.8	21	1.1	16.2	0.5	NT	NT
	+SPD	2.8	NT[c]	0.6	19.7	0.5	NT	NT
aga	Minimal	2.8	26	0.6	17.0	0.7	<0.1	−
	+ARG	5.0	1459	<0.1	2.0	1.0	0.6	+
TP-138	Minimal	8.1	0	<0.1	1.1	0.2	<0.1	−
	+SPD	2.8	0	<0.1	19.7	0.8	NT	NT

[a] Supplements were added to 1 m*M* final concentration.
[b] Abbreviations: ODC, ornithine decarboxylase; PUT, putrescine; SPD, spermidine; SPM, spermine; CAD, cadaverine; APC, aminopropylcadaverine, ARG, arginine.
[c] NT, not tested; − = absent; + = present.

into scintillation vials with 1.5 ml of water and 10 ml of scintillation fluid (1 part Triton X-100 : 3 parts toluene; the toluene contains 5 g of 2,5-diphenyloxazole per liter).

Characteristics of Ornithine-Deprived *aga* Cultures

Growth. Cultures of *aga* grown in arginine-supplemented media have a doubling time of 5–6 hr for as long as they have been maintained (ca. 24 hr). It is likely that growth would continue at this rate indefinitely, because stationary cultures continue increasing linearly in dry weight for 10 days.[1] In contrast, cultures of an ornithine decarboxylase-deficient strain (*spe-1*) grown on minimal medium grow more slowly in exponential culture (doubling time = 7–16 hr) for 48 hr and are unable to sustain growth in the long term in stationary cultures.[10] It is likely that after initial depletion of normal polyamines, the two kinds of culture owe their difference to the formation of cadaverine in arginine-grown *aga* cells, a compound that *spe-1* cultures cannot form[5] (see below).

Pools. Steady-state cultures of wild type and of the *spe-1* strain grown in minimal medium, and of the *aga* strain grown in arginine, differ in their ornithine and polyamine pools (see the table). Both polyamine-deprived cultures have little or no putrescine, and their low spermidine content can be accounted for by the amount in the inoculum. The two cultures differ in that cadaverine and aminopropylcadaverine are found in arginine-grown *aga* cultures, whereas neither is detectable in *spe-1* cultures. The

[10] K. J. McDougall, J. Deters, and J. Miskimin, *Antonie van Leeuwenhoek* **43,** 143 (1977).

two unusual polyamines are not found in wild-type or in ornithine-sufficient *aga* cultures. If lysine is added to arginine-grown *aga* cultures, a substantial enhancement of the cadaverine pool is seen. This does not happen if lysine is added to starving *spe-1* cultures.

If the *aga* strain is grown in minimal medium, its polyamine pools are quite normal. Upon the addition of arginine to such cultures, the synthesis of putrescine and spermidine stops completely after about an hour, and this is correlated with the exhaustion of ornithine.[2] The putrescine present at the time of arginine addition persists for many hours, as though it were sequestered from use by spermidine synthetase.[2]

Ornithine Decarboxylase. Wild-type and the *aga* strain, when grown in minimal medium, have similar specific activities of ornithine decarboxylase (see the table). The arginine-grown *aga* strain, in contrast, has greatly elevated activity (50- to 100-fold) of the enzyme. The *spe-1* culture has no activity under any condition tested. The high level of ornithine decarboxylase in ornithine-deprived *aga* cultures, combined with the lack of ornithine, allows the weak lysine decarboxylase activity of this enzyme to be expressed *in vivo;* cadaverine therefore appears in such cultures.

Restoration of Ornithine to Ornithine-Deprived aga Cultures. The most useful characteristics of the ornithine-deprived *aga* culture in our hands are the polyamine-depleted cytoplasm and the derepressed ornithine decarboxylase. An example of the use of this situation is shown in Fig. 1, where ornithine is restored to a starved *aga* culture. Rapid putrescine and spermidine synthesis ensue, with polyamine formation ceasing at the time that spermidine synthesis returns to normal.[2] We have speculated that until spermidine levels are normal, all of it is tightly bound within the cell. When the binding sites are saturated, free spermidine appears which can exert a negative effect (as yet undefined) upon the highly derepressed ornithine decarboxylase.[2] The pattern of polyamine restoration is highly reminiscent of that seen in mammalian cells after relief of methylglyoxal bisguanylhydrazone (MGBG) treatment.[11]

Mutant Selection. Polyamine-sufficient cells, when suddenly unable to make polyamines, are able to grow for some time before the polyamine requirement is expressed. For this reason, Whitney and Morris[3] grew arginaseless cells of yeast on arginine to deprive them of polyamines before they were able to isolate the first series of polyamine mutants in that organism. Our own mutant (*spe-1*) of *N. crassa* was isolated by the same rationale. In both cases, it is probable that the preconditioning of cells reduced the "phenotypic lag" and allowed polyamine mutants to escape the negative selection against the wild type.

[11] J. Jänne, L. Alhonen-Hongisto, P. Seppänen, and E. Höltta, *Adv. Polyamine Res.* **3**, 85 (1981).

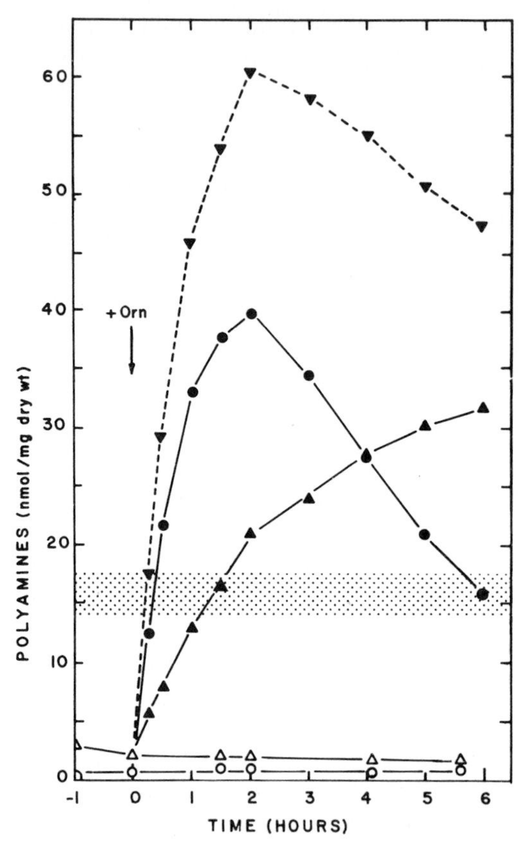

FIG. 1. Putrescine and spermidine pools of *aga* upon restoration of ornithine to an arginine-grown culture. Cells are grown from inoculation in medium containing 1 m*M* L-arginine. At time zero, a control culture continued growth (○, △) and one culture was supplemented with 5 m*M* L-ornithine (●, ▲, ▼). Symbols: putrescine (○, ●); spermidine (△, ▲); sum of putrescine and spermidine (▼) in treated culture. The horizontal stippled band shows the normal range of spermidine concentrations in wild-type cultures. Taken from Paulus and Davis[2] with permission.

Conclusion

A state of ornithine deprivation is in principle achievable in any culture by a combination of metabolic or mutational blocks in ornithine biosynthesis and in arginine catabolism. Cells in this state can be used to study regulation of ornithine decarboxylase and other enzymes, cadaverine synthesis, flux of ornithine into polyamines, and the properties of polyamine-depleted cytoplasm. Cells deprived of polyamines are favorable starting material for isolation of polyamine mutants.

Acknowledgments

Research was supported by grants from National Institutes of Health (AM-20083) and the American Cancer Society (BC-366).

[17] Cloning of the *Escherichia coli* Genes for the Biosynthetic Enzymes for Polyamines[1–3]

By CELIA WHITE TABOR, HERBERT TABOR, EDMUND W. HAFNER, GEORGE D. MARKHAM, and STEPHEN M. BOYLE

A. Principle[4]

The genes for the biosynthetic enzymes involved in polyamine biosynthesis[5] were identified in a bank of *E. coli* DNA that had been cloned into ColE1 plasmids by Clarke and Carbon.[6] The clones containing the genes of interest were then grown to obtain the plasmid DNA. This DNA was cut by restriction endonucleases and cloned into a high copy-number plasmid vector. Strains of *E. coli* transformed with these hybrid DNAs produce many times the wild-type concentration of the gene product and are useful for the preparation of large amounts of the homogeneous pro-

[1] E. W. Hafner, C. W. Tabor, and H. Tabor, *Fed. Proc., Fed. Am. Soc. Exp. Biol.* **38,** 395 (1979).

[2] G. D. Markham, E. W. Hafner, C. W. Tabor, and H. Tabor, *J. Biol. Chem.* **255,** 9082 (1980).

[3] S. M. Boyle, G. D. Markham, E. W. Hafner, J. M. Wright, H. Tabor, and C. W. Tabor, in preparation. This contribution includes restriction maps for the *speA, speB, speC,* and *metK* genes.

[4] The specific details for each step in the procedure are not given in this chapter, since Vol. 68 of this series contains such details for all the methods needed for cloning a desired gene. (For another collection of methods, see T. Maniatis, E. F. Fritsch, and J. Sambrook, "Molecular Cloning." Cold Spring Harbor Laboratory, Cold Spring Harbor, New York, 1982.)

The methods that we have used are essentially the same or very similar to those described in Vol. 68, which are listed below.

a. Preparation of competent bacteria for transformation: Vol. 68 [16], p. 252.
b. Transformation of cells: Vol. 68 [16], p. 252.
c. Preparation of plasmids (small or large scale): Vol. 68 [16], pp. 264, 265; [17], pp. 269–272. For the rapid purification of DNA, see Vol. 68 [17], p. 271.
d. Amplification of plasmid DNA: Vol. 68 [16], pp. 264, 265.
e. DNA joining methods: Vol. 68 [1], pp. 16–18. We have followed the directions provided by the supplier of the ligases.
f. Restriction endonucleases for cleaving DNA: Vol. 68 [2], pp. 27–41. We have followed the specific directions provided by the supplier of each endonuclease prepara-

ISBN 0-12-181994-9

tein and for studies of the regulation of the enzyme. After amplification of the plasmids, the DNA can be isolated in large amounts for sequencing studies or for use in *in vitro* protein synthesis.

B. Identification of Cells Containing Plasmids Carrying the Genes for Adenosylmethionine Synthetase (*metK*), Arginine Decarboxylase (*speA*), and Agmatine Ureohydrolase (*speB*)

The Clarke–Carbon collection[6] was grown overnight in grids on LB[7] agar containing colicin E1.[8,9] The plasmids from these cells were then transferred to strain EWH313[10] by mixing the two strains on another LB

tion. It is important to destroy the endonuclease by a 2-min immersion in boiling water before adding a ligase.

g. Agarose gel electrophoresis of DNA and detection of DNA on gels: Vol. 68 [9], pp. 152–176; [16], pp. 265, 266.
h. Elution of DNA from agarose gels: Vol. 68 [10], pp. 176–182.
i. Acrylamide gel electrophoresis of DNA: Vol. 68 [16], pp. 265, 266.
j. Elution of DNA from polyacrylamide gels: Vol. 68 [16], pp. 266, 267.
k. Cloning vehicles: Vol. 68 [1], pp. 5–11.
l. Assay of expression of gene function by localization of proteins in denaturing polyacrylamide gels: Vol. 68 [38], pp. 517, 518. We routinely used 1-mm-thick slab gels for this purpose and increased the percentage of acrylamide from 10 to 25% for the assay of small protein molecules. A stacking gel of 5% acrylamide was used.
m. Expression of gene function in minicells: Vol. 68 [36], pp. 495–497. Although the use of minicells for the study of bacteriophage-coded proteins is described, similar techniques can be used for plasmid-coded proteins.

[5] The *speD* gene is located at 2.7 min on the *E. coli* chromosomal map. *speB* and *speA* are at about 62.8 min, *metK* is at 63.1 min, and *speC* is at 63.4 min. See also this volume [12] for the identification of the various genes; see also footnote 3 and W. K. Maas, *Mol. Gen. Genet.* **119,** 1 (1972).

[6] L. Clarke and J. Carbon, *Cell* **9,** 91 (1976). See also this series, Vol. 68 [27]. pLC is the abbreviation for the plasmids from this collection. The number of the plasmid designates the position in the storage plates of the collection. The first number represents the plate number and the second number represents the position on the plate.

[7] See this series, Vol. 17A [1], p. 5 for the composition of Vogel–Bonner medium; for the composition of LB medium, see J. H. Miller, "Experiments in Molecular Genetics," p. 433. Cold Spring Harbor Laboratory, Cold Spring Harbor, New York, 1974.

[8] For the preparation of colicin E1, see this series, Vol. 68 [16], p. 264. See also J. A. Spudich, V. Horn, and C. Yanofsky, *J. Mol. Biol.* **53,** 49 (1970). Colicin E1 kills all cells that do not have ColE1 plasmids and is needed to prevent loss of this rather unstable plasmid.

[9] The stamping technique was described by B. Weiss and C. Milcarek, this series, Vol. 27 [16], p. 180. See also this volume [12].

[10] Strain EWH313 has the genotype F^-, *hisG1 lacY1 gal-6 rpsL tsx-64* Δ(*glc speC*) Δ(*speA speB*), and has a near-absolute requirement for polyamines for growth if leucine and threonine are added to the medium (unpublished data). It seems likely that the strain that we have isolated recently [HT414, *J. Bacteriol.* **147,** 702 (1981)], which has an absolute requirement for amines for growth, would also be suitable for this purpose.

plate with the stampers.[9] After growth the resultant grid was transferred to plates containing minimal media[7] plus the auxotrophic requirements plus colicin E1 and streptomycin. After full growth, the grid was transferred again onto the same medium plus leucine (100 μg/ml) and threonine (100 μg/ml). Those colonies that grew well presumably had received plasmids with the genes for putrescine biosynthesis[10]; i.e., either *speC* alone, or *speA* plus *speB,* or all three. After purification by single-colony isolation, these isolates were tested for *speA* and *speC* by the methods listed in this volume [12, 18]. Five colonies were found to have the *speA* gene (and presumably the *speB* gene), but not *speC;* namely, the colonies that had received the plasmids from positions 2-5, 5-8, 5-14, 12-37, and 27-37 of the Clarke–Carbon collection. (Position 20-5 contained the *speC* gene; see Section D).

Four of these plasmids were then tested for the presence of the *metK* gene by transferring the plasmids from the respective Clarke–Carbon donors into a *metJ* recipient strain (EWH205). The presence of the $metK^+$ gene was shown by measurement of adenosylmethionine synthetase *in vitro,* as described elsewhere in this volume [35]. All four of the plasmids tested (2-5, 5-8, 5-14, 27-37) contained the *metK* gene in addition to the *speA* and *speB* genes.

C. Insertion of *metK, speA,* and *speB* Genes into a High Copy-Number Plasmid

1. DNA was isolated[4] from pLC2-5 of the Clarke–Carbon collection, and treated with *Eco*RI restriction endonuclease. The DNA fragments were then ligated with *Eco*RI-treated DNA from pBR322. The ligated DNA was used to transform a *metK E. coli* strain (DM22).[11] One transformant (pK5) had high levels of adenosylmethionine synthetase, arginine decarboxylase, and agmatine ureohydrolase when assayed *in vivo* or *in vitro,* i.e., about 5–10 times the wild-type level.

2. To prepare a plasmid containing only the *metK* gene, DNA was isolated from the pK5 plasmid and was treated with *Pst*I restriction endonuclease. The fragments were then religated and the ligated DNA was used to transform the *metK* strain DM22. One transformant (pK8) showed increased levels of adenosylmethionine synthetase, but not of arginine decarboxylase or of agmatine ureohydrolase. The insert carrying the *metK* gene is about 1.5 kb long.

[11] DM22 is an *hsd* derivative of *E. coli* K12 strain EWH154 [E. W. Hafner, C. W. Tabor, and H. Tabor, *J. Bacteriol.* **132,** 832 (1977)]. The genotype of EWH154 is *metK86 serA25* Δ(*speC glc-1*) *rpsL25 thr-1 leu-6 thi-1 proA2 hisC3 metG87.*

3. To prepare a plasmid containing the *metK* and *speA* genes, but not the *speB* gene, we treated the DNA from plasmid pK5 with *Sal*I endonuclease. The DNA was religated and used to transform *E. coli* strain DM22. One transformant (pK11) was found to have lost the tetracycline resistance marker of pK5 because the *Sal*I fragment had been reversed when religation occurred. The DNA of pK11 was isolated and cut with *Bal*I. The fragments were then religated, and the ligated DNA was used to transform the *metK* strain DM22. One transformant was found (pK12) that had high adenosylmethionine synthetase (*metK*) and arginine decarboxylase (*speA*) activity, but no agmatine ureohydrolase (*speB*) (in contrast to the parent plasmid pK11). The insert carrying the *speA* and *metK* genes was about 4 kb in length.

D. Identification of Cells in the Clarke–Carbon Collection Containing Plasmids Carrying the Gene for Ornithine Decarboxylase (*speC*)[1,3]

The procedure was essentially the same as described in Section B for the *metK* gene, except that, after growth in minimal medium (supplemented with the auxotrophic requirements, colicin E1, and streptomycin), ornithine decarboxylase activity was assayed either *in vivo* as described in this volume [12], or *in vitro*, as described in this volume [12, 18]. One clone from the Clarke–Carbon collection, 20-5,[10] was found to have a high ornithine decarboxylase activity.

E. Insertion of the *speC* Gene into a High Copy-Number Plasmid

DNA was isolated[4] from plasmid pLC20-5 and treated with the restriction endonuclease *Pst*I.[4] The DNA fragments were then ligated to *Pst*I-treated pBR322. The ligated DNA was used to transform a *speC E. coli* strain (HT289).[12] One transformant (pODC1) had high levels of ornithine decarboxylase when tested *in vitro*.[13] When this transformant was tested for arginine decarboxylase, agmatine ureohydrolase, or adenosylmethionine synthetase activities, no increase over the wild-type level was found. The insert carrying the ornithine decarboxylase gene was 3.2 kb long.

[12] H. Tabor, C. W. Tabor, M. S. Cohn, and E. W. Hafner, *J. Bacteriol.* **147,** 702 (1981). The genotype of HT289 is Δ(*speA speB*) Δ(*speC glc*) *hsd sup thr leu thi*.

[13] See this volume [12, 18].

F. Identification of Cells in the Clarke–Carbon Collection Containing Plasmids That Carry the Gene for Adenosylmethionine Decarboxylase (*speD*)

The mating procedure was essentially the same as that described in Section B for the *metK* gene, except that the recipient organisms were mutants lacking the *speD* gene.[14] After full growth on minimal plates containing the auxotrophic requirements plus colicin E1 and streptomycin, adenosylmethionine decarboxylase activity was assayed *in vivo* as described in this volume [12]. One colony, 37-29, had high adenosylmethionine decarboxylase activity.

G. Insertion of the *speD* Gene into a High Copy-Number Plasmid[14]

1. DNA was isolated from plasmid pLC37-29 and treated with restriction endonuclease *Pst*I.[4] The DNA fragments were ligated to *Pst*I-treated pBR322 and the ligated DNA was used to transform a *speD E. coli* strain (D18).[15] One colony was found that had a high adenosylmethionine decarboxylase activity. This colony contained a new plasmid, pSPD1, which had an 8 kb insert in pBR322.

2. Plasmid pSPD1 was cut with the restriction endonuclease *Sal*I and was religated into a new pBR322, which had been cut with *Sal*I. The ligated DNA was used for transformation. A new plasmid pSPD3 was found that contained a 6 kb insert.

3. The plasmid pSPD3 was cut with *Pvu*II and the DNA was religated. The ligated DNA was used for transformation. One colony was found that had a high adenosylmethionine decarboxylase activity (pSPD5). The new plasmid had a 2.8 kb insert and the colony carrying this plasmid produces about 15 times the wild-type activity.

[14] Strains containing Δ*speD* were first described by C. W. Tabor, H. Tabor, and E. W. Hafner, *J. Biol. Chem.* **253,** 3671 (1978); see also this volume [12]. The recipient strain used for section G (strain HT383) contains the Δ*speD* and *hsd* genes [G. D. Markham, C. W. Tabor, and H. Tabor, *J. Biol. Chem.* **257,** 12063 (1982)].

Section IV

Ornithine Decarboxylase, Arginine Decarboxylase, Lysine Decarboxylase

A. Enzyme Assays and Preparations

Articles 18 through 30

B. Enzyme Inhibitors

Articles 31 through 34

[18] Biosynthetic and Biodegradative Ornithine and Arginine Decarboxylases from *Escherichia coli*

By DAVID R. MORRIS and ELIZABETH A. BOEKER

L-Ornithine → putrescine + CO_2
L-Arginine → agmatine + CO_2

Under appropriate culture conditions, some strains of *Escherichia coli* are able to synthesize two sets of ornithine and arginine decarboxylases.[1] One set, produced during growth on neutral minimal medium, forms a network for regulation of putrescine biosynthesis that responds to the availability of substrates from the pathway of arginine biosynthesis.[1,2] These two enzymes are referred to as the biosynthetic ornithine decarboxylase (bODC) and the biosynthetic arginine decarboxylase (bADC) and have been found in all strains of *E. coli* tested. bODC[3] and bADC[4] have both been purified to homogeneity.

When cultures of *E. coli* are grown at low pH in the presence of ornithine or arginine, degradative ornithine decarboxylase (dODC) and degradative arginine decarboxylase (dADC), respectively, are induced. The physiological role of these enzymes appears to be to regulate environmental pH.[1] Homogeneous preparations of both dODC[5] and dADC[6] have been obtained. Although most *E. coli* strains can produce dADC, dODC is of quite limited occurrence.[3] The purification of dADC was described in this series,[7] and the methods are updated in this chapter. The purification of a related enzyme, the inducible lysine decarboxylase from *E. coli*, is described in this volume.[8]

Assay Methods

Principle. All four enzymes are conveniently assayed using commercially available labeled substrate. The $^{14}CO_2$ evolved during the course of

[1] D. R. Morris and R. H. Fillingame, *Annu. Rev. Biochem.* **43,** 303 (1974).
[2] D. R. Morris, W. H. Wu, D. Applebaum, and K. L. Koffron, *Ann. N.Y. Acad. Sci.* **171,** 968 (1970).
[3] D. M. Applebaum, J. C. Dunlap, and D. R. Morris, *Biochemistry* **16,** 1580 (1977).
[4] W. H. Wu and D. R. Morris, *J. Biol. Chem.* **248,** 1687 (1973).
[5] D. Applebaum, D. L. Sabo, E. H. Fischer, and D. R. Morris, *Biochemistry* **14,** 3675 (1975).
[6] E. A. Boeker, E. H. Fischer, and E. E. Snell, *J. Biol. Chem.* **244,** 5239 (1969).
[7] E. A. Boeker and E. E. Snell, this series, Vol. 17B, p. 657.
[8] E. A. Boeker and E. H. Fischer, this volume [30].

ISBN 0-12-181994-9

the reaction is trapped on filter paper impregnated with an amine,[9] in this case 2-aminoethanol.

Reagents

bODC

Substrate solution: 8.9 m*M* L-[1-^{14}C]ornithine (ca. 2.5×10^4 dpm/μmol) in a buffer containing 120 m*M* *N*-2-hydroxyethylpiperazine-*N*′-2-ethanesulfonic acid (pH 8.25), 48 μ*M* pyridoxal phosphate, and 2.0 m*M* dithiothreitol

Enzyme diluent: buffer A (see Purification)

bADC

Substrate solution: 8.4 m*M* uniformly labeled L-[^{14}C]arginine (ca. 1.5×10^5 dpm/μmol) in a buffer containing 120 m*M* *N*-2-hydroxyethylpiperazine-*N*′-2-ethanesulfonic acid (pH 8.4), 48 μ*M* pyridoxal phosphate, 4.8 m*M* $MgSO_4$, and 1.2 m*M* dithiothreitol

Enzyme diluent: buffer A (see Purification)

dODC

Substrate solution: identical to that described above for bODC except for the substitution of 120 m*M* 3-(*N*-morpholino)propanesulfonic acid (pH 7.0) as buffer

Enzyme diluent: buffer C (see Purification)

dADC

Substrate solution: 30 m*M* uniformly labeled L-[^{14}C]arginine (ca. 1.5×10^5 dpm/μmol) in a buffer containing 200 m*M* sodium acetate (pH 5.2) and 60 μ*M* pyridoxal phosphate

Enzyme diluent: 200 m*M* sodium acetate (pH 5.2) containing 60 μ*M* pyridoxal phosphate and 3 mg of bovine serum albumin per milliliter

All assays

2-Aminoethanol–2-methoxyethanol (1 : 1)

Trichloroacetic acid, 100% (w/v)

Procedure. The substrate solution (0.25 ml) is placed in the bottom of a stoppered 15 × 85 mm culture tube and warmed at 37° for 5 min. A 1 × 2.5 cm piece of Whatman No. 1 filter paper is fluted, placed crosswise in the upper part of the tube, and impregnated with 20 μl of the aminoethanol solution. The enzyme is diluted, and 50 μl (containing enough enzyme to use no more than 10% of the substrate) is added. After 15 min, each assay is stopped by adding 20 μl of the trichloroacetic acid solution. After an additional 15 min, the filter paper is removed with tweezers, added to 10 ml of a nonaqueous scintillation fluid, and counted.

[9] D. R. Morris and A. B. Pardee, *Biochem. Biophys. Res. Commun.* **20,** 697 (1965).

The assays can be standardized internally by using a large excess of the enzyme in question in order to release and trap CO_2 from all the substrate present. This gives an effective specific radioactivity for the substrate that takes into account the efficiency of both counting and CO_2 trapping. The overall assay efficiency is usually about 50%. A unit of enzyme activity is defined as the quantity that gives 1 μmol of CO_2 liberated per minute under these assay conditions.

Purification Procedures

The purification procedures for bODC and bADC begin with the same crude extract. Since specific culture conditions are required for induction of dODC and dADC, separate batches of cells are required for these purification procedures.

bODC[3]

Cell Growth. Escherichia coli strain UW44 (ATCC 27549)[5] is grown at 37°, with aeration, in medium 63[10] supplemented with trace elements,[11] 0.2% glucose, and 0.05% yeast extract. After growth to late log phase (A_{540} = 1.2), the cells are harvested, washed with 0.9% NaCl, and stored as a paste at −20°. The cell yield is generally 150–200 g wet weight from 100 liters of culture.

Step 1. Crude Extract. One kilogram of cell paste is suspended in 3 liters of buffer A [50 m*M* potassium phosphate (pH 6.0), 1 m*M* dithiothreitol, 5 m*M* $MgSO_4$, and 40 μ*M* pyridoxal phosphate] and disrupted in 250-ml portions by sonic oscillation for 20 min with a Branson S-75 sonifier at a setting of 5.5 A. The suspension is cooled in a −15° ice–salt bath during this procedure, and is then centrifuged at 16,000 *g* for 50 min. The pooled supernatant fractions are diluted with buffer A to A_{280} = 116.

Step 2. Streptomycin Precipitation. For every 100 ml of the diluted crude extract, 7.5 ml of a freshly prepared 10% (w/v) streptomycin sulfate solution is added dropwise with stirring. After addition of the streptomycin, the preparation is stirred for 30 min in an ice bath and centrifuged at 12,500 *g* for 40 min.

Step 3. Ammonium Sulfate Fractionation. Buffer B (identical to buffer A except that the dithiothreitol is replaced by 15 m*M* 2-mercaptoethanol) is saturated with ammonium sulfate at room temperature and then stored at 5°. The ammonium sulfate solution is added dropwise with stir-

[10] G. N. Cohen and H. V. Rickenberg, *Ann. Inst. Pasteur, Paris* **91,** 693 (1956).
[11] B. N. Ames, B. Garry, and L. Herzenberg, *J. Gen. Microbiol.* **22,** 369 (1960).

ring at 0° to the supernatant fluid from step 2. After each step below, the preparation is stirred for 30 min and then centrifuged at 12,500 *g* for 30 min. The material precipitating between 0 and 25% saturation is discarded. The precipitate forming between 25 and 39% saturation is suspended in 1.6 liters of 50 m*M* potassium phosphate buffer (pH 8.0) containing dithiothreitol, $MgSO_4$, and pyridoxal phosphate at the concentrations specified for buffer A. This 25–39% fraction is used for step 4 of the bADC purification. The fraction between 30 and 50.5% saturation, containing bODC, is suspended in 2 liters of buffer A and used for step 4 of this purification.

Step 4. Heat Treatment. The ammonium sulfate fraction (30–50.5%) is dialyzed for 24 hr against three 20-liter changes of 200 m*M* potassium phosphate buffer (pH 6.3) supplemented as is buffer B. The pyridoxal phosphate concentration is then increased to 0.44 m*M*, and the solution is stirred for 2 hr. The preparation is divided into 30-ml portions and heated at 62° for 10 min in 25 × 150 mm Pyrex culture tubes with periodic shaking. The portions are cooled rapidly to 0° and combined. After centrifuging at 16,000 *g* for 25 min, the supernatant solutions are pooled and dialyzed against three 20-liter changes of buffer B.

Step 5. Hydroxyapatite Chromatography. A column (5 × 20 cm) of hydroxyapatite (Bio-Rad Laboratories) is washed with 5 volumes of 50 m*M* potassium phosphate (pH 6.0) and 3 volumes of buffer B. The material from step 4 is applied and the column is washed with 800 ml of buffer B. The column is eluted with a 2-liter gradient from 50 to 250 m*M* potassium phosphate (pH 6.0) supplemented as is buffer A. The fractions containing bODC (ca. 150 m*M* buffer) are pooled and concentrated to 20 ml using an Amicon ultrafiltration cell with a PM-10 membrane. This preparation is dialyzed twice against 1 liter of 0.4 *M* sodium acetate (pH 5.6) containing 40 μM pyridoxal phosphate and 15 m*M* 2-mercaptoethanol.

Step 6. Sephadex G-200 Gel Filtration. Glycerol is added to the material from step 5 to a final concentration of 10%. The solution is layered onto the top of the gel bed (5 × 51 cm). The bODC activity elutes after approximately 400 ml of buffer has passed through the column. The active fractions are pooled, concentrated by ultrafiltration, and dialyzed twice for 6 hr against 1 liter of 50 m*M* sodium succinate buffer (pH 5.9) containing 40 μM pyridoxal phosphate and 15 m*M* 2-mercaptoethanol.

Step 7. DEAE-Sephadex A-25 Chromatography. A column (3.5 × 26 cm) of DEAE-Sephadex A-25 is equilibrated with the buffer used for dialysis. The enzyme is applied to the column and eluted with a 2-liter linear gradient from 50 to 200 m*M* sodium succinate (pH 5.9) supplemented as above for dialysis. The active fractions (ca. 80 m*M* buffer) are pooled and concentrated by ultrafiltration.

Summary. Table I summarizes a typical purification from 1 kg of cells. This preparation is estimated to be at least 85% pure. No decarboxylase activity toward arginine or lysine could be detected.[3]

bADC[4]

The growth of the cells and the first two purification steps are identical to those described for dODC. At step 3, bADC precipitates between 25 and 39% saturation. This is used for bADC purification. Buffers A and B are described in the bODC purification.

Step 4. Ammonium Sulfate Fractionation at pH 8. An aqueous solution of ammonium sulfate is saturated at 5° and adjusted to pH 8.0 with concentrated ammonium hydroxide (pH measured at a 1 : 10 dilution). The 25–39% fraction from step 3, dissolved as described in pH 8.0 buffer, is fractionated with the pH 8 ammonium sulfate solution. The precipitate that forms between 35 and 53% saturation is collected by centrifugation at 10,000 *g* for 20 min, dissolved in 500 ml of buffer A, and dialyzed first against 10 liters of buffer A for 10 hr and then against 4 liters of 250 m*M* potassium phosphate (pH 6.0) supplemented as described for buffer A. After at least 5 hr, the preparation is centrifuged at 20,000 *g* for 30 min.

Step 5. Heat Treatment. The material from step 4 is supplemented with pyridoxal phosphate to a final concentration of 0.5 m*M*. This solution is split into 50-ml portions that are heated individually, with gentle swirling, in 600-ml glass round-bottomed flasks in a 62° water bath. After 10 min in the bath, the flask is rapidly chilled in ice water. After centrifugation at 20,000 *g* for 30 min, the supernatant solution is dialyzed against

TABLE I
PURIFICATION OF bODC

Step	Fraction	Protein (mg)	Specific activity (units/mg)	Recovery (%)
1	Crude extract	79,000	0.02	100
2	Streptomycin sulfate supernatant solution	70,900	0.02	83[a]
3	Ammonium sulfate, 39–50.5% saturation	20,400	0.09	96
4	Heat treatment	6,020	0.19	64
5	Hydroxyapatite chromatography	345	2.1	41
6	Sephadex G-200 chromatography	63	5.8	20
7	DEAE-Sephadex A-25 chromatography	1.8	99	10

[a] The enzyme activity was slightly inhibited in the streptomycin sulfate supernatant solution.

two 10-liter changes of 50 m*M* potassium succinate (pH 5.85) supplemented as specified for buffer A.

Step 6. DEAE-Sephadex A-25 Chromatography. A column (4 × 35 cm) of DEAE-Sephadex A-25 is equilibrated with the 50 m*M* potassium succinate (pH 5.85) buffer and the sample from step 5 is applied. After washing the column with 200 ml of the equilibrating buffer, the enzyme is eluted with a 3-liter linear gradient from 0.1 to 0.4 *M* of this buffer. Those fractions with bADC activity (ca. 0.25 m*M* buffer) are pooled and concentrated by ultrafiltration. This material is then dialyzed overnight against 200 volumes of buffer A.

Step 7. Hydroxyapatite Chromatography. A hydroxyapatite (Bio-Rad Laboratories) column (3 × 15 cm) is washed with at least 5 volumes of buffer A. The enzyme solution from step 6 is applied and the column is eluted with a 1200-ml linear gradient of 50 to 250 m*M* potassium phosphate (pH 6.0). The active fractions (ca. 150 m*M* buffer) are pooled and concentrated by ultrafiltration.

Crystallization. The enzyme from step 7 will readily crystallize from 35% saturated ammonium sulfate (pH 6.0) at 5°. This does not produce an increase in specific enzymatic activity and leads to aggregation of the protein.[4] Crystallization is therefore not recommended for routine purification.

Summary. The purification of bADC is summarized in Table II. The final preparation routinely shows multiple activity peaks after electrophoresis under nondenaturing conditions. Gel electrophoresis in the presence of 0.1% sodium dodecyl sulfate gives, reproducibly, two bands with relative migrations corresponding to molecular weights of 70,000–72,000 and 75,000–76,000.[4] The preparation did not decarboxylate ornithine or lysine.

TABLE II
PURIFICATION OF bADC

Step	Fraction	Protein (mg)	Specific activity (units/mg)	Recovery (%)
1	Crude extract	94,500	0.01	100
2	Streptomycin sulfate supernatant solution	83,600	0.01	94
3	pH 6 ammonium sulfate, 25–39% saturation	12,600	0.07	83
4	pH 8 ammonium sulfate, 35–53% saturation	1,900	0.39	76
5	Heat treatment	1,400	0.51	72
6	DEAE-Sephadex A-25 chromatography	45	12.6	57
7	Hydroxyapatite chromatography	27	16.4	44

dODC[5]

Cell Growth. An inoculum culture (1 liter) is grown in medium 63[10] supplemented with trace elements,[11] 0.2% glucose, and 0.05% yeast extract at 37° with aeration. The inoculum suspension is added to 10 liters of induction medium [1% nutrient broth, 0.05% $(NH_4)_2SO_4$, 0.1% NaCl, 0.1% K_2HPO_4, 0.05% sodium citrate, and 0.8% L-ornithine, adjusted to pH 5.2] and allowed to grow without aeration for 7 hr at 37°. The harvested cells are washed with 0.9% NaCl and used immediately for purification.

Step 1. Crude Extract. The washed cells are suspended in 200 ml of buffer C [50 m*M* potassium phosphate (pH 6.7) supplemented with 0.4 m*M* pyridoxal phosphate and 10 m*M* 2-mercaptoethanol]. The cell suspension is disrupted as described for the bODC purification and centrifuged at 35,000 *g* for 30 min.

Step 2. Protamine Sulfate Precipitation. A 2% protamine sulfate solution is prepared at room temperature in buffer C and added dropwise to the crude extract with stirring [volume added (ml) = 0.025 × extract volume (ml) × A_{260}]. The A_{260} varied between 43.6 and 48.5 for various preparations. The preparation is stirred for 30 min and then centrifuged at 35,000 *g* for 30 min. The supernatant solution is slowly adjusted to pH 7.0 with 0.15 *M* NH_4OH.

Step 3. Ammonium Sulfate Fractionation. A saturated solution of ammonium sulfate (at 5°) is prepared in 50 m*M* potassium phosphate (pH 7.0) and adjusted to pH 7.0 with concentrated NH_4OH. The preparation from step 2 is fractionated with the saturated ammonium sulfate solution; dODC is found in the 20–30% saturation fraction. The precipitates are collected by centrifugation for 20 min at 35,000 *g*; the enzyme-containing precipitate is dissolved in 10 ml of buffer C.

Purification Summary. The purification of dODC is summarized in Table III. The resulting preparation had no detectable decarboxylase ac-

TABLE III
PURIFICATION OF dODC

Step	Fraction	Protein (mg)	Specific activity (units/mg)	Recovery (%)
1	Crude extract	1000	9.3	100
2	Protamine sulfate supernatant solution	690	13.1	97
3	Ammonium sulfate, 20–30% saturation	66	130	92

tivity toward arginine and lysine and showed a single band on electrophoresis under either denaturating or nondenaturing conditions.[5]

dADC[7]

A procedure suitable for the production of 50–100 mg of dADC was described in Vol. 17B.[7] That procedure has now been modified to yield 0.5–1 g of pure enzyme. The relevant modifications are described below.

Cell Growth. The inoculum culture of *E. coli* B is grown at 37° on a chemically defined medium (medium I in Table I of Boeker and Snell[7]) in three stages: four 25-ml cultures for 11 hr and four 250-ml cultures for 12 hr, both on a rotary shaker, and a 10-liter carboy culture, vigorously aerated for 8 hr. The final inoculum culture is added to a fermentor containing 90 liters of a sterile, enriched medium (medium II in Table I of Boeker and Snell[7]) or is divided among four 40-liter carboys, each containing 22 liters of the enriched medium. In either case, the cells are grown with very slight aeration for ca. 12 hr at 30° and collected with a Sharples centrifuge.

Purification. Unless specifically stated, the conditions are as previously described.[7] The cells are disrupted by sonic oscillation, and the resulting preparation is heated to 60° for 2 min and subjected to ammonium sulfate fractionation and dialysis.[7] Attempts to disrupt the cells more efficiently, either by using a French pressure cell or by grinding with glass beads, produced an extract that could not be purified successfully. In the final purification step, the dialyzed ammonium sulfate fraction is applied to a 4 × 50 column of Whatman DE-52. The column is washed with 50 m*M* sodium succinate (pH 5.8), and the enzyme is eluted with an 8-liter linear gradient from 50 to 200 m*M* sodium succinate (pH 5.8). The final specific activity should be greater than 350 units/mg.

TABLE IV
COMPARISON OF THE BASIC AMINO ACID DECARBOXYLASES OF *Escherichia coli*

Enzyme	Optimum pH	Turnover number[a]	Substrate K_m[b]	Cofactors	Subunit ($M_r \times 10^{-3}$)	Number of subunits
bODC	8.3	157	5.6	PLP	81	2
dODC	6.9	173	3.6	PLP	80	2
bADC	8.4	18	0.03	PLP, Mg^{2+}	74	4
dADC	5.2	547	0.65	PLP	82	2 or 10
LDC	5.7	1330	1.5	PLP	80	2 or 10

[a] Picomoles per second (μmol pyridoxal phosphate)$^{-1}$.
[b] Millimolar.

Properties

The catalytic and structural properties of the four enzymes described here and a related enzyme, lysine decarboxylase (LDC), have been studied in some detail[3–6,12–20] and have been reviewed.[1,7,8,21] A number of their properties are compared in Table IV. Except for the two ornithine decarboxylases, the enzymes differ considerably in catalytic properties, with a range of 3 units in pH optimum and a span of two orders of magnitude in both turnover number and K_m. The three biodegradative enzymes are optimally active below neutrality, consonant with their physiological role in pH regulation, and both biosynthetic enzymes show optimal activity above pH 8. All five enzymes require pyridoxal phosphate for activity and bind one coenzyme molecule per subunit. In addition, bADC requires magnesium ion for covalent interaction of the coenzyme with the enzyme and for formation of the active tetramer.

Although there is no evidence for immunological cross reaction among any of these five enzymes, four have rather similar structural properties (Table IV). It seems that the fundamental structure of bODC, dODC, dADC, and LDC is a dimer of 160,000 molecular weight. At low pH or high ionic strength, the dimers of dADC and LDC associate to a decamer; indeed, the decamer is the active form. On the other hand, neither ornithine decarboxylase dimer shows any tendency to associate under conditions that give optimal activity. The pyridoxal phosphate binding sites of dODC, dADC, and LDC have been investigated by reduction with $NaBH_4$ and sequence analysis of the resulting phosphopyridoxal peptides (Table V). All three peptides have a histidine residue on the amino side of the phosphopyridoxal lysine and an extremely hydrophobic sequence on the carboxyl side. Aside from these two general features, the sequence of the peptide from bODC bears little resemblance to the other two decarboxylases. On the other hand, the homology between the sequences of the dADC and LDC peptides is most impressive. Out of 12 overlapping

[12] W. H. Wu and D. R. Morris, *J. Biol. Chem.* **248,** 1696 (1973).

[13] S. L. Blethen, E. A. Boeker, and E. E. Snell, *J. Biol. Chem.* **243,** 1671 (1968).

[14] E. A. Boeker and E. E. Snell, *J. Biol. Chem.* **243,** 1678 (1968).

[15] E. A. Boeker, E. H. Fischer, and E. E. Snell, *J. Biol. Chem.* **246,** 6676 (1971).

[16] E. A. Boeker, *Biochemistry* **17,** 258 (1978).

[17] E. A. Boeker, *Biochemistry* **17,** 263 (1978).

[18] S. M. Nowak and E. A. Boeker, *Arch. Biochem. Biophys.* **207,** 110 (1981).

[19] D. L. Sabo, E. A. Boeker, B. Byers, H. Waron, and E. H. Fischer, *Biochemistry* **13,** 662 (1974).

[20] D. L. Sabo and E. H. Fischer, *Biochemistry* **13,** 670 (1974).

[21] E. A. Boeker and E. E. Snell, *in* "The Enzymes" (P. D. Boyer, ed.), 3rd Ed., Vol. 6, p. 217. Academic Press, New York, 1972.

TABLE V
PYRIDOXAL PHOSPHATE BINDING SITES OF THE BIODEGRADATIVE DECARBOXYLASES

LDC:	. . . Glu-Thr-Glu-Ser-Thr-His-(ε-Pxy)Lys-Leu-Leu-Ala-Ala-Phe
dADC:	Ala-Thr-His-Ser-Thr-His-(ε-Pxy)Lys-Leu-Leu-Asn-Ala-Leu . . .
dODC:	Val-His-(ε-Pxy)Lys-Gln-Gln-Ala-Gly-Gln

residues, 8 are identical, and two of the four differences, Glu → Ala and Phe → Leu, can be accounted for by single base changes in the corresponding codons.

The dimer–decamer interconversion of dADC has been studied in considerable detail.[6,13–20] Under normal assay conditions (0.2 *M* sodium acetate, pH 5.2), the enzyme exists as the decamer (molecular weight 820,000, $s^{\circ}_{20,w}$ = 23.3 S). The decamer dissociates to 5 dimers (molecular weight 162,000, $s^{\circ}_{20,w}$ = 8.0 S) if the pH is increased to 6.5 or more and the sodium ion concentration is decreased to 40 m*M* or less. The decamer is re-formed under any of the following conditions: decrease of the pH to less than 6.0, increase of sodium ion concentration to greater than 80 m*M*, addition of a divalent cation, addition of the substrate or substrate analog. Intermediate species (tetramer, hexamer, and octamer) can be observed in the ultracentrifuge when association is induced by the substrate or an analog, but not when it occurs under the other conditions. When the initial rate of dissociation of the decamer is analyzed kinetically, using a chemical assay and a stopped-flow spectrophotometer, the empirical rate law is consistent with an ordered reaction in which the dissociation of a proton and the binding of a sodium ion precede an irreversible (on this time scale) dissociation step. The dimer has no measurable enzymatic activity when assayed at very low protein concentrations. As the concentration of dADC is increased from 0.02 to 10 μg/ml, activity appears and increases to that of the decamer. The time course of the appearance of activity under these conditions is sigmoidal. Thus, it seems that the dimer is inactive and that activity appears only when the dADC concentration is high enough for the substrate to induce association to a higher molecular weight form.

[19] Ornithine Decarboxylase[1] (*Saccharomyces cerevisiae*)

By ANIL K. TYAGI, CELIA W. TABOR, and HERBERT TABOR

$$\text{Ornithine} \xrightarrow{-CO_2} \text{putrescine}$$

Assay Method

Principle. $^{14}CO_2$ enzymatically produced by the decarboxylation of L-[1-^{14}C]ornithine is collected on filter paper impregnated with saturated barium hydroxide solution.

Reagents

Tris-HCl buffer: 0.025 *M* (pH 7.6) containing 5 m*M* dithiothreitol, 1 m*M* $MgCl_2$, 0.1 m*M* pyridoxal 5-phosphate, and 0.1 m*M* EDTA
L-[*carboxy*-^{14}C]Ornithine (7.23 mCi/mmol)
Barium hydroxide solution saturated at room temperature
KH_2PO_4, 1.0 *M*

Procedure. The assays[2] are performed at 37° in Tris-HCl buffer, 0.025 *M* (pH 7.6) containing 5 m*M* dithiothreitol, 1 m*M* $MgCl_2$, 0.1 m*M* pyridoxal 5-phosphate, and 0.1 m*M* EDTA. The assay mixture (in a scintillation vial) contains enzyme and buffer in a final volume of 0.4 ml. The reaction is started by adding 0.5 μCi (in 10 μl) of L-[*carboxy*-^{14}C]ornithine (7.23 mCi/mmol). The vials are closed tightly with caps containing a piece of Whatman No. 3 filter paper (1.5 × 1.5 cm) impregnated with 50 μl of saturated barium hydroxide solution.[3] The vials are incubated at 37° for 20 min, at which time the reaction is stopped by adding 250 μl of 1.0 *M* KH_2PO_4 solution. After further incubation (with shaking) at room temperature for 45 min, trapped $Ba^{14}CO_3$ on the filter paper is determined by scintillation spectrometry. Under these conditions, production of $^{14}CO_2$ is linear with respect to time and enzyme concentration.

Definition of Unit and Specific Activity. One unit of enzyme is defined as the amount of enzyme protein that liberates 1 μmol of $^{14}CO_2$ from L-[1-^{14}C]ornithine in 1 hr under the standard conditions of the method described above. Protein is routinely estimated by the method of Brad-

[1] EC 4.1.1.17; L-ornithine carboxy-lyase.
[2] A. K. Tyagi, C. W. Tabor, and H. Tabor, *J. Biol. Chem.* **256,** 12156 (1981).
[3] H. Tabor, C. W. Tabor, and E. Hafner, *J. Bacteriol.* **128,** 485 (1976).

ISBN 0-12-181994-9

ford.[4] Purified enzyme protein is measured by the method of Lowry *et al.*[5] Bovine serum albumin is used as a standard in both cases. Specific activity is expressed as units of enzyme activity per milligram of protein.

Growth of Cells

A medium containing 2% dextrose, 0.7% yeast nitrogen base without amino acids (Difco), 0.005% adenine, and 0.025% threonine is adjusted to pH 7.0 with K_2HPO_4. The *spe*2-4 strain of *Saccharomyces cerevisiae*,[6] which lacks *S*-adenosylmethionine decarboxylase and has a high activity of ornithine decarboxylase, is grown in this medium in a 100-liter fermentor at 28° with vigorous aeration. The cells are harvested at $A_{650\ nm}$ of ~5.0.

Purification Procedures

Two methods of purification of ornithine decarboxylase are presented below.

Method 1[2]

This was the initial procedure used in our laboratory for the purification of this enzyme and did not include phenylmethylsulfonyl fluoride (PMSF). The enzyme was purified to homogeneity by this procedure and has a molecular weight of 68,000.

The cells (300 g), grown and harvested as described above, are suspended in 3 volumes of 0.025 *M* Tris-chloride, pH 7.6, containing 5 m*M* dithiothreitol, 1 m*M* $MgCl_2$, and 0.1 m*M* EDTA (buffer A) and are broken by a single pass through a French press at 18,000 psi. The resulting suspension is centrifuged for 30 min at 20,000 *g*.

Ammonium Sulfate Fractionation. For each 100 ml of supernatant, 10.8 g of "enzyme grade" ammonium sulfate is added with stirring. The precipitate is discarded, and additional ammonium sulfate (21.4 g/100 ml) is added to the supernatant solution. The resulting precipitate, which contains most of the activity, is dissolved in 150 ml of buffer A and dialyzed overnight against 15 liters of buffer A with two changes of buffer.

DE-52 Chromatography. The dialyzed enzyme is applied to a DE-52 column (3.9 × 22 cm) preequilibrated with 0.05 *M* NaCl in buffer A. The

[4] M. M. Bradford, *Anal. Biochem.* **72,** 248 (1976).

[5] O. H. Lowry, N. J. Rosebrough, A. L. Farr, and R. J. Randall, *J. Biol. Chem.* **193,** 265 (1951).

[6] M. S. Cohn, C. W. Tabor, and H. Tabor, *J. Bacteriol.* **134,** 208 (1978).

column is washed with 1 liter of equilibrating buffer, and ornithine decarboxylase is eluted with a 550-ml linear gradient of 0.05 *M* NaCl to 0.35 *M* NaCl in buffer A. The active fractions are pooled and solid ammonium sulfate is added to 55% saturation. After stirring for 30 min, the precipitate is collected by centrifugation at 25,000 *g* for 15 min and dissolved in 8 ml of buffer A.

Sephacryl S-200 Chromatography. The concentrated enzyme is loaded onto a Sephacryl S-200 column (2.5 × 90 cm), which is preequilibrated with buffer A. Ornithine decarboxylase is eluted with the same buffer at a flow rate of 16 ml/hr. The peak fractions containing the enzyme are pooled and subjected to ammonium sulfate precipitation (55% saturation). The pellet obtained after the centrifugation is dissolved in 30 ml of buffer A containing 0.5 *M* $(NH_4)_2SO_4$.

Phenyl-Sepharose Chromatography. The final purification is achieved on a column of phenyl-Sepharose CL-4B (Pharmacia) (1.5 × 10 cm) that has been equilibrated with buffer A containing 0.5 *M* $(NH_4)_2SO_4$. After loading the active fraction from the Sephacryl column, the column is washed with 100 ml of 0.5 *M* $(NH_4)_2SO_4$ in buffer A, followed by 500 ml of 0.25 *M* $(NH_4)_2SO_4$ in buffer A. Ornithine decarboxylase is then eluted with a linear reverse gradient of 0.25 *M* to 0.0 *M* $(NH_4)_2SO_4$ in 500 ml of buffer A. The column is washed with an additional 200 ml of buffer A containing no $(NH_4)_2SO_4$. Fractions near the end of the gradient that contain ornithine decarboxylase are pooled, concentrated in an Amicon pressure cell with an XM-50 membrane, and stored at −80°.

A typical purification is summarized in the table.

PURIFICATION OF ORNITHINE DECARBOXYLASE FROM *Saccharomyces cerevisiae*[a]

Step	Protein (mg)	Specific activity (μmol/mg hr^{-1})	Yield of activity (%)	Purification (fold)
Cell-free extract	7900	0.02	100	1
Ammonium sulfate	4240	0.05	120	2
DE-52 cellulose	300	0.45	79	21
Sephacryl S-200	41	2.50	60	120
Phenyl-Sepharose	2	31	38	1480

[a] The data refer to 300 g wet weight of yeast cells as the starting material. Activity of ornithine decarboxylase is measured at pH 7.6 in the presence of 175 μM ornithine and 100 μM pyridoxal 5-phosphate at 37°. At the pH optimum (pH 8.0) the specific activity of the pure enzyme is 42 μmol/mg per hour.

Method 2

We have found that if PMSF is present during the extraction of yeast all the enzyme in the crude extract has a molecular weight of 86,000 (with a small amount of a 81,000 M_r form), and there is no 68,000 M_r form (as measured by SDS acrylamide electrophoresis). By using an antibody column we have been able to obtain a preparation that is 40% pure, but we have not yet developed a method for the purification of this 86,000 M_r preparation to homogeneity. During the procedure some of the 86,000 M_r form is converted to the 68,000 M_r form. This purification procedure is described below.

The cells are grown and harvested as described above but are suspended in buffer A containing 1 mM PMSF. The extract is made by passing the suspension through a French press at 18,000 psi at 4°; the resulting suspension is centrifuged for 30 min at 20,000 g (4°).

Preparation of Antibody Affinity Column. Purified ornithine decarboxylase (50 μg in 500 μl) obtained by Method 1 is emulsified with an equal volume of complete Freund's adjuvant and injected subcutaneously, distributed equally among 10 sites in a male rabbit weighing 2 kg. One month later a second injection of 50 μg of enzyme in complete Freund's adjuvant is given in the same way. Fifteen days after the second injection 50 μg of ornithine decarboxylase (in 500 μl) is given subcutaneously distributed equally among 10 sites after emulsifying the enzyme solution with an equal volume of incomplete Freund's adjuvant. Two weeks after the third injection, about 40 ml of blood are removed from the external ear vein, and the serum is separated and processed as described by Harboe and Ingild[7] up to the final acetate dialysis. The lipoproteins precipitated during dialysis are removed by centrifugation. After dialyzing the supernatant solution against 0.1 M NaCl containing 15 mM sodium azide, the supernatant is stored at 4°.

Anti-ornithine decarboxylase IgG can be coupled to CNBr-activated Sepharose 4B by the procedure described by Livingston *et al.*[8] and the instructions provided by Pharmacia Fine Chemicals. CNBr-activated Sepharose 4B (Pharmacia Fine Chemicals) is washed with 1 mM HCl and then suspended in coupling buffer (0.1 M $NaHCO_3$, pH 8.3). Anti-ornithine decarboxylase IgG is dialyzed against the coupling buffer at a concentration of 5 mg/ml. The protein solution is mixed with the gel suspension at a Sepharose : protein ratio of 33 : 1 and the contents are shaken on a rotary shaker for 16 hr at 4°. Under these conditions more

[7] N. Harboe and A. Ingild, *Scand. J. Immunol.* **2,** 161 (1973).

[8] D. M. Livingston, E. M. Scolnick, W. P. Parks, and G. J. Todaro, *Proc. Natl. Acad. Sci. U.S.A.* **69,** 393 (1972).

than 90% of added protein is coupled to Sepharose 4B. Excess protein is washed off with coupling buffer, and the remaining active groups are blocked by reacting with 30 ml of 1 *M* ethanolamine solution (pH 8.0) for 3 hr at room temperature. Unreacted blocking reagent is removed by washing with 0.1 *M* sodium acetate (pH 4.0) containing 0.5 *M* NaCl, followed by coupling buffer. The conjugated Sepharose is then packed into a column 1.5 × 7 cm.

Affinity Chromatography of Ornithine Decarboxylase on Antibody Column. All operations are performed at 0–4°. Samples of fresh crude extract, prepared as described above, are applied to the column equilibrated with buffer A containing 1 m*M* PMSF. Samples may be applied at a rate of 85 ml/hr, and the eluate is recycled through the column at the same rate. In this way 200–300 ml of crude extract (8–10 mg of protein per milliliter) can be loaded on the column with more than 95% binding of ornithine decarboxylase to the column. The column is then washed with buffer A containing 1 *M* NaCl (300–400 ml) at a flow rate of 80 ml/hr. Finally, the enzyme is eluted by step elution with a small volume of 4.5 *M* $MgCl_2$ in buffer A, and the active fractions are dialyzed thoroughly against buffer A to remove $MgCl_2$. The enzyme after dialysis is concentrated in an Amicon pressure cell with a XM-50 membrane. This procedure results in 40% pure ornithine decarboxylase. During this procedure some of the 86,000 M_r form is converted to the 68,000 M_r form. The final preparation is a mixture of these two forms and a 34,000 M_r form. In addition there is a small amount of the 81,000 M_r form.

Properties of the Enzyme

The properties of the enzyme that we studied are those of a pure 68,000 M_r form. The enzyme follows Michaelis–Menten kinetics from pH 6.5 to pH 9.0 with the V_{max} at pH 8.0; it requires pyridoxal phosphate for activity and is stable for at least 10 weeks at −80°. The enzyme is specific for ornithine. It is inhibited reversibly by putrescine and α-methylornithine and irreversibly by the suicide inactivator α-difluoromethylornithine. The enzyme is inactive in the absence of thiols and can be activated by thiols. Among the thiol compounds dithiothreitol is the most effective. Some nucleotides stimulate the enzyme to a moderate extent.[2]

[20] Ornithine Decarboxylase and the Ornithine Decarboxylase-Modifying Protein of *Physarum polycephalum*[1]

By JOHN L. A. MITCHELL

The acellular slime mold *Physarum polycephalum* is a primitive eukaryote of considerable value in ornithine decarboxylase research. It maintains very high concentrations of the active state of this enzyme, compared with mammalian tissues, and it can be easily manipulated and inexpensively mass cultured. The activity of this enzyme in *Physarum* changes rapidly in response to the same stimuli that affect mammalian ornithine decarboxylase. Unlike the mammalian systems, however, total ornithine decarboxylase enzyme protein appears to be quite stable in *Physarum,* with activity modulation produced by a reversible posttranslational modification in the enzyme protein that renders the enzyme inactive in the *in vivo,* but not the *in vitro,* environment.[2-4] The physical and kinetic properties of the active and modified ornithine decarboxylase forms,[5] and of the polyamine-stimulated enzyme that catalyzes this posttranslational protein modification,[6] are studied to elucidate the mechanism of regulation of polyamine biosynthesis in this system and yield insight by analogy to the comparable biosynthetic control in mammalian tissues.

Organism and Culture Conditions

Physarum polycephalum, subline M3C (VIII), was grown in 15-liter batches in a 28-liter New Brunswick fermentor (Model MF 128S).[7] The semidefined medium of Daniel and Baldwin[8] was inoculated with 300 ml

[1] See also G. D. Kuehn, V. J. Atmar, and G. R. Daniels, this volume [21].
[2] J. L. A. Mitchell, S. N. Anderson, D. D. Carter, M. J. Sedory, J. F. Scott, and D. A. Varland, *Adv. Polyamine Res.* **1,** 39 (1978).
[3] J. L. A. Mitchell, D. D. Carter, and J. A. Rybski, *Eur. J. Biochem.* **92,** 325 (1978).
[4] J. L. A. Mitchell and G. E. Kottas, *FEBS Lett.* **102,** 265 (1979).
[5] J. L. A. Mitchell and D. D. Carter, *Biochim. Biophys. Acta* **483,** 425 (1977).
[6] J. L. A. Mitchell, T. A. Augustine, and J. M. Wilson, *Biochim. Biophys. Acta* **657,** 257 (1981).
[7] E. N. Brewer, S. Kuraiski, J. C. Garver, and F. M. Strong, *Appl. Microbiol.* **12,** 161 (1964).
[8] J. W. Daniel and H. H. Baldwin, *Methods Cell Physiol.* **1,** 9 (1964).

ISBN 0-12-181994-9

of log-phase microplasmodia stock cultures and allowed to grow for 4 days at 28° with constant stirring at 100 rpm and aeration at about 4 liters per minute. Resulting microplasmodia were removed from the culture medium by centrifugation at 400 *g* for 60 sec and washed once with 40 m*M* NaCl. The washed material was then homogenized in acetone at 0° followed by centrifugation at 600 *g* for 5 min. This acetone precipitate could be stored in sealed bottles for at least 3 months at −20°.

Assay

Ornithine decarboxylase was monitored during purification by placing 100 μl of enzyme solution into 25-ml Erlenmeyer flasks containing 1.9 ml of 20 m*M* HEPPS [4-(2-hydroxyethyl)-1-piperazine propanesulfonic acid] buffer (pH 8.4) containing 0.5 m*M* dithiothreitol, 0.5 m*M* Na_2EDTA, and 100 μ*M* pyridoxal 5′-phosphate. The addition of 0.1 μmol of L-[1-^{14}C]ornithine (0.01 μCi, Amersham/Searle) initiated the reaction. The assay flasks were sealed with rubber stoppers suspending center wells that contained 0.01 ml of 1.0 *M* Hyamine hydroxide in methanol and agitated at 26°. After 60 min, 1 ml of 2 *M* citric acid was injected, and after an additional 30 min the center wells were removed and the absorbed radioactive CO_2 was counted in toluene-based scintillation fluid. One unit of enzyme activity was defined as the amount that catalyzes the release of 1 μmol of CO_2 per minute of assay. Both the A form and the B form of the enzyme were determined in the standard assay.

Purification of Ornithine Decarboxylase

A typical purification is shown in the table.[9] Acetone pellets containing approximately 25 g of protein were suspended by Sorvall homogenization in 5 liters of ice cold 0.05 *M* potassium phosphate (pH 7.4) containing 0.5 m*M* Na_2EDTA, 15 m*M* 2-mercaptoethanol, 20 μ*M* pyridoxal 5′-phosphate, and 0.12% (w/v) NaN_3. Insoluble material was pelleted by a 5-min centrifugation at 300 *g*. The resulting enzyme suspension was filtered through cheesecloth to remove a slimy foam and stirred into a suspension of 1600 ml of DEAE-cellulose equilibrated in the same buffer. This slurry was then poured into a column (31 cm by 10.5 cm) that contained a 500-ml bed of this resin. After packing to a bed volume of 2.1 liters, this column was washed with an additional 2 liters of the same buffer, and finally the enzyme was rapidly eluted by the application of 3 liters of buffer containing 0.2 *M* KCl. The enzymatically active fractions of this elution (about

[9] J. M. Wilson and J. L. A. Mitchell, in preparation.

PURIFICATION OF ORNITHINE DECARBOXYLASE

Step	Total protein (mg)	Total activity (units)	Specific activity (units/mg)	Relative purification (fold)	Yield (%)
Harvest	25,900	103	0.004	1	100
Extraction of acetone pellet	6,500	72.7	0.007	1.8	71
DEAE-cellulose	3,330	41.7	0.01	3.1	41
Sephacryl S-300	710	27.5	0.04	9.7	27
DEAE-Sephacel	14.0	19.7	1.40	350	19
Ultrogel AcA-34	3.8	12.6	3.26	820	12
Hydroxyapatite	0.12	2.42	19.8	5000	2.4

250 ml) were pooled and concentrated to 50 ml using Amicon hollow-fiber (HlX-50) membrane filtration.

The concentrated enzyme fraction was applied to a Sephacryl S-300 column (60 × 5.0 cm) preequilibrated in buffer without KCl, and eluted in 20 ml fractions at 50 ml/hr. The most active 4–5 fractions were combined and applied to a column (8.0 × 2.3 cm) of DEAE-Sephacel that had been equilibrated with the same buffer except that the 2-mercaptoethanol had been replaced with 5.0 m*M* dithiothreitol. This column was washed at 40 ml/hr with two column volumes of buffer. The bound enzyme was eluted with a linear, 0 to 0.2 *M* KCl gradient in 600 ml of buffer, and the eluate was collected in 7.0-ml fractions. The most active fractions were pooled (90 ml) and concentrated to 8 ml on an Amicon PM-10 ultrafiltration membrane.

The concentrated enzyme sample was then applied to a column of Ultrogel AcA-34 (2.5 × 90 cm) and eluted at 18 ml/hr into 7.0-ml fractions. The most active fractions were pooled (35 ml) and concentrated to 2.0 ml by ultrafiltration on an Amicon PM-10 membrane.

This material was quite stable at either 4° or −20° in the presence of 30% glycerol and was sufficiently clean for most enzyme kinetic and protein modification studies. Further purification was achieved, as shown in the table, by dialyzing the above concentrated enzyme against 0.09 *M* potassium phosphate buffer (pH 7.2) containing 5.0 m*M* dithiothreitol, 20% v/v glycerol, and 20 μ*M* pyridoxal 5′-phosphate, and applying the sample to a column (1.15 × 7.0 cm) of hydroxyapatite at 25 ml/hr. Unadsorbed material, which included the A form of this enzyme, was eluted with 40 ml of the buffer and collected in 1-ml fractions. The peak fractions were again concentrated by ultrafiltration and stored in 30% glycerol at −20°.

Enzyme Form Separation

The purification illustrated in the table was of cell material containing predominantly A-form enzyme. When the culture pellets were washed with 100 m*M* NaCl instead of 40 m*M*, the cellular enzyme was predominantly of the B form.[4] The subsequent purification of this enzyme form was exactly the same as above except that the B form bound to the hydroxyapatite column and was eluted by a 0.09–0.2 *M* phosphate gradient of 100 ml.[6] The recovery of B-form ornithine decarboxylase from this column was quite low (40–60%), and this material rapidly lost activity during subsequent storage at −20°. Small amounts (approximately 5.0×10^{-3} units) of mixed A and B forms of this enzyme were separated with much greater recovery on a heparin-Sepharose CL-6B (Pharmacia) column as follows. One milliliter of this matrix was equilibrated with 0.05 *M* potassium phosphate buffer (pH 7.2) containing 0.5 m*M* EDTA, 2 m*M* dithiothreitol, 20 μM pyridoxal 5′-phosphate, 1% glycerol, and 0.01% Triton X-100. Of the 1 ml of mixed enzyme applied to the column, the A form was not retained whereas 100% of the B form bound and was eluted at about 0.3 *M* in a 20-ml gradient from 0 to 0.8 *M* NaCl at the flow rate of 2.5 ml/hr.

Estimation of Purity

Aliquots of the purified ornithine decarboxylase (5–10 μg of protein) were dialyzed extensively against distilled water at 4°, lyophilized, and eventually suspended in 100° SDS sample buffer (62.5 m*M* Tris, pH 6.8, 1% 2-mercaptoethanol, 0.5% SDS, and 0.25 *M* sucrose). These samples along with appropriate molecular weight standards were electrophoresed in vertical slab 10% acrylamide gels (Bio-Rad Laboratories, Model 220) at 30 mA constant current and eventually were stained with Coomassie Brilliant Blue R-250. Most of the stained material appeared as a single band at M_r 52,000; however, light bands were also apparent at M_r 48,000, 68,000, and in the region of 85,000.

To facilitate the positive identification of the active site-containing subunit of this enzyme, we employed ^{14}C-labeled α-difluoromethylornithine (special order from Amersham/Searle). This ornithine analog is recognized by the enzyme and in the course of reaction is covalently attached to the enzyme's active site, rendering the enzyme inactive.[10] Purified enzyme (0.52 unit) in 0.01 *M* potassium phosphate buffer (pH 8.0)

[10] B. W. Metcalf, P. Bey, C. Danzin, M. J. Jung, P. Casara, and J. P. Vevert, *J. Am. Chem. Soc.* **100,** 2551 (1978).

with 5.0 m*M* dithiothreitol, 10% glycerol, and 10 m*M* KCl was mixed with 100 nmol of [^{14}C]difluoromethylornithine and 10 nmol of pyridoxal phosphate in a total volume of 0.10 ml. After 30 min at 26° no active enzyme remained. Excess reactants were removed by dialysis against distilled water, and 35-μl samples were analyzed by electrophoresis as above. The stained gels were dried onto stiff filter paper, and subsequent autoradiography showed incorporation of ^{14}C only in the 52,000 M_r band.

Various measured amounts of purified enzyme were inactivated by [^{14}C]difluoromethylornithine as above, and the ^{14}C label attached to trichloroacetic acid-precipitable material was quantitated using calculations similar to those of Pritchard *et al.*[11] Knowing the specific activity of the labeled inhibitor to be 60 μCi/μmol and assuming one active site per M_r 52,000, the average number of units of enzyme activity per mole of enzyme were calculated. This method indicated that pure ornithine decarboxylase from this organism should have a specific activity of 48.8 units/mg, implying that our final preparation in the table was 41% pure, active enzyme.

Enzyme Properties

Physarum A-form and B-form ornithine decarboxylase appear in the same fractions when eluted from gel filtration columns with buffers containing 0.3 *M* NaCl.[6] They also demonstrate the same molecular weight by SDS–polyacrylamide gel electrophoresis after labeling with [^{14}C]difluoromethylornithine. Isoelectric focusing of active enzyme, however, shows the A form to have an isoelectric point of pH 5.5 while the B form is 5.9.[2] These are about 5.4 and 5.7, respectively, in the presence of 9 *M* urea.[9] The pH optima of these forms also differ when assayed in the HEPPS buffer system, peaking at pH 8.0 and 8.4, respectively, for the A and B forms.[4] At low ornithine (i.e., 0.05 m*M*) and high coenzyme (i.e., 100 μM) levels the peak for the A form broadens considerably and is approximately 10% higher at pH 8.4 than at pH 8.0.

These enzyme forms are quite similar in their K_m for ornithine at 0.08 and 0.09 m*M*, respectively, for the A and B forms measured at pH 8.4 using 50 μM coenzyme. The enzyme forms differ widely in their affinity for pyridoxal 5′-phosphate with K_m^{PLP} of 0.05 μM for the A form at its pH optimum (8.0) and 12.8 μM for the B form at its optimum.[12] These values are somewhat lower than observed in the presence of borate or phosphate assay buffers.[5]

[11] M. L. Pritchard, J. E. Seely, H. Pösö, L. S. Jefferson, and A. E. Pegg, *Biochem. Biophys. Res. Commun.* **100,** 1597 (1981).

[12] J. L. A. Mitchell, R. A. Yingling, and G. K. Mitchell, *FEBS Lett.* **131,** 305 (1981).

Assay of Ornithine Decarboxylase-Modifying (A to B) Protein

The modification of the A-form enzyme to produce the B enzyme state was assayed in 0.2-ml reaction mixes that contained either crude homogenate or subsequent column fractions in 0.02 *M* HEPPS buffer (pH 8.0) with 0.5 m*M* EDTA, 1.0 m*M* spermidine, 2.5 m*M* dithiothreitol, and about 6.0×10^{-3} units of partially purified A-form enzyme.[6] Immediately after mixing, 0.02-ml aliquots were removed and duplicate assays were performed both for A-form and for total enzyme activity. These assays were repeated after 1 hr at 25°, and the loss in the A-form enzyme, in comparison to the stable total activity, was used to indicate enzyme modification. A-form activity before and after this incubation period was assayed as described for total (A plus B) activity except that the buffer was at pH 8.0 and only 0.5 μM pyridoxal 5′-phosphate was used. This assay of ornithine decarboxylase modification was not quantitative in crude cellular fractions owing to an apparent instability of this modifying protein under these conditions and the variable contribution of cellular A-form and B-form enzyme. After partial purification preparations of this modifying protein were quite stable when 20% polyethylene glycol was added to the reaction mix, and resulting enzyme modifications were linear with respect to time and substrate (A-form enzyme) concentrations. One unit of conversion activity is defined as the amount that catalyzes the modification of 1 unit of A-form enzyme to B-form enzyme in 1 min.

Preparation of the Ornithine Decarboxylase-Modifying Protein

Microplasmodia from 500 ml of exponential culture were centrifuged (300 *g* for 60 S) and the pellet (about 50 ml packed cell volume) was washed in 23° distilled water for 2 min. This material was again pelleted and homogenized by sonication in 1 liter of 0.02 *M* HEPPS buffer (pH 8.0) containing 0.5 m*M* EDTA and 3 m*M* dithiothreitol. The precipitate produced between 40 and 70% saturation with $(NH_4)_2SO_4$ at 4° was resuspended in 30 ml of this buffer at 0.05 *M* NaCl and desalted and depigmented on a column of Sephadex G-25 (200-ml bed volume). The unabsorbed fraction was then diluted to 500 ml using the same buffer (also with 0.05 *M* NaCl) and applied to a 50-ml DEAE-Sephacel column. After washing with 100 ml of buffer, the conversion factor activity was eluted with a 0.05 to 0.25 *M* NaCl linear gradient (800 ml), where 20-ml fractions were collected at about 50 ml/hr. Samples (0.1 ml) of each fraction were dialyzed against the buffer without salt for 2 hr in a Microdialyzer (Bethesda Research Laboratories, Inc.), and 0.09 ml of the desalted fractions was assayed for enzyme modification activity. The active fractions (about

0.15 *M* NaCl) were pooled and the protein was precipitated by 70% saturation with $(NH_4)_2SO_4$. The pellet was resuspended in 10 ml of buffer containing 0.3 *M* NaCl and applied to an AcA-44 column (2.5 × 84 cm). Buffer plus 0.3 *M* NaCl was run through this column at 15 ml/hr, and 7-ml fractions were collected. The peak of enzyme activity was determined on desalted fractions as indicated above, and the 4 peak fractions were combined, concentrated by Amicon PM-10 ultrafiltration to 2 ml, and stored at −20° in buffer containing 30% glycerol. The average yield of this procedure was 0.068 units of ornithine decarboxylase-modifying activity.

Characteristics of the Ornithine Decarboxylase-Modifying Protein

The factor catalyzing this enzyme modification was shown to be a protein of M_r 30,000–35,000.[6] The reaction was stimulated by glycerol and polyethylene glycol and strongly inhibited by ionic strengths greater than 0.05 *M*. This reaction requires either spermine or spermidine (apparent K_ms of 0.12 and 0.35 m*M*, respectively) or much greater concentrations of putrescine (apparent $K_m > 5$ m*M*).[13]

Comments

The chemical difference between these two forms of the *Physarum* enzyme, and thus the mechanism of action of the protein that catalyzes this modification, is still not known. The modifying protein is not thought to be a transglutaminase, as putrescine is a poor substrate and ^3H-labeled polyamine used as substrate did not end up covalently bound to the product (B-form enzyme). This reaction does not have the ATP requirement of a protein kinase, and alkaline phosphatase does not promote conversion between these forms. Clearly this enzyme modification is distinct from the inactivation of this enzyme by either transglutaminase or protein kinase, which are discussed elsewhere in this volume.

Acknowledgments

This work was supported by Grants 81-16 from the American Cancer Society, Illinois Division, 17949-06 from NIH, and RR 07176 from the Biomedical Research Support Grant Program, NIH.

[13] J. L. A. Mitchell, G. K. Mitchell, and D. D. Carter, *Biochem. J.* **205,** 551 (1982).

[21] Polyamine-Dependent Protein Kinase from the Slime Mold *Physarum polycephalum*[1]

By Glenn D. Kuehn, Valerie J. Atmar, and Gary R. Daniels

(EC 2.7.1.37, ATP : protein phosphotransferase)

ATP^{4-} + ornithine decarboxylase enzyme →

ADP^{3-} + H^{+} + phosphoornithine decarboxylase

Ornithine decarboxylase (OrnDCase) [L-ornithine carboxy-lyase, EC 4.1.1.17] catalyzes the first, and rate-limiting, reaction of polyamine biosynthesis in all eukaryotic species.[2] The enzyme is of interest from a regulatory point of view because the turnover (appearance and disappearance) of its catalytic activity ranks among the most rapid changes found among enzymes from eukaryotic organisms as a result of administration of numerous chemical and physical stimuli.[2,3] To account for such rapid changes in activity, many mechanisms have been proposed for the control of OrnDCase.[4] These proposals have included: (*a*) control at the level of RNA synthesis; (*b*) translational control by polyamines; (*c*) antizyme induction or release by polyamines; and (*d*) transitions between an active and a less active form. A conservative appraisal of the available evidence for these proposals supports a stronger case for some type of posttranscriptional control for OrnDCase as opposed to a transcriptional control. Moreover, there is increasing evidence that the polyamines spermidine and spermine exert a key regulatory role in some type of negative feedback process.[5-7] The negative control is antagonized by putrescine.

In general, the regulation of the catalytic function of biosynthetic enzymes through posttranslational, reversible phosphorylation by protein kinases and phosphoprotein phosphatases has been well documented.[8]

[1] See also J. L. A. Mitchell, this volume [20].

[2] D. R. Morris and R. H. Fillingame, *Annu. Rev. Biochem.* **43,** 303 (1974).

[3] U. Bachrach, *in* "Polyamines in Biomedical Research" (J. M. Gaugas, ed.), p. 81. Wiley, New York, 1980.

[4] P. P. McCann, *in* "Polyamines in Biomedical Research" (J. M. Gaugas, ed.), p. 109. Wiley, New York, 1980.

[5] T. J. Paulus and R. H. Davis, *J. Bacteriol.* **145,** 14 (1981).

[6] P. S. Mamont, A. M. Joder-Ohlenbusch, M. Nussli, and J. Grove, *Biochem. J.* **196,** 411 (1981).

[7] V. J. Atmar and G. D. Kuehn, *Proc. Natl. Acad. Sci. U.S.A.* **78,** 5518 (1981).

[8] P. Cohen, "Recently Discovered Systems of Enzyme Regulation by Reversible Phosphorylation. Molecular Aspects of Cellular Regulation," Vol. 1. Elsevier/North-Holland Biomedical Press, Amsterdam, 1981.

ISBN 0-12-181994-9

Below, we describe the preparation of a protein kinase from nucleoli or nuclei of *Physarum polycephalum* that phosphorylates OrnDCase. The phosphorylation reaction is dependent on spermidine and spermine. Putrescine antagonizes activation of the protein kinase by spermidine and spermine. Phosphorylation of OrnDCase inhibits its capacity to catalyze decarboxylation of L-ornithine.

Apart from the interesting specificity of this protein kinase toward phosphorylation of OrnDCase, is the important implication that its discovery offers for investigating the regulation of biosynthetic enzymes in general metabolism. This is a unique example of control of the first enzyme in a biosynthetic pathway by a protein kinase that is, in turn, modulated by the immediate end products of the pathway. This phenomenon may extend to other enzymes in metabolism.

Assay Method

Principle. To assay polyamine-dependent protein kinase, it is essential to establish that transfer of [^{32}P]$HOPO_3^{2-}$ from [γ-^{32}P]ATP to ornithine decarboxylase is spermidine- and spermine-dependent. The assay involves measurement of acid-stable [^{32}P]$HOPO_3^{2-}$ incorporated into ornithine decarboxylase after terminating the protein kinase reaction with a cold solution of bovine serum albumin containing EDTA.

Reagents

Incubation reagent containing 0.3 *M* β-glycerol phosphate disodium salt, 30 m*M* $Mg(CH_3CO_2)_2 \cdot 4H_2O$, 3 m*M* ethyleneglycol-bis(β-aminoethyl ether) *N,N'*-tetraacetic acid (EGTA). After preparation, the pH is adjusted to 6.0 with HCl (25°).

[γ-^{32}P]ATP Tris salt reagent, 100 m*M*, with specific radioactivity of approximately 5×10^9 cpm/μmol or 2250 mCi/mmol. This is prepared by a published procedure.[9]

Stopping reagent containing 100 m*M* EDTA, 50 mg/ml bovine serum albumin

Spermidine–spermine solution containing 3 m*M* spermidine (free base) and 3 m*M* spermine (free base)

$HClO_4$, solutions of 10 and 5%

Tris-HCl, 0.05 *M*, containing 0.3 *M* and 1 *M* NaCl. After preparation, the pH is adjusted to 7.5 with HCl (25°)

Scintillation cocktail containing 1 liter of toluene, 1 liter of 2-ethoxyethanol, and 8 g of Omnifluor

[9] I. M. Glynn and J. B. Chappell, *Biochem. J.* **90,** 147 (1964).

Procedure. For assay of the enzyme from *P. polycephalum* and three additional cell types including bovine spermatozoa, rat liver nuclei, and Ehrlich ascites tumor cells, incubation mixtures at pH 6.8 contain 50 μl of incubation reagent, 10 μl of [γ-^{32}P]ATP Tris salt reagent, 10 μl of spermidine–spermine solution, 10 μl of 0.70 M NaCl containing 20–100 μg of OrnDCase, and 40 μl of enzyme preparation in appropriate buffer (see Purification section). The OrnDCase substrate protein is prepared by AG3X4A ion-exchange chromatography as previously described.[10] All components except [γ-^{32}P]ATP are preincubated for 5 min at 30°, and the solution of [γ-^{32}P]ATP is added at zero time. After 20 min, the reaction is stopped by adding 15 μl of stopping reagent and placing the reaction tube on ice. Zero-time control reactions are treated similarly except that [γ-^{32}P]ATP is added to the reaction mixture after stopping reagent. A sample of 0.1 ml from each reaction is immediately spotted uniformly over the area of a disk, 2.5 cm diameter, of Whatman 3 MM paper. These disks are presoaked in distilled water to remove loose fibers and are dried at room temperature prior to spotting of reaction mixtures. The spotted disks are washed successively in a beaker for 30 min in 10% $HClO_4$ at 5°, again for 30 min in 10% $HClO_4$ at 5°, and finally for 30 min in 5% $HClO_4$ at room temperature. One liter of $HClO_4$ solution is supplied for a maximum of 30 disks. After these washings, each disk is blotted with absorbent paper and dried at room temperature or under a heat gun. Each dried disk is counted for ^{32}P-labeled phosphate content in nonaqueous scintillation cocktail. Concurrent with all experiments, the radioactivity of the [γ-^{32}P]ATP used is determined.

One enzyme unit catalyzes the incorporation of 1 nmol of [γ-^{32}P]$HOPO_3^{2-}$ into OrnDCase per minute at 30° under the conditions described. Specific activity is units per milligram of protein.

Purification from *Physarum polycephalum*

Growth of Microplasmodia. The culture of *P. polycephalum* is a subculture of strain M_3cV originally obtained from Dr. H. P. Rusch at the University of Wisconsin. It has been maintained on a glucose–tryptone–yeast extract medium[11] at New Mexico State University since 1970 with transfer every 3 days. Cultures may be obtained by request to G. D. Kuehn.

Isolation of Nucleoli and Nuclei from Microplasmodia. Nucleoli have been isolated from synchronized macroplasmodia of *P. polycephalum*

[10] G. D. Kuehn, H. U. Affolter, V. J. Atmar, T. Seebeck, U. Gubler, and R. Braun, *Proc. Natl. Acad. Sci. U.S.A.* **76,** 2541 (1979).

[11] B. Chin and I. A. Bernstein, *J. Bacteriol.* **96,** 330 (1968).

grown as plate cultures.[11,12] However, these methods do not yield sufficient amounts of nucleoli for enzyme purification trials. The following procedure describes a method for isolating nucleoli from microplasmodia grown in shake cultures. The method yields adequate quantities for isolating polyamine-dependent protein kinase. Successful isolation of nucleoli from microplasmodia is critically dependent on reducing the cell-associated calcium ion concentration to a level that labilizes the nuclei, such that they will lyse during homogenization in a Waring blender. This is accomplished by washing the microplasmodia in buffered EDTA solution prior to homogenization and conducting the initial homogenization in the absence of supplemented calcium ion.

Microplasmodia are grown as shake cultures for 48 hr in 100 ml of growth medium contained in 500-ml Erlenmeyer flasks. The microplasmodia in 10 flasks are collected by centrifugation in 1-liter cups at 1500 *g* for 2 min. The growth medium is decanted and discarded. All succeeding procedures are conducted near 4°. All adjustments of pH for Tris-HCl buffers are performed at room temperature. Fifty milliliters of wet-packed microplasmodia are suspended in 400 ml of cold Tris-HCl buffer, 50 m*M*, containing 2 m*M* disodium EDTA, pH 7.5. The suspension is immediately centrifuged for 3 min at 1500 *g*. After discarding the wash solution, a second washing is repeated as described above. The wet-packed plasmodial mass is partitioned into 5-ml portions. Each 5-ml sample is processed by the following protocol. Microplasmodia are homogenized in 195 ml of Tris-HCl buffer, 10 m*M*, containing 250 m*M* sucrose and 0.5% Triton X-100, pH 7.5. The suspension is homogenized at 120 V for 1 min using a 1-liter cup and a Waring blender. The homogenate is immediately transferred to a second 1-liter cup and is rehomogenized for an additional 30 sec at 120 V after addition of 2 ml of 1 *M* $CaCl_2$. The suspension is allowed to settle for 10 min. Excess foam is removed by suction. The homogenate is filtered through two layers of Kleen Test Milk filters (Kleen Test Products, Inc., Milwaukee, Wisconsin). The filtrate is centrifuged at 4800 *g* for 15 min. After discarding the supernatant fraction, the nucleoli from two homogenizations (5 ml of original microplasmodia each) are combined and suspended in 20 ml of Tris-HCl, 10 m*M*, containing 250 m*M* sucrose, 10 m*M* $CaCl_2$, pH 7.5. The nucleoli are uniformly dispersed with four strokes in a 30-ml Dounce homogenizer with a Teflon pestle. The suspension is transferred to a 50-ml conical polycarbonate centrifuge tube. After addition of 6 ml of Percoll (Pharmacia Fine Chemicals Co.), each tube is inverted three times to mix the suspension. The suspension is centrifuged

[12] H. U. Affolter, K. Behrens, T. Seebeck, and R. Braun, *FEBS Lett.* **107,** 340 (1979).

for 15 min at 1500 *g*. Nucleoli sediment to the bottom of the tube where they are recovered after removing the top slime layer and the buffered Percoll medium with a water aspirator. The nucleolar pellet is washed once in 20 ml of the Tris–sucrose buffer containing 10 m*M* $CaCl_2$. After centrifugation at 1500 *g* for 15 min, the buffer solution is decanted and the pelleted nucleoli are used immediately for isolation of polyamine-dependent protein kinase. Purity of the nucleolar preparation can be determined by phase contrast microscopy. Preparations routinely contain less than 5% contamination by unlysed nuclei.

Nuclei, in large quantities, can be isolated from microplasmodia of *P. polycephalum* by the method of Mohberg and Rusch.[13]

Crude Extract Preparation and BioRex 70 Treatment. All procedures are conducted near 4°. Isolated nucleoli or nuclei are suspended in an equal volume of Tris-HCl, 50 m*M*, containing 0.3 *M* NaCl, pH 7.5. The suspension is passed three times through a prechilled French pressure cell at 1400 kg/cm^2. The homogenate is centrifuged at 100,000 *g* for 2 hr. The following is a modification of a procedure previously published that has proved to be effective in fractionating nuclear protein kinases.[14] BioRex 70 (Na^+ form, 400 mesh, Bio-Rad Laboratories, Richmond, California) is added to the supernatant fraction to a ratio of about 20 mg of BioRex 70 per milligram of protein. The suspension is stirred slowly for 30 min, then is centrifuged for 15 min at 6000 *g*. The supernatant solution is retained, and the treatment with BioRex 70 is repeated two more times. The final supernatant solution is dialyzed into 50 m*M* Tris-HCl containing 0.3 *M* NaCl, pH 7.5. The dialyzed solution of nuclear or nucleolar proteins is next fractionated by phosphocellulose column chromatography.

Preparation of Phosphocellulose Column. Ten grams of phosphocellulose (Sigma, fine mesh, product No. C2258) is suspended in 300 ml of 0.5 *N* NaOH and stirred for 30 min at room temperature. The fines are removed by suction, and the slurry is filtered in a Büchner funnel with Whatman No. 1 filter paper. The cake is washed on the funnel with distilled water until the pH of the filtrate is about 8. The phosphocellulose cake is transferred to a clean beaker and stirred with 300 ml of 0.5 *N* HCl for 30 min. The cellulose is again filtered and washed as above until the pH of the filtrate is about 4. A final treatment with 0.5 *N* NaOH is carried out as described above. The cellulose is suspended in 50 m*M* Tris-HCl containing 0.3 *M* NaCl, pH 7.5, and the pH is adjusted to 7.5 using 6 *N* HCl. The suspension is stirred gently for 3–4 hr at room temperature, and the pH is readjusted to 7.5 as needed. Phosphocellulose washed in this

[13] J. Mohberg and H. P. Rusch, *Exp. Cell Res.* **66,** 305 (1971).

[14] V. M. Kish and L. J. Kleinsmith, *Methods Cell Biol.* **19,** 101 (1978).

manner should be used within 1 week for optimum separation of protein kinases. A 0.9 × 15 cm column of the phosphocellulose is prepared and is washed in the cold (4°) with 100 ml of 50 m*M* Tris-HCl–0.3 *M* NaCl (pH 7.5).

Isolation of Polyamine-Dependent Protein Kinase. The fraction of nuclear or nucleolar proteins, containing up to 70 mg of protein, is applied to the phosphocellulose column. The column is eluted with 50 m*M* Tris-HCl–0.3 *M* NaCl (pH 7.5) until the $A_{280\ nm}$ of the effluent is less than 0.1. The eluent is then changed to a 300-ml, linear gradient from 0.3 to 1.0 *M* NaCl in 50 m*M* Tris-HCl, pH 7.5. Fractions of 3 ml are collected. Polyamine-dependent protein kinase is consistently eluted from the phosphocellulose column when the gradient reaches about 0.7 *M* NaCl. The tubes that contain the protein kinase are quickly selected, by identifying the few fractions containing 0.7 *M* NaCl, with a conductivity meter. By assaying the ten preceding and ten succeeding fractions around the fractions with 0.7 *M* NaCl, the protein kinase activity is found.

Detection of the protein kinase exploits its novel property of autophosphorylation in the absence of OrnDCase.[7,15] Thus, all column fractions suspected of containing the protein kinase are assayed twice. One assay is conducted in the presence, and a second assay in the absence, of spermidine–spermine. No OrnDCase substrate protein is added to these assays. Two different fractions containing protein kinase activity are routinely observed among the gradient fractions containing 0.7 *M* NaCl. An early fraction is observed that is inhibited by spermidine–spermine. This fraction contains the polyamine-dependent protein kinase, free of its substrate, OrnDCase. In this state, the protein kinase autophosphorylates itself and the autophosphorylation reaction is inhibited by spermidine–spermine. If provided OrnDCase substrate, the protein kinase will phosphorylate the 70,000 M_r subunit of OrnDCase in a reaction that is spermidine- and spermine-dependent.

A later fraction, eluting immediately after the polyamine-dependent protein kinase from phosphocellulose, exhibits little protein kinase activity in the absence of spermidine–spermine but is markedly activated by spermidine–spermine.[7,15] This fraction contains the polyamine-dependent protein kinase copurified with its 70,000-M_r substrate protein, OrnDCase.[7,15] If the protein kinase assays are conducted at NaCl concentrations greater than approximately 150 m*M*, the polyamine-dependent protein kinase reaction is readily detected. If assays are conducted below 150 m*M* NaCl, little or no protein kinase activity is observed. Similarly, OrnDCase activity is detected under the high salt condition, but not at

[15] G. R. Daniels, V. J. Atmar, and G. D. Kuehn, *Biochemistry* **20,** 2525 (1981).

concentrations much below 150 mM NaCl. At low salt, the polyamine-dependent protein kinase and OrnDCase form a complex that exhibits neither of the two catalytic activities in low salt solutions.[7]

Notes on Fractionation Procedures and General Properties of the Protein Kinase. Polyamine-dependent protein kinase activity cannot be demonstrated in crude extracts prepared from nuclei or nucleoli. It is only after partial purification through BioRex 70 treatment that the activity can be detected. It is assumed that a high phosphoprotein phosphatase activity is responsible for this effect because Na_2MoO_4 is required to observe the activity in crude extract preparations.[15] Thus, a purification summary table cannot be presented.

The protein kinase preparations from phosphocellulose chromatography are stable for weeks when stored at refrigerator temperature in buffered 0.7 M NaCl solution. After about 3 weeks, the enzyme that copurifies with OrnDCase begins slowly to lose its capacity to be activated by spermidine–spermine.

Putrescine antagonizes the capacity of spermidine–spermine to activate the protein kinase reaction. The mechanism for how this is accomplished is unknown.

No modifiers, previously recognized to influence other protein kinases, have been found to influence this protein kinase except the polyamines.[15]

The specific activity of the purified polyamine-dependent protein kinase is consistently in the vicinity of 10–15 units/mg of protein when OrnDCase is the protein substrate. The only endogenous substrate yet identified for the protein kinase is OrnDCase. The Michaelis constants for ATP and Mg^{2+} are 10 μM and 53 μM, respectively. The protein kinase will phosphorylate casein at approximately 0.1 the rate observed with OrnDCase. The casein kinase activity is not activated by spermidine–spermine.

Fractions from phosphocellulose chromatography in which the protein kinase and OrnDCase copurify frequently contain more OrnDCase than protein kinase.

Electrophoretograms of purified preparations of the polyamine-dependent protein kinase yield a single protein zone in both 8 and 12.5% SDS–polyacrylamide gels.

Purification from Other Eukaryotic Sources

The procedures described have also been successful for purifying polyamine-dependent protein kinase from rat liver nuclei, bovine spermatozoa,[16] nuclei from Ehrlich ascites tumor cells as well as the cytoplasmic compartment of this tumor cell line.[17]

[16] V. J. Atmar, G. D. Kuehn, and E. R. Casillas, *J. Biol. Chem.* **256,** 8275 (1981).
[17] V. Sekar, V. J. Atmar, and G. D. Kuehn, *Biochem. Biophys. Res. Commun.* **106,** 305 (1982).

[22] Ornithine Decarboxylase (Rat Liver)

By SHIN-ICHI HAYASHI and TAKAAKI KAMEJI

L-Ornithine → putrescine + CO_2

Assay Method

Principle. The enzyme activity is determined by measuring the rate of formation of $^{14}CO_2$ from [1-^{14}C]ornithine.

Reagents

Tris-HCl buffer, 0.1 *M*, pH 7.4
Pyridoxal phosphate, 0.5 m*M*
Dithiothreitol, 25 m*M*; freshly prepared
Substrate solution: L-ornithine hydrochloride, 10 m*M*, containing either L-[1-^{14}C]ornithine, 1.25 μCi/ml, or DL-[1-^{14}C]ornithine, 2.5 μCi/ml
KOH, 10%
HCl, 6 *N*
Scintillation fluid: 4 g of PPO and 0.1 g of POPOP in 700 ml of toluene and 300 ml of ethyl alcohol

Procedure.[1] The standard assay mixture contains 1 ml of Tris-HCl buffer, 0.1 ml of pyridoxal phosphate, 0.1 ml of dithiothreitol, 0.1 ml of substrate solution, and enzyme solution in a final volume of 2.5 ml. Incubation is carried out at 37° for 30 min in an airtight 25-ml conical flask with a center well containing a piece of filter paper moistened with 0.05 ml of 10% KOH. The reaction is stopped by injection of 0.5 ml of 6 *N* HCl, and the mixture is shaken for more than 20 min. The filter paper is then transferred to a vial, and radioactivity is determined after addition of 4 ml of scintillation fluid. Values are corrected for the counts in blanks without enzyme.

[1] T. Noguchi, Y. Aramaki, T. Kameji, and S. Hayashi, *J. Biochem.* (*Tokyo*) **85,** 953 (1979).

ISBN 0-12-181994-9

Comments. Ornithine decarboxylase is mainly located in the cytosol.[2] Enzyme solution should be free from particulate fractions that may cause indirect decarboxylation of ornithine unrelated to the enzyme activity. Enzyme solution should be dialyzed if it contains any inhibitory substance, such as DL-α-hydrazino-δ-aminovaleric acid,[3,4] NaCl,[5] or ornithine at high concentration. In later stages of enzyme purification, the assay is performed in the presence of 0.2 mg of bovine serum albumin per milliliter to prevent enzyme denaturation.

Definition of Unit and Specific Activity. One unit of enzyme activity is defined as the amount catalyzing the formation of 1 nmol of CO_2 per hour. Specific activity is expressed as units per milligram of protein. Up to step 4 of purification, protein is measured by the method of Lowry *et al.*[6] with bovine serum albumin as a standard. At steps 5 and 6, the enzyme sample is subjected to electrophoresis by the method of Laemmli[7] in parallel with various amounts of bovine serum albumin as a standard. Protein bands are stained with 0.2% Coomassie Brilliant Blue R250, and absorbance is recorded at 600 nm with a densitometer for protein determination.[8]

Preparation of Affinity Adsorbents. Pyridoxamine phosphate (PMP)-Sepharose 4B and PMP-succinyldiaminodipropylamine (arm)-Sepharose 4B are prepared as described by Miller *et al.*[9] Heparin-Sepharose 4B is prepared as follows. Seventy-five milliliters of Sepharose 4B activated by cyanogen bromide[10] is washed and suspended in 75 ml of 0.1 *M* $NaHCO_3$, pH 9.8. The suspension is mixed with 150 mg of heparin sodium salt (159 IU/mg) and stirred gently at 4° overnight. Then the gel is washed with 1 *M* NaCl followed by 0.1 *M* $NaHCO_3$, pH 9.8, and suspended in 75 ml of 0.1 *M* $NaHCO_3$ containing 1 *M* monoethanolamine, pH 9.8, and stirred gently for 4 hr at room temperature to mask unreacted active groups.

[2] A. E. Pegg and H. G. Williams-Ashman, *in* "Polyamines in Biology and Medicine" (D. R. Morris and L. J. Marton, eds.), p. 3. Dekker, New York, 1981.

[3] H. Inoue, Y. Kato, M. Takigawa, K. Adachi, and Y. Takeda, *J. Biochem.* (*Tokyo*) **77,** 879 (1975).

[4] T. Sawayama, H. Kinugasa, and H. Nishimura, *Chem. Pharm. Bull.* **24,** 326 (1976).

[5] M. F. Obenrader and W. F. Prouty, *J. Biol. Chem.* **252,** 2860 (1977).

[6] O. H. Lowry, N. J. Rosebrough, A. L. Farr, and R. J. Randall, *J. Biol. Chem.* **193,** 265 (1951).

[7] U. K. Laemmli, *Nature* (*London*) **227,** 680 (1970).

[8] T. Kameji, Y. Murakami, K. Fujita, and S. Hayashi, *Biochim. Biophys. Acta* **717,** 111 (1982).

[9] J. V. Miller, Jr., P. Cuatrecasas, and E. B. Thompson, *Biochim. Biophys. Acta* **276,** 407 (1972).

[10] P. Cuatrecasas, *J. Biol. Chem.* **245,** 3059 (1970).

Purification Procedure[8]

Treatment of Rats for Enzyme Induction. One hundred and fifty male Sprague–Dawley rats weighing 200–300 g are injected intraperitoneally with 150 mg of thioacetamide per kilogram 20 hr before sacrifice and with 75 mg of DL-α-hydrazino-δ-aminovaleric acid per kilogram, a competitive inhibitor of ornithine decarboxylase,[3,4] 6 hr before sacrifice. The latter treatment increases the enzyme level more than twofold.[3,11]

Step 1. Preparation of Crude Extract. Rats are decapitated and their livers are quickly removed and cooled. All subsequent procedures are performed at 0–4°. Livers are homogenized with two volumes of homogenizing medium (0.2 *M* sucrose in 25 m*M* Tris-HCl, 1 m*M* dithiothreitol, and 0.1 m*M* EDTA, pH 7.5) in a Dounce-type all-glass homogenizer. The homogenate is centrifuged at 15,000 g_{max} for 15 min, and the supernatant is centrifuged at 107,000 g_{max} for 60 min.

Step 2. DEAE-Cellulose Column Chromatography. The crude extract is directly applied to a column of DEAE-cellulose (4 × 60 cm) that has been equilibrated with 25 m*M* Tris-HCl, pH 7.5, containing 1 m*M* dithiothreitol and 0.1 m*M* EDTA (buffer A). The column is washed with the same buffer and then the enzyme is eluted with a linear gradient of NaCl in buffer A (0 to 0.5 *M*, 6 liters). Active fractions are pooled and brought to 60% saturation with solid ammonium sulfate. The precipitate is collected by centrifugation, dissolved in 25 m*M* Tris-HCl, pH 7.2, containing 1 m*M* dithiothreitol and 0.1 m*M* EDTA (buffer B), and dialyzed against the same buffer overnight.

Step 3. Affinity Chromatography on PMP-arm-Sepharose. The enzyme solution is applied to a column of PMP-arm-Sepharose 4B (2.2 × 55 cm) that has been equilibrated with buffer B. The column is washed with 1 liter of the same buffer and then the enzyme is eluted from the column with a pH gradient obtained by placing buffer B containing 0.4 *M* NaCl and 0.1% Tween 80 (500 ml) in a mixing chamber and 50 m*M* Tris base containing 0.4 *M* NaCl, 1 m*M* dithiothreitol, and 0.1% Tween 80 (500 ml) in a reservoir. One milliliter of 0.25 *M* Tris-HCl, pH 7.0, is added to each collecting tube beforehand to prevent enzyme denaturation by alkaline pH. Active fractions (12 ml each) are collected, concentrated in an Amicon ultrafiltration apparatus, and dialyzed against buffer B containing 0.1% Tween 80.

Step 4. First Affinity Chromatography on Heparin-Sepharose. The enzyme solution is applied to a column of heparin-Sepharose (1.2 × 24 cm) that has been equilibrated with buffer B containing 0.1% Tween 80. The column is washed with the same buffer, and then the enzyme is eluted

[11] S. I. Harik, M. D. Hollenberg, and S. H. Snyder, *Mol. Pharmacol.* **10,** 41 (1974).

with a linear gradient of NaCl in buffer B containing 0.1% Tween 80 (0 to 0.3 *M*, 400 ml). Active fractions are pooled, concentrated by ultrafiltration, and dialyzed against 50 m*M* Tris-HCl, pH 7.2, containing 1 m*M* dithiothreitol, 0.2 m*M* EDTA, and 0.1% Tween 80 (buffer C).

Step 5. Affinity Chromatography on PMP-Sepharose. The enzyme solution is applied to a column of PMP-Sepharose (1.5 × 10 cm) that has been equilibrated with buffer C. The column is washed with the same buffer and then the enzyme is eluted with a linear gradient of pyridoxal phosphate in the same buffer (0 to 0.2 m*M*, 200 ml). Active fractions are pooled and concentrated by ultrafiltration.

Step 6. Second Affinity Chromatography on Heparin-Sepharose. The enzyme solution is directly applied to a column of heparin-Sepharose (1.5 × 5 cm) that has been equilibrated with buffer C containing 50 μM pyridoxal phosphate (buffer D). The column is washed with the same buffer and then the enzyme is eluted with a linear gradient of NaCl in buffer D (0 to 0.15 *M*, 100 ml). Active fractions are pooled and concentrated by ultrafiltration.

Comments on Purification. A summary of the purification is shown in the table. It may be easier to carry out the procedure up to step 2 twice on half scale and then combine the preparations for the subsequent purification. At the time of liver excision, it is necessary to eliminate pancreatic tissue completely since the enzyme appears to be quite sensitive to pancreatic proteases. At each chromatographic step, the enzyme is eluted as a single peak and therefore most of the active fractions are collected. The last step increases the purity of the enzyme from about 80% (step 5) to near homogeneity, as judged by sodium dodecyl sulfate (SDS)–polyacrylamide gel electrophoresis. The slight decrease in specific activity at the last step may be due to denaturation of the enzyme because of its extremely low protein concentration.

PURIFICATION OF ORNITHINE DECARBOXYLASE FROM RAT LIVER

Step	Volume (ml)	Total protein (mg)	Specific activity (units/mg)	Purification (fold)	Yield (%)
1. Crude extract	3395	99,300	3.1	1	100
2. DEAE-cellulose	132	4,430	31	10	44
3. PMP[a]-arm-Sepharose	62	9.8	11,000	3,600	36
4. First heparin-Sepharose	4.5	0.86	120,000	38,000	33
5. PMP-Sepharose	1.1	0.051	1,200,000	390,000	20
6. Second heparin-Sepharose	0.75	0.025	1,100,000	350,000	8.1

[a] PMP, Pyridoxamine phosphate.

Properties[8,12]

Purity. The enzyme is shown to be nearly homogeneous by polyacrylamide disc gel electrophoresis both in the absence and in the presence of SDS and also by Ouchterlony double immunodiffusion analysis.

Molecular Weight. The apparent molecular weight of ornithine decarboxylase is estimated to be about 105,000 by polyacrylamide gel electrophoreses at several different gel concentrations.[13] The enzyme is composed of two identical subunits each with a molecular weight of 50,000, as judged by SDS–polyacrylamide gel electrophoresis.

Isoelectric Point. The enzyme has a single isoelectric point at pH 4.1 as determined by chromatofocusing.

Stability. The purified enzyme is relatively stable in the presence of dithiothreitol, pyridoxal phosphate, and Tween 80 unless very dilute. No loss of activity is observed on storage at −80° for 2 weeks. Even when incubated at 52° in the presence of 2 m*M* dithiothreitol, 0.1 m*M* pyridoxal phosphate, and 0.1 mg/ml of bovine serum albumin, the enzyme activity remains unchanged for 45 min. The yield of purified enzyme is apparently improved by addition of Tween 80[14] at step 3 and thereafter.

Abbreviations used: PMP, pyridoxamine phosphate; arm, succinyldiaminodipropylamine; SDS, sodium dodecyl sulfate; PPO, 2,5-diphenyloxazole; POPOP, 2,2′-*p*-phenylenebis(5-phenyloxazole).

[12] T. Kameji and S. Hayashi, *Biochim. Biophys. Acta* **705,** 405 (1982).
[13] J. L. Hedrick and A. J. Smith, *Arch. Biochem. Biophys.* **126,** 155 (1968).
[14] T. Kitani and H. Fujisawa, *Eur. J. Biochem.* **119,** 177 (1981).

[23] Ornithine Decarboxylase (Mouse Kidney)

By JAMES E. SEELY and ANTHONY E. PEGG

Ornithine decarboxylase (EC 4.1.1.17) catalyzes the conversion of L-ornithine to putrescine and CO_2.[1] It will also act on L-lysine, forming cadaverine, but the K_m for this substrate is 100 times higher than for ornithine.[2] Mouse kidney is a good source of this enzyme after treatment with androgens, having a specific activity 100–200 times that of regenerating rat liver.[2,3]

[1] A. E. Pegg and H. G. Williams-Ashman, *Biochem. J.* **108,** 533 (1968).
[2] A. E. Pegg and S. M. McGill, *Biochim. Biophys. Acta* **568,** 416 (1979).
[3] J. E. Seely, H. Pösö, and A. E. Pegg, *Biochem. J.* **206,** 311 (1982).

METHODS IN ENZYMOLOGY, VOL. 94

ISBN 0-12-181994-9

Assay Method

Principle. The activity is assayed by following $^{14}CO_2$ release from L-[1-^{14}C]ornithine.

Reagents

Tris-HCl buffer, 0.5 *M*, pH 7.5
Dithiothreitol, 25 m*M*
Pyridoxal 5′-phosphate, 2 m*M*
L-Ornithine, 20 m*M*
L-[1-^{14}C]Ornithine, 12.5 μCi/ml (specific activity 57 mCi/mmol)
Enzyme, in 25 m*M* Tris-HCl, pH 7.5 containing 2.5 m*M* dithiothreitol and 0.1 m*M* EDTA
Hyamine hydroxide, 1 *M*
Sulfuric acid, 5 *M*

Procedure. An assay mixture is made up containing 5 μl of 20 m*M* L-ornithine, 10 μl of L-[1-^{14}C]ornithine (57 mCi/mmol), 5 μl of 2 m*M* pyridoxal phosphate, 12.5 μl of 25 m*M* dithiothreitol, 2.5 μl of 0.5 *M* Tris-HCl (pH 7.5), and enzyme (0.1–3 units) in a total volume of 0.25 ml. The reaction is carried out in test tubes closed with rubber stoppers carrying polypropylene wells (Kontes Glass Company, Vineland, New Jersey) containing 0.2 ml of 1 *M* Hyamine hydroxide. The assay is started by the addition of the enzyme solution, and the tubes are incubated at 37° for 30 min in a shaking water bath. The reaction is then stopped by the injection of 0.3 ml of 5 *M* sulfuric acid through the rubber cap. The acid releases $^{14}CO_2$ from the assay medium, and, after a further 15 min at 37° to ensure complete absorption of the $^{14}CO_2$ in the Hyamine hydroxide, the well is removed, placed in 5 ml of a toluene-based scintillation fluid, and counted in a liquid scintillation spectrometer. Blank tubes are set up in which the sulfuric acid is added before the enzyme or the enzyme is omitted.

Definition of Unit. One unit is defined as the amount of enzyme releasing 1 nmol of $^{14}CO_2$ per 30 min at 37°.

Purification Procedure

Step 1. Induction of Enzyme and Preparation of Extracts. Approximately 100 male Crl : CD-1 mice (Charles River Breeding Laboratories, Wilmington, Massachusetts) are treated with testosterone propionate (100 mg/kg) by subcutaneous injection of a solution of 4 mg/ml in sesame seed oil. Three days later the mice are killed by cervical dislocation and the kidneys are homogenized at 0–4° in 2 volumes of buffer A (25 m*M* Tris-HCl, pH 7.5, containing 0.1 m*M* EDTA and 2.5 m*M* dithiothreitol). The homogenate is centrifuged at 100,000 *g* for 45 min.

Step 2. Fractionation with Ammonium Sulfate. Proteins precipitating between 30 and 50% saturation with ammonium sulfate are collected by centrifugation at 12,000 *g* for 15 min, dissolved in a volume of buffer A to give a final protein concentration of 25–50 mg/ml, and dialyzed overnight against 50 volumes of buffer A.

Step 3. DEAE-Cellulose Chromatography. The dialyzed fraction is loaded onto a DEAE-cellulose column (2.5 × 25 cm) that has been equilibrated with buffer A. The column is washed with 125 ml of buffer A followed by 125 ml of buffer A containing 0.08 *M* NaCl. Enzyme is eluted using a linear gradient (300 ml) of 0.08 to 0.3 *M* NaCl in buffer A. Elution of enzyme activity occurs at a NaCl concentration of about 0.2 *M*. The active fractions are pooled, concentrated 10-fold by ultrafiltration using a Diaflo PM-10 membrane, and dialyzed overnight against 50 volumes of buffer B (25 m*M* Tris-HCl, pH 7.5, containing 0.1 m*M* EDTA, 2.5 m*M* dithiothreitol, and 0.02% Brij 35). The use of Brij 35 is essential to maintain good enzyme activity, as the purified enzyme is extremely unstable when the detergent is not present.[4]

Step 4. Affinity Chromatography on Pyridoxamine 5′-Phosphate-Agarose. An affinity absorbent[4,5] is prepared by the reaction of pyridoxamine 5′-phosphate with Affi-Gel 10 (Bio-Rad Laboratories, Richmond, California). The Affi-Gel 10 is packaged fully swollen in ethanol, 25 ml per vial. The slurry is transferred to a fritted glass funnel, the ethanol is drained, and the moist cake is washed with 3 bed volumes of cold 2-propanol followed by 3 bed volumes of cold distilled water, using gentle suction to aid filtration. The gel is scraped from the funnel and placed in a clean bottle containing 14.2 mg of pyridoxamine 5′-phosphate in 12.5 ml of 0.1 *M* HEPES buffer, pH 7.5. The coupling reaction is allowed to proceed for 4 hr at 4° with rotational mixing of the suspension. Unreacted groups on the gel are then coupled to ethanolamine by adding 25 ml of 1 *M* ethanolamine HCl, pH 8, and allowing the reaction to continue for an additional hour at 4°. The gel is transferred to a fritted glass funnel and washed to remove unbound ligand. The funnel is washed successively with 300-ml portions of ice-cold 1 *M* NaCl, distilled water, and 25 m*M* Tris-HCl, pH 7.5, containing 0.1 m*M* EDTA and 0.02% Brij 35. The coupled gel is stored in an equal volume of 25 m*M* Tris-HCl, pH 7.5, containing 0.1 m*M* EDTA, 0.02% Brij 35, and 0.02% sodium azide. Prior to use, the gel is equilibrated with several column volumes of buffer B.

The enzyme from step 3 is applied to a column (1.5 × 7.5 cm) of the agarose-pyridoxamine 5′-phosphate at a flow rate of 1.5 ml/hr. The

[4] M. L. Pritchard, J. E. Seely, H. Pösö, L. S. Jefferson, and A. E. Pegg, *Biochem. Biophys. Res. Commun.* **100,** 1597 (1981).

[5] The preparation and use of this absorbent is modified from R. J. Boucek and K. J. Lembach, *Arch. Biochem. Biophys.* **184,** 408 (1977).

PURIFICATION OF MOUSE KIDNEY ORNITHINE DECARBOXYLASE

Fraction	Total protein (mg)	Total units	Specific activity (units/mg)	Purification (fold)	Yield (%)
Supernatant	4550	654,000	144	1	100
Dialyzed ammonium sulfate precipitate	1700	464,000	273	2	71
DEAE-cellulose eluate	246	366,000	1,307	9	57
Pyridoxamine 5′-phosphate-agarose eluate	0.61	235,000	384,000	2,670	36
Gel filtration	0.088	131,000	1,490,000	10,400	20

column is washed with 2 column volumes of buffer B at 1.5 ml/hr followed by 4 column volumes of buffer B at 100–150 ml/hr. (Under these conditions less than 10% of the initial activity fails to remain attached to the column.) The enzyme is then eluted with buffer B containing 10 μM pyridoxal phosphate at a flow rate of 1.5 ml/hr. Fractions containing activity are concentrated to a volume of 1–2 ml by ultrafiltration.

Step 5. Gel Filtration. The enzyme fraction from step 4 is applied to a column (1.6 × 65 cm) of Ultrogel AcA-34 equilibrated with buffer B containing 10 μM pyridoxal 5′-phosphate. Elution is carried out with the same buffer at a flow rate of 10 ml/hr, and fractions of 1.5 ml are collected. Those containing enzyme activity (usually fractions 48–60) are pooled and concentrated by ultrafiltration.

A summary of the purification procedure is shown in the table. The protein determination was carried out by the method of Bradford[6] using aldolase as a standard. As discussed in this volume,[7] the final specific activity agrees well with that predicted from the titration of the enzyme with the irreversible inhibitor α-difluoromethylornithine.

Properties

The enzyme is very unstable unless maintained in the presence of dithiothreitol, pyridoxal 5′-phosphate, and Brij 35. In the presence of buffer B containing 10 μM pyridoxal 5′-phosphate, it was stored at −40° for 3 weeks with no loss of activity. It has an M_r of about 100,000 and consists of two subunits of M_r 53,000.[8] Activity requires pyridoxal 5′-phosphate.

[6] M. M. Bradford, *Anal. Biochem.* **72,** 248 (1976).

[7] J. E. Seely, H. Pösö, and A. E. Pegg, this volume [32].

[8] J. E. Seely, H. Pösö, and A. E. Pegg, *Biochemistry* **21,** 3394 (1982).

[24] Ornithine Decarboxylase (Germinated Barley Seeds)

By DIMITRIOS A. KYRIAKIDIS, CHRISTOS A. PANAGIOTIDIS, and JOHN G. GEORGATSOS

The contribution of ornithine decarboxylase (EC 4.1.1.17) (ODC) to polyamine biosynthesis in plant cells was thought to be insignificant since ODC activity in most plant tissues was found to be much lower than that of arginine decarboxylase.[1,2] However, reports have shown high levels of ODC activity in rapidly proliferating plant cells[3] and great similarities between plant and mammalian ODC.[4]

In the previous investigations, including the reports cited above, ODC activity was assayed in the 10,000 *g* supernatant. We have reported that in barley seeds germinated for more than 90 hr 75% of ODC activity is located in the nucleus, tightly bound to chromatin, with the remaining 25% in the cytosol.[5]

$$\underset{\text{Ornithine}}{H_2NCH_2CH_2CH_2\underset{^{14}COOH}{\underset{|}{C}}HNH_2} \xrightarrow{ODC} \underset{\text{Putrescine}}{H_2NCH_2CH_2CH_2CH_2NH_2} + {}^{14}CO_2$$

Assay Method

Principle. The assay depends upon the measurement of $^{14}CO_2$ released from DL-[1-^{14}C]ornithine.

Reagents

Tris-HCl buffer: 50 m*M* Tris-HCl, (pH 8.5) at 20°, 0.3 m*M* EDTA, 50 μM pyridoxal phosphate, and 5 m*M* dithiothreitol
DL-[1-^{14}C]Ornithine, specific activity 50 mCi/mmol
Enzyme, 0.5–2.0 unit per assay
Trichloroacetic acid, 10% (w/v)
NCS (tissue solubilizer) diluted 1 : 1 with toluene

[1] M. J. Montague, J. W. Koppenbring, and E. G. Jaworski, *Plant Cell Physiol.* **62,** 430 (1978).
[2] T. A. Smith, *Phytochemistry* **18,** 1447 (1979).
[3] Y. M. Heimer, Y. Mizrahi, and U. Bachrach, *FEBS Lett.* **104,** 146 (1979).
[4] Y. M. Heimer and Y. Mizrahi, *Biochem. J.* **201,** 373 (1982).
[5] C. A. Panagiotidis, J. G. Georgatsos, and D. A. Kyriakidis, *FEBS Lett.* **146,** 193 (1982).

METHODS IN ENZYMOLOGY, VOL. 94

ISBN 0-12-181994-9

Procedure. The reaction is carried out in 17 × 100 mm polystyrene tubes sealed with polyethylene caps.[6] A $\frac{1}{4}$-inch filter paper disk, impregnated with 25 μl of NCS (Amersham) diluted 1 : 1 with toluene and transfixed with an 18-gauge syringe needle through the cap, is present during the incubation. The reaction mixture contains 50 μl of 50 m*M* Tris pH 8.5, 0.3 m*M* EDTA, 50 μ*M* pyridoxal phosphate, 5 m*M* dithiothreitol (buffer A), 5 μl of enzyme, and 5 μl of DL-[1-^{14}C]ornithine (0.023 μmol). The mixture is incubated at 37° for 60 min, then stopped by the injection of 0.1 ml of 10% (w/v) trichloroacetic acid into each tube through the syringe needle; the incubation at 37° is continued for an additional hour. The liberated $^{14}CO_2$ is determined by placing the paper disk into 3 ml of scintillation fluid and counting in a Packard Tri-Carb liquid scintillation spectrometer.

Definition of Unit and Specific Activity. One unit of ODC activity is defined as the amount of enzyme that releases 1 nmol of $^{14}CO_2$ in 1 hr under the conditions of the experiment. Specific activity is defined as units per milligram of protein. Protein is determined by the method of Ross and Schatz[7] using bovine serum albumin as a standard.

Purification Procedure

Germination of Barley Seeds. Barley seeds (*Hordeum vulgare* var. Beca) are germinated for 90 hr in the presence of gibberellic acid (5×10^{-5} *M*) and 10 hr in the presence of gibberellic acid (5×10^{-5} *M*) and actinomycin D (2 μg/ml) as previously described by Panagiotidis *et al.*[5] Ornithine decarboxylase activity is usually induced fourfold by gibberellic acid and superinduced two- to threefold by actinomycin D.[5]

Preparation of Chromatin and Cytosolic Fraction. Barley seeds germinated for 100 hr (1100 *g*) are homogenized in a Waring blender for 1 min at low speed and 2 min more at high speed in two volumes of isotonic grinding medium (0.25 *M* sucrose in buffer A). The homogenate is filtered through four layers of gauze and the filtrate is centrifuged at 4000 *g* for 30 min. Chromatin is isolated from the pellet using the method of Bonner.[7] The 4000 *g* supernatant fraction is centrifuged at 100,000 *g* for 1 hr; this supernatant solution is used as the source of cytosolic ODC.

[6] J. S. Heller, K. Y. Chen, D. A. Kyriakidis, W. F. Fong, and E. S. Canellakis, *J. Cell. Physiol.* **96,** 225 (1978).

[7] J. Bonner, *in* "Plant Biochemistry" (J. Bonner and J. E. Varner, eds.), 3rd Ed., p. 37. Academic Press, New York, 1976.

Purification of Chromatin-Bound ODC

All the subsequent manipulations are carried out at 0–4°.

Step 1. Preparation of Crude Chromatin. The chromatin is suspended in 120 ml of 0.25 *M* sucrose in buffer A and centrifuged at 4000 *g* for 30 min. This treatment is repeated twice.

Step 2. Extraction of Chromatin-Bound ODC. Chromatin of step 1 is suspended in 80 ml of buffer A; the suspension is frozen and thawed once, stirred for 45 min, and centrifuged at 100,000 *g* for 1 hr. After this treatment more than 50% of the ODC activity is released. The remaining activity is recovered after repeating this 2 or 3 more times.

Step 3. DEAE-BioGel A Chromatography. A DEAE-BioGel A column of 2.5 × 27 cm is packed without pressure and equilibrated with buffer A. The supernatant of step 2 is applied to the column. The column is thoroughly washed with buffer A, and the enzyme is eluted with a 700-ml linear gradient containing 0 to 0.3 *M* NaCl in buffer A. Fractions with 10-fold purification are pooled, diluted five times with buffer A, and used in the next step.

Step 4. TEAE-Cellulose Chromatography. A triethylaminoethyl-cellulose column (2.0 × 9.5 cm) is equilibrated with buffer A; the enzyme solution of step 3 is applied, and the column is washed with 150 ml of buffer A followed by 150 ml of 0.1 *M* NaCl in buffer A. The enzyme is eluted with 150 ml of 0.25 *M* NaCl in buffer A, and the active fractions are pooled and concentrated to 5.6 ml by Amicon ultrafiltration (PM-10 filter).

Step 5. Sephadex G-200 Chromatography. The enzyme solution from step 4 is applied to a Sephadex G-200 (2.8 × 60 cm) column previously equilibrated with buffer A. The enzyme is eluted with buffer A. The active fractions are pooled and concentrated to 4.8 ml by Amicon ultrafiltration (PM-10); 10% (v/v) glycerol is added, and the enzyme is stored at −20°.

Results of the overall purification procedure are summarized in Table I.

Purification of Cytosolic ODC

Step 1. Crude 100,000 g Supernatant. The 100,000 *g* supernatant (1200 ml) from 1100 g of germinated barley seeds is used for the purification of cytosolic enzyme.

Step 2. DEAE-BioGel A Chromatography. The supernatant from step 1 is divided in three portions and applied to three DEAE-BioGel A columns of 2.5 × 30 cm equilibrated with buffer A. The columns are

TABLE I
PURIFICATION OF CHROMATIN-BOUND ORNITHINE DECARBOXYLASE[a]

Step	Volume (ml)	Protein (mg)	Total activity (units)	Specific activity (units/mg)
1. Crude chromatin[b]	600	2340	42,100	18.0
2. Chromatin extract	170	820	40,000	49
3. DEAE-BioGel A	150	55	22,000	400.0
4. TEAE-cellulose	5.6	6	20,000	3,330.0
5. Sephadex G-200	4.8	0.8	16,000	20,000.0

[a] Overall yield, 38%; overall purification, 1110-fold.
[b] The number of each step corresponds to that in the text.

washed with 250 ml of buffer A, and the enzyme is eluted from each column with 200 ml of 0.15 *M* NaCl in buffer A. The active fractions from the three columns are combined, diluted fourfold with buffer A, and used in the next step.

Step 3. TEAE-Cellulose Chromatography. The enzyme solution (550 ml) of step 2 is applied on a (triethylaminoethyl)cellulose (2.0 × 30 cm) column equilibrated with buffer A. The column is washed with 300 ml of buffer A, and the enzyme is eluted with a 300-ml linear gradient of 0 to 0.3 *M* NaCl in buffer A. The active fractions are pooled and concentrated to 6 ml by Amicon ultrafiltration (PM-30).

Step 4. BioGel P-100 Chromatography. Six milliliters of enzyme solution of step 3 are applied to a BioGel P-100 (2.8 × 68 cm) column previously equilibrated with buffer A; buffer A is used for elution of the enzyme. The active fractions are pooled and concentrated to 4.5 ml by Amicon ultrafiltration (PM-30); glycerol 10% (v/v) is added, and the solution is stored at −20°.

Results of the overall purification procedure are summarized in Table II.

Properties

Storage of the crude cytosolic preparation at 0° for more than 10 hr results in 80–90% decrease of ODC activity, whereas the chromatin preparation loses only 10% of ODC activity. A 10% decrease of ODC activity is obtained when the partially purified enzymes (after the TEAE-cellulose

TABLE II
PURIFICATION OF CYTOSOLIC ORNITHINE DECARBOXYLASE[a]

Step	Volume (ml)	Protein (mg)	Total activity (units)	Specific activity (units/mg)
1. Crude 100,000 *g* supernatant[b]	1200	15,800	20,500	1.3
2. DEAE-BioGel A	550	1,590	15,500	9.7
3. TEAE-cellulose	6	460	12,600	28
4. BioGel P-100	4.5	98	9,900	101

[a] Overall yield, 48%; overall purification, 77-fold.
[b] The number of each step corresponds to that in the text.

column) are allowed to remain at room temperature for 5 hr; at 0° both enzymes at this stage of purification appear to be stable. Repeated freezing-thawing in the presence of 10% v/v glycerol causes 15–20% loss in both activities.

Kinetics. Both partially purified enzymes obey classical Michaelis–Menten kinetics with a K_m of 1 mM for the chromatin ODC and of 2.2 mM for the cytosolic ODC.

Inhibitors and Activators. The two enzymes require pyridoxal phosphate and dithiothreitol for maximum enzymatic activity.

[25] Immunocytochemical Localization of Ornithine Decarboxylase[1]

By LO PERSSON, ELSA ROSENGREN, FRANK SUNDLER, and ROLF UDDMAN

Principle

The immunocytochemical localization of ornithine decarboxylase (L-ornithine carboxy-lyase, EC 4.1.1.17) requires a specific and potent antiserum against the enzyme. The steps involved in the purification of the enzyme to be used for immunization have been described previously.[2]

[1] See also this volume [26] for another procedure for the localization of ornithine decarboxylase in tissues. For other preparations of mammalian ornithine decarboxylase see this volume [22] and [23].
[2] L. Persson, *Acta Chem. Scand., Ser. B* **B35,** 451 (1981).

ISBN 0-12-181994-9

Enzyme antibody (rabbit) + Enzyme in tissue → Antigen- -antibody complex + Fluorescent anti-rabbit antibody → Final complex

IMMUNOFLUORESCENCE TECHNIQUE

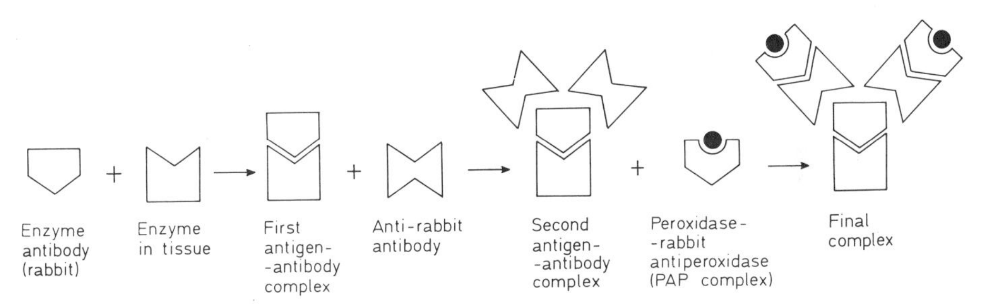

IMMUNOPEROXIDASE (PAP) TECHNIQUE

FIG. 1. Immunofluorescence and immunoperoxidase techniques.

Also the production and characterization of antibodies against ornithine decarboxylase have been described.[3] Once having obtained an ornithine decarboxylase antiserum, the enzyme can be localized cytochemically by the indirect immunofluorescence method of Coons *et al.*[4] (Method I) or by the peroxidase-antiperoxidase (PAP) procedure of Sternberger[5] (Method II) (see Fig. 1). In both techniques the first step consists of the binding of a specific antibody (usually raised in rabbit) to ornithine decarboxylase to the antigen present in the tissue section. In the indirect immunofluorescence method (Method I) the site of the antigen–antibody reaction is then demonstrated by the use of a fluorochrome-labeled antibody against rabbit immunoglobulins[6] (which is usually raised in sheep or goats). In the PAP technique (Method II) this second antibody is unlabeled; this antibody is used in excess and will therefore bind to the primary antibody with only one of its two binding sites. The tissue section is then incubated

[3] L. Persson, *Acta Chem. Scand., Ser. B* **B36,** 685 (1982).

[4] A. H. Coons, E. H. Leduc, and J. M. Connolly, *J. Exp. Med.* **102,** 49 (1955).

[5] L. A. Sternberger, "Immunocytochemistry." Prentice-Hall, Englewood Cliffs, New Jersey, 1974.

[6] Available commercially from Miles Laboratories, Inc.

with a soluble complex of peroxidase and antiperoxidase[6] (raised in the same species as the primary antibody) that will bind to the free binding site of the second antibody. The peroxidase activity is revealed by the method of Graham and Karnovsky.[7]

Both techniques have been employed for the immunocytochemical localization of ornithine decarboxylase in kidney of testosterone-treated mice,[8] ovary of gonadotropin-treated rats,[9] and rat placenta,[10] i.e., tissues known to contain high ornithine decarboxylase activity.

Reagents

Formaldehyde, 4% in 0.1 *M* phosphate buffer, pH 7.2, containing 5 m*M* dithiothreitol (perfusion solution)
Sucrose, 5% in 0.1 *M* phosphate buffer, pH 7.2
Phosphate-buffered saline (PBS): 0.9% NaCl in 0.01 *M* phosphate buffer, pH 7.5; with or without 0.25% Triton X-100
Rabbit antiserum against ornithine decarboxylase diluted in PBS containing 0.25% human serum albumin and 0.25% Triton X-100
Fluorescein-conjugated anti-rabbit IgG
Unconjugated anti-rabbit IgG
PAP complex; peroxidase coupled to rabbit antibodies raised against horseradish peroxidase
Tris · HCl buffer, 0.05 *M*, pH 7.6
3,3′-Diaminobenzidine tetrahydrochloride
Hydrogen peroxide

Procedure

The animals are deeply anesthetized and perfused with cold formaldehyde solution. Specimens from tissues to be investigated are removed and immersed overnight in the perfusion solution. After thorough rinsing in buffer containing 5% sucrose, the specimens are frozen on Dry Ice and sectioned on a cryostat at 10–15 μm.

In the immunofluorescence procedure, sections are incubated with ornithine decarboxylase antiserum in the appropriate dilution (depending on the antiserum and the tissue) for 3 hr at room temperature. After thorough rinsing in PBS containing 0.25% Triton X-100, the sections are

[7] R. C. Graham and M. J. Karnovsky, *J. Histochem. Cytochem.* **14,** 291 (1966).
[8] L. Persson, E. Rosengren, and F. Sundler, *Biochem. Biophys. Res. Commun.* **104,** 1196 (1982).
[9] L. Persson, E. Rosengren, and F. Sundler, *Histochemistry* **75,** 163 (1982).
[10] L. Persson, E. Rosengren, F. Sundler, and R. Uddman, *Adv. Polyamine Res.* (in press).

incubated with fluorescein-conjugated anti-rabbit IgG, diluted 1 : 20, for 30 min at room temperature. The sections are then rinsed in PBS containing 0.25% Triton X-100 and mounted in phosphate-buffered glycerin (PBS and glycerin, 1 : 1). The sections are examined in a fluorescence microscope equipped with filters selected to give peak excitation at 490 nm.

In the PAP procedure the sections are exposed to the ornithine decarboxylase antiserum, used in several times higher dilution than in the immunofluorescence method, for 18 hr at 4°. After rinsing in PBS containing 0.25% Triton X-100, the sections are incubated with unlabeled anti-rabbit IgG, diluted 1 : 30, for 30 min at room temperature. The sections are then rinsed and incubated with PAP complex in the appropriate dilution (usually 1 : 100 to 1 : 300) for 30 min at room temperature. After rinsing in 0.05 *M* Tris buffer, pH 7.6, the peroxidase activity is visualized by incubation with a solution of 0.06% 3,3′-diaminobenzidine tetrahydrochloride and 0.01% hydrogen peroxide in 0.05 *M* Tris buffer, pH 7.6, for 1 hr at room temperature. The sections are then rinsed in water, dehydrated in a series of ethanol solutions and xylene, and finally mounted in Permount (Fischer Scientific Co., Fair Lawn, New Jersey) and examined using an ordinary light microscope.

Several different controls have been devised for immunocytochemistry.[5] These controls include sections exposed to antiserum inactivated by the addition of the pure antigen. However, in this case it must be emphasized that such a control does not exclude the possibility of immunoreactivity due to the presence of small amounts of protein contaminants in the purified enzyme preparations. Hence, additional control experiments must be carried out to establish that the immunoreactivity reflects the presence of ornithine decarboxylase.[11]

[11] For further discussion, see J. Rossier, *Neuroscience* **6**, 989 (1981).

[26] Autoradiographic Localization of Ornithine Decarboxylase

By Ian S. Zagon, James E. Seely, and Anthony E. Pegg

Ornithine decarboxylase activity has been detected in extracts from many different mammalian cells, other animals, plants, lower eukaryotes, and prokaryotes.[1] In many cases the enzyme activity shows a remarkable

[1] A. E. Pegg and H. G. Williams-Ashman, *in* "Polyamines in Biology and Medicine" (D. R. Morris and L. J. Marton, eds.), p. 3. Dekker, New York, 1981.

ISBN 0-12-181994-9

fluctuation in response to a variety of trophic stimuli, and there has been much interest in the possibility that the induction of ornithine decarboxylase activity is required for growth.[2,3] Many of the mammalian tissues in which ornithine decarboxylase activity has been shown to be highly inducible are complex mixtures of different cell types. In rodents, the ornithine decarboxylase activities of different organs differ by as much as three orders of magnitude.[4,5] It is probable that the basal levels and the induced activities of different cell types within the same tissue would also vary greatly. Direct methods for localizing and quantitating the amount of enzyme protein present in tissue sections are, therefore, needed.

A second reason for development of methodology for the histological detection of ornithine decarboxylase is that there are no definitive studies of the subcellular localization of this enzyme. Most biochemical assays made with homogenates fractionated by differential centrifugation have indicated that the bulk of the enzyme is present in the cytosol.[1,2] However, nuclear ornithine decarboxylase activity has been reported,[1,3,6] and the possibility that the enzyme leaks out into the soluble fraction of the homogenate during processing cannot be totally excluded. The extraordinarily rapid turnover of ornithine decarboxylase protein[1-3] also raises the question of whether some of the enzyme or degradation fragments might be present in lysosomes. A specific method for localization capable of high resolution is needed to answer these questions. At present, two possible approaches have been shown to be useful: measurements using specific antibodies for the enzyme[7] and autoradiographic localization using radioactive α-difluoromethylornithine (DFMO).[4]

Principle

α-Difluoromethylornithine is an enzyme-activated irreversible inhibitor[8] that forms a covalent linkage to ornithine decarboxylase when incubated with the enzyme under conditions that allow the enzyme to decarboxylate this drug.[9] α-Difluoromethylornithine penetrates readily into

[2] J. Jänne, H. Pösö, and A. Raina, *Biochim. Biophys. Acta* **473,** 241 (1978).

[3] D. H. Russell, *Pharmacology* **20,** 117 (1980).

[4] A. E. Pegg, J. E. Seely, and I. S. Zagon, *Science* **217,** 68 (1982).

[5] J. E. Seely, H. Pösö, and A. E. Pegg, *Biochem. J.* **206,** 311 (1982).

[6] A. J. Bitonti and D. Couri, *Biochem. Biophys. Res. Commun.* **99,** 1040 (1981).

[7] L. Persson, E. Rosengren, and F. Sundler, *Biochem. Biophys. Res. Commun.* **104,** 1196 (1982).

[8] B. W. Metcalf, P. Bey, C. Danzin, M. J. Jung, P. Casara, and J. P. Vevert, *J. Am. Chem. Soc.* **100,** 2551 (1978).

[9] M. L. Pritchard, J. E. Seely, H. Pösö, L. S. Jefferson, and A. E. Pegg, *Biochem. Biophys. Res. Commun.* **100,** 1597 (1981).

cells when administered to animals and is not metabolized significantly in other pathways. Therefore, the only cellular macromolecule that becomes labeled when radioactive DFMO is administered is ornithine decarboxylase.[10] If animals are treated with the labeled DFMO and tissues are fixed and washed to remove unbound drug, the residual radioactivity represents that bound to ornithine decarboxylase and can be used to localize the enzyme. It is possible that the entry of DFMO into some cells occurs more extensively than into others; this could produce an artifactual result based on the fact that the enzyme is labeled more completely in cells with the higher concentration of DFMO. However, this possibility can be ruled out by using a dose of DFMO that inactivates virtually all the enzyme activity in the tissue sample under study. There is a stoichiometric attachment of DFMO to the enzyme which then becomes irreversibly inactivated and cannot bind further drug.[9] Therefore, all molecules of the enzyme that are inactivated become labeled; when all of the activity is lost, all of the enzyme is labeled.

Methodology

In order to illustrate the method, the localization of ornithine decarboxylase in mouse kidney by using [5-^{3}H]DFMO is described below. This procedure can be used with minor variations for other tissues in which *in vivo* labeling can be carried out.

In preliminary experiments with unlabeled DFMO, it was found that treatment with 1 mg/kg reduced renal ornithine decarboxylase by 90% within 30–60 min. Therefore, the mice were treated with 1 mg (1 mCi) of [5-^{3}H]DFMO per kilogram of body weight by intraperitoneal injection of a solution in 0.9% (w/v) NaCl and killed 30 min later. Radioactive DFMO suitable for this procedure is now commercially available from New England Nuclear (Boston, Massachusetts) at a specific activity of 15–22 Ci/mmol and can be diluted with unlabeled DFMO to the desired specific activity.

The mice were anesthetized with ether, and tissues were fixed by cardiac perfusion with 10% neutral buffered formalin at an air pressure of 120 mm Hg. Tissues were removed and further fixed and washed by immersion for at least 3 days with the fixative changed 2–3 times daily. After fixation, tissues were embedded in polyester wax and sectioned at 10 μm. In order to investigate the regional localization of the enzyme within the kidney, sections were exposed to LKB ultrafilm (^{3}H), (purchased from LKB Instruments, Inc., Gaithersburg, Maryland) in an X-ray

[10] J. E. Seely, H. Pösö, and A. E. Pegg, *J. Biol. Chem.* **257**, 7549 (1982).

cassette for 4 weeks at 4°. The films were developed for 5 min in D-19 (Kodak). The developed films were viewed to provide a qualitative assessment of the localization of the enzyme. A quantitative assessment could be obtained from densitometric readings using a Tobias densitometer (Tobias Associates, Inc., Model TCS, Ivyland, Pennsylvania). The developed film showed clearly that the enzyme was present primarily in the cortex, with much lower levels in the medulla (Fig. 1A). Additional evidence that this radioactivity does represent ornithine decarboxylase was obtained by treating the mice with cycloheximide (20 mg/kg) to block protein synthesis 6 hr before giving the labeled drug. Since ornithine decarboxylase turns over very rapidly, this reduced the enzyme activity by up to 95% and greatly decreased the radioactivity that became fixed in the tissue (Fig. 1B).

More detailed resolution was obtained by coating histological sections

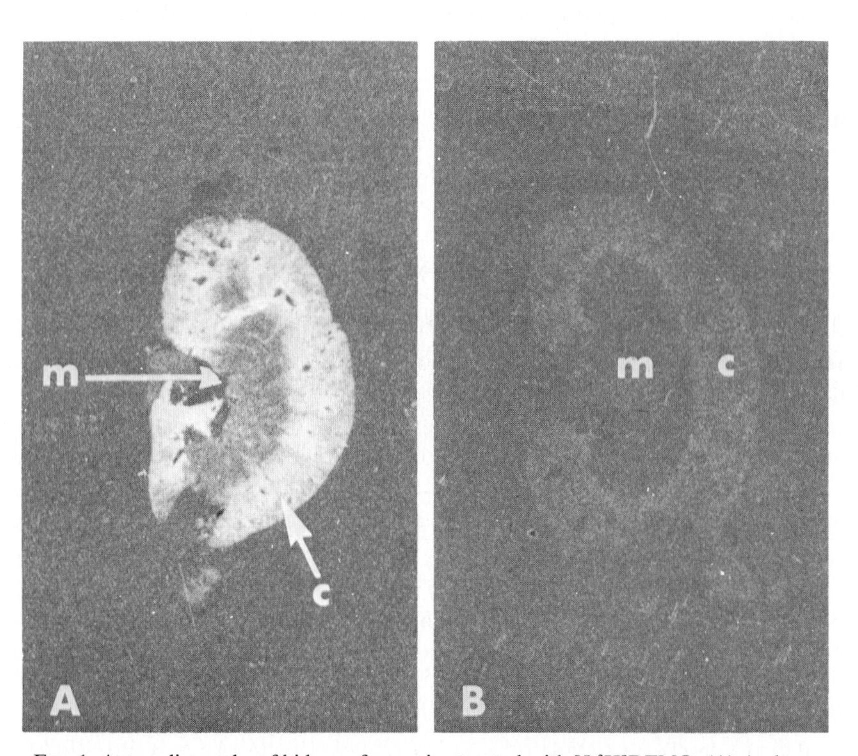

FIG. 1. Autoradiographs of kidneys from mice treated with [5-^{3}H]DFMO. (A) Androgen-treated mouse kidney with marked incorporation noted in the cortex (c), but not in the medulla (m). (B) Cycloheximide-treated mouse kidney demonstrating a noticeable reduction in DFMO activity (compare with A), but differences between the cortex (c) and medulla (m) are still apparent. ×4.5.

with Kodak nuclear track emulsion (NTB-2) (Rochester, New York). These sections were incubated for 6 weeks at 4°, developed in D-19 (Kodak) and stained with hematoxylin. In addition to qualitative observation of the distribution of silver grains, the results were quantitated by counting the number of grains over a given area using a light microscope with 630 × or 1000 × magnification. The kidney sections showed many grains corresponding to the proximal tubules (Fig. 2A) with lower amounts over glomeruli, Bowman's capsules, distal tubules, and collecting ducts. The distribution of grains in the sections from cycloheximide-treated rats (Fig. 2B) was similar, but markedly decreased. The majority of the ornithine decarboxylase in the proximal tubule cells was in the cytoplasm, but some grains clearly derive from the nucleus, indicating that some ornithine decarboxylase is present there as well.

This method can be used for other tissues exactly as described above and has also been carried out for the mouse kidney using [5-^{14}C]DFMO with exactly similar results.[4] The [5-^{14}C]DFMO (60 Ci/mol) was purchased from Amersham/Searle (Arlington Heights, Illinois) and used undiluted at a dose of 1 mg/kg. The 10-μm sections were exposed to Lo-Dose Mammography Film (DuPont, Wilmington, Delaware) in an X-ray cassette for 4–6 weeks at 4°, and films were developed in a Kodak X-OMAT processor. Histological sections were treated as were those exposed to [5-^{3}H]DFMO. The major limiting factor is the low amount of ornithine decarboxylase present in the tissues and the specific activity of the radioactive DFMO. The mouse kidney contains up to 4900 units (1 unit is 1 nmol of CO_2 released per 30 min) per gram wet weight of ornithine decarboxylase, and excellent autoradiographs can be obtained with labeled DFMO of 60 Ci/mol. Most mammalian tissues have enzyme levels much less than this. Based on our studies of other tissues and kidneys in which the enzyme activity is reduced by treatment with cycloheximide, an approximate calculation of the specific activity of DFMO needed can be made. This suggests that a specific activity of DFMO of 12 Ci/mmol would be needed for tissues having activities of 1 unit per gram. This can be achieved using tritiated DFMO, but it is not yet known whether the very high levels of radioactivity that would be needed can be given while maintaining the same specificity as for labeling only ornithine decarboxylase. Also, it is not feasible to maintain such a high specific activity and give intact animals doses of 1 mg/kg, which are needed for complete loss of activity. This problem may be avoided by labeling *in vitro*. For this procedure, fresh tissue slices (200 μm) were obtained using a Vibratome (Lancer Series 1000, St. Louis, Missouri) and collected in phosphate-buffered saline at 4° (Flexi-Cool, FTS Systems, Stone Ridge, New York). The sections were immediately incubated under 95% O_2:5%

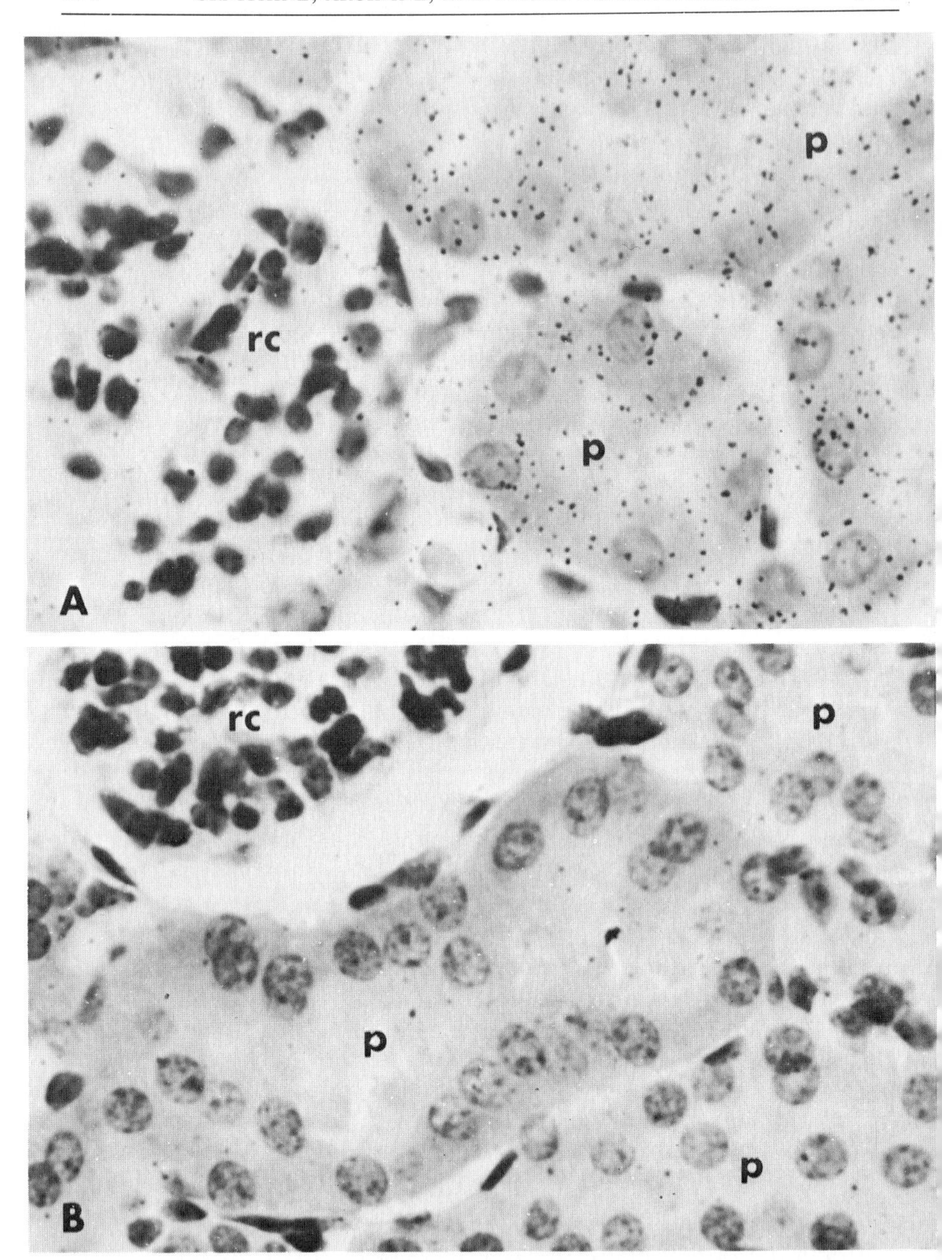

FIG. 2. Histological autoradiographs from androgen-treated kidneys of mice treated with [5-^{3}H]DFMO. (A) A section of cortex showing extensive labeling of the proximal convoluted tubules (p) and much less activity over the renal corpuscle (rc). (B) A similar section of cortex from a cycloheximide-treated mouse. ×850.

CO_2 with Krebs–Henseleit buffer containing 10 μM [5-^{3}H]DFMO for 30–60 min. These sections were then fixed in 10% formalin, washed, embedded in polyester wax, sectioned (10 μm), and analyzed as described above.

Advantages and Limitations

The major advantage of this method is that it is simple, does not require biochemical reagents such as monospecific antibodies, and provides extremely good resolution and accurate quantitation. The covalent attachment of DFMO to protein in the tissues provides a distinct advantage in permitting the use of conventional methodology for light microscopy, thereby ensuring optimal preservation of structure. Similarly, standard techniques for preparation for electron microscopic visualization and for autoradiographic analysis can also be undertaken with the promise of maximizing anatomical details. Results obtained at high magnification are quite adequate to determine the distribution of the enzyme within the subcellular organelles of a single cell. The procedure may also be valuable for following the degradation of ornithine decarboxylase, since it has been shown that the inactivated enzyme with DFMO attached is degraded *in vivo* at the same rate as the native enzyme.[10]

The greatest drawback of this procedure is that incorporation of label from DFMO into ornithine decarboxylase requires the exposure of the enzyme under conditions in which it is enzymatically active. Furthermore, only one molecule of DFMO can be bound per molecule of ornithine decarboxylase subunit. As discussed above, the enzyme is present in some cells in very small amounts (equivalent to only a few hundred molecules per cell), and, therefore, a high specific activity of DFMO is needed in order to incorporate sufficient radioactivity even when all of the enzyme is inactivated. Ornithine decarboxylase is a very labile enzyme, and it is unlikely that sufficient activity would be retained in materials stored for a long period of time. Therefore, only the exposure to the drug of freshly isolated tissue sections or *in vivo* administration of the DFMO are likely to be successful.

The present method cannot be used to quantitate "cryptic inactive" forms of ornithine decarboxylase, and there is evidence that such forms may exist under some circumstances.[11-13] Any inactive enzyme would not be detected at all by this approach, but would be amenable to study by

[11] E. S. Canellakis, D. Viceps-Madore, D. A. Kyriakidis, and J. S. Heller, *Curr. Top. Cell. Regul.* **155,** 155 (1979).

[12] D. H. Russell, *Med. Biol.* **59,** 286 (1981).

[13] A. K. Tyagi, C. W. Tabor, and H. Tabor, *J. Biol. Chem.* **256,** 12156 (1981).

immunological techniques.[7] The use of the two methods in parallel may, therefore, provide additional information, since the latter would indicate total enzyme protein and the former would focus on only the active fraction.

The present method bears a superficial resemblance to that proposed by Gilad and Gilad[14] in which a complex of DFMO with rhodamine or biotin was used to react with the enzyme. It is not clear how such complexes react with the enzyme, since ornithine decarboxylase does not recognize substrates with the amino groups blocked[15] and the reaction between the enzyme and these complexes apparently occurred in sections in which the tissue had been fixed with formaldehyde and glutaraldehyde. No information on the chemical structures of the complexes, the stoichiometry of their reaction with ornithine decarboxylase, or the specificity of the reaction are available. For these reasons, the use of radioactive DFMO of which the purity is readily determined, that is known to react in equimolar amounts,[9,16] and that labels only the enzyme as indicated by polyacrylamide gel electrophoresis of tissue extracts exposed to the drug,[10] is preferable.

[14] G. M. Gilad and V. H. Gilad, *J. Histochem. Cytochem.* **29,** 687 (1981).

[15] P. Bey, C. Danzin, V. Van Dorsselaer, P. S. Mamont, M. Jung, and C. Tardif, *J. Med. Chem.* **21,** 50 (1978).

[16] J. E. Seely, H. Pösö, and A. E. Pegg, *Biochemistry* **21,** 3394 (1982).

[27] Arginine Decarboxylase (Oat Seedlings)

By TERENCE A. SMITH

$$\underset{\text{Arginine}}{NH_2\underset{\underset{COOH}{|}}{C}HCH_2CH_2CH_2NH\overset{\overset{NH}{\|}}{C}NH_2} \rightarrow \underset{\text{Agmatine}}{NH_2CH_2CH_2CH_2CH_2NH\overset{\overset{NH}{\|}}{C}NH_2} + CO_2$$

Arginine decarboxylase (EC 4.1.1.19; L-arginine carboxy-lyase) and its product agmatine have been detected in many higher plants and bacteria, but very rarely in animals.[1] In oats, arginine decarboxylase is particularly active and the present account concerns the arginine decarboxylase from this source. Arginine decarboxylases have also been purified and characterized from *Lathyrus sativus* and *Oryza sativa* (rice).[2]

[1] T. A. Smith, *Phytochemistry* **18,** 1447 (1979).

[2] S. Ramakrishna and P. R. Adiga, *Eur. J. Biochem.* **59,** 377 (1975); M. M. Choudhuri and B. Ghosh, *Agric. Biol. Chem.* **46,** 739 (1982).

METHODS IN ENZYMOLOGY, VOL. 94

ISBN 0-12-181994-9

Assay[3]

Isotopic Method[1,4]

Reagents

L-Arginine 25 mM
L-[U-^{14}C]arginine, 1 μCi/ml
Tris · HCl buffer, 0.1 M, pH 7.5 at 30°
NaOH, 0.1 M
HCl, 1 M
$CaCl_2$, 0.2 M
Extract containing 50–250 pkat of arginine decarboxylase

Procedure. The assay is conducted in Conway units. The enzyme in 2.3 ml of buffer is placed in the outer compartment. The center well contains 0.2 ml of 0.1 M NaOH. Carrier and labeled arginine (0.1 ml of each) are added to the outer compartment, and the units are sealed and incubated at 30° for 1 hr. HCl (0.5 ml) is then added to the outer compartment, and the units are resealed. The CO_2 is allowed to distill at room temperature for 45 min. $CaCl_2$ (0.1 ml) is then added to the center well, and the suspension is transferred quantitatively to tissue-lined planchettes for counting. Blanks are used to estimate the volatile ^{14}C impurities distilling from the labeled arginine.[4]

In this assay [carboxy-labeled ^{14}C]arginine would be preferable, since $^{14}CO_2$ may be released, by the combined action of arginase and urease, from the guanidino group when using [U-^{14}C]arginine.

Spectrophotometric Method[4]

Principle. Agmatine, the product of arginine decarboxylase, is oxidized by diamine oxidase. The hydrogen peroxide formed during this oxidation is estimated by the peroxidase–guaiacol system as a chromogen with λ_{max} at 470 nm.

Reagents

Tris buffer, 0.1 M, pH 7.5, as measured at 30°
Pea seedling amine oxidase, 100 nkat/ml; prepared by method of Smith[4]
Peroxidase, Sigma type II (200 purpurogallin units/mg, 1 mg/ml)
L-Arginine, 25 mM
Guaiacol, 25 mM
Extract containing 50–250 pkat of arginine decarboxylase.

[3] Assays for arginine decarboxylase are described also in this volume [18].
[4] T. A. Smith, *Anal. Biochem.* **92,** 331 (1979).

Procedure. In an optically matched cylindrical spectrophotometric cell (1.2 × 12.5 cm) are placed Tris buffer (2.5 ml) amine oxidase (0.1 ml), peroxidase (0.1 ml), guaiacol (0.1 ml), and extract (0.1 ml). Greater dilutions of enzyme can be accommodated by reducing the volume of Tris buffer and increasing the volume of the extract to a total volume of 2.9 ml. The tube is preincubated in a water bath at 30° for 2 min and then placed in a spectrophotometer with a thermostated cuvette holder at 30°. On addition of arginine (0.1 ml) and mixing, the increase in absorbance at 470 nm is determined on a recorder (full scale, normally 0.2 absorbance unit). Experiments have shown that the use of fluorogenic peroxidase substrates[5] gives a considerably increased sensitivity. The method cannot be applied to the estimation of enzymes requiring thiol reagents (e.g., ornithine decarboxylase), since the —SH group is oxidized preferentially by the peroxidase.

Purification

Material. Oat seedlings *Avena sativa* cv. 'Black Supreme' are grown for 21 days in sand in polyethylene pots with diurnal illumination (16 hr/day, 10 klux, 24° in light, 19° in dark). The plants are watered daily with a nutrient medium containing 2 m*M* Na_2SO_4, 1.5 m*M* $MgSO_4$, 4 m*M* $CaCl_2$, 0.33 m*M* Na_2HPO_4, 4 m*M* $NaNO_3$, 4 m*M* $(NH_4)_2SO_4$, Fe EDTA, and micronutrients. These plants are grown without potassium, since activity is enhanced 2- to 5-fold when this element is deficient.[6] All subsequent operations are performed at 0–5° unless otherwise specified.

Step 1. Extraction. Leaves are macerated in 2 volumes of 50 m*M* Na_2HPO_4; the extract is filtered under pressure through nylon cloth and frozen for 18 hr (crude extract).

Step 2. Ammonium Sulfate Fractionation. On thawing, 200 g of $(NH_4)_2SO_4$ are dissolved per liter. The precipitate is discarded after centrifugation at 2000 *g* for 15 min. $(NH_4)_2SO_4$ is then added to increase the concentration to 500 g per liter. The precipitate obtained on centrifuging at 2000 *g* for 15 min is dispersed in Tris · HCl buffer (0.1 *M*, pH 6.7) and dialyzed against this buffer for 18 hr.

Step 3. Acetone Fractionation. Acetone at −15° (50 ml per 100 ml of enzyme solution) is then added with stirring at 0°, the extract is centrifuged at 2000 *g* for 10 min, and the precipitate is discarded. Further acetone (150 ml per 100 ml of enzyme solution) at −15° is added with

[5] K. Zaitsu and Y. Ohkura, *Anal. Biochem.* **109,** 109 (1980).

[6] T. A. Smith, *Phytochemistry* **2,** 241 (1963).

stirring at 0°, and the precipitate recovered by centrifugation at 5000 *g* for 5 min is dissolved in Tris · HCl buffer (0.1 *M*, pH 7). The extract is dialyzed against this buffer and centrifuged at 15,000 *g* for 5 min.

Step 4. Gel Filtration. This enzyme preparation (usually 10 ml) is applied to a column (3.2 × 70 cm, bed volume 570 ml) of BioGel A-1.5m equilibrated with 50 m*M* Tris-HCl buffer (0.1 *M*, pH 7) containing 0.1 *M* KCl. The column is eluted with this buffer, and 5-ml fractions are collected at 10-min intervals. The active fractions are combined and concentrated by ultrafiltration on a Diaflo PM-30 membrane. Two fractions (A, M_r 195,000; B, M_r 118,000) are obtained at this stage.

Step 5. DEAE-Cellulose Chromatography. Fraction A, which has a greater activity than fraction B, is applied to a 1.9 × 7 cm column of DEAE-cellulose (Whatman DE-23) equilibrated with Tris · HCl (50 m*M*, pH 6.7). The enzyme is eluted with a linear gradient of KCl (0–1 *M*), total volume 400 ml; fractions (5 ml) are collected at 10-min intervals. Active fractions are pooled and concentrated by ultrafiltration. At this stage polyacrylamide gel electrophoresis shows that activity is associated with a band that comprises 80–90% of the total protein.

Typical purifications obtained in three experiments are summarized in the table.

PURIFICATION OF ARGININE DECARBOXYLASE FROM POTASSIUM-DEFICIENT OAT SEEDLINGS

Step	Volume (ml)	Total protein (mg)	Total activity (nkat)	Specific activity (nkat/mg protein)	Purification factor	Yield (%)
Experiment 1						
Crude extract	2,050	10,500	697	0.066	1	100
$(NH_4)_2SO_4$ fractionation	170	1,750	765	0.43	6.7	106
Acetone fractionation	74	880	606	0.69	11.5	86
Experiment 2						
Acetone fractionation	10	60	170	2.8	1 (40)	100
Gel filtration[a]	83	4.9	161	32.5	12 (464)	95
Experiment 3						
Gel filtration[a]	53	0.8	26.5	33.3	1 (475)	100
DEAE-cellulose[a]	66	0.04	9.9	247	7.4 (3530)	37

[a] Results refer to fraction A. The elution volume of this fraction on gel filtration was 310 ml, and on DEAE cellulose 130 ml. Reprinted with permission from reference 1.

Properties

L-Arginine appears to be the only substrate for the enzyme. D-Arginine inhibits 60% at 1 m*M*, and L-canavanine at 1 m*M* inhibits the enzyme by 50%. Agmatine and related amines had relatively little direct effect on the enzyme. NSD 1055 (4-bromo-3-hydroxybenzyloxyamine dihydrogen phosphate) is a powerful inhibitor, suggesting that the enzyme is pyridoxal phosphate dependent. Pyridoxal phosphate stimulates activity of the purified enzyme by 34% at 0.8 m*M*. The pH optimum of the enzyme is 7–7.5, and the K_m is 3×10^{-5} *M*.

Intact oat leaves stored at −15° for 8 months lost 89% of their activity, but purified preparations obtained on gel filtration are stable for 3 months at 4°. Fractions obtained after DEAE-cellulose chromatography lose 50% of their activity on storage for 1 week at −15°.

[28] Lysine Decarboxylase[1] (*Escherichia coli* B)[2]

By ELIZABETH A. BOEKER and EDMOND H. FISCHER

L-Lysine → cadaverine + CO_2

Three highly basic amino acids—ornithine, arginine, and lysine—are decarboxylated by extracts of *Escherichia coli*.[3] Ornithine and arginine are each decarboxylated by two types of enzymes: a degradative, inducible enzyme[4–9] with an acidic pH optimum, and a biosynthetic, constitutive enzyme[10,11] with a neutral optimum. Lysine is known to be decarboxylated only by the first type of enzyme.[12,13] All five enzymes have been

[1] EC 4.1.1.18; L-Lysine carboxy-lyase.
[2] For Donna Sabo, in memory.
[3] E. F. Gale, *Adv. Enzymol.* **6,** 1 (1940).
[4] S. L. Blethen, E. A. Boeker, and E. E. Snell, *J. Biol. Chem.* **243,** 1671 (1968).
[5] E. A. Boeker and E. E. Snell, *J. Biol. Chem.* **243,** 1678 (1968).
[6] E. A. Boeker, E. H. Fischer, and E. E. Snell, *J. Biol. Chem.* **244,** 5239 (1969).
[7] E. A. Boeker, E. H. Fischer, and E. E. Snell, *J. Biol. Chem.* **246,** 6776 (1971).
[8] E. A. Boeker, *Biochemistry* **17,** 258 (1978).
[9] S. M. Nowak and E. A. Boeker, *Arch. Biochem. Biophys.* **207,** 110 (1981).
[10] D. M. Applebaum, J. C. Dunlap, and D. R. Morris, *Biochemistry* **16,** 1580 (1977).
[11] W. H. Wu and D. R. Morris, *J. Biol. Chem.* **248,** 1687 (1973).
[12] D. L. Sabo, E. A. Boeker, B. Byers, H. Waron, and E. H. Fischer, *Biochemistry* **13,** 662 (1974).
[13] D. L. Sabo and E. H. Fischer, *Biochemistry* **13,** 670 (1974).

METHODS IN ENZYMOLOGY, VOL. 94

ISBN 0-12-181994-9

purified to homogeneity.[4,10–12,14] The inducible enzymes for arginine and lysine form a particularly interesting pair: while their polypeptide chains clearly differ, their gross physical properties are very nearly identical. All the evidence to date suggests that these two enzymes derived from a common ancestor. The inducible arginine decarboxylase from *E. coli* B was described in Vol. 17B of this series. We describe here the enzyme that acts on lysine. All the information presented here results from the extensive characterization carried out by Donna Sabo.[12,13]

Assay

Lysine decarboxylase is readily assayed by trapping and counting the $^{14}CO_2$ produced from labeled lysine on filter paper.[15] The assay can be standardized internally.

Reagents

Sodium acetate buffer, 0.2 *M*, pH 5.7, containing 60 μM pyridoxal phosphate
L-Lysine · HCl, 30 m*M*, U-^{14}C (ca. 10^5 dpm/μmol) in the above buffer
Bovine serum albumin, 3 mg/ml, in the above buffer
2-Aminoethanol–2-methoxyethanol, 1 : 1
Trichloroacetic acid, 100% (w/v)

Procedure. The lysine solution (0.25 ml) is placed in the bottom of a stoppered 15 × 85 mm culture tube and incubated at 37° for 5 min. A 1 × 2.5 cm piece of Whatman No. 1 filter paper is fluted, placed crosswise in the upper part of the tube, and soaked with 20 μl of aminoethanol–methoxyethanol. The enzyme is diluted in the serum albumin solution, and 50 μl (containing enough decarboxylase to produce 0.01 to 0.2 μmol of CO_2 per minute) is added. After 15 min, the assay is stopped with 20 μl of trichloroacetic acid, and the tube is rapidly restoppered. After an additional 15 min, the filter paper is removed with tweezers, added to 10 ml of any good, nonaqueous scintillant solution, and counted.

The assay is linear until about 50% of the substrate has been consumed. It can be standardized internally by using a large excess of enzyme in order to release and trap all the CO_2 from the substrate. This gives an effective specific radioactivity for the lysine that takes into account the efficiency of both counting and CO_2 trapping. The overall assay efficiency is usually about 50%.

[14] D. M. Applebaum, D. L. Sabo, E. H. Fischer, and D. R. Morris, *Biochemistry* **14,** 3675 (1975).

[15] D. R. Morris and A. B. Pardee, *Biochem. Biophys. Res. Commun.* **20,** 697 (1965).

TABLE I
GROWTH MEDIA FOR *Escherichia coli* B

Component	Medium I	Medium II
$(NH_4)_2SO_4$	0.4%	0.4%
NaCl	0.1%	0.1%
K_2HPO_4	0.1%	0.1%
$MgSO_4$	0.1%	0.01%
Na_3 citrate	0.05%	0.05%
Glucose	0.1%	1.0%[a]
L-Lysine · HCl	0%	1.0%
H_3PO_4 to a final pH of:	6.8	5.5
Sterilize at 120° for:	30 min	60 min

[a] The glucose is sterilized separately.

Purification

Growth of E. coli B. An inoculum culture is grown from agar slants in three steps: two 50-ml portions for 24 hr and two 500-ml portions for 15 hr, in flasks, and one 10-l portion for 9 hr, in a carboy. The cultures are grown at 37°, in medium I (Table I), using vigorous aeration, either by shaking or through a sintered-glass sparger. The final culture is added to 90 liters of medium II (Table I) in a 100-liter fermentor and allowed to grow without aeration for 9 hr at 37°. The cells (100–150 g) are collected in a large Sharples centrifuge. These conditions produce very high levels of lysine decarboxylase. Induction approximately parallels growth. The specific activity of the disrupted cells (see below) begins to decrease early in stationary phase. A summary of the remaining purification steps is given in Table II.

Preparation of Crude Extract. Freshly harvested cells are suspended in four volumes of 0.1 *M* sodium succinate buffer, pH 6.2, containing

TABLE II
PURIFICATION OF LYSINE DECARBOXYLASE[a]

Step	Volume (ml)	Protein (g)	Specific activity (μmol/min/mg)	Yield (1%)
Crude extract	750	14.9	23	(100)
Heat treatment	650	2.8	110	86
$(NH_2)_2SO_4$ fractionation	44	1.6	330	150
Acid precipitation	22	0.4	1000	130

[a] From 190 g of *Escherichia coli* B.

10 mM 2-mercaptoethanol, 1 mM EDTA, and 0.1 mM pyridoxal phosphate and disrupted in 200-ml portions by sonic oscillation for 20 min at 0°. The suspension is centrifuged for 30 min at 16,000 g.

Heat Treatment. The cloudy supernatant solution is divided into 200-ml aliquots, heated to 70° by swirling in an 80° water bath, maintained at this temperature for 5 min, chilled, and centrifuged for 10 min at 16,000 g.

$(NH_4)_2SO_4$ Fractionation. Enough of a saturated solution of ammonium sulfate in 0.1 M sodium succinate, pH 6.2, is added to the clear, chilled yellow supernatant solution to make it 45% in $(NH_4)_2SO_4$. After 30 min at 0°C, the solution is centrifuged at 16,000 g for 20 min. The supernatant solution is then brought to 55% saturation, allowed to stand for several hours, and centrifuged. The yellow precipitate is dissolved in a small volume of 0.1 M sodium succinate, pH 6.2, and centrifuged.

Acid Precipitation. The supernatant solution is dialyzed for 16 hr at 4°C against 100 volumes of 50 mM sodium succinate, pH 5.2. The precipitate is then collected by centrifugation at 30,000 g for 15 min, washed twice with several volumes of dialysis buffer, and dissolved in a small volume of 0.1 M sodium succinate, pH 6.0, containing 1 mM EDTA, by warming briefly. It can be stored at −10° for long periods by making the buffer 20% in glycerol.

Properties

This preparation of lysine decarboxylase appears to be essentially homogeneous when subjected to acrylamide gel electrophoresis, either with or without sodium dodecyl sulfate, and when subjected to isoelectric focusing. Arginine decarboxylase is partially induced during the growth of *E. coli* and has very similar physical properties, but it is denatured in the heat step and remains in solution during acid precipitation. Antibodies prepared against the final lysine decarboxylase preparation do not cross-react, in Ouchterlony double-diffusion experiments, with the inducible arginine, ornithine, or glutamate decarboxylases. Lysine decarboxylase also has less than 0.1% activity against the substrates of these enzymes.

In addition to lysine (K_m = 1.5 mM), the enzyme will also decarboxylate S-aminoethyl-L-cysteine (K_m = 3.4 mM) and δ-hydroxylysine (K_m = 7.0 mM) at rates of 15 and 25%, respectively, of that for lysine. The optimum pH is 5.7. The absorption spectrum has maxima at 278 and 422 nm and is independent of pH from 9.0 to 6.0, which is as low as it is possible to go before the enzyme starts to precipitate. $A^{1\%}_{280\ nm}$ is 13.3. The partial specific volume, calculated from the amino acid composition, is 0.739 ml/g.

The lysine decarboxylase polypeptide chain has a molecular weight of

80,000, as determined both by equilibrium ultracentrifugation in 6 *M* guanidine · HCl and by gel electrophoresis on sodium dodecyl sulfate. Since this produces a single band, and since the N-terminal sequence and an internal sequence are unique for 7 and 15 residues, respectively, it seems clear that there is but one type of polypeptide. This cannot be separated from that of arginine decarboxylase by sodium dodecyl sulfate gel electrophoresis.

In a potassium phosphate buffer with a pH of 8.0 and an ionic strength of 0.2, lysine decarboxylase sediments as a homogeneous dimer ($s^0_{20,w}$ = 7.8 S, M_r = 154,000–156,000). At pH 7.0 and μ 0.1, the enzyme sediments as a homogeneous decamer ($s^0_{20,w}$ = 22.2 S, M_r = 730,000–800,000). Further decreasing the pH to 6.0 and increasing μ to 0.5 produces a heterogeneous aggregate with a very high average molecular weight. All three forms can be visualized in the electron microscope. The dimer is somewhat elongated and has a slight waist. The decamer is a pentamer of dimers. The aggregates are long stacks of decamers in which alternate spacings are clearly more distinct.[12]

Based on gel filtration experiments under pseudo-assay conditions, and by analogy to arginine decarboxylase, it is nearly certain that the dimer is inactive and the decamer active. There is no information about the aggregate. A striking feature of the purification procedure is the activation that occurs during the $(NH_4)_2SO_4$ precipitation. Although it has not been demonstrated directly, it seems likely that this is due to association of the dimer to the decamer concomitant with the increase in ionic strength.

As suggested by the absorption spectrum and the reaction itself, lysine decarboxylase is a pyridoxal phosphate-dependent enzyme. The coenzyme can be removed by dialysis against 0.1 *M* cysteine and 0.05 *M* mercaptoethanol is K_2HPO_4, pH 6.0, μ 0.1. Of the activity, 95% is restored when 1.0–1.2 mol of pyridoxal phosphate are added per 80,000 g. The apoenzyme shows the same self-association behavior as the holoenzyme, but is somewhat less heat stable. $NaBH_4$ reduction inactivates the holoenzyme; peptides isolated from the reduced enzyme have been used to determine a 15-residue sequence around the pyridoxal phosphate binding site with the following structure[13]

Val-Ile-Tyr-Glu-Thr-Glu-Ser-Thr-His-
(ε-Pyridoxyl)Lys-Leu-Leu-Ala-Ala-Phe

[29] Antizyme and Antizyme Inhibitor of Ornithine Decarboxylase (Rat Liver)

By SHIN-ICHI HAYASHI and KAZUNOBU FUJITA

Ornithine decarboxylase antizyme[1–5] is a protein inhibitor that specifically inactivates ornithine decarboxylase by forming a complex with it. The antizyme is induced in rat liver and in various cultured cells by administration of some natural polyamines or their analogs. We have discovered another protein factor in rat liver, antizyme inhibitor,[6,7] that specifically inactivates antizyme, apparently by forming a complex with it. Antizyme inhibitor reactivates ornithine decarboxylase in the inactive enzyme–antizyme complex and therefore can be used to assay the amount of the inactive complex.

Since both antizyme and antizyme inhibitor act in a time-independent manner, their activities are assayed by measuring inhibition of ornithine decarboxylase activity and reversal of the inhibition, respectively, without any preincubation.

Antizyme

Assay Method

Principle. Antizyme activity is determined by measuring inhibition of ornithine decarboxylase activity.

Reagents

Tris-HCl buffer, 0.1 M, pH 7.4
Pyridoxal phosphate, 0.5 mM
Dithiothreitol, 25 mM; freshly prepared

[1] W. F. Fong, J. S. Heller, and E. S. Canellakis, *Biochim. Biophys. Acta* **428,** 456 (1976).
[2] J. S. Heller, W. F. Fong, and E. S. Canellakis, *Proc. Natl. Acad. Sci. U.S.A.* **73,** 1858 (1976).
[3] J. S. Heller, K. Y. Chen, D. A. Kyriakidis, W. F. Fong, and E. S. Canellakis, *J. Cell. Physiol.* **96,** 225 (1978).
[4] J. S. Heller and E. S. Canellakis, *J. Cell. Physiol.* **107,** 209 (1981).
[5] P. P. McCann, C. Tardif, J. Hornsperger, and P. Böhlen, *J. Cell. Physiol.* **99,** 183 (1979).
[6] K. Fujita, Y. Murakami, and S. Hayashi, *Biochem. J.* **204,** 647 (1982).
[7] K. Fujita, Y. Murakami, T. Kameji, S. Matsufuji, K. Utsunomiya, R. Kanamoto, and S. Hayashi, *Adv. Polyamine Res.* **4** (in press).

ISBN 0-12-181994-9

Bovine serum albumin, 5 mg/ml

Substrate solution: L-ornithine hydrochloride (10 mM) containing either L-[1-^{14}C]ornithine (1.25 μCi/ml) or DL-[1-^{14}C]ornithine (2.5 μCi/ml)

Ornithine decarboxylase, dialyzed against 25 mM Tris-HCl (pH 7.5) containing 1 mM dithiothreitol and 0.1 mM EDTA (buffer A), about 1000 units/ml

Antizyme, dialyzed against buffer A

KOH, 10%

HCl, 6 N

Scintillation fluid: 4 g of 2,5-diphenyloxazole (PPO) and 0.1 g of 2,2′-*p*-phenylenebis(5-phenyloxazole) (POPOP) in 700 ml of toluene and 300 ml of ethyl alcohol

Preparation of Ornithine Decarboxylase. The enzyme is partially purified from rat liver by DEAE-cellulose chromatography (step 2) and pyridoxamine phosphate–succinyldiaminodipropylamine–Sepharose affinity chromatography (step 3) as described elsewhere in this volume,[8] except that the final preparation is dialyzed against buffer A.

Procedure. The standard assay mixture contains 1 ml of Tris-HCl buffer, 0.1 ml of pyridoxal phosphate, 0.1 ml of dithiothreitol, 0.1 ml of bovine serum albumin, 0.1 ml of substrate solution, 10–30 units of ornithine decarboxylase, and antizyme in a final volume of 2.5 ml (a). Antizyme is omitted from the control assay mixture (b). Ornithine decarboxylase activity is measured as described in this volume.[8] Antizyme activity is calculated as $b - a$.

Definition of Unit and Specific Activity. One unit of antizyme activity is defined as the amount inhibiting one unit of ornithine decarboxylase activity, which in turn is defined as the amount forming 1 nmol of CO_2 per hour. Specific activity is expressed as units per milligram of protein.

Comments on Assay. The amount of antizyme added to the assay mixture should not give more than 75% inhibition. Inhibition beyond this range is nonlinear with the amount of antizyme added. Attention should be paid to the conditions of the antizyme assay, since both ornithine decarboxylase and antizyme activities may be affected by various factors. Both preparations should be dialyzed before assay to remove possible low-molecular-weight inhibitors of the enzyme, such as sodium and potassium chloride (>0.1 M),[9] ornithine, and arginine, which may be converted to ornithine by arginase. Both preparations should be free from proteases that easily inactivate ornithine decarboxylase and also from

[8] S. Hayashi and T. Kameji, this volume [22].

[9] M. F. Obenrader and W. F. Prouty, *J. Biol. Chem.* **252,** 2860 (1977).

phosphatases that hydrolyze pyridoxal phosphate. Furthermore, the ornithine decarboxylase preparation should be free from both inactivated enzyme and antizyme inhibitor. It is recommended that several criteria of antizyme activity be examined, such as its stoichiometric and time-independent action, heat lability, and sensitivity to antizyme inhibitor (see Figs. 1 and 3).

Purification Procedure[6]

Step 1. Induction of Antizyme and Preparation of Liver Extract. Four Sprague–Dawley rats weighing about 300 g are given two intraperitoneal injections of 1 mmol of 1,3-diaminopropane per kilogram with a 2-hr interval between injections. Food should be removed from animal cages before injections to prevent dietary induction of ornithine decarboxylase.[10] Animals are killed by decapitation between 1 and 3 hr after the second injection. Livers are quickly excised and chilled on ice. All subsequent procedures are performed at 0–4°. Livers are homogenized with 2 volumes of buffer A containing 0.2 *M* sucrose in a Dounce-type all-glass homogenizer. The homogenate is centrifuged at 100,000 g_{max} for 60 min.

Step 2. DEAE-Cellulose Column Chromatography. The crude extract is directly applied to a column of DEAE-cellulose (2 × 25 cm) that has been equilibrated with buffer A. The column is washed with the same buffer, and then antizyme is eluted with a linear gradient of NaCl in buffer A (0.05 to 0.6 *M*, 400 ml). Active fractions are pooled and brought to 70% saturation with solid ammonium sulfate. The precipitate is collected by centrifugation at 10,000 g_{max} for 15 min, dissolved in a small volume of 50 m*M* Tris-HCl (pH 7.5) containing 50 m*M* NaCl, 1 m*M* dithiothreitol, and 0.1 m*M* EDTA (buffer B), and dialyzed against the same buffer.

Step 3. Sephadex G-75 Gel Filtration. The antizyme solution is applied to a column of Sephadex G-75 (2.2 × 70 cm) that has been equilibrated with buffer B and eluted with the same buffer. Active fractions are pooled and dialyzed against buffer A.

A summary of the partial purification is shown in the table. More extensive purification has been reported by Canellakis *et al.*[11]

Properties

Molecular Weight. The apparent molecular weight of antizyme from rat liver is 24,500[6] or 25,500,[2] as estimated by gel-filtration analysis.

[10] T. Noguchi, Y. Aramaki, T. Kameji, and S. Hayashi, *J. Biochem.* (*Tokyo*) **85,** 953 (1979).

[11] E. S. Canellakis, D. Viceps-Madore, D. A. Kyriakidis, and J. S. Heller, *Curr. Top. Cell. Regul.* **15,** 155 (1979). For the preparation of antizymes from *Escherichia coli,* see this volume [30].

PARTIAL PURIFICATION OF ANTIZYME FROM RAT LIVER[a]

Step	Total protein (mg)	Total activity (units)	Specific activity (units/mg)	Purification (fold)	Yield (%)
1. Crude extract	4610	3800	0.82	1	100
2. DEAE-cellulose	390	3450	8.8	11	90
3. Sephadex G-75	34	2050	60	72	54

[a] From Fujita *et al.*[7]

Stability. Antizyme is stable for at least 2 days at 0° and can be stored for several months at −80° without any loss of activity. The activity gradually decreases, however, on repeated freezing and thawing. Antizyme is completely inactivated by heat treatment at 59° for 30 min or by treatment with trypsin or chymotrypsin,[2] but not with RNase.[2]

Mode of Action. Antizyme inhibits ornithine decarboxylase in a time-independent manner. Thus, in the presence of antizyme, the enzyme reaction proceeds linearly at a reduced rate (Fig. 1). This excludes the possibility that antizyme is an enzyme, such as protease, inactivating ornithine decarboxylase. Several lines of evidence indicate that antizyme inhibits ornithine decarboxylase by forming a complex with it. First, its inhibitory action is stoichiometric (Fig. 3). Second, gel-filtration analysis shows that equivalent amounts of antizyme and ornithine decarboxylase disappear upon mixing the two and reappear on treatment with a high concentration of salts.[2] Third, the inactive complex of ornithine decarboxylase and antizyme, determined with the aid of antizyme inhibitor, is eluted slightly before free ornithine decarboxylase on gel-filtration analysis.[6]

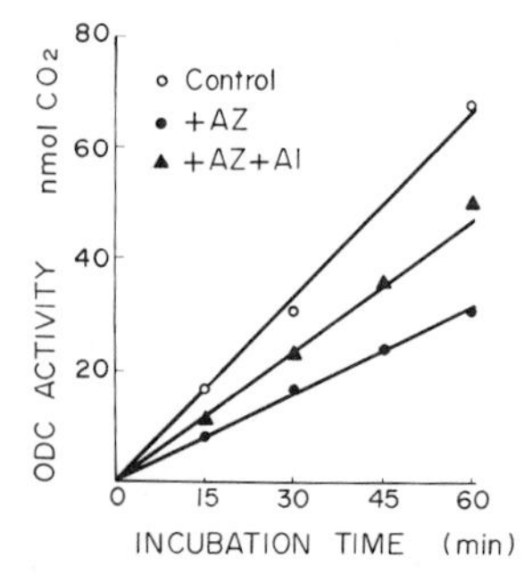

FIG. 1. Time course of the ornithine decarboxylase (ODC) reaction in the presence of antizyme (AZ) and antizyme plus antizyme inhibitor (AI). The enzyme reaction was allowed to proceed at 37° without any preincubation. From Fujita *et al.*[6]

Occurrence. Antizyme activity is not detected in cell extracts containing active ornithine decarboxylase. It is induced by administration of a relatively high concentration of various diamines or polyamines, and appears only after complete disappearance of ornithine decarboxylase activity. Such induction of antizyme activity has been reported in various tissues and cells of higher animals, including rat liver,[2,6] chicken liver,[12] rat thyroid,[13] H-35 rat hepatoma cells,[1] HTC cells,[5] primary cultures of rat hepatocytes,[7] L1210 mouse leukemia cells,[2] and mouse neuroblastoma cells.[2] It also occurs in *Escherichia coli*[11] and germinating barley seeds.[14]

Antizyme is located in the cytosol fraction, and appears to turn over rapidly with a half-life comparable with that of ornithine decarboxylase.[2]

Antizyme Inhibitor

Assay Method

Principle. Antizyme inhibitor is assayed by measuring reversal of inhibition of ornithine decarboxylase caused by antizyme.

Reagents

Tris-HCl buffer, 0.1 *M*, pH 7.4
Pyridoxal phosphate, 0.5 m*M*
Dithiothreitol, 25 m*M*; freshly prepared
Bovine serum albumin, 5 mg/ml
Substrate solution: L-ornithine hydrochloride (10 m*M*) containing either L-[1-^{14}C]ornithine (1.25 μCi/ml) or DL-[1-^{14}C]ornithine (2.5 μCi/ml)
Ornithine decarboxylase (step 3),[8] dialyzed against buffer A, about 1000 units/ml
Antizyme (step 2), dialyzed against buffer A, about 500 units/ml
Antizyme inhibitor, dialyzed against buffer A
KOH, 10%
HCl, 6 *N*
Scintillation fluid: 4 g of 2,5-diphenyloxazole (PPO) and 0.1 g of 2,2′-*p*-phenylenebis(5-phenyloxazole) (POPOP) in 700 ml of toluene and 300 ml of ethyl alcohol

[12] M. A. Grillo, S. Bedino, and G. Testore, *Int. J. Biochem.* **11,** 37 (1980).
[13] Y. Friedman, S. Park, S. Levasseur, and G. Burke, *Biochim. Biophys. Acta* **500,** 291 (1977).
[14] D. A. Kyriakidis, *Adv. Polyamine Res.* **4** (in press).

Procedure. The standard assay mixture contains 1 ml of Tris-HCl buffer, 0.1 ml of pyridoxal phosphate, 0.1 ml of dithiothreitol, 0.1 ml of bovine serum albumin, 0.1 ml of substrate solution, 15–30 units of ornithine decarboxylase, 10–20 units of antizyme, and antizyme inhibitor in a final volume of 2.5 ml (*a*). Three control assay mixtures are prepared: one without antizyme inhibitor (*b*), one without both antizyme and antizyme inhibitor (*c*), and one without both ornithine decarboxylase and antizyme (*d*). The third control may be omitted if the antizyme inhibitor preparation is known not to contain any ornithine decarboxylase activity. Ornithine decarboxylase activity is measured as described in this volume.[8] Antizyme inhibitor activity is calculated as $a - b - d$.

Definition of Unit and Specific Activity. One unit of antizyme inhibitor activity is defined as the amount reversing the inhibition of 1 unit of ornithine decarboxylase activity by antizyme. Specific activity is expressed as units per milligram of protein.

Comments. The amount of antizyme added to the assay mixture should not give more than 75% inhibition, as described before. Moreover, the amount of antizyme inhibitor added should not give more than 80% reversal of the inhibition by antizyme. Beyond this range reversal is nonlinear with the amount of antizyme inhibitor added. Assay of antizyme inhibitor requires similar precautions to those described for antizyme assay.

It is impossible at present to assay the activity of antizyme inhibitor accurately in crude liver extracts containing high activity of ornithine decarboxylase.

Preparation Procedure

Step 1. Treatment of Rats and Preparation of Liver Extract. Twenty Sprague–Dawley rats weighing about 210 g are injected intraperitoneally with 150 mg of thioacetamide per kilogram 20 hr before sacrifice and 75 mg of DL-α-hydrazine-δ-aminovaleric acid per kilogram 6 hr before sacrifice. The treatment is the same as that used to induce hepatic ornithine decarboxylase.[8] Animals are killed by decapitation, and livers are quickly excised and chilled on ice. All subsequent procedures are performed at 0–4°. Livers are homogenized with 2 volumes of buffer A containing 0.2 *M* sucrose in a Dounce-type all-glass homogenizer. The homogenate is centrifuged at 100,000 g_{max} for 60 min.

Step 2. DEAE-Cellulose Column Chromatography. The crude extract is directly applied to a column of DEAE-cellulose (2.2 × 45 cm) that has been equilibrated with buffer A, and the column is washed with the same buffer. Antizyme inhibitor is eluted with a linear gradient of NaCl in buffer A (0 to 0.5 *M*, 760 ml). Active fractions are pooled and brought to 70% saturation with solid ammonium sulfate. The precipitate is collected by centrifugation at 10,000 g_{max} for 15 min, dissolved in buffer A, and dialyzed against the same buffer.

Comments on Preparation. As shown in Fig. 2, antizyme inhibitor is eluted from a DEAE-cellulose column as a single peak (0.30 to 0.37 *M* NaCl) that is separated almost completely from that of ornithine decarboxylase (0.2 to 0.3 *M* NaCl).

Because antizyme inhibitor activity cannot be assayed accurately in crude extracts, quantitative values for purification and recovery rates are not shown.

Properties[6,7]

Molecular Weight. The apparent molecular weight of antizyme inhibitor is about the same as that of ornithine decarboxylase, i.e., 105,000 as estimated by gel-filtration analysis.

Stability. Although antizyme inhibitor is stable for at least 1 day at 0°, it is gradually inactivated on storage at −18° and rapidly inactivated on repeated freezing and thawing.

General Properties. Antizyme inhibitor is inactivated completely by heat treatment at 70° for 15 min and also by treatment with insolubilized trypsin. Antizyme inhibitor activity is not affected at all by treatment with rabbit antiserum raised against nearly homogeneous ornithine decarboxylase.

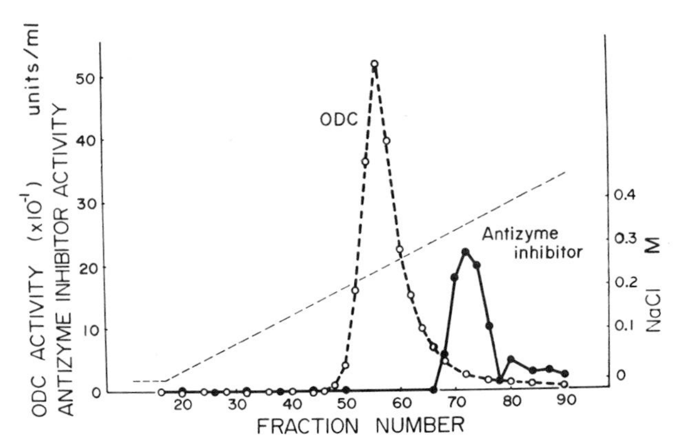

FIG. 2. Separation of antizyme inhibitor from ornithine decarboxylase (ODC) by DEAE-cellulose column chromatography. The light dashed line indicates NaCl concentration. From Fujita *et al.*[6]

Mode of Action. Antizyme inhibitor inhibits antizyme, apparently by forming a complex with it. It also reactivates ornithine decarboxylase in the inactive enzyme–antizyme complex, apparently by releasing free enzyme. These actions are supported by the following observations. Antizyme inhibitor acts in a time-independent and stoichiometric manner (Figs. 1 and 3). Gel-filtration analysis shows that equivalent amounts of antizyme and antizyme inhibitor are consumed when the two are mixed. Antizyme inhibitor exhibits the same activity whether it is added to antizyme before or after ornithine decarboxylase. Furthermore, when a mixture of ornithine decarboxylase and antizyme inhibitor is titrated with antizyme, the enzyme is inhibited by antizyme only after all the antizyme inhibitor activity present has been neutralized (Fig. 3). These results indicate that antizyme inhibitor binds to antizyme with much higher affinity than does ornithine decarboxylase.

Occurrence. At present, antizyme inhibitor has been found in liver extracts from rats in which hepatic ornithine decarboxylase had been induced either by thioacetamide and DL-α-hydrazino-δ-aminovaleric acid or by feeding a high-protein diet. Its activity is not detected in liver extracts from starved rats, containing no ornithine decarboxylase activity. Antizyme inhibitor does not coexist with antizyme. Antizyme inhibitor is located in the cytosol fraction.

Use of Antizyme Inhibitor To Assay the Inactive Complex of Ornithine Decarboxylase and Antizyme. In studies on the physiological role of antizyme, it is essential to have a simple and reliable method for assay of the

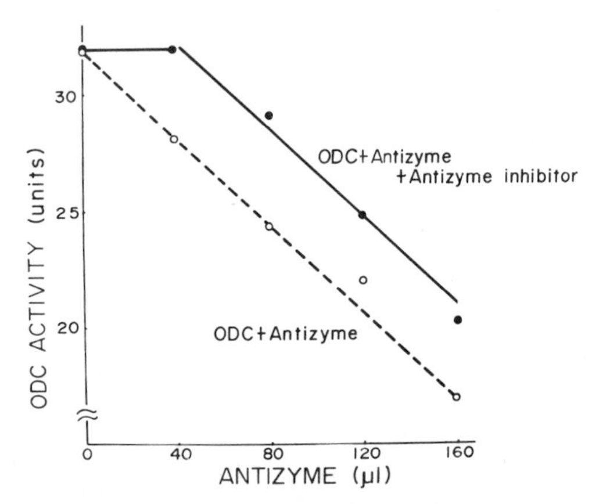

FIG. 3. Titration of ornithine decarboxylase (ODC) with antizyme in the presence and in the absence of a fixed amount of antizyme inhibitor. Increasing amounts of antizyme were added to a series of flasks containing a fixed amount of ornithine decarboxylase with or without a fixed amount of antizyme inhibitor. All mixtures were adjusted to the same volume with buffer A and assayed for enzyme activity. From Fujita *et al.*[6]

amount of inactive ornithine decarboxylase–antizyme complex. Antizyme inhibitor can be used for this purpose. Thus, the amount of the inactive complex is determined as the amount of ornithine decarboxylase activity that can be fully reactivated by antizyme inhibitor. Usually 1.4 equivalents of antizyme inhibitor suffice for complete reactivation when there is no free antizyme activity.

[30] Purification of Ornithine Decarboxylase Antizymes[1] (*Escherichia coli*)

By DIMITRI A. KYRIAKIDIS, JOHN S. HELLER, and EVANGELOS S. CANELLAKIS

The addition of putrescine, the end product of the ornithine decarboxylase (ODC; EC 4.1.1.17) reaction, or of polyamines, to cultures of mammalian cells as well as their administration to rats, stimulates the appearance of a protein inhibitor of eukaryotic ODC.[1a,2] The name ODC-antizyme was proposed for such protein inhibitors.[2] A previous report from this laboratory[3] has shown the presence of ODC antizyme in the *Escherichia coli* ODC mutant MA255 (speB, speC),[4] with properties similar to those of the mammalian ODC antizyme and whose level of activity is increased by the presence of putrescine.[3] This article describes the purification to homogeneity of an acidic antizyme protein and of two basic antizyme proteins isolated from the MA255 strain of *E. coli*. The present method has been developed as a rapid way of purifying these macromolecular inhibitors of ornithine decarboxylase from *E. coli*, as opposed to the original method using milder conventional procedures.[3]

Assay Method

Principle. Whereas the ornithine decarboxylase assay measures the $^{14}CO_2$ produced in the conversion of DL-[1-^{14}C]ornithine to putrescine, the ODC antizyme assay measures the inhibition of a partially purified prepa-

[1] For the preparation of an antizyme from rat liver, see this volume [29].

[1a] W. F. Fong, J. S. Heller, and E. S. Canellakis, *Biochim. Biophys. Acta* **428,** 456 (1976).

[2] J. S. Heller, W. F. Fong, and E. S. Canellakis, *Proc. Natl. Acad. Sci. U.S.A.* **73,** 1858 (1976).

[3] D. A. Kyriakidis, J. S. Heller, and E. S. Canellakis, *Proc. Natl. Acad. Sci. U.S.A.* **75,** 4699 (1978). See also this volume [24].

[4] S. Cunningham-Rundles and W. K. Maas, *J. Bacteriol.* **124,** 791 (1975).

ISBN 0-12-181994-9

ration of ornithine decarboxylase activity by *E. coli* extracts having antizyme activity.

Reagents

Tris · HCl buffer, 50 m*M*, pH 8.3, at 37°, 0.1 m*M* EDTA, 0.05 m*M* pyridoxal phosphate, and 2.5 m*M* dithiothreitol (freshly prepared)
DL-[1-^{14}C]Ornithine, 6.6 m*M*, 7.7 mCi/mmol
Ornithine decarboxylase, 1–2 units (specific activity, 6700 units/mg). The enzyme is partially purified from *E. coli* AB1203 as previously described.[3]
NCS (Amersham) diluted 1 : 1 with toluene

Procedure. The reaction mixtures are prepared in 11 × 75 mm test tubes, immersed in ice, in a final volume of 0.07 ml containing 0.05 ml of Tris-HCl buffer and 0.005 ml of ornithine decarboxylase. Varying amounts of sample ranging from 0.001 ml to 0.01 ml are added, followed by 0.005 ml of DL-[1-^{14}C]ornithine. The test tube is capped with a stopper penetrated by a 1½-inch, 21 gauge needle that pierces the center of a ¼-inch paper disk impregnated with 20 μl of the NCS solution.[5] The reaction is incubated for 30 min at 37°, stopped by injection of 0.1 ml of 10% trichloroacetic acid through the needle, and incubated further for 1 hr. The paper disks are placed in Betafluor and counted in a Beckman LS 7000 liquid scintillation counter. These experiments are performed at concentrations of ornithine below the K_m of ornithine decarboxylase. Because of the high K_m of the *E. coli* ornithine decarboxylase (K_m = 5.6 m*M*),[6] it is economically not feasible to work at saturating substrate levels.

Definition of Unit and Specific Activity. One unit of ornithine decarboxylase activity is defined as the amount of enzyme that catalyzes the release of 1 nmol of $^{14}CO_2$ during 60 min of incubation. Values for the 30-min assays are normalized to 1 hr. One unit of antizyme activity is defined as the amount of antizyme that will inhibit 1 unit of ornithine decarboxylase activity.

Purification Procedures

Step 1. Growth and Lysis of Cells. Antizyme is prepared from *E. coli* MA255 grown in three 8-liter carboys in Luria medium[7] containing 0.2% dextrose, 0.5 m*M* putrescine, and 0.5 m*M* spermidine after reducing the NaCl concentration to 0.5 g/liter. The cells are harvested in late log phase with a refrigerated Sharples centrifuge, providing a total yield of approxi-

[5] J. S. Heller and E. S. Canellakis, *J. Cell. Physiol.* **107,** 209 (1981).
[6] D. M. Applebaum, J. C. Dunlap, and D. R. Morris, *Biochemistry* **16,** 1580 (1977).
[7] S. E. Luria and J. W. Burrous, *J. Bacteriol.* **74,** 461 (1957).

mately 100 g of cells. The cells are suspended in 400 ml of 50 mM Tris · HCl (pH 8.2 at 25°) containing 200 μg of lysozyme per milliliter. The suspension is incubated for 1 hr at room temperature, cooled to 4°, and sonicated with a Bronson Sonifier cell disruptor 200 (output level 7 for periods of 2–5 min) while being kept in an ice–salt bath. The sonication step ensures the complete lysis of *E. coli* and substantially decreases the viscosity of the solution, allowing centrifugation at 10,000 g (10 min). The pellet is resuspended, in the cold, in 200 ml of 50 mM Tris-HCl (pH 8.2 at 25°) stirred for $\frac{1}{2}$ hr, and centrifuged as above; the supernatant solutions are combined. All subsequent steps are performed at 0–4°.

Step 2. Acid Fractionation. The 10,000 g supernatant solution is adjusted to pH 1.8 by the slow addition of concentrated HCl, stirred for 60 min, and centrifuged at 10,000 g for 10 min. This step removes the bulk of the proteins and leaves the acidic and the basic antizyme(s) in solution. The acidic antizyme is precipitated from the acid supernatant solution by the slow dropwise addition of concentrated perchloric acid to a final concentration of 0.26 M perchloric acid (step 3A). The solution is further stirred for 30 min to complete the precipitation of the acidic antizyme. The suspension is centrifuged at 10,000 g. The basic antizyme(s) remains in solution. The pellet containing the acidic antizyme is dissolved in approximately 100 ml of 0.02 M HCl (pH 1.8, solution A). The two solutions, containing the basic and acidic antizymes, respectively, are dialyzed against three 16-liter volumes of solution A over an 18-hr period. The perchloric acid-soluble supernatant solution (containing the basic antizymes) is concentrated by ultrafiltration (YM5 membrane, Amicon) to 20–30 ml. Over a series of many experiments, the relative proportion of inhibitory activity found in the solutions containing the acidic and basic antizymes varied from approximately 1 : 4 to 1 : 10. This may be the result of subtle variations in growth conditions and the time of harvest.

Purification of the Acidic Antizyme

Step 4A. DEAE-BioGel A Chromatography. A DEAE-BioGel A column (5 × 75 cm) is equilibrated with solution A. The dialyzed solution of the acidic antizyme (approximately 100 ml), clarified by centrifugation, is applied to the column and further eluted with solution A. The peak of the antizyme-containing fractions usually appears at 0.53 column volume. The antizyme fractions are collected and concentrated to approximately 2 ml by ultrafiltration (PM-10 filter, Amicon).

Step 5A. Isoelectric Focusing. Isoelectric focusing is performed in 10% polyacrylamide gels using ampholytes of pH range 3–10.[8] Riboflavin

[8] C. W. Wrigley, this series, Vol. 22, p. 559.

TABLE I
PURIFICATION OF ACIDIC ANTIZYME

Fractionation step	Volume (ml)	Total protein[a] (mg)	Total units $\times 10^{-3}$	Specific activity (units/mg protein) $\times 10^{-3}$
1. Crude supernatant	500	6600	—[b]	<0.01
2. HCl supernatant	460	890	145	0.16
3A. Perchloric acid pellet	100	291	21	0.07
4A. DEAE-BioGel A	2.0	31	19	0.63
5A. Isoelectric focusing	0.2	0.56	2.4	4.4

[a] Protein concentrations were determined by the method of M. Bradford [*Anal. Biochem.* **72,** 248 (1976)] using bovine serum albumin as a standard.

[b] This fraction contains excessive amounts of compounds that activate ornithine decarboxylase. Therefore, little if any inhibitory activity can be detected.

5′-phosphate is used for the polymerization of the gel according to Brackenridge and Bachelard.[9] The upper and lower chambers contain 0.01 *M* H_3PO_4 and 0.02 *M* NaOH, respectively. Sucrose (10%, w/v) and ampholytes (1%, v/v; pH 3–10) are added to the antizyme sample. This is added to the gel at the anode, and electrofocusing is started with reverse polarity at 0.5 mA/tube for 1 hr, the power supply is switched to 2 mA/tube, and the electrophoresis is continued overnight. The gels are cut in 1-mm slices and ground; the antizyme is extracted with 50 m*M* Tris · HCl buffer (pH 8.2 at 25°). Results of the overall purification procedure are summarized in Table I.

Purification of the Basic Antizyme

Step 4B. CM-BioGel A Chromatography. CM-BioGel A preequilibrated with 50 m*M* Tris-HCl (pH 7.6, 4°), 0.1% Brij 58 and 10% glycerol (buffer B) is packed into a (2.5 × 40 cm) column. The dialyzed solution of the basic antizyme from step 3B is adjusted to pH 7.4 with 0.1 *M* Tris base, supplemented to 0.1% Brij 58 and 10% glycerol, and applied to the column. The column is eluted with (*a*) 100 ml of buffer B; (*b*) 100 ml of buffer B containing 0.1 *M* NaCl; and (*c*) a total 1-liter linear gradient 0.1 *M* to 0.45 *M* NaCl in buffer B. The active fractions, which elute at an NaCl concentration of approximately 0.35–0.4 *M*, are pooled, concentrated by ultrafiltration (YM5, Amicon) to approximately 8–10 ml, and dialyzed against solution A containing 0.1% Brij 58 and 10% glycerol.

[9] C. J. Brackenridge and H. S. Bachelard, *J. Chromatogr.* **41,** 242 (1969).

Step 5B. Gel Electrophoresis. The gel electrophoresis system for basic proteins as described by Panyim and Chalkley[10] is used for the final purification step of the basic antizyme(s). Two 0.2-mm-thick slab gels are prepared. One of these is to be used as a support gel and contains half the total polyacrylamide gel volume of the other; both are preelectrophoresed at 5 mA overnight. Approximately 1 ml of step 4B material (0.38 mg) containing pyronin-Y as a "tracking dye" is loaded on the larger of the two gels. Electrophoresis is performed at 10 mA to 15 mA for 6 hr. Two major protein bands can be detected, one at an R_f 0.35–0.55 relative to the tracking dye, the other at an R_f 0.80 relative to the tracking dye. These two antizyme bands are eluted into solution A by a modification of the method of Otto and Snejdarkova,[11] as follows.

One glass plate is removed from the gel, and a section of gel with an R_f 0.35–0.55 relative to the tracking dye is removed and transferred to the top of the support slab gel. It is important to place the gel section so that no bubbles are trapped in the interface of the two gels. Several drops of glycerol are placed on the gel, the glass plate is replaced, and the gel is clamped onto the electrophoresis apparatus. One milliliter of solution A containing 0.1% Brij 58 and 50% glycerol is layered on the gel followed by 2 ml of 2 *M* NaCl, and the remaining space is filled with electrophoresis buffer.

To obtain the second band of basic antizyme, the region of the original gel between R_f 0.55 and 0.80 relative to the tracking dye is discarded, leaving the bottom section of gel attached to the original plate. The glass plate is replaced, and the gel is clamped onto the electrophoresis apparatus. One milliliter of solution A containing 0.1% Brij 58 and 50% glycerol is also layered on this gel, followed by 2 ml of 2 *M* NaCl while the remaining space is filled with electrophoresis buffer as above.

Both gels are then electrophoresed with reverse polarity for 2 hr at 5–10 mA. The 2 *M* NaCl and solution A fractions are removed and dialyzed overnight against solution A containing 0.1% Brij 58 and 10% glycerol. In each case, 95% of the antizyme activity is extractable. Results of a typical purification are summarized in Table II.

Properties

Purity

Acidic ODC Antizyme. The active fraction of step 5A migrates as a single band on SDS–PAGE[12] (10% polyacrylamide). Isoelectric point is 3.8.

[10] S. Panyim and R. Chalkley, *Arch. Biochem. Biophys.* **130,** 337 (1969).
[11] M. Otto and M. Snejdarkova, *Anal. Biochem.* **111,** 111 (1981).
[12] U. K. Laemmli, *Nature (London)* **222,** 680 (1970).

TABLE II
PURIFICATION OF BASIC ANTIZYME(S)

Fractionation step	Volume (ml)	Total protein (mg)	Total units $\times 10^{-3}$	Specific activity (units/mg protein $\times 10^{-3}$)
1. Crude supernatant	500	6600	—[a]	<0.01
2. HCl supernatant	460	890	145	0.16
3B. Perchloric acid supernatant	470	600	124	0.20
4B. CM-BioGel A[b]	8.1	4.5	93	20.4
5B. Gel electrophoresis				
R_f 0.35–0.55 Band I	1.5	0.8	37.2	45
$R_f > 0.80$ Band II	2.1	1.2	44.5	36

[a] This fraction contains excessive amounts of compounds that activate ornithine decarboxylase. Therefore, little, if any, inhibitory activity can be detected.

[b] Protein concentrations for steps 4B and 5B were determined fluorometrically using *o*-phthaldialdehyde as described by E. C. Butcher and O. H. Lowry [*Anal. Biochem.* **76,** 502 (1976)].

Basic ODC Antizyme. The active fractions of step 5B migrate as two distinct single bands on polyacrylamide gel electrophoresis in urea at pH 2.2[10] (15% polyacrylamide). In addition, each band migrates as a single band on SDS–PAGE[12] (15% polyacrylamide). Isoelectric points above 9.5.

Stability

Acidic ODC Antizyme. The acidic antizyme is stable at 0° and loses 20–30% activity when heated at 100° for 3 min.

Basic ODC Antizyme. There has been little loss of activity of either purified protein when stored in solution A containing 0.1% Brij 58 and 10% glycerol at −70° for 3 months. At 4° in 50 m*M* Tris · HCl (pH 7.6) there occurs approximately 50% loss in activity in 24 hr. The addition of 0.1% Brij 58 and 10% glycerol to the 50 m*M* Tris · HCl (pH 7.6) solution reduces the loss to 5–10% in 24 hr, while storage at −70° at this pH results in 50–60% loss in activity in 1–2 weeks.

Molecular Weight

Acidic Antizyme. The molecular weight of the acidic antizyme assessed by SDS–PAGE[12] (10% polyacrylamide) was found to be 49,500. In order to detect the protein, which apparently is soluble in acetic acid, we found it necessary to stain or destain the gels in 10% (w/v) trichloroacetic acid. To remove the SDS, the gels are shaken for approximately 18 hr in a 10% (w/v) trichloroacetic acid solution with at least 3–4 changes of this solution. Coomassie Brilliant Blue R 250 is then dissolved in water (0.05% w/v); (50% w/v) trichloroacetic acid is slowly added to a final concentration of 10% (w/v). The mixture is then stirred for 15 min, sonicated for 5 min in a Branson Sonifier (power 6, pulse), and filtered; the gels are stained in the above solution for 6–8 hr. The gels are destained in 10% trichloroacetic acid (w/v) solution.

Basic Antizyme. Basic proteins are known to bind to Sephadex and Bio-Rad molecular exclusion gels, preventing accurate estimates of molecular weight by this method.[13] The minimum molecular weights estimated by SDS–PAGE[12] are 11,000 and 9000 for the two basic antizymes (5BI and 5BII, respectively) (corrected for the mobility of basic proteins as described by Panyim and Chalkley[14]).

Distribution

ODC antizyme activity has been found in *E. coli*, rat liver, a variety of mammalian cell cultures, plant cells, and barley seeds.

Acknowledgments

The authors are indebted to Dr. B. J. Bachmann for the generous supplies of the *E. coli* strains and to Dr. K. Brooks Low for assistance and helpful discussions. We thank also R. C. Rostomily for excellent technical assistance. This work was supported by NIH Grant 2R01GM26559.

[13] K. R. Sommer and R. Chalkley, *Biochemistry* **13,** 1022 (1974).
[14] S. Panyim and R. Chalkley, *J. Biol. Chem.* **246,** 7557 (1971).

[31] Synthesis of Irreversible Inhibitors of Polyamine Biosynthesis

By PHILIPPE BEY, PATRICK CASARA, JEAN-PAUL VEVERT, and BRIAN METCALF

Ornithine decarboxylase, the rate-determining enzyme for the biosynthesis of polyamines in eukaryotic cells is subject to irreversible inhibition of the "suicide" type[1] by fluorinated and acetylenic analogs of the enzyme substrate ornithine and of the product of enzymatic decarboxylation

[1] C. Walsh, *Horiz. Biochem. Biophys.* **3,** 36 (1977).

ISBN 0-12-181994-9

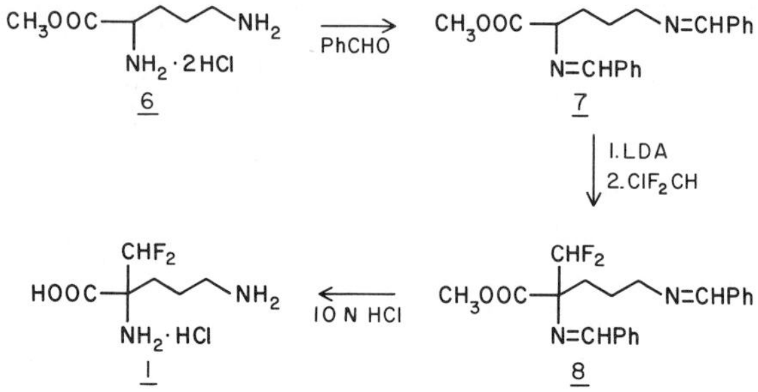

CH_3OOC–C($NH_2 \cdot 2HCl$)–$CH_2CH_2CH_2NH_2$ (6) $\xrightarrow{PhCHO}$ CH_3OOC–C(N=CHPh)–$CH_2CH_2CH_2$N=CHPh (7) $\xrightarrow{1. LDA;\ 2. ClF_2CH}$ CH_3OOC–C(CHF_2)(N=CHPh)–$CH_2CH_2CH_2$N=CHPh (8) $\xrightarrow{10\ N\ HCl}$ HOOC–C(CHF_2)($NH_2 \cdot HCl$)–$CH_2CH_2CH_2NH_2$ (1)

SCHEME 1. Synthesis of α-difluoromethylornithine (**1**).

putrescine.[2-4] Syntheses of the representative examples α-difluoromethylornithine (**1**), α-monofluoromethylputrescine (**2**), α-ethynylornithine (**3**), and α-ethynylputrescine (**4**) are discussed here. The proposed mechanisms of inactivation of ornithine decarboxylase have been presented elsewhere. In addition, the conversion of α-difluoromethylornithine to α-difluoromethylarginine[5] (**5**), an irreversible inhibitor of arginine decarboxylase,[6] will be described.

α-Difluoromethylornithine[5] (**1**, Scheme 1)

To a suspension of ornithine methyl ester dihydrochloride (**6**) (65 g, 0.3 mol) and freshly distilled benzaldehyde (63.6 g, 0.6 mol) in dichloromethane (200 ml) cooled to 0° and magnetically stirred was added slowly a solution of triethylamine (60.6 g, 0.6 mol) in dichloromethane (70 ml). The reaction mixture was then stirred overnight at room temperature. After removal of the solvent, the residue was taken up in anhydrous ether and the insoluble material was filtered off. The filtrate was washed with water, then brine, dried over $MgSO_4$, and concentrated to give methyl 2,5-bis(benzylideneamino)pentanoate (**7**) (80.8 g). Recrystallization from pentane afforded the analytical sample; mp 42°.

To a solution of lithium diisopropylamide (0.9 mol) in tetrahydrofuran cooled to −78° and magnetically stirred under nitrogen was added a solution of **7** (261 g, 0.81 mol) in 1.5 liters of tetrahydrofuran. At the end of the addition, the cooling bath was replaced by a warm water bath. When the

[2] B. W. Metcalf, P. Bey, C. Danzin, M. J. Jung, P. Casara, and J. P. Vevert, *J. Am. Chem. Soc.* **100,** 2251 (1978).

[3] P. Bey, *in* "Enzyme-Activated Irreversible Inhibitors" (N. Seiler, M. J. Jung, and J. Koch-Weser, eds.), p. 27. Elsevier/North-Holland, Amsterdam, 1978.

[4] C. Danzin, P. Casara, N. Claverie, and B. W. Metcalf, *J. Med. Chem.* **24,** 16 (1980).

[5] P. Bey, J. P. Vevert, V. Van Dorsselaer, and M. Kolb, *J. Org. Chem.* **44,** 2732 (1979).

[6] A. Kallio, P. P. McCann, and P. Bey, *Biochemistry* **20,** 3163 (1981).

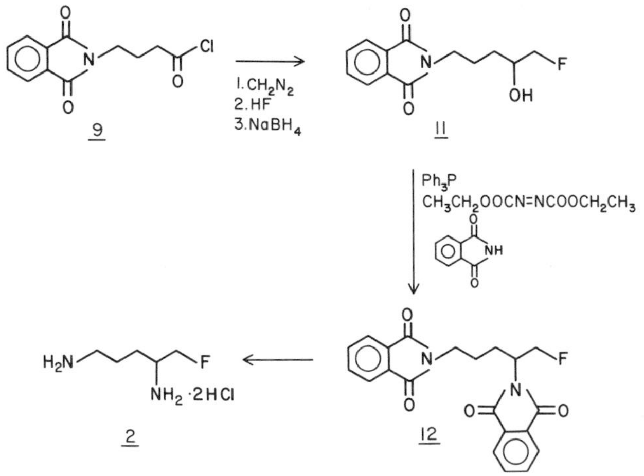

SCHEME 2. Synthesis of α-monofluoromethylputrescine (**2**).

temperature of the reaction mixture reached 40°, the nitrogen inlet was disconnected and replaced by a balloon. The reaction mixture was then saturated with chlorodifluoromethane by passing a rapid stream of chlorodifluoromethane through the solution (saturation was obtained when the balloon started to expand) maintained at a temperature of 40–50°. After stirring for 1 hr under Freon atmosphere, the reaction mixture was quenched with brine and extracted with ether to give 294 g (97%) of crude methyl 2-difluoromethyl-2,5-bis(benzylideneamino)pentanoate (**8**) as an oil unstable to distillation and to chromatography.

The crude **8** was stirred at room temperature for 3 hr with 1 *N* HCl (400 ml). The aqueous layer was washed with chloroform (3 × 100 ml), then refluxed overnight with 10 *N* HCl (400 ml). On cooling, the solution was washed with chloroform (3 × 100 ml) and concentrated to dryness under reduced pressure. The residue (50 g) was dissolved in water (80 ml), then triethylamine (approximately 25 ml) was added until pH 3.5. Charcoal (approximately 0.5 g) was then added, and the mixture was heated at 60° for 2 hr. After filtration, ethanol (320 ml) was added to the filtrate, which on cooling gave α-difluoromethylornithine (**1**) as the hydrochloride monohydrate (28.8 g), mp 183°, after recrystallization from ethanol–water.

α-Monofluoromethylputrescine (2, Scheme 2)

To a magnetically stirred solution of diazomethane in ether[7] (110 ml of approximately 0.8 *M* solution) at 0° under a nitrogen atmosphere was

[7] J. P. Moore and D. E. Reed, *Org. Synth.* **41,** 16 (1961).

added during 40 min a solution of 4-phthalimidobutyryl chloride[8] (**9**, 5.2 g, 20.6 mmol) in ether (76 ml). On completion of the addition, the mixture was stirred for $1\frac{1}{4}$ hr at room temperature. This solution was then added dropwise to the hydrogen fluoride–pyridine reagent (40 ml, Aldrich) contained in a polyethylene Erlenmeyer flask cooled in an ice–salt bath. On completion of the addition the solution was stored at room temperature for $1\frac{1}{2}$ hr then poured onto approximately 150 g of ice. The ethereal layer was separated and washed with aqueous bicarbonate and brine then dried ($MgSO_4$). The solution was then concentrated under reduced pressure and the residue (4.1 g) was recrystallized from ether–pentane to afford 1-fluoro-5-*N*-phthalimidopentan-2-one (**10**), mp 92°.

To the ketone **10** (550 mg, 2.2 mmol) in tetrahydrofuran (5 ml) and methanol (5 ml) at −20° was added a cold (−20°) solution of sodium borohydride (28 mg, 0.81 mmol) in tetrahydrofuran (5 ml) and methanol (5 ml) by syringe. After 15 min at −20° the solution was acidified by the addition of 2 *N* HCl and allowed to warm to room temperature. The mixture was concentrated at room temperature under reduced pressure, then extracted with chloroform. The chloroform layer was washed with brine then dried ($MgSO_4$) and concentrated. The residue was recrystallized from ether to afford 1-fluoro-5-*N*-phthalimidopentan-2-ol (**11**, 474 mg), mp 85°.

To the alcohol **11** (264 mg, 1.0 mmol) in tetrahydrofuran (1.5 ml) was added phthalimide (170 mg, 1.0 mmol) in tetrahydrofuran (5 ml), triphenylphosphine (302 mg, 1.1 mmol) in tetrahydrofuran (1 ml), and diethylazidodicarboxylate (201 mg, 1.1 mmol) in tetrahydrofuran (1 ml). The mixture was stirred under nitrogen for 2 hr at room temperature, then concentrated to dryness. The residue was then chromatographed on silica gel (50 g) in ethyl acetate–hexane (1 : 1). Elution with this solvent gave the bisphthalimide **12**, which was recrystallized from tetrahydrofuran–ether to afford 283 mg, mp 112°. A suspension of **12** (3.1 g) in concentrated hydrochloric acid (140 ml) was heated under reflux for 3 days. On cooling to 4° the precipitated phthalic acid was filtered off and the filtrate was concentrated to 20 ml under reduced pressure then refiltered and concentrated to dryness. The residue was boiled in isopropanol (40 ml) for 30 min, then concentrated under reduced pressure. This operation was repeated three times, then the residue was recrystallized three times from ethanol to afford the dihydrochloride of α-fluoromethylputrescine (**2**, 1.1 g), mp 154°.

[8] S. Gabriel and J. Colman, *Ber. Dtsch. Chem. Ges.* **41**, 513 (1908).

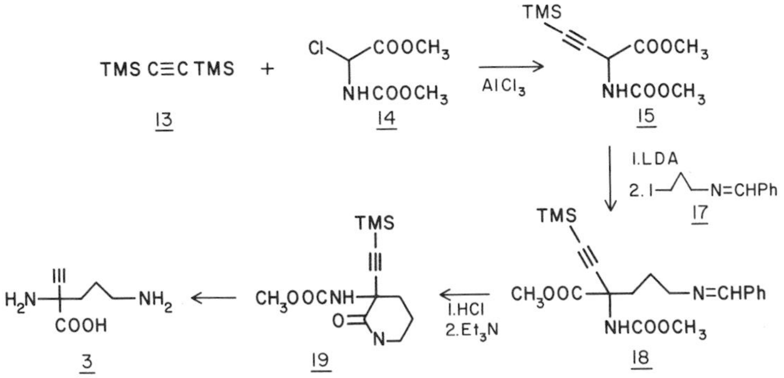

SCHEME 3. Synthesis of α-ethynylornithine (**3**).

α-Ethynylornithine (3, Scheme 3)[4]

To a cooled (0°) solution of methyl α-chloro-*N*-carbomethoxyglycinate[9] (**14**, 91 g, 0.5 mol) and bis(trimethylsilyl)acetylene (**13**, 95 g, 0.56 mol) in freshly distilled dichloromethane (500 ml) was added aluminum trichloride (75 g, 0.56 mol) in small portions. After the addition was completed, the reaction mixture was maintained at 20° overnight. The solution was washed with water (3 × 200 ml), dried, and distilled to give **15** (55 g, 45%), bp 90° (0.15 mm). To a solution of 1,3-bromopropylamine hydrobromide (22.5 g, 0.1 mol) and benzaldehyde (10.6 g, 0.1 mol) in dichloromethane (150 ml) was added triethylamine (10.1 g). After standing for 2 hr at room temperature, the solution was washed with water (2 × 100 ml) and dried ($MgSO_4$); it yielded by distillation 1-bromo-3-benzaldiminopropane (**16**, 18.5 g, 82%), by 90° (0.05 mm).

A solution of **16** (22.5 g, 0.1 mol) and NaI (15 g, 0.1 mol) in anhydrous tetrahydrofuran (200 ml) was heated under reflux overnight. The reaction mixture was cooled to room temperature, and the solid formed was filtered off. The resulting iodo compound **17** was used without further purification in the next step.

To a cooled (−78°) solution of lithium diisopropylamide (0.02 mol) in tetrahydrofuran–hexamethylphosphoramide (100 ml, 9 : 1) was added a solution of **15** (2.45 g, 10 mmol) in tetrahydrofuran (5 ml) under N_2. After 1 hr at −78°, a solution of **17** (2.75 g, 10 mmol) in tetrahydrofuran (5 ml) was added. The reaction mixture was stirred at −78° for 2.5 hr and then hydrolyzed with 1 *N* AcOH (100 ml). After extraction with ether, the intermediary imine **18** was hydrolyzed with 1 *N* HCl (100 ml). The benzal-

[9] Z. Bernstein and D. Ben-Ishai, *Tetrahedron* **33,** 881 (1977).

dehyde was extracted with ether, and the aqueous phase was evaporated and dried under vacuum. The residue was dissolved in methanol (100 ml), and triethylamine was added (1.5 ml, 10 mmol). After 12 hr at room temperature, the methanol was evaporated and ether (100 ml) was added. The solution was washed with water (3 × 100 ml) and dried ($MgSO_4$). After evaporation of the solvent, the residue was chromatographed on a silica gel column (50 g). Elution with ether afforded 1.2 g (45%) of the lactam **19**, mp 146°.

A solution of **19** (1.42 g, 5.3 mmol) and trimethylsilyl iodide (1.5 ml, 10 mmol) in chloroform (5 ml) was heated under reflux for 1 hr and then 1 ml of methanol was added slowly. After evaporation, the residue was hydrolyzed with 2 *N* NaOH (7.5 ml) overnight. The aqueous solution was diluted with water (100 ml), washed with chloroform (2 × 50 ml), acidified with 1 *N* HCl, and washed with chloroform (2 × 50 ml). After concentration, the aqueous phase, adjusted to pH 6 with 1 *N* NH_4OH solution, was passed through an Amberlite IR 120, H^+ form, resin column. The ninhydrin-positive fractions eluted with 1 *N* NH_4OH were collected and evaporated to give 0.8 g of crude α-ethynylornithine (**3**). The solid was dissolved in 1 *N* HCl (10 ml), and the residue obtained after evaporation was dissolved in EtOH (10 ml). The monohydrochloride hydrate of **3** was obtained by addition of triethylamine and recrystallization in EtOH–H_2O (9:1) (600 mg, 59%), mp 171°.

α-Ethynylputrescine (**4**, Scheme 4)[2]

A solution of propargylamine (**20**, 26.1 g, 0.47 mol) and benzaldehyde (52 g, 0.49 mol) in benzene (150 ml) was treated with $MgSO_4$ (20 g); the mixture was stirred at room temperature for 30 min, then filtered. Excess water was removed via azeotropic distillation, and the solution was concentrated on the rotorvapor, then distilled. The imine **21** (55.5 g, 85%), bp 107–110° water pump pressure, was collected.

To a mechanically stirred solution of the imine **21** (43.5 g, 0.30 mol) in tetrahydrofuran (400 ml) at 0° was added, during 30 min, ethylmagnesium bromide (285 ml of a 1.12 mol solution, 0.316 mol). After 30 min at 0°, the resulting solution was treated with a solution of trimethylsilyl chloride (32.4 g, 0.30 mol) in tetrahydrofuran (100 ml), the addition taking 45 min. The solution was stirred at 0° for a further $1\frac{1}{2}$ hr, then treated with brine. The organic phase was separated and washed well with brine (8 × 100 ml), then dried and concentrated on a rotorvapor. The residue was distilled to afford **22** (52.2 g, 89%), bp 92–110°, 0.6 mm.

n-Butyllithium (195 ml, 2.15 mol in hexane, 0.4 mol) was added to a

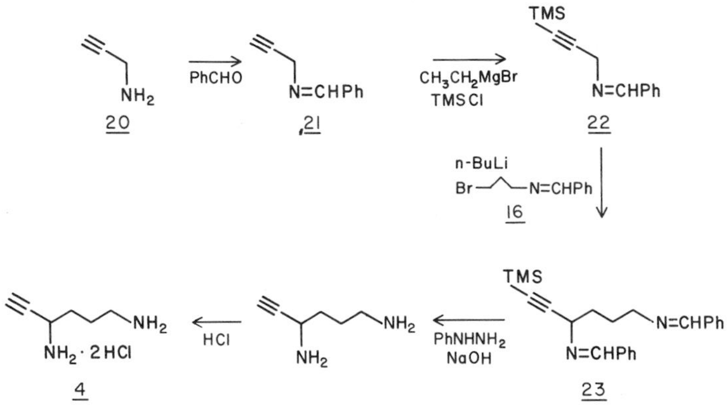

SCHEME 4. Synthesis of α-ethynylputrescine (**4**).

stirred solution of **22** (86.4 g, 0.4 mol) in anhydrous tetrahydrofuran (2.4 g) at −78°. The mixture was stirred for 10 min and then treated with a solution of **16** (90.4 g, 0.4 mol) in anhydrous tetrahydrofuran (100 ml). After stirring at −78° for a further 3 hr the mixture was allowed to reach room temperature, washed with water (2 × 400 ml), and dried ($MgSO_4$). The residue (152.8 g) of crude **23** after evaporation of the tetrahydrofuran was treated with benzene (50 ml), pentane (400 ml), and phenylhydrazine (91 g). After stirring for 1 hr at room temperature, the mixture was allowed to stand for 1½ hr in a refrigerator before being filtered to remove benzaldehyde phenylhydrazone. The residue (120 g) after evaporation of the solvent was dissolved in a mixture of concentrated hydrochloric acid (80 ml) and water (50 ml); the solution was clarified by filtration and extraction with dichloromethane (3 × 100 ml). Evaporation of the aqueous solution gave a residue that was strongly alkalinized by the addition of 40% sodium hydroxide (100 ml) and sodium hydroxide pellets added until two phases were obtained. The aqueous phase was extracted with dichloromethane; the organic phases were combined and dried, then distilled to give a liquid bp 58°, 0.2 mm (19.5 g) consisting of α-ethynylputrescine and a little phenylhydrazine (~12%). The crude base was converted to its monohydrochloride by the addition of 1 *N* hydrochloric acid (156 ml); the solution was clarified by extraction with dichloromethane (3 × 50 ml), a further 156 ml of 1 *N* hydrochloric acid was added, and the solution was evaporated to dryness. Crystallization of the residue (27 g) from methanol gave 5-hexyne-1,4-diamine (**4**) as the dihydrochloride (11 g), mp 170°. A further 4.1 g was obtained by concentration of the mother liquors.

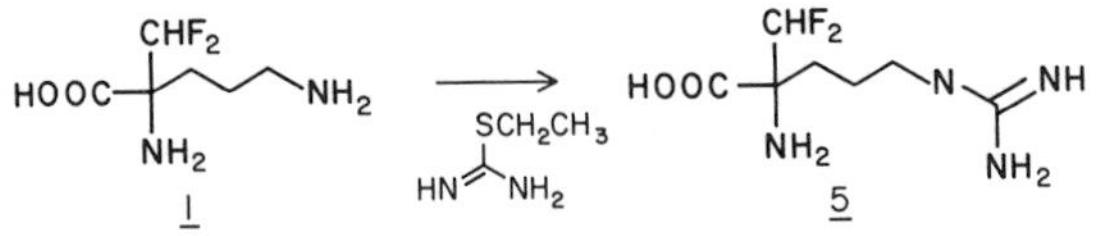

CHF_2 HOOC— —NH_2 NH_2 1 → (SCH_2CH_3, HN=C—NH_2) → CHF_2 HOOC— —N— NH NH_2 NH_2 5

SCHEME 5. Synthesis of α-difluoromethylarginine (**5**).

α-Difluoromethylarginine[5] (5, Scheme 5)

To a solution of α-difluoromethylornithine **1** (5 g, 21.1 mmol) in 8.5 ml of sodium hydroxide was added ethylthiouronium hydrobromide (7.2 g, 42.2 mmol). The pH of the solution was adjusted to 10.5 with 2 *N* sodium hydroxide, where it was maintained for 4 days. The reaction mixture was then neutralized to pH 7 with 1 *N* hydrochloric acid and concentrated *in vacuo*. The residue was passed on an Amberlite IR 120 H^+-form resin column. Elution with 2 *N* ammonium hydroxide afforded 2.3 g of **5** (50%) which was recrystallized from aqueous ethanol to afford crystals, mp 257°.

[32] Labeling and Quantitation of Ornithine Decarboxylase Protein by Reaction with α-[5-^{14}C]Difluoromethylornithine

By JAMES E. SEELY, HANNU PÖSÖ, and ANTHONY E. PEGG

Principle

α-Difluoromethylornithine is a potent, irreversible inhibitor of ornithine decarboxylase and is thought to act via an enzyme-activated "suicide" mechanism.[1,2] Inactivation occurs because the inhibitor is decarboxylated by the enzyme, generating an intermediate carbanionic species which, with the loss of a fluorine, alkylates a nucleophilic site (probably lysine) on the enzyme. Such inactivation, therefore, proceeds with stoichiometric attachment of the difluoromethylornithine derivative to the enzyme. By using radioactive inhibitor, it is possible to label the enzyme specifically; by quantitation of the amount of binding needed for complete

[1] B. W. Metcalf, P. Bey, C. Danzin, M. J. Jung, P. Casara, and J. P. Vevert, *J. Am. Chem. Soc.* **100,** 2551 (1978).

[2] P. Bey, *in* "Enzyme-Activated Irreversible Inhibitors" (N. Seiler, M. J. Jung, and J. Koch-Weser, eds.), p. 27. Elsevier/North-Holland, Amsterdam, 1978.

ISBN 0-12-181994-9

inactivation, the number of enzyme molecules can be calculated.[3,4] Ornithine decarboxylase must be catalytically active in order for binding to occur via this mechanism. Therefore, the binding assay is carried out under conditions similar to those used in the determination of enzyme activity.

Reagents

Buffer A: 0.025 *M* Tris · HCl, pH 7.5, containing 2.5 m*M* dithiothreitol and 0.1 m*M* EDTA
Pyridoxal 5′-phosphate, 2 m*M*
L-Ornithine, 0.24 *M*
DL-α-[5-^{14}C]Difluoromethylornithine, 12.5 μCi/ml (60 mCi/mmol)
Perchloric acid, 1 *M*
Ethanol–chloroform–ether, 2 : 1 : 1 (v/v/v)
Sodium hydroxide, 0.1 *M*

Procedure

Exposure to 5 μM DL-α-[5-^{14}C]difluoromethylornithine for 60 min at 37° under the standard assay conditions is sufficient for complete inactivation and maximal labeling of the enzyme.[3-5] Extracts containing ornithine decarboxylase activity are dialyzed overnight against 200 volumes of buffer A to remove any endogenous ornithine. A total volume of 0.48 ml containing 0.4 ml of the enzyme extract, 0.01 ml of 2 m*M* pyridoxal phosphate, and 0.0114 ml of [5-^{14}C]difluoromethylornithine is placed in small centrifuge tubes and incubated for 60 min at 37°. Proteins are then precipitated by addition of 2 ml of 1 *M* perchloric acid followed by centrifugation at 12,000 *g* for 5 min. The pellet is washed twice with 2-ml portions of 1 *M* perchloric acid followed by one wash with 2 ml of ethanol–chloroform–ether (2 : 1 : 1) and a final wash with 2 ml of ether. Each wash is followed by centrifugation at 12,000 *g* for 10 min to remove the protein from the wash solution. The protein is dried in air for 2 hr and dissolved in 1 ml of 0.1 *N* NaOH at 100°, cooled, mixed with 10 ml of ACS II liquid scintillation fluid (Amersham/Searle, Arlington Heights, Illinois) and counted in a liquid scintillation spectrometer at an efficiency of 67%.

[3] M. L. Pritchard, J. E. Seely, H. Pösö, L. S. Jefferson, and A. E. Pegg, *Biochem. Biophys. Res. Commun.* **100,** 1597 (1981).
[4] A. E. Pegg, I. Matsui, J. E. Seely, and H. Pösö, *Med. Biol.* **59,** 327 (1981).
[5] J. E. Seely, H. Pösö, and A. E. Pegg, *Biochem. J.* **206,** 311 (1982).

Any nonspecific binding can be corrected for by including in the incubation mixture a large excess of L-ornithine which protects the enzyme from interaction with the DL-α-[5-^{14}C]difluoromethylornithine. The assay tubes should, therefore, include a control that contains 60 μl of 0.24 *M* L-ornithine, and the radioactivity present in this tube is subtracted from the binding found in the complete assay.

Applications

The binding of labeled α-difluoromethylornithine can be used to assess the purity of ornithine decarboxylase preparations and to determine the number of active molecules of the enzyme present in tissue extracts obtained from various physiological states. Such extracts from rat liver after various treatments affecting ornithine decarboxylase activity, including partial hepatectomy and treatment with thioacetamide, carbon tetrachloride, or cycloheximide, gave a constant binding of 26 fmol of inhibitor bound per unit of enzyme activity[5] (1 unit represents 1 nmol of CO_2 released per 30 min). The radioactively labeled rat ornithine decarboxylase has an approximate M_r of 100,000 on gel filtration and a subunit M_r of 55,000 as determined by polyacrylamide gel analysis under denaturing conditions.[3] Assuming that both subunits can bind a molecule of the inhibitor for full inactivation, these results suggest that the specific activity to be expected for homogeneous rat liver ornithine decarboxylase is about 7.5×10^5 units per milligram of protein and that, even after induction by thioacetamide or partial hepatectomy, an 800,000-fold purification would be needed to achieve this value, since the activity of initial extracts is 0.9 units per milligram.[3,4]

Similar experiments with mouse extracts have indicated that mouse kidney ornithine decarboxylase is slightly smaller than the rat enzyme, with a M_r of about 53,000 for the subunits on electrophoresis under denaturing conditions.[6] The mouse enzyme binds 13.5 fmol of inhibitor per unit of enzyme activity, indicating that homogeneous mouse kidney ornithine decarboxylase should have a specific activity of 1.53×10^6 units per milligram of protein.[5] We have purified the enzyme to this specific activity starting from extracts of androgen-induced mouse kidney, which have an activity of 1.44 units per milligram.[6]

Other potential uses for the specific labeling of ornithine decarboxylase by reaction with radioactive α-difluoromethylornithine include studies of the enzyme structure and active center, since the labeled protein (and the peptide containing the labeled drug after proteolytic digestion)

[6] J. E. Seely and A. E. Pegg, this volume [23].

can be compared from various sources; studies of the degradation of the enzyme protein *in vivo,* since we have observed that the labeled protein is degraded at the same rapid rate as the enzyme activity is lost when protein synthesis is inhibited[7]; and autoradiographic localization of the enzyme.[8] The latter applications depend on the absolute specificity of labeling *in vivo*. This has as yet been investigated only in the mouse kidney; in this organ, only a single labeled protein is found when extracts from mice treated with DL-α-[5-^{14}C]difluoromethylornithine are analyzed by polyacrylamide gel electrophoresis under denaturing conditions.[7] This protein was identical to that labeled *in vitro* when purified ornithine decarboxylase was allowed to react with the drug.[7,8]

[7] J. E. Seely, H. Pösö, and A. E. Pegg, *J. Biol. Chem.* **257,** 7549 (1982).
[8] A. E. Pegg, J. E. Seely, and I. S. Zagon, *Science* **217,** 68 (1982).

[33] Methods for the Study of the Treatment of Protozoan Diseases by Inhibitors of Ornithine Decarboxylase

By PETER P. MCCANN, CYRUS J. BACCHI, WILLIAM L. HANSON, HENRY C. NATHAN, SEYMOUR H. HUTNER, and ALBERT SJOERDSMA

Polyamine metabolism has been suggested[1,2] as a target for chemotherapy of parasitic protozoan disease, particularly as drug–polyamine interactions have been demonstrated in African trypanosomes.[2,3] DL-α-Difluoromethylornithine (DFMO) (MDL 71,782), a specific irreversible inhibitor of the first step in polyamine biosynthesis [i.e., formation of putrescine from ornithine by ornithine decarboxylase (EC 4.1.1.17)], cured mice infected with a virulent, rodent-passaged strain of *Trypanosoma brucei brucei*.[4] This parasite is closely related to the trypanosomes that cause human sleeping sickness and it causes a form of the disease in domestic animals. The drug, remarkably nontoxic, was completely effective when administered in drinking water of the host mice. Furthermore, DFMO is potently anticoccidial against infections of *Eimeria tenella* in

[1] S. S. Cohen, *Science* **205,** 964 (1979).
[2] C. J. Bacchi, *J. Protozool.* **28,** 20 (1981).
[3] H. C. Nathan, C. J. Bacchi, T. T. Sakai, D. Rescigno, D. Stumpf, and S. H. Hutner, *Trans. R. Soc. Trop. Med. Hyg.* **75,** 394 (1981).
[4] C. J. Bacchi, H. C. Nathan, S. H. Hutner, P. P. McCann, and A. Sjoerdsma, *Science* **210,** 332 (1980).

ISBN 0-12-181994-9

chickens and markedly reduces the cecal lesions that often lead to severe blood loss and death.[5,6] In both types of protozoal infections, reversal of the therapeutic effects of DFMO can be seen with coadministration of polyamines that bypass the enzymatic block.[5-7]

Trypanosoma b. brucei Infections in Mice

Materials

Organisms. The EATRO 110 (East African Trypanosome Research Organization) isolate was obtained from the American Type Culture Collection (Rockville, Maryland 20852) and was syringe-passaged in 20- to 30-g outbred Swiss–Webster male and female mice (Royal Hart Laboratories, New Hampton, New York 10958). This pleomorphic strain produces a rapid fatal parasitemia in 4–7 days (blood counts 0.5–1.5 $\times 10^9$/ml).

Inoculum. Mouse blood (2–3 ml) was aspirated from ventricles of animals with a near-terminal infection (72–96 hr) and suspended in 8 ml of 0.09 *M* Tris-buffered 5 m*M* saline plus 2% (w/v) glucose, pH 7.8. Suspensions were then centrifuged at 3800 *g* for 10 min (International Model HN). The upper buffy coat of the pellet containing the trypanosomes was removed by syringe and resuspended in buffer. Trypanosomes (per mm^3) were counted in a Neubauer hemacytometer, and suspensions were diluted with buffer to a final count of $\sim 10^6$/ml. Suspensions were used immediately after dilution and agitated frequently to promote aeration, since the organisms are intensely aerobic. Groups of five 20- to 25-g mice (mixed males and females) were then inoculated intraperitoneally (ip) with 0.25 ml of the Tris–saline–glucose buffer containing 5 $\times 10^5$ organisms per animal.[8]

Administration of DFMO and Coadministration of Polyamines

The DFMO as the monohydrochloride (M_r 236.6) is very water soluble (436 g/liter); it is obtainable from Merrell Dow Research Center (Cincinnati, Ohio 45215). The inhibitor was given as a percentage solution (g/100 ml) in the drinking water (usually tap water) of infected animals;

[5] W. L. Hanson, M. M. Bradford, W. L. Chapman, Jr., V. B. Waits, P. P. McCann, and A. Sjoerdsma, *Am. J. Vet. Res.* **43,** 1651 (1982).

[6] P. P. McCann, C. J. Bacchi, W. L. Hanson, G. D. Cain, H. C. Nathan, S. H. Hutner, and A. Sjoerdsma, *Adv. Polyamine Res.* **3,** 97 (1981).

[7] C. J. Bacchi, H. C. Nathan, S. H. Hutner, P. P. McCann, and A. Sjoerdsma, *Biochem. Pharmacol.* **31,** 2833 (1982).

[8] H. C. Nathan, K. V. M. Soto, R. Moreira, L. Chunosoff, S. H. Hutner, and C. J. Bacchi, *J. Protozool.* **26,** 657 (1979).

TABLE I
RESULTS OF TREATMENT OF *Trypanosoma brucei brucei* INFECTIONS IN MICE

Drug, treatment regimen	Total dose of DFMO (mg)	Average survival (days)
None	0	0
DFMO[a]		
2.0%, 6 days	600	>30
2.0%, 3 days	300	>30
1.0%, 6 days	300	>30
1.0%, 3 days	150	>30
0.75%, 3 days	112.5	22.6
0.5%, 3 days	75	4.0
0.25%, 3 days	37.5	2.5
0.1%, 3 days	15	0
2.0%, 3 days		
+ Spermine · 4 HCl, 50 mg/kg	300	7.7
+ Spermidine · 3 HCl, 100 mg/kg	300	14.1
+ Putrescine · 2 HCl, 500 mg/kg	300	13.4

[a] DFMO, DL-α-Difluoromethylornithine.

it was first administered 24 hr after infection and continued for 3–6 days. The total dose of DFMO administered (mg) was calculated on the basis of a daily intake of 5 ml of water per 25-g mouse per day. Water intake of the animals will vary with the housing conditions and should be determined accordingly. The drinking water containing DFMO was constantly available. Polyamines (i.e., putrescine, spermidine, spermine), all as hydrochlorides (Sigma Chemical Co., St. Louis, Missouri 63178), were administered as a single ip injection (0.2 ml) in aqueous solution daily for 3 days concurrent with DFMO treatment.[7]

The results in Table I are expressed as average survival time (days) beyond death of control animals. Cages were checked daily at approximately the same time of day. Groups of animals receiving DFMO as a 1 or 2% solution were completely cured of infection, as indicated by survival times >30 days beyond that of untreated animals. Parasites disappeared from the blood 5 days after treatment was begun, and Giemsa-stained tail blood smears taken 30 days after death of controls were negative. Healthy mice injected with brain suspensions (sometimes the site of cryptic trypanosome infection[9]) from cured animals failed to develop parasitemia.

[9] F. W. Jennings, D. D. Whitelaw, P. H. Holmes, H. G. B. Chizyuka, and G. M. Urquhart, *Int. J. Parasitol.* **9**, 381 (1979).

Polyamines, administered alone to infected animals, do not affect the usual lethal course of infection. The doses of polyamines described in Table I have no toxic effects on uninfected mice (outbred Swiss–Webster) used as controls.

Eimeria tenella Infections in Chickens

Materials

Organisms. The Wisconsin strain of *Eimeria tenella* isolate can be obtained from W. L. Hanson, Department of Parasitology, College of Veterinary Medicine, University of Georgia (Athens, Georgia 30602). It rapidly multiplies by schizogony in the cecal mucosal cells, resulting in a rapid increase in parasite number and damage to the mucosal epithelium, often with severe blood loss and death.

Inoculum. Eimeria tenella is normally maintained in host chickens and oocysts obtained from heavily infected chickens. Groups of nine, healthy, 2-week-old male chickens were given a sublethal dose of 1×10^5 sporulated oocysts via gavage. The infection time was 5 days, at the end of which chickens were killed with CO_2 (3–5 min in 100% CO_2) and necropsied; cecal lesions were evaluated by the standard lesion-scoring procedure, which has a range of 0 (no lesions) to 4.0 (cecal walls greatly distended with blood or large blood cores).[10]

TABLE II
RESULTS OF TREATMENT OF *Eimeria tenella* INFECTIONS IN CHICKENS

Drug, treatment regimen	Days of treatment	Mean lesion scores
None	—	3.50
DFMO[a]		
2.0%	−1 to +5	0
1.0%	−1 to +5	0
0.5%	−1 to +5	0.10
0.25%	−8 to +5	0.44
0.125%	−8 to +5	0
0.0625%	−8 to +5	0.66
0.5% + putrescine · 2 HCl, 300 mg/kg	−1 to +5	4.00
Amprolium, 0.012%	−2 to +5	0.30

[a] DFMO, DL-α-Difluoromethylornithine.

[10] W. M. Reid, *Am. J. Vet. Res.* **30,** 447 (1969).

Administration of DFMO and Coadministration of Polyamines

The DFMO was given, as above, in the drinking water of infected animals and was administered either 1 day or 8 days before infection and continued for the 5-day course of the infection.[5,6] Drinking water containing DFMO was constantly available. Putrescine (· 2 HCl) was administered as a single ip injection (300 mg/kg per day) on the day of infection and the four following days.[5,6]

The typical data in Table II are given as mean cecal lesion scores for treated and untreated chickens. When DFMO was started 1 day before infection and continued for 5 days, a concentration of 0.5% DFMO was as effective as 0.012% Amprolium, a standard anticoccidial drug. Even lower doses of DFMO, i.e., 0.0625%, were as active against *E. tenella*, provided treatment was initiated 8 days prior to infection (day 0) and continued thereafter. Birds cured by DFMO were immune to reinfection by *E. tenella* when challenged 1 week after completion of therapy. Putrescine, administered alone to infected control animals, had no effect on lesions.

[34] Methods for the Study of the Experimental Interruption of Pregnancy by Ornithine Decarboxylase Inhibitors: Effects of DL-α-Difluoromethylornithine

By JOHN R. FOZARD and MARIE-LOUISE PART

A sharp rise in uterine–deciduomal ornithine decarboxylase (ODC; EC 4.1.1.17) activity is a characteristic feature of the period immediately following implantation in early mammalian embryogenesis.[1] Direct evidence of an essential role for this enzyme has been obtained using DL-α-difluoromethylornithine (MDL 71782; DFMO), a highly selective, irreversible inhibitor of ODC.[2] When administered during early gestation DFMO arrested embryonal development in mouse, rat, and rabbit.[1,3] We describe here our procedure for the consistent interruption of pregnancy in the mouse with DFMO. We further indicate how the principles employed may be generalized to other mammalian species.

[1] J. R. Fozard, M.-L. Part, N. J. Prakash, J. Grove, P. J. Schechter, A. Sjoerdsma, and J. Koch-Weser, *Science* **208,** 505 (1980).

[2] B. W. Metcalf, P. Bey, C. Danzin, M. J. Jung, P. Casara, and J. P. Vevert, *J. Am. Chem. Soc.* **100,** 2551 (1978).

[3] J. R. Fozard, M.-L. Part, N. J. Prakash, and J. Grove, *Eur. J. Pharmacol.* **65,** 379 (1980).

ISBN 0-12-181994-9

Principles

Ornithine decarboxylase has a biological half-life of only 10–20 min.[4] It follows, therefore, that even with an irreversible inhibitor it is necessary to maintain adequate tissue concentrations of the agent to ensure continued inhibition of the enzyme. In practice, DFMO is rapidly cleared both from the circulation ($t_{1/2} \sim 20$ min) and from deciduomata of the mouse. Repeated drug administration is therefore necessary if sustained inhibition of the enzyme is to be achieved. Our previous experiments[1,3] have established days 5–8 as the period during early murine gestation when deciduomal putrescine biosynthesis increases rapidly to a peak. This is also the period during embryonal development most sensitive to inhibition by DFMO.[3] Based on these considerations, the contragestational effect of DFMO and its relationship to dose is most consistently demonstrated after regular administration of the drug during days 5–8 of gestation.

Experimental Procedure

Animals. CDA, HAM ICR albino mice of proved fertility, weighing 30–45 g at the start of the experiments, are used. They are housed at 22° and have standard laboratory chow and water *ad libitum*. They are maintained on a 12-hr light–dark schedule with light from 22:00 to 10:00. For mating, two females are transferred to the home cage of a singly housed male between 08:00 and 09:00. Animals are inspected daily thereafter between 08:00 and 09:00 for the presence of a vaginal plug; the day of the discovery of the plug is designated day 1 of gestation. Of a series of 311 mice proved to have mated, 274 were subsequently found to be pregnant (88%).

Drug Administration. Sustained enzyme inhibition by DFMO can be achieved by incorporating it in the drinking water or by subcutaneous injection every 6 hr. The DFMO may be obtained from the Centre de Recherche Merrell International, Strasbourg. To prepare the drinking fluids, DFMO is dissolved at the appropriate concentrations (see below) in distilled water and placed in 300-ml plastic drinking bottles with nondrip spouts (Ehret GmbH, Emmendingen, Federal Republic of Germany) for administration to the animals. Solutions for subcutaneous injection are prepared in sterile saline at concentrations chosen so that the required dose is given when 10 ml per kilogram of body weight is injected.

[4] D. V. Maudsley, *Biochem. Pharmacol.* **28,** 153 (1979).

Mated females are housed 5–10 per cage. Treatment with DFMO is started at 12:00 on day 5, at which time either the sole drinking fluid is changed from water to a DFMO solution or the first of 16 six-hourly subcutaneous injections of DFMO is given. Treatment is maintained for 96 hr until the end of day 8. Control animals receive water to drink throughout this time or injections of saline.

Contragestational effects are assessed by killing the animals 24 hr before the expected time of parturition on day 18 of gestation. The uteri are removed, pinned to a dissecting board, and slit longitudinally along the side opposite to the mesometrium. Live and dead fetuses, placentas, and resorption nodules are removed, counted, blotted, and weighed.

Dose–Response Relationship. Concentrations of DFMO in the drinking water of 0.125, 0.5, or 2% provide each mouse with approximately 250, 1000, and 3000 mg of DFMO per kilogram per 24-hr period (Table I). With the two lower doses, contragestational effects are manifested as a dose-related decline in the number of viable fetuses per mated female and in an increase in the number of resorption nodules (Table I). After the highest concentration of DFMO, none of the animals show any sign of pregnancy when autopsied on day 18 of gestation. This reflects the complete resorption or loss from the uterus of the arrested embryo and surrounding decidual tissue.[1] Qualitatively similar effects are observed when DFMO is given by subcutaneous injection at doses of 50 and 200 mg/kg every 6 hr on days 5–8 of gestation (Table I).

TABLE I
CONTRAGESTATIONAL EFFECTS OF DL-α-DIFLUOROMETHYLORNITHINE (DFMO) IN THE MOUSE

Treatment during days 5–8	Dose of DFMO (mg/kg per 24 hr)	Fetuses per mouse	Resorption nodules per mouse	Number of mice
Control	0	10.0 ± 1.0	3.4 ± 0.7	17
In drinking water				
0.125%	259[a]	7.2 ± 1.4	3.5 ± 0.9	6
0.5%	989[a]	2.0 ± 1.0[b]	10.4 ± 1.5[b]	7
2.0%	3070[a]	0[b]	0[b]	19
By injection every 6 hr				
50 mg/kg	200	5.5 ± 1.5[b]	8.8 ± 1.5[b]	8
200 mg/kg	800	0[b]	0[b]	10

[a] Based on an average body weight of 40 g.
[b] Significantly different from control value, $p < 0.05$.

TABLE II
COMPARISON OF DEVELOPMENTAL STAGES OF MOUSE, RAT, RABBIT, SHEEP, MONKEY, AND MAN[a]

Standard stages (Witschi)[b]	Identification of stages	Gestation age (days)					
		Mouse	Rat	Rabbit	Sheep	Monkey	Man
3	Four cells	1.5	3	0.5	1.5	1.5	2
7–8[c]	Beginning of implantation	4–5	6	7	10	9	6.5
12[c]	Primitive streak	7	8.5	6.5	13	19	19
16[c]	13–20 Somite embryo	9	10.5	9	17	25	27
25	End of embryonic period	11	12.5	10	21	28	36
35–36	Birth	19	22	32	150	164	267

[a] Adapted from Table 26 in "Biology Data Handbook." Fed. Am. Soc. Exp. Biol., Washington D.C., 1964.
[b] From Witschi.[5]
[c] Stages 7–16 were the critical period of murine gestation.

Generalization to Other Species

The critical period of murine gestation that we have identified as being susceptible to inhibition by DFMO covers standard stages 8–16 of mammalian development as defined by Witschi.[5] For comparison, some representative stages of other common species including man are presented in Table II. Extrapolating from the mouse, we administered DFMO via the drinking water to rats and rabbits during standard stages 8–16 of development. In each case we obtained complete arrest of embryonic development.[3] Thus, the principles underlying the method of interrupting pregnancy in the mouse generalize to at least two other species. Whether they hold in those species (sheep, monkey, man) where stages 8–16 extend over a much longer time span remains to be determined.

[5] E. Witschi, "The Development of Vertebrates." Saunders, Philadelphia, 1956.

Section V

Adenosylmethionine Synthetase (Methionine Adenosyltransferase) and Adenosylmethionine Decarboxylase

A. Enzyme Assays and Preparations
Articles 35 through 39

B. Enzyme Inhibitors
Articles 40 and 41

[35] *S*-Adenosylmethionine Synthetase (Methionine Adenosyltransferase) (*Escherichia coli*)[1–3]

By GEORGE D. MARKHAM, EDMUND W. HAFNER, CELIA WHITE TABOR, and HERBERT TABOR

$$\text{L-Methionine} + \text{ATP} + H_2O + E \rightarrow [E \cdot \text{AdoMet} \cdot \text{PPP}_i] \quad (1)$$

$$[E \cdot \text{AdoMet} \cdot \text{PPP}_i] \rightarrow E + \text{AdoMet} + P_i + PP_i \quad (2)$$

$$\text{L-Methionine} + \text{adenosine-}\overset{\alpha,\beta,\gamma}{\text{PPP}} + H_2O \rightarrow \text{AdoMet} + \overset{\alpha,\beta}{PP_i} + \overset{\gamma}{P_i} \quad (3)$$

Assay

Principle. The enzyme converts L-[$^{14}CH_3$]methionine and ATP to the strongly basic sulfonium compound *S*-adenosylmethionine. This product is adsorbed selectively to squares of cation-exchange paper. The radioactivity on the paper is then counted.[4]

Reagents

Tris-chloride, 1 *M*, pH 8.0
KCl, 1 *M*
$MgCl_2$, 0.2 *M*
Adenosine triphosphate, 0.1 *M*
L-[$^{14}CH_3$]Methionine, 0.012 *M* (specific activity 1.8 mCi/mmol)
EDTA (Na^+), 0.05 *M*, pH 8.0

Procedure. The incubation mixture contains 5 μl of 1 *M* Tris-HCl, pH 8.0, 5 μl of 1 *M* KCl, 5 μl of 0.2 *M* $MgCl_2$, 5 μl of 0.1 *M* ATP, 5 μl of 12 m*M* L-[$^{14}CH_3$]methionine, 15 μl of water, and 5 μl of enzyme. The incubation is at 25° for 5–30 min. The reaction is stopped by adding 0.1 ml of 50 m*M* EDTA, pH 8.0. One-tenth milliliter of the incubation mixture is then placed on a 3 × 3 cm square of Whatman IRC-50 cation-exchange paper or on Whatman P-81 phosphocellulose paper (numbered with pencil).[5]

[1] G. D. Markham, E. W. Hafner, C. W. Tabor, and H. Tabor, *J. Biol. Chem.* **255,** 9082 (1980).

[2] EC 2.5.1.6. The gene for this enzyme is designated *metK*.

[3] AdoMet, *S*-Adenosylmethionine.

[4] Other assays for adenosylmethionine formation can also be used (see this volume [8, 9], and this series, Vol. 17B [190, 191]).

[5] For some experiments in which labeled nucleotides are used, we have found that background counts could be minimized by pretreating the ion-exchange filters with 25 m*M* sodium tripolyphosphate and drying them at room temperature before use.

ISBN 0-12-181994-9

The papers from several assays are spread on a Büchner funnel and washed with 4 liters of distilled water. The filters are then placed in vials with Aquasol or Omnifluor scintillation fluid and counted.

Definition of Unit. One unit of enzyme activity is defined as the amount of enzyme producing 1 μmol of adenosylmethionine per minute under the above conditions.[6]

Purification

Growth of Culture. For the purification of AdoMet synthetase, a strain of *E. coli* K12 was developed that contained a *metK* plasmid in a *metJ* background. The *metK* plasmid was found in the Clarke–Carbon collection in position 27–37 and was originally detected as a plasmid containing *speA speB* (markers adjoining *metK*). This plasmid was transferred by standard techniques into EWH205; the resultant strain was designated EWH205/pLC 27-37 (*thyA argG6 metJ47 rpsL* rif^R *recA*).[8]

Cells for the enzyme preparation are grown with aeration at 37° in LB medium[9] to stationary phase, harvested, and stored at −70°. Under these growth conditions, the AdoMet synthetase levels in the plasmid-containing *metJ* strain are 80-fold greater than in a $metJ^+$ strain without this plasmid, 20-fold greater than in a $metJ^+$ strain that does carry this plasmid, and 5-fold greater than the enzyme level in a $metJ^-$ strain that does not carry the plasmid.

Fifty (50) grams of cells are suspended in 200 ml of 0.1 *M* Tris-HCl, pH 8.0, containing 3 μ*M* phenylmethylsulfonyl fluoride, and the cells are broken by a single pass through a French press at 15,000 psi.[10] To the resultant solution a fresh 5% solution of streptomycin sulfate is added to a final concentration of 1%. The viscous solution is stirred for 10 min and then centrifuged for 15 min at 15,000 *g*.

Ammonium Sulfate Fractionation. For each 100 ml of the supernatant solution, 22 g of ammonium sulfate are added; after stirring for 15 min the

[6] Adenosylmethionine synthetase preparations have a tripolyphosphatase activity which is stimulated by adenosylmethionine.[7] With purified preparations of the enzyme, this tripolyphosphatase activity can be assayed with inorganic tripolyphosphate as the substrate and a colorimetric assay for the amount of inorganic phosphate formed.

[7] S. H. Mudd, *J. Biol. Chem.* **238,** 2156 (1963).

[8] This strain may be obtained from the authors. We have also developed a plasmid with a much smaller DNA insert that still contains the *metK* gene, but the strain containing this plasmid has not been used for the purification procedure. (See this volume [17].)

[9] LB medium contains 10 g of Bacto-tryptone, 5 g of Bacto-yeast extract, and 10 g of NaCl per liter.

[10] All purification steps are performed at 0–4°. All buffers contain 10% glycerol, 1 m*M* EDTA, and 0.1% mercaptoethanol.

solution is centrifuged. For each 100 ml of the supernatant solution 12 g of ammonium sulfate are added; after stirring for 15 min the precipitated enzyme is collected by centrifugation. The precipitate is dissolved in 150 ml of 10 m*M* Tris-HCl, pH 8.0, containing 0.75 *M* $(NH_4)_2SO_4$.[11]

Phenyl-Sepharose Chromatography. The enzyme is then loaded onto a column of phenyl-Sepharose (Pharmacia, 2.5 × 30 cm) that has been equilibrated with 10 m*M* Tris-HCl, pH 8.0, 0.75 *M* $(NH_4)_2SO_4$. After loading, the column is washed with 2 volumes of the equilibration buffer, and the AdoMet synthetase is eluted with a reverse gradient of 0.75 to 0 *M* ammonium sulfate in 600 ml of 10 m*M* Tris-HCl, pH 8.0. Fractions with AdoMet synthetase activity are pooled, concentrated in an Amicon pressure cell with a XM-50 membrane, and dialyzed against 50 volumes of 50 m*M* Tris-HCl, pH 8.0.

DEAE-52 Cellulose. The dialyzed enzyme is loaded onto a 2.5 × 20 cm column of DEAE-52 cellulose (Whatman) that has been equilibrated with the dialysis buffer. After washing the loaded column with 2 column volumes of the dialysis buffer, AdoMet synthetase is eluted with a linear gradient of 0 to 350 m*M* KCl in 400 ml of 50 m*M* Tris-HCl, pH 8.0. Fractions containing AdoMet synthetase activity are pooled and concentrated to 5–10 ml in an Amicon pressure cell or a collodion bag apparatus.

Sephacryl S-300. The concentrated enzyme is then loaded onto a Sephacryl S-300 column (2.5 × 90 cm, flowing at 20 ml/hr) that has been equilibrated with 50 m*M* Tris-HCl, pH 8.0, and 50 m*M* KCl buffer. AdoMet synthetase is eluted as the major protein peak. Final purification is obtained by chromatography on a column of DEAE-Sephadex A-50 (1.6 × 10 cm) that has been equilibrated with the 50 m*M* Tris-HCl, pH 8.0, and 50 m*M* KCl buffer. AdoMet synthetase is eluted with a linear gradient of 50 to 300 m*M* KCl in 250 ml. The pure enzyme is concentrated to approximately 5 mg/ml and stored at −70°. A typical purification is presented in the table.[1,12]

Properties

Substrate Specificity. 3′-Deoxy-ATP is a good substrate for adenosylmethionine synthetase, as is 8-bromo-ATP, the A isomer of ATPβS, and the fluorescent analog formycin triphosphate. Selenomethionine can substitute for methionine as a substrate.

[11] The enzyme at this stage is suitable for the preparation of labeled adenosylmethionine, as described in this series, Vol. 17B [190]. It can also be used to prepare substrate quantities of adenosylselenomethionine (T. C. Stadtman, personal communication).

[12] Small amounts of enzyme can be crystallized by a vapor diffusion (hanging drop) method [G. L. Gilliland, G. D. Markham, and D. R. Davies, *Proc. Am. Crystallogr. Assoc.* **8,** 43 (1980)].

PURIFICATION OF ADENOSYLMETHIONINE SYNTHETASE[a,b]

Step	Volume (ml)	Protein[c] (mg/ml)	Total activity (μmol/min)	Specific activity (μmol/min/mg)	Yield (%)	Purification (fold)
Streptomycin sulfate supernatant	258	64.0	100	5.9×10^{-3}	100	1.0
Ammonium sulfate fractionation (redissolved)	150	50.0	84	1.1×10^{-2}	84	1.9
Phenyl-Sepharose (dialyzed)	61	16.6	75	7.4×10^{-2}	75	12.5
DEAE-52 cellulose	72	2.4	73	4.2×10^{-1}	73	71.0
Sephacryl S-300	20	3.4	44	6.5×10^{-1}	44	110.0
DEAE-Sephadex A-50	47	1.2	35	7.5×10^{-1}	35	126.0

[a] Preparation of AdoMet synthetase from 50 g of *E. coli* K12, strain EWH205/pLC 27-37.

[b] The purified enzyme migrates as a single band in polyacrylamide gel electrophoresis both in the presence and the absence of sodium dodecyl sulfate. The purified enzyme is stable indefinitely at −70° and for at least several hours at 25°.

[c] During the enzyme purification, protein concentrations are estimated by assuming that a 1 mg/ml solution has an absorbance at 280 nm of 1.0. For purified AdoMet synthetase, an extinction coefficient, $A_{280\,nm}^{1\%,\,1\,cm}$ of 13 is determined by drying to constant weight at 100° *in vacuo*.

Molecular Weight. The molecular weight of native adenosylmethionine synthetase is 180,000 (by gel filtration). The subunit molecular weight is 43,000.

Cation Requirements. Adenosylmethionine synthetase requires a divalent cation for activity, such as Mg^{2+}, Mn^{2+}, or Co^{2+}. For maximal activation the concentration of the cation must be at least equal to the ATP concentration. The enzyme is markedly stimulated by monovalent cations.

Inhibitors. The enzyme is inhibited competitively with respect to ATP by adenosylmethionine, adenyl-5′-ylimidodiphosphate (AMP-PNP), and tripolyphosphate. *S*-Carbamoylcysteine is a competitive inhibitor with respect to methionine, but noncompetitive with respect to ATP. PP_i and P_i are noncompetitive inhibitors for both ATP and methionine.

Tripolyphosphatase Activity.[6] The tripolyphosphatase activity requires a divalent cation, a monovalent cation, and adenosylmethionine for optimal activity. When the nucleotide analog AMP-PNP is used, the formation of AdoMet can be dissociated from the tripolyphosphatase activity[1,13] (i.e., AdoMet and imidotriphosphate are formed).

[13] G. D. Markham, *J. Biol. Chem.* **256,** 1903 (1981).

[36] Fractionation of Methionine Adenosyltransferase[1] Isozymes (Rat Liver)

By JERALD L. HOFFMAN

S-Adenosylmethionine (AdoMet) is important as a direct metabolic donor of methyl and α-amino-*n*-butyryl groups. More pertinent to this volume is the role of AdoMet (after decarboxylation) as an aminopropyl donor for spermidine and spermine synthesis. AdoMet is synthesized from methionine (Met) and ATP in a reaction catalyzed by methionine adenosyltransferase (MAT) as follows:

$$\text{Met} + \text{ATP} \rightarrow \text{AdoMet} + \text{PP}_i + \text{P}_i$$

The enzyme is also commonly known as AdoMet synthetase; the systematic name is ATP:L-methionine-*S*-adenosyltransferase (EC 2.5.1.6).

Although MAT has been known for almost 30 years, homogeneous preparations have only recently been obtained. Chiang and Cantoni[2] purified MAT from yeast and obtained two closely related forms, but yields were quite low. Markham *et al.*[3] purified the enzyme from a recombinant strain of *E. coli* derepressed for MAT synthesis and carrying additional *E. coli* MAT genes in a plasmid. This strain gave high yields on the order of 1 mg per gram of cells.

We have given a preliminary report on the first purification of a mammalian (rat liver) MAT.[4] In the course of this work it became apparent that rat liver contains three forms of MAT. The following method permits separation of these three forms and purification of one to homogeneity.

Assay Method

Principle. Methionine adenosyltransferase is assayed by measuring the conversion of [^{35}S]methionine to [^{35}S]AdoMet. The strongly cationic AdoMet is isolated by spotting reaction mixtures on cellulose phosphate paper disks and washing off unreacted methionine with ammonium formate buffer.

[1] See also this volume [35] and [64].

[2] P. K. Chiang and G. L. Cantoni, *J. Biol. Chem.* **252,** 4506 (1977).

[3] G. D. Markham, E. W. Hafner, C. W. Tabor, and H. Tabor, *J. Biol. Chem.* **255,** 9082 (1980).

[4] J. L. Hoffman and G. L. Kunz, *Fed. Proc., Fed. Am. Soc. Exp. Biol.* **39,** Abstr. 448 (1980).

ISBN 0-12-181994-9

Reagents and Supplies

Assay mixture: 300 mM KCl, 15 mM $MgSO_4$, 4 mM dithiothreitol, 100 mM N-2-hydroxyethylpiperazine-N'-2-ethanesulfonic acid (HEPES), 10 mM ATP, and 50 μM [^{35}S]methionine (specific activity approximately 400 Ci/mol). The pH is adjusted to 7.5 with KOH prior to addition of ATP and methionine.

Ammonium formate, 0.1 M, adjusted to pH 3.0 with formic acid

Disks (2.3 cm in diameter) of Whatman P-81 phosphocellulose paper marked in pencil for identification

Liquid scintillation fluid: toluene-diluted Liquifluor is used routinely, but any comparable preparation would suffice.

Procedure. To 50 μl of assay mixture is added 50 μl of MAT-containing sample. Enzyme dilution or incubation time is adjusted in order that no more than 20% of [^{35}S]methionine is converted to AdoMet. Reaction mixtures are incubated at 37° and terminated by spotting 80 μl on phosphocellulose paper disks that are then placed in 0.1 M ammonium formate (total volume about 10 ml per disk). Once all the disks in a set of assays are collected, the ammonium formate is decanted and the disks are washed twice more with 0.1 M ammonium formate, once with ethanol, and once with ether. The ethanol and ether washes should each consist of about 10 ml per disk and may be used for up to 10 batches. Disks are allowed to dry in air and the [^{35}S]AdoMet on each is determined by counting under 5 ml of liquid scintillation fluid. The energy spectrum of ^{35}S is so similar to that of ^{14}C that ^{14}C push-button or module settings commonly found on counters may be used. Uniform counting efficiencies of near 25% are routinely obtained. Blank mixtures containing water instead of enzyme should be carried through the entire procedure. Such blanks should have less than 500 cpm as compared to about 500,000 cpm applied in the original reaction mixture. Standardization is most simply obtained by spotting, drying (without washing), and counting 50 μl of the assay mixture.

Purification

Reagents and Supplies

Homogenizing buffer: 250 mM sucrose, 2 mM dithiothreitol, 10 mM $MgSO_4$, 1 mM EDTA, 25 mM HEPES, adjusted to pH 7.5 with KOH

Buffer A: 2 mM dithiothreitol, 25 mM HEPES, adjusted to pH 7.5 with KOH

Buffer B: 0.3 *M* KCl in buffer A
Buffer C: 2 m*M* dithiothreitol, 10 m*M* *N,N*-bis(2-hydroxyethyl)glycine (Bicine), adjusted to pH 8.5 with KOH
Buffer D: 40% (v/v) dimethyl sulfoxide (DMSO) in buffer C
Buffer E: 150 m*M* KCl, 5 m*M* $MgSO_4$ in buffer A
DEAE-Sephacel column (Pharmacia Fine Chemicals): 1 ml bed volume for each gram of liver; packed and equilibrated with buffer A
Phenyl-Sepharose column (Pharmacia Fine Chemicals): 0.5 ml bed volume per gram of liver; packed and equilibrated with buffer B
Sephacryl S-300 column (Pharmacia Fine Chemicals): 2.5 × 90 cm; packed and equilibrated with buffer E
Ultrafiltration equipment having a membrane with a molecular weight cutoff of about 30,000. We generally use Amicon YM-30 membranes.

Procedure. All operations are done in the cold (0–5°) unless stated otherwise. Female Sprague–Dawley rats weighing 200–300 g are used, although other strains would be satisfactory. Animals are anesthetized with ether, decapitated, and allowed to bleed before livers are removed. Livers are weighed, minced, and then disrupted in 4 ml of homogenizing buffer per gram of liver using a glass homogenizer with a motor-driven Teflon pestle. The homogenate is centrifuged at 100,000 *g* for 1 hr.

The resulting supernatant is applied to the DEAE-Sephacel column. The column is washed with buffer A until the $A_{280\ nm}$ is near zero, at which time a linear gradient from buffer A to buffer B is initiated. The total gradient volume should be 10 times the column bed volume. The flow rate may be up to 40 ml/cm^2 per hour. Collected fractions should be assayed for $A_{280\ nm}$ and for MAT activity. The MAT elutes during the gradient at 0.15 to 0.2 *M* KCl on the trailing edge of the broad $A_{280\ nm}$ peak. There is slight separation of the MAT forms from each other, and so the fractions saved for the next step should include the entire peak of MAT activity, not just the major fractions.

The pooled DEAE fraction should be mixed with solid KCl to elevate the KCl concentration by an additional 0.2 *M* before application to the phenyl-Sepharose column. This column is eluted stepwise with five bed volumes each of buffers B, C, and then D at a flow rate of 40 ml/cm^2 per hour. Fractions are assayed for $A_{280\ nm}$ and MAT activity. A peak of MAT activity is eluted at each step (counting sample application plus buffer B as one step), which we designate MAT-I, -II, and -III in order of elution. MAT-I approximates 15%, MAT-II is 5%, and MAT-III is 80% of the total activity recovered.

Each MAT peak is separately handled from this point. After concentration to 5 ml by ultrafiltration, the MATs are further purified by gel filtration chromatography on the Sephacryl S-300 column, eluting at 100 ml/hr with buffer E and assaying for $A_{280\ nm}$ and MAT. This step yields homogeneous MAT-III, and there is generally an obvious $A_{280\ nm}$ peak coinciding with the MAT activity. MAT-I will have a more complex $A_{280\ nm}$ profile, while so little protein remains in the MAT-II preparation that it may be difficult to measure. The MATs are concentrated to 10 ml by ultrafiltration, divided into 1-ml portions, and stored frozen for further studies.

Notes. The following schedule has been found to be convenient. With all columns and buffers ready, homogenates are prepared about 11:00 A.M. the first day. The ultracentrifuged supernatant is then applied to the DEAE column in time to complete the wash with buffer A and initiate the gradient for an overnight run. Assay of these fractions can be finished sufficiently early the next morning to enable completion of the phenyl-Sepharose column step the second day. If the sample plus buffer B eluent from the phenyl-Sepharose column is assayed immediately after elution, MAT-I can be located, concentrated, and run on the Sephacryl S-300 column overnight after the second day. MAT-II can be run on this column the third day, and MAT-III run overnight after the third day.

Results. A typical purification is shown in Table I. Particularly noteworthy are the excellent yields of total activity and of homogeneous (on polyacrylamide gel electrophoresis) MAT-III. The yield of MAT-III approximates 1 mg per liver for rats of 200–300 g, which bodes well for further chemical, physical, and immunochemical studies. MAT-I and MAT-II are not completely purified by this technique, since both preparations show 4–6 proteins on polyacrylamide gel electrophoresis.

Properties

The properties of the three MATs so obtained are summarized in Table II. We find the MAT-I, -II, -III designation useful, since this is not only the order of elution from phenyl-Sepharose, but also indicates the relative order of molecular weights from high to low. MAT-III appears to be composed of 2 identical subunits. The kinetic properties of the three forms are quite distinct. MAT-II has the lowest K_m (Met), is negatively cooperative as indicated by the Hill coefficient of 0.7, and is inhibited weakly by DMSO and very strongly by AdoMet. MAT-I has an intermediate K_m (Met), has no cooperativity, is not affected by DMSO, and is weakly inhibited by AdoMet. MAT-III has strong positive cooperativity as evidenced by a Hill coefficient of 1.9. This prevents determination of K_m, but $S_{0.5}$ is estimated at 205 μM Met by inspection of the Michaelis–

TABLE I
PURIFICATION OF RAT LIVER METHIONINE ADENOSYLTRANSFERASES (MAT-I, MAT-II, AND MAT-III)

Step	Volume (ml)	Activity (units[a]/ml)	Protein (mg/ml)	Specific activity (units/mg)	Yield (%)	Purification (fold)
Homogenate[b]	103	91	22.9	4.0	100	1
DEAE-Sephacel	95	82	2.5	33	83	8.3
Phenyl-Sepharose						
I	163	4.2	0.11	40	7.3	10.0
II	45	4.6	0.067	69	2.2	17.4
III	138	41	0.068	610	61	154
				Total	70.5	
Sephacryl S-300						
I	36	10.4	0.14	74	4.0	18.8
II	36	3.4	0.006	610	1.3	153
III	40	97	0.077	1260	42	318
				Total	47.3	

[a] Unit = nanomoles of AdoMet per hour at 5 mM ATP, 25 μM Met.
[b] Three rat livers, 27 g total weight.

Menten curve. Most unusual is the strong activation of MAT-III by both DMSO and AdoMet.

Fractionation and kinetics studies on rat tissues other than liver generally show the presence of only MAT-II. This indicates a unique ability for the liver, having MAT-III, to maintain high rates of AdoMet synthesis in the presence of high levels of Met and AdoMet.

TABLE II
PROPERTIES OF METHIONINE ADENOSYLTRANSFERASES (MAT-I, MAT-II, AND MAT-III)

MAT	M_r[a]	Subunits[b]	K_m or *$S_{0.5}$ (μM Met)	Hill coefficient	Effect[c] of DMSO	Effect[c] of AdoMet
I	208,000	—	42	1.0	0	−
II	120,000	—	5.4	0.7	−	−−−
III	97,000	2 × 47,000	*205	1.9	+++	++

[a] From calibrated gel filtration chromatography.
[b] From polyacrylamide gel electrophoresis in sodium dodecyl sulfate.
[c] 0, no effect; −, inhibition; +, activation. DMSO, Dimethyl sulfoxide; AdoMet, S-adenosylmethionine.

Its ready availability in high yields by the above procedure, its ability to respond to high concentrations of Met, and the lack of inhibition by its product AdoMet make MAT-III valuable for the preparation *in vitro* of AdoMet in either nonradioactive form or with various specific radioactive labels. Such AdoMet preparations are required for different chromatographic or enzymatic studies, several of which are related to polyamine biochemistry.

[37] *S*-Adenosylmethionine Decarboxylase (*Escherichia coli*)

By GEORGE D. MARKHAM, CELIA WHITE TABOR, and HERBERT TABOR[1,2]

S-Adenosyl-L-[^{14}COOH]methionine (AdoMet) → $^{14}CO_2$ + decarboxylated *S*-adenosylmethionine

Assay Method

Principle. The $^{14}CO_2$ formed from *S*-adenosyl-L-[^{14}COOH]methionine by the enzyme is trapped on filter paper and counted.

Reagents

HEPES/KOH, 1 *M*, pH 7.4
KCl, 1 *M*
EDTA, 10 m*M*
$MgCl_2$, 1 *M*
Adenosyl-L-[^{14}COOH]methionine, 1 m*M* (0.9 mCi/mmol)

Procedure. The assay mixture (in a scintillation vial) contains 20 μl of 1 *M* HEPES/KOH buffer, pH 7.4, 20 μl of 1 *M* KCl, 4 μl of 10 m*M* EDTA, 4 μl of 1 *M* $MgCl_2$, 16 μl of 1 m*M* [^{14}COOH]AdoMet (0.9 mCi/mmol), and 0.33 ml of H_2O. Enzyme (usually 10 μl) is added, and the vials are rapidly closed with caps into which a piece of Whatman No. 3 filter paper wet with 50 μl of 1 *M* Hyamine hydroxide has been inserted. Assay incubations, 1–10 min in duration, are performed at 25°. The assay is terminated by addition of 0.4 ml of 1 *M* KH_2PO_4. After 40 min at 37°, trapped $^{14}CO_2$ on the filter papers is determined by scintillation counting in Aquasol. One unit of enzyme activity is defined as the amount of enzyme that releases 1 μmol of $^{14}CO_2$ per minute under these assay conditions.

[1] G. D. Markham, C. W. Tabor, and H. Tabor, *J. Biol. Chem.* **257,** 12063 (1982).
[2] Another purification from a standard strain of *E. coli* was reported earlier from this laboratory [R. B. Wickner, C. W. Tabor, and H. Tabor, *J. Biol. Chem.* **245,** 2132 (1970)].

METHODS IN ENZYMOLOGY, VOL. 94

ISBN 0-12-181994-9

Growth of Organism. *Escherichia coli* strain HT 383/pSPD1 (*thi pro hsd lac gal ara*) carries a tetracycline-resistant hybrid plasmid that contains the *E. coli* gene for adenosylmethionine decarboxylase.[3] This organism produces eight times as much adenosylmethionine decarboxylase as the wild-type parent strain.[1] The cells are grown in a 50-liter vat at 37° with aeration to late log phase in minimal medium[4] containing 0.01% proline, 0.001% thiamin, 0.2% glucose, and 0.03 mg of oxytetracycline per milliliter. The cells are harvested with a Sharples centrifuge and stored frozen at −70° until used.

Purification Procedure

All steps of the purification are carried out at 0–4°. All buffers contain 0.5 m*M* EDTA and 0.1% 2-mercaptoethanol.

Step 1. Preparation of the Extract and Streptomycin Precipitation. Forty (40) grams of cells are suspended in 200 ml of 10 m*M* Tris-HCl, 0.1 m*M* phenylmethylsulfonyl fluoride, pH 8.0. The cells are broken by a single pass through a French press at 15,000 psi. To the resultant extract a 5% solution of streptomycin sulfate is added to a final concentration of 1%. After stirring for 15 min, the solution is centrifuged for 30 min at 15,000 *g*.

Step 2. MGBG-Sepharose[5] Affinity Chromatography. To the supernatant solution solid KCl is added to a final concentration of 0.6 *M*; 1 *M* $MgCl_2$ is added to a final concentration of 10 m*M*, and the pH of the extract is adjusted to 8.0 with 5 *N* KOH. The solution is then loaded onto a column (5 × 10 cm) of MGBG-Sepharose that has been equilibrated with 20 m*M* Tris-HCl, 10 m*M* $MgCl_2$, and 0.6 *M* KCl, pH 8.0. After loading, the column is washed with the equilibration buffer until no protein is

[3] This strain is available from the authors.

[4] H. J. Vogel and D. M. Bonner, *J. Biol. Chem.* **218,** 97 (1956). See also this series, Vol. 17A, p. 5.

[5] Synthesis of MGBG-Sepharose: Pegg and Pösö (see this volume [39]) introduced the use of a methylglyoxal bis(guanylhydrazone)-Sepharose (MGBG-Sepharose) column for the purification of liver adenosylmethionine decarboxylase. We have used a comparable adsorption step, although the MGBG-Sepharose preparation is different.[1] The MGBG is coupled to epoxy-activated Sepharose 6B by gently shaking 15 g of swollen gel (prepared as directed by the manufacturer) in 0.1 *M* MGBG, adjusted to pH 9.0 with KOH, for 24 hr at 37°. The reacted gel is washed with water and with 2 *M* NaCl until no ultraviolet absorbance from MGBG is detected in the eluate. The gel is stored at 4° in 0.02% NaN_3. The amount of MGBG bound to the gel (ca 3 μmol per milliliter of gel) is estimated from the absorbance at 325 nm of resin that is solubilized by boiling in 0.1 *N* NaOH; the extinction coefficient for MGBG in 0.1 *N* NaOH is $3.35 \times 10^4\ M^{-1}\ cm^{-1}$.

PURIFICATION OF AdoMet DECARBOXYLASE FROM *Escherichia coli* K12 STRAIN HT383/pSPD1[a]

Step	Volume (ml)	Protein (mg)	Total activity (μmol/min)	Specific activity (μmol/mg/min)	Yield (%)	Purification (fold)
Extract	255	3270	17.1	0.005	100	1
MGBG-Sepharose	140	13	10.7	0.82	64	160
DEAE-Cellulose	62	10	8.2	0.88	48	170

[a] Forty grams, wet weight.

detected in the eluate (ca 500 ml). AdoMet decarboxylase is then eluted by washing the column with 20 m*M* potassium phosphate and 0.6 *M* KCl, pH 7.0. The enzyme, which is estimated by electrophoresis to be approximately 90% pure, is concentrated to 10 ml with an Amicon pressure cell (XM-50 membrane) and diluted with 100 ml of 50 m*M* potassium phosphate, pH 7.4.

Step 3. DE-52 Cellulose Chromatography. The enzyme solution is then loaded onto a column of DE-52 cellulose (1.2 × 20 cm) that has been equilibrated with 50 m*M* potassium phosphate, pH 7.4. The enzyme activity is eluted near the middle of a linear gradient of 0.05 to 0.4 *M* KCl in 200 ml of 50 m*M* phosphate buffer, pH 7.4. The resultant enzyme, which is electrophoretically homogeneous, is concentrated to ca 5 mg/ml in an Amicon pressure cell and stored at 4°, where it is stable for at least several months.

A typical purification is presented in the table.

Properties

The native enzyme has a molecular weight of 108,000 and and contains 6 subunits of M_r 17,000. Each subunit has one covalently attached pyruvate which is essential for activity. The enzyme also requires divalent cations for activity; Mg^{2+}, Mn^{2+}, and Ca^{2+} are about equally active.

The enzyme is inactivated by phenylhydrazine. The enzyme is also inactivated by $NaCNBH_3$, but this inactivation occurs only in the presence of substrate or product and Mg^{2+}, resulting in the binding of the product, decarboxylated adenosylmethionine, to the enzyme. Decarboxylated adenosylmethionine and MGBG also inhibit the enzyme; this inhibition is competitive with respect to adenosylmethionine.

[38] *S*-Adenosylmethionine Decarboxylase (*Saccharomyces cerevisiae*)[1]

By MURRAY S. COHN, CELIA WHITE TABOR, and HERBERT TABOR

S-Adenosyl-L-[^{14}COOH]methionine → decarboxylated adenosylmethionine + $^{14}CO_2$[1]

Assay Method

Principle. The assay depends upon the measurement of the $^{14}CO_2$ released from *S*-adenosyl-L-[^{14}COOH]methionine.

Reagents

Tris-HCl, 1 *M*, pH 7.2
Dithiothreitol, 0.1 *M*
EDTA, 0.01 *M*
Putrescine dihydrochloride, 0.1 *M*
Bovine serum albumin, 10 mg/ml
Hyamine hydroxide, 1 *M*
KH_2PO_4, 1 *M*
S-Adenosyl-L-[^{14}COOH]methionine, 0.2 m*M* (54.6 μCi/μmol)

Procedure. The reaction is carried out in screw-capped scintillation vials. The incubation mixture (in a vial) contains 0.04 ml of 1 *M* Tris-hydrochloride buffer, pH 7.2, 4 μl of 0.1 *M* dithiothreitol, 4 μl of 0.01 *M* EDTA, 10 μl of 0.1 *M* putrescine dihydrochloride, 4 μl of bovine serum albumin (10 mg/ml), water, and enzyme in a total volume of 0.4 ml. The vial is incubated for 5 min at 37°. Then 1 nmol (5 μl) of *S*-adenosyl-L-[^{14}COOH]methionine (54.6 μCi/μmol, New England Nuclear) is added to the reaction mixture, and 20 μl of Hyamine hydroxide are placed on a 1.8 cm square of Whatman No. 3 filter paper wedged in the cap. The cap is closed immediately and the vial is incubated at 37° for 30 min. The reaction is stopped by the addition of 0.2 ml of 1.0 *M* KH_2PO_4. The vial is immediately closed again and shaken for 1 hr at room temperature to facilitate the release of $^{14}CO_2$. The paper containing the trapped $^{14}CO_2$ is transferred to another vial containing 5 ml of Aquasol for determination of radioactivity in a liquid scintillation counter. If less than 10% of the substrate has been depleted, the assay is linear with time and enzyme concentration. One unit of activity is defined as the amount of enzyme catalyzing the formation of 1 nmol of $^{14}CO_2$ per minute.

[1] M. S. Cohn, C. W. Tabor, and H. Tabor, *J. Biol. Chem.* **252,** 8212 (1977).

METHODS IN ENZYMOLOGY, VOL. 94

ISBN 0-12-181994-9

Purification of *S*-Adenosylmethionine Decarboxylase

Preparation of Crude Extract. Saccharomyces cerevisiae (*a ade3 leu1 pep4*) is grown in a 50-liter fermentor with vigorous aeration, to late logarithmic phase at 30° in 2% dextrose, 2% Bacto-peptone, 1% Bacto-yeast extract, and 0.04% adenine sulfate. The pH is automatically maintained at 6.25 with 4 *N* NaOH. The cells (approximate wet weight 800 g from 50 liters) are harvested in a refrigerated Sharples centrifuge and are then suspended in an equal volume of 50 m*M* Tris-HCl (pH 7.2), containing 1.0 m*M* dithiothreitol, 0.1 m*M* EDTA, and 2.5 m*M* putrescine dihydrochloride (buffer I). The suspension is passed through a Gaulin homogenizer at 12,000 psi four times with cooling to 4° after each pass. The homogenate is centrifuged at 16,000 *g* for 30 min, and the pH of the supernatant fluid is immediately adjusted to 7.2 with 4 *N* KOH (fraction I). The temperature is maintained between 2 and 4° in all the following steps.

Protamine Sulfate. A volume of 4% protamine sulfate in buffer I equal to 20% of the crude extract volume is added dropwise to fraction I over a period of 15 min with constant stirring. Stirring is continued for 30 min, the mixture is then centrifuged at 16,000 *g* for 30 min, and the supernatant fluid is collected (fraction II).

DEAE-52 Cellulose Chromatography. A DEAE-52 column (5 × 50 cm) is equilibrated with 100 m*M* Tris-HCl (pH 7.2) containing 1.0 m*M* dithiothreitol, 2.5 m*M* putrescine dihydrochloride, and 0.1 m*M* EDTA (buffer II). The conductivity of fraction II is lowered to that of a 100 m*M* solution of Tris-HCl by dilution with a solution of 1.0 m*M* dithiothreitol, 2.5 m*M* putrescine dihydrochloride, and 0.1 m*M* EDTA (usually 1 ml of diluting buffer per 2 ml of enzyme solution). The sample is then applied to the column, followed by a wash with one column volume of buffer II. A linear gradient, consisting of 1 liter of buffer II and 1 liter of 220 m*M* Tris-HCl (pH 7.2) containing 1.0 m*M* dithiothreitol, 2.5 m*M* putrescine dihydrochloride, and 0.1 m*M* EDTA, is applied. The peak of enzymatic activity emerges at approximately 170 m*M* Tris-HCl, and the most active fractions are pooled (fraction III).

Hydroxyapatite Chromatography. Fraction III is applied to a 2.5 × 30 cm column of hydroxyapatite that has been equilibrated with buffer I, and the column is then washed with one column volume of buffer I. A linear gradient (500 ml of buffer I and 500 ml of buffer I containing 100 m*M* potassium phosphate, pH 7.2) is applied to the column. The peak of activity emerges at approximately 40 m*M* potassium phosphate, and the most active fractions are pooled (fraction IV).

PURIFICATION OF S-ADENOSYLMETHIONINE DECARBOXYLASE

Fraction	Step	Volume (ml)	Total protein (mg)	Total activity[a] (units)	Specific activity[b]	Yield (%)	Purification (fold)
I	Crude extract[c]	1,200	85,000	760	0.009	100	1.0
II	Protamine sulfate	1,200	28,500	700	0.025	92	2.7
III	DEAE-cellulose	1,900	7,300	550	0.076	73	8.4
IV	Hydroxyapatite	51	570	275	0.49	36	54
V	MGBG-Sepharose I	9.9	7.6	215	28	28	3,100
VI	MGBG-Sepharose II	2.5	1.7	180	106	24	11,800

[a] One unit = 1 nmol of $^{14}CO_2$ released per minute.
[b] Units per milligram of protein.
[c] The purification is carried out on the cells obtained from 50 liters of early stationary-phase yeast (800 g of paste).

MGBG-Sepharose I.[2] A column (5 × 25 cm) is equilibrated with buffer I, and fraction IV is applied. The column is then washed with 500 ml each of buffer I and buffer I containing 0.3 *M* NaCl. The enzymatic activity is then eluted by buffer I containing 0.3 *M* NaCl and 1.0 m*M* MGBG.[4] Activity emerges at the front of the MGBG peak, which is monitored by its absorbance at 277 nm. Enzyme fractions are pooled and concentrated to 10 ml with an Amicon ultrafiltration unit containing a PM-10 ultrafilter, and the concentrated sample is dialyzed against 1 liter of buffer I overnight (fraction V).

MGBG-Sepharose II. The fraction V enzyme is applied to a second column (2.5 × 40 cm), which is equilibrated and eluted as described above except that washes are only of 200 ml volume. The active fractions are pooled, concentrated to 2.5 ml, and dialyzed as before. This enzyme preparation (fraction VI) is stored at 2–4°.

A typical purification is shown in the table.

Properties

The pure enzyme is homogeneous upon electrophoresis. The native enzyme has a molecular weight of 88,000. Upon SDS–acrylamide electrophoresis, a subunit of M_r 41,000 is found.

[2] MGBG-Sepharose is prepared as described by Pegg.[3] After use, the MGBG-Sepharose is transferred to a sintered-glass funnel (500 ml packed volume) and is regenerated by four 1-liter washes of 1 *M* NaCl, followed by two 1-liter washes of distilled water.

[3] A. E. Pegg, *Biochem. J.* **141,** 581 (1974). See also this volume [39].

[4] MGBG, Methylglyoxal bis(guanylhydrazone).

The enzyme is stable during storage at 2° for 3 months, but loses activity rapidly at 37° or upon freezing at −70°. Bovine serum albumin at 100 μg/ml stabilizes the activity at 37°. The enzyme is inactivated by reduction with sodium borohydride, and by other carbonyl-binding reagents. The carbonyl group, which is essential for activity, is due to the presence of 1 mol of covalently linked pyruvate per mole of subunit. Pyridoxal phosphate does not serve as a cofactor. The yeast adenosylmethionine decarboxylase resembles other eukaryote adenosylmethionine decarboxylases in that putrescine, but not Mg^{2+}, is required for activity. The enzyme is inhibited by MGBG.[1]

[39] *S*-Adenosylmethionine Decarboxylase (Rat Liver)[1]

By Anthony E. Pegg and Hannu Pösö

This enzyme (EC 4.1.1.50) catalyzes the conversion of *S*-adenosylmethionine to *S*-5′-deoxyadenosyl-(5′)-3-methylthiopropylamine (decarboxylated *S*-adenosylmethionine) and CO_2. Preparation of the bacterial enzyme from *Escherichia coli,* which is Mg^{2+} dependent, has been described in an earlier volume.[2] The mammalian enzyme is strongly activated by putrescine.[3]

Assay Method

Principle. The activity is most conveniently assayed by following $^{14}CO_2$ release from *S*-adenosyl-L-[*carboxy*-^{14}C]methionine.

Reagents

Sodium phosphate buffer, 0.5 *M*, pH 7.5
Dithiothreitol, 25 m*M*
Putrescine dihydrochloride, 15 m*M*, adjusted to pH 7.0
S-Adenosyl-L-methionine, 4 m*M*
S-Adenosyl-L-[*carboxy*-^{14}C]methionine, 20 μCi/ml (specific activity, 56 mCi/mmol)
Enzyme: 0.1 to 0.001 unit is used for assay.

[1] For the preparation of this enzyme from other sources, see this volume [37] and [38].
[2] R. B. Wickner, C. W. Tabor, and H. Tabor, this series, Vol. 17B, p. 647.
[3] A. E. Pegg and H. G. Williams-Ashman, *J. Biol. Chem.* **244,** 682 (1969).

ISBN 0-12-181994-9

Procedure. Place 25 μl of 0.05 *M* phosphate buffer, 12.5 μl of 25 m*M* dithiothreitol, 50 μl of 15 m*M* putrescine, 12.5 μl of 4 m*M* unlabeled *S*-adenosylmethionine, 5–10 μl of *S*-adenosyl-L-[*carboxy*-^{14}C]methionine (0.1–0.2 μCi), enzyme, and water (final volume 250 μl) in small test tubes. Reactions are initiated by addition of the enzyme, and the tubes are immediately closed with rubber stoppers that carry a polypropylene well (stoppers and wells are supplied by the Kontes Glass Company, Vineland, New Jersey). The wells contain 200 μl of 1 *M* Hyamine hydroxide. After incubation for 30–120 min, 0.2 ml of 5 *N* sulfuric acid is injected through the rubber stopper in order to stop the reaction and to release $^{14}CO_2$ from the medium. The measurements are carried out at 37°. After a further 30 min of incubation (to ensure complete absorption of $^{14}CO_2$), the well is removed and placed together with its contents into a scintillation vial containing 5 ml of toluene-based scintillation fluid and counted in a liquid scintillation spectrometer. A reagent blank is made up by adding 0.2 ml of 5 *N* sulfuric acid to the assay tube before adding the enzyme.

Definition of Unit. One unit of activity is defined as the amount of enzyme releasing 1 nmol of $^{14}CO_2$ per minute at 37°. Specific activity is expressed as units per milligram of protein.

Purification Procedure

Step 1. Induction of Enzyme and Preparation of Extracts. The basal level of adenosylmethionine decarboxylase is low in rat tissues. A convenient way to increase it is to treat the rats with methylglyoxal bis-(guanylhydrazone) (MGBG[4]), which increases activity at least 20-fold in rat liver.[5] MGBG obtained as the dihydrochloride salt is dissolved in 0.9% NaCl at a concentration of 40 mg/ml and injected at a dose of 80 mg/kg 20–24 hr prior to the death of the animals. Livers from 35 male rats (weighing 400–500 g) are homogenized with two volumes of ice-cold 25 m*M* sodium phosphate, pH 7.5, 2.5 m*M* dithiothreitol, 2.5 m*M* putrescine, and 0.1 m*M* EDTA (buffer A). All steps are carried out at 4°. The homogenate is then centrifuged at 105,000 *g* for 30 min.

Step 2. Fractionation with Ammonium Sulfate. Proteins precipitated between 35 and 65% saturation with ammonium sulfate are collected, dissolved in 100 ml of buffer A, and dialyzed for 18 hr against 50 volumes of the same buffer.

Step 3. DEAE-Cellulose Chromatography. The dialyzed solution (final volume, 200 ml) is applied to a column (5 × 45 cm) of DEAE-cellulose

[4] The correct chemical name for the substance widely known as methylglyoxal bis(guanylhydrazone) and abbreviated here as MGBG is 1,1′-[(methylethanediylidene)dinitrilo]diguanidine. It is available commercially as the dihydrochloride salt.

previously equilibrated with buffer A without EDTA. The column, run at a flow rate of 80 ml/hr, is then washed with 200 ml of 0.1 *M* NaCl in buffer A and eluted with a linear gradient of 0.1 to 0.3 *M* NaCl in the same buffer (total volume, 2 liters). Fractions of 20 ml are collected and those fractions containing enzyme activity are pooled (the enzyme comes out at a salt concentration of 0.17–0.19 *M* NaCl) and made 70% saturated with ammonium sulfate. The protein precipitate is collected, dissolved in 40 ml of 10 m*M* Tris-HCl, pH 7.5, 2.5 m*M* putrescine, 2.5 m*M* dithiothreitol, and 0.1 m*M* EDTA (buffer B), and the solution is dialyzed against 100 volumes of the same buffer for 15 hr.

Step 4. MGBG-Sepharose Column Chromatography. In this final step the enzyme is purified to homogeneity by using its high affinity for MGBG.[5] The inhibitor is linked to 6-aminohexanoic acid-Sepharose 4B (Pharmacia Fine Chemicals, Uppsala, Sweden) as follows: 15 g dry weight of 6-aminohexanoic acid–Sepharose 4B is swollen in 1 liter of 0.5 *M* NaCl and then washed with 4 liters of 0.5 *M* NaCl by leaving it for 30 min with 500 ml of NaCl eight times. The resin is washed in distilled water with six changes (500 ml for 30 min) to remove NaCl and then suspended in as small a volume of distilled water as possible (about 80 ml) containing 2.5 g of MGBG; the pH is adjusted to 5.5 with 1 *M* NaOH. The mixture is stirred slowly, and 3.5 g of 1-ethyl-3-(3-dimethylaminopropyl)carbodiimide dissolved in 5 ml of water is added dropwise. The solution is shaken gently for 15 min, and then the pH is checked and adjusted if necessary to ensure that it is between 5 and 6. After this, 1 g more of the carbodiimide is added and the solution is shaken gently for 48 hr at room temperature. It should not be stirred with a magnetic stirrer, since this breaks up the gel. The pH is checked periodically and maintained at 5.5. The Sepharose is packed in a column, washed with 1 *M* NaCl in 10 m*M* Tris-HCl, pH 8.0 (three times with 1 liter), and stored in 0.5 *M* NaCl in 10 m*M* Tris-HCl, pH 7.5, at 4° before use.

A column (1.6 × 18 cm) is packed with the Sepharose linked to MGBG and washed with 500 ml of buffer B. The dialyzed DEAE-column eluate (final volume, 55 ml) is then applied to the column at a flow rate of 45 ml/hr, and the column is then washed with 200 ml of buffer B followed by 250 ml of buffer B containing 0.5 *M* NaCl. Adenosylmethionine decarboxylase is then eluted from the column by buffer B containing 0.5 *M* NaCl and 1 m*M* MGBG. Fractions (10 ml) are collected and assayed for enzyme activity by adding 5-μl samples to the standard assay mixture lacking unlabeled adenosylmethionine. The concentration of MGBG in the eluate

[5] A. E. Pegg, *Biochem. J.* **141,** 581 (1974).

PURIFICATION OF RAT LIVER ADENOSYLMETHIONINE DECARBOXYLASE

Fraction	Total protein (mg)	Total units	Specific activity (units/mg)	Purification (fold)	Yield (%)
Supernatant	21,400	1600	0.08[a]	1	100
Dialyzed $(NH_4)_2SO_4$ precipitate	8,200	1680	0.21	3	105
DEAE-cellulose eluate	260	1200	4.60	61	75
MGBG-Sepharose eluate	1.1	700	630	8430	43

[a] Activity as measured after dialysis to remove MGBG.

is measured by using the absorbance at 283 nm (ε equals 38,400 at pH 1). The liver enzyme usually elutes in fractions 3–5 after addition of MGBG to the elution buffer. The enzyme is then concentrated to a volume of about 1 ml using an Amicon ultrafiltration cell, and the volume is made up to 10 ml with buffer B. The solution is again concentrated to 1 ml, and this procedure is repeated five times to remove MGBG. Any remaining traces of MGBG are removed by dialysis for 48 hr against 1 liter of the same buffer.

A typical purification is summarized in the table.

Properties

The purified enzyme is stable for at least 6 weeks at 0–2° when putrescine is present. Storage in the presence of putrescine is essential, since the activity is lost rapidly without putrescine (80% in 14 hr at 2°). The enzyme has an M_r of 68,000, and the preparation gives a single band upon electrophoresis in the presence of sodium dodecyl sulfate, corresponding to an M_r of 32,500, suggesting that the enzyme has two subunits. The preparation gives a single band in isoelectric focusing having an isoelectric point of pH 5.7.[6] Liver adenosylmethionine decarboxylase resembles that from *E. coli*[2] in that it contains a covalently bound pyruvate prosthetic group,[7,8] but differs in that it is activated by putrescine,[3] whereas the bacterial enzyme requires Mg^{2+}.[2] The K_m for *S*-adenosylmethionine is about 50 μM in the presence of 2.5 m*M* putrescine and at pH 7.5.[6]

[6] H. Pösö and A. E. Pegg, *Biochemistry* **21,** 3116 (1982).

[7] A. A. Demetriou, M. S. Cohn, C. W. Tabor, and H. Tabor, *J. Biol. Chem.* **253,** 1684 (1978).

[8] A. E. Pegg, *FEBS Lett.* **84,** 33 (1977).

Other Sources

Putrescine-activated adenosylmethionine decarboxylases has also been purified from rat liver by Demetriou *et al.*,[7] as well as from yeast,[9,10] from mouse liver and mammary gland,[11] and from bovine liver.[12] This enzyme has also been purified from rat psoas muscle by a procedure similar to that described above.[6] However, the psoas enzyme differs from that of liver in its isoelectric point (5.3), degree of activation by putrescine, K_m for *S*-adenosylmethionine, and sensitivity to inhibition by MGBG.[6] The psoas enzyme is more strongly inhibited by MGBG and can readily be separated from the liver enzyme by the chromatography on MGBG-Sepharose described above, since it comes out later than the liver enzyme (by about three fractions). The greater sensitivity to inhibition by MGBG and the higher concentration of MGBG present in the later fractions, which contain the psoas enzyme, makes it essential to dialyze the fractions to remove inhibitor before assay in order to detect the enzyme.[6] Therefore, with preparations from other mammalian sources in which the form(s) of *S*-adenosylmethionine decarboxylase present are not known, it would be desirable to check all fractions from the MGBG-Sepharose column after removal of the inhibitor.

Preparation of Antibodies to Rat Liver *S*-Adenosylmethionine Decarboxylase

Antibodies to the purified rat liver enzyme were raised in New Zealand White female rabbits.[13] The rabbits were injected subcutaneously at multiple sites with 0.5 mg of enzyme emulsified in Freund's complete adjuvant. Starting 2 weeks after this treatment, three booster injections of 0.1 mg of enzyme in incomplete adjuvant were given; the rabbits were bled 10–30 days after the last injection. Immunoglobulins were prepared by fractionation with ammonium sulfate, and this antiserum was monospecific for *S*-adenosylmethionine decarboxylase by criteria of Ouchterlony double diffusion and immunoelectrophoresis.[13] The antiserum could be used for immunotitration of adenosylmethionine decarboxylase activity in ultracentrifuged tissue extracts. The extracts in 0.1 m*M* dithiothreitol, 2.5 m*M* putrescine, and 25 m*M* sodium phosphate buffer,

[9] H. Pösö, R. Sinervirta, and J. Jänne, *Biochem. J.* **151,** 67 (1975).

[10] M. S. Cohn, C. W. Tabor, and H. Tabor, *J. Biol. Chem.* **252,** 8212 (1977).

[11] T. Sakai, C. Hori, K. Kano, and T. Oka, *Biochemistry* **18,** 5441 (1979).

[12] C. E. Seyfried, O. E. Oleinik, J. L. Degen, K. Resing, and D. R. Morris, *Biochim. Biophys. Acta* **716,** 169 (1982).

[13] A. E. Pegg, *J. Biol. Chem.* **254,** 3249 (1979).

pH 7.0 (buffer B), were incubated with varying amounts of the antiserum in a total volume of 0.15 ml overnight at 4°. The tubes were then centrifuged at 10,000 rpm for 15 min, and the residual *S*-adenosylmethionine decarboxylase activity was measured as described above. The amount of antibody needed to inactivate 1 unit of enzyme was determined over the portion of the curve from 0 to 60% inhibition, which was linear with increasing addition of antiserum. Approximately 2 μg of immunoglobulin was needed per unit of enzyme activity, and this value was constant for a number of conditions under which the activity was altered, indicating that the alterations were due to changes in the amount of enzyme protein.[13]

The antiserum was also suitable for the precipitation of labeled adenosylmethionine decarboxylase, which was prepared by reduction of the enzyme at its pyruvate prosthetic group with tritiated sodium borohydride (205 Ci/mol). The enzyme (0.2 mg/ml) was incubated in 0.5 *M* $NaHCO_3$ with 1 mCi of NaB^3H_4 per milliliter (added in 0.1 ml of 50 m*M* NaOH) for 1 hr at room temperature. The solution was then dialyzed for 6 hr five times against 2 liters each time of 0.5 *M* $NaHCO_3$, lyophilized, and redissolved in buffer B. This material contained about 3260 dpm per microgram of protein and could be precipitated quantitatively with the antiserum on incubation as described above followed by addition of goat antirabbit IgG or protein A. This can be used as the basis of a radioimmunoassay for *S*-adenosylmethionine decarboxylase.

Antisera specific for mouse *S*-adenosylmethionine decarboxylase have also been prepared against the mouse[11] and bovine[12] liver enzymes.

[40] Inhibitors of *S*-Adenosylmethionine Decarboxylase

By ANTHONY E. PEGG

Three classes of inhibitors of *S*-adenosylmethionine decarboxylase are available: (*a*) methylglyoxal bis(guanylhydrazone) (MGBG[1]) and congeners, which are reversible inhibitors; (*b*) 1,1′-[(methylethanediylidene)-dinitrilo]bis(3-aminoguanidine) (MBAG[1]) which is an irreversible inhibitor; and (*c*) nucleosides related to *S*-adenosylmethionine and its decarboxylated derivative.

[1] The abbreviations used are MGBG for methylglyoxal bis(guanylhydrazone) (the correct name for this substance is 1,1′-[(methylethanediylidene)dinitrilo]diguanidine), MBAG for 1,1′-[(methylethanediylidene)dinitrilo]bis(3-aminoguanidine), and EGBG for ethylglyoxal bis(guanylhydrazone).

ISBN 0-12-181994-9

Methylglyoxal bis(guanylhydrazone) and Congeners

Synthesis. MGBG is commercially available as the hydrochloride salt (Aldrich Chemical Company, Milwaukee, Wisconsin). It is readily synthesized by the addition of 50 mmol of pyruvaldehyde (7 ml of 43% solution) to 60 mmol of aminoguanidine hydrochloride dissolved in 10 ml of 1 *N* HCl and 15 ml of ethanol. After 1 hr at 4° and a further 1 hr at 40°, during which time the mixture is stirred constantly, the reaction mixture is poured into 300 ml of acetone. The precipitate (90% yield) is collected by filtration, washed with cold acetone and ether, and dried *in vacuo*. A scaled-down method for this synthesis can be used to prepare the labeled compound from [^{14}C]aminoguanidine.[2] Similar synthetic methods using other substituted glyoxals and aminoguanidine to yield homologs in the methylglyoxal moiety have been published.[3,4]

Inhibition. As first reported by Williams-Ashman and Schenone,[5] prokaryotic putrescine-activated *S*-adenosylmethionine decarboxylases are strongly inhibited by MGBG. The inhibition is reversible and appears to be competitive with *S*-adenosylmethionine and uncompetitive with putrescine, but, owing to the complex kinetics for decarboxylation of *S*-adenosylmethionine, which varies with pH and the presence of the activator putrescine, most results have been expressed as the concentration of MGBG giving 50% inhibition under standard conditions. Enzymes from rat liver, prostate, and psoas muscle are inhibited by 31, 55, and 68%, respectively, by 1 μM MGBG at saturating putrescine and *S*-adenosylmethionine concentrations (Table I). The ethyl- or dimethylglyoxal homologs of MGBG are more potent inhibitors, but substitution of a butyl group reduces the inhibitory potency. Increasing the number of carbon atoms between the aminoguanidine moieties also greatly reduces the inhibition. Addition of methyl groups to the internal nitrogen atoms slightly increases potency, but addition to the terminal nitrogens greatly reduces it (Table I). *S*-Adenosylmethionine decarboxylase from psoas is more sensitive to inhibition by MGBG and all derivatives tested than is the liver enzyme.

[2] V. T. Oliverio and C. Denham, *J. Pharm. Sci.* **52,** 202 (1963).

[3] E. G. Podrebarac, W. H. Nyberg, F. A. French, and C. C. Cheng, *J. Med. Chem.* **6,** 283 (1963).

[4] F. Baiocchi, C. C. Cheng, W. J. Haggerty, L. R. Lewis, T. K. Liao, W. H. Nyberg, D. E. O'Brien, and E. G. Podrebarac, *J. Med. Chem.* **6,** 431 (1963).

[5] H. G. Williams-Ashman and A. Schenone, *Biochem. Biophys. Res. Commun.* **46,** 288 (1972).

[6] A. Corti, C. Dave, H. G. Williams-Ashman, E. Mihich, and A. Schenone, *Biochem. J.* **139,** 351 (1974).

[7] H. Pösö and A. E. Pegg, *Biochemistry* **21,** 3116 (1982).

At present, there is no convincing explanation for the strong inhibitory action of MGBG toward adenosylmethionine decarboxylases. Although the drug is more active against the putrescine-activated *S*-adenosylmethionine decarboxylases of eukaryotes when these are in the presence of putrescine, MGBG and its ethyl homolog are inhibitory to the enzyme from *E. coli,* which is Mg^{2+}-activated, and from the slime mold *Physarum polycephalum,* which requires no cofactors (Table II). The inhibition of these enzymes requires 20- to 50-fold higher concentrations but is still in the micromolar range and suggests a strong interaction between the enzyme and these inhibitors. Affinity chromatography on MGBG-Sepharose is very useful for the purification of mammalian and yeast *S*-adenosylmethionine decarboxylases[8,9] and should also work for the bacterial and slime mold enzymes.[9a] MGBG is not a potent inhibitor of transmethylases using *S*-adenosylmethionine, but inhibition of the decarboxylase is competitive with the substrate.[6,10] Guanethidine, which also contains an aminoguanidine group, is also claimed to inhibit prostate *S*-adenosylmethionine decarboxylase activity,[11] but it is much less active than MGBG. Aminoguanidine and diaminoguanidine do not inhibit decarboxylation at 1 m*M* concentrations.[5,12]

Specificity. Although MGBG is a potent inhibitor of adenosylmethionine decarboxylase, it is not totally specific, since it is an even more potent inhibitor of diamine oxidase, for which the K_i is 0.1 μM.[12]

Activity in Vivo. MGBG is taken up into mammalian cells by an active transport system[13] and is an effective inhibitor of spermidine synthesis *in vivo.*[14–16] However, the ability of cells to concentrate MGBG by an active transport process is such that very high intracellular contents are achieved,[15] and it is unclear whether the consequences of exposure to MGBG, such as reduced cellular growth rate and mitochondrial damage, are due to the effects on polyamine metabolism.[16] Since the active transport system also transports polyamines, there is competition for uptake, and reversal of inhibitory effects of MGBG by spermidine is not convincing evidence for these effects being mediated via polyamine depletion.

[8] A. E. Pegg, *Biochem. J.* **141,** 581 (1974).

[9] A. E. Pegg and H. Pösö, see this volume [38] and [39].

[9a] See also this volume [37].

[10] A. E. Pegg, *J. Biol. Chem.* **253,** 539 (1978).

[11] E. M. Johnson and A. S. Taylor, *Biochem. Pharmacol.* **29,** 113 (1980).

[12] A. E. Pegg and S. M. McGill, *Biochem. Pharmacol.* **27,** 1625 (1978).

[13] J. L. Mandel and W. F. Flintoff, *J. Cell. Physiol.* **97,** 335 (1978).

[14] C. E. Seyfried and D. R. Morris, *Cancer Res.* **39,** 4861 (1979).

[15] P. Seppänen, L. Alhonen-Hongisto, and J. Jänne, *Eur. J. Biochem.* **110,** 7 (1980).

[16] C. W. Porter, C. Dave, and E. Mihich, *in* "Polyamines in Biology and Medicine" (D. R. Morris and L. J. Marton, eds.), p. 407. Dekker, New York, 1981.

TABLE I

INHIBITION OF RAT PROSTATE, LIVER, AND PSOAS *S*-ADENOSYLMETHIONINE DECARBOXYLASE ACTIVITY BY MGBG AND RELATED COMPOUNDS

Inhibitor	Formula	Concentration	Percentage inhibition of enzyme from		
			Prostate[a]	Liver[b]	Psoas[b]
Methylglyoxal bis-(guanylhydrazone)	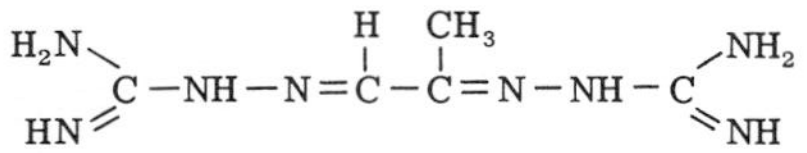 $H_2N(HN{=})C-NH-N{=}C(H)-C(CH_3){=}N-NH-C({=}NH)NH_2$	1 μM	55	31	68
Ethylglyoxal bis(guanyl-hydrazone)	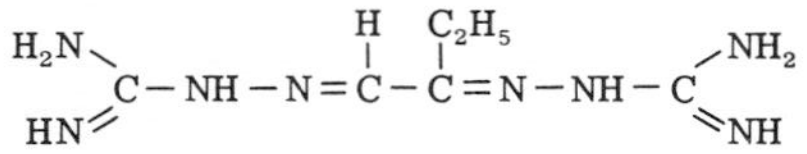 $H_2N(HN{=})C-NH-N{=}C(H)-C(C_2H_5){=}N-NH-C({=}NH)NH_2$	1 μM	90	80	95
Dimethylglyoxal bis(guanylhydrazone)	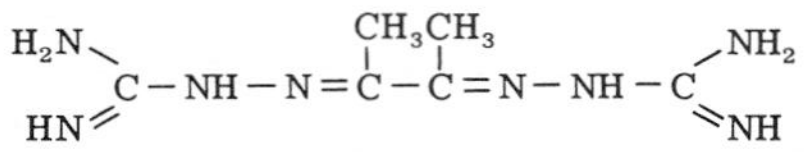 $H_2N(HN{=})C-NH-N{=}C(CH_3)-C(CH_3){=}N-NH-C({=}NH)NH_2$	1 μM	80	85	95
n-Butylglyoxal bis(guanylhydrazone)	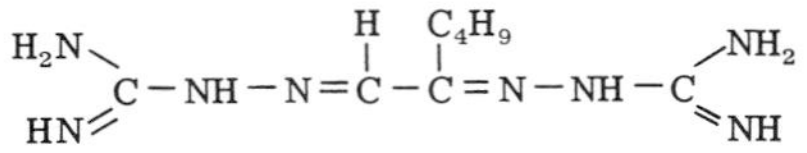 $H_2N(HN{=})C-NH-N{=}C(H)-C(C_4H_9){=}N-NH-C({=}NH)NH_2$	10 μM	ND	25	50

Propane dialdehyde bis(guanylhydrazone)	$(H_2N)(HN{=})C-NH-N{=}C(H)-CH_2-C(H){=}N-NH-C({=}NH)(NH_2)$	1 m*M*	35	15	25
Pentane dialdehyde bis(guanylhydrazone)	$(H_2N)(HN{=})C-NH-N{=}C(H)-[CH_2]_3-C(H){=}N-NH-C({=}NH)(NH_2)$	1 m*M*	75	30	50
Di-*N*‴-methylglyoxal bis(guanylhydrazone)	$(CH_3-NH)(HN{=})C-NH-N{=}C(H)-C(CH_3){=}N-NH-C({=}NH)(NHCH_3)$	10 μM	10	ND[c]	ND
Di-*N*″-methylglyoxal bis(guanylhydrazone)	$(H_2N)(HN{=})C-N(CH_3)-N{=}C(H)-C(CH_3){=}N-N(CH_3)-C({=}NH)(NH_2)$	1 μM	80	ND	ND

[a] From Corti *et al.*[6]
[b] From Pösö and Pegg.[7]
[c] ND, Not determined.

TABLE II
Inhibition of Rat Liver, *Escherichia coli*, Yeast, and Slime Mold *S*-Adenosylmethionine Decarboxylases by MGBG and EGBG

Source of enzyme	Concentration (μM) needed for 50% inhibition	
	MGBG	EGBG
Escherichia coli (assayed plus Mg^{2+})	105	15
Yeast (*Saccharomyces cerevisiae*) assayed plus putrescine	1.7	0.3
Physarum polycephalum	30	8
Rat liver (assayed plus putrescine)	1.8	0.3
Rat liver (assayed without putrescine)	45	NT[a]

[a] NT, Not tested.

The ethyl and dimethyl homologs of MGBG are much less effective *in vivo,* presumably owing to reduced rate of uptake.[15,16] MGBG stabilizes adenosylmethionine decarboxylase against intracellular degradation[17] and in this way produces a paradoxical increase in the amount of the enzyme it inhibits.

Measurement of MGBG Concentrations. MGBG is readily quantitated by its absorbance in the ultraviolet, which has a maximum of 283 nm at pH 1 (ε = 38,400) and at 325 nm at pH 11 (ε = 33,500). It can be assayed in biological samples either by high-performance liquid chromatography[18] or by an enzymatic assay.[19]

1,1′-[(Methylethanediylidene)dinitrilo]bis(3-aminoguanidine)

Synthesis. MBAG is prepared[4] by the dropwise addition of 50 mmol of pyruvaldehyde (7 ml of 43% solution) to 100 mmol of *N,N′*-diaminoguanidine dissolved in 175 ml of 0.1 *N* H_2SO_4 at 50°. The mixture containing a heavy precipitate is stirred for 1 hr; the precipitate is collected by centrifugation, washed with cold water, and recrystallized from water containing a few drops of sulfuric acid (yield 60%).

Inhibition in Vitro. MBAG inhibits mammalian *S*-adenosylmethionine decarboxylase irreversibly, and activity could not be restored by exten-

[17] A. E. Pegg, *J. Biol. Chem.* **254,** 3249 (1979).

[18] M. G. Rosenblum and T. L. Loo, *J. Chromatogr.* **183,** 363 (1980).

[19] P. Seppänen, L. Alhonen-Hongisto, H. Pösö, and J. Jänne, *FEBS Lett.* **111,** 99 (1980). See also this volume [41].

sive dialysis.[10,20] Inactivation was much more rapid in the presence of putrescine and proceeded as a first-order reaction showing a rate saturation effect at high concentrations of the inhibitor.[10] The apparent K_i was 66 μM, and the rate constant for inactivation was 0.2 min^{-1}. Either S-adenosylmethionine or MGBG protected the enzyme from irreversible inactivation by MBAG. These results are consistent with a model for inactivation in which MBAG binds reversibly to the enzyme at the same site as MGBG and then reacts with the bound pyruvate cofactor[21] to cause permanent inactivation.[10] A variety of adducts between pyruvate derivatives and hydrazones were synthesized in the hope that these would be more potent inhibitors of the enzyme, but all were inactive.[22]

MBAG also inactivates the adenosylmethionine decarboxylases from yeast (50% inactivation produced by exposure to 7 μM for 1 hr, a rate similar to that for the rat prostate enzyme), *E. coli* (50% inactivation requires 1 mM MBAG for 1 hr), and *Physarum polycephalum* (50% inhibition requires 25 μM MBAG for 1 hr).

Specificity. MBAG is also a strong inhibitor of diamine oxidase.[12] Inhibition is competitive, but the K_i is 0.02 μM, and a single dose of 50 mg/kg inhibited activity for 72 hr in treated rats.

Inhibition in Vivo. MBAG inhibits polyamine synthesis *in vivo* in rat tissues[10] and in cultured rodent cells, but it is no more potent than MGBG and is more toxic.[23] This is probably due to a combination of slower uptake, the chemical reactivity of the hydrazine groups of MBAG, which may react slowly with many cellular constituents, and the rapid turnover of adenosylmethionine decarboxylase, which permits new enzyme synthesis to replace that which is inactivated.[17]

Nucleosides Related to S-Adenosylmethionine

A large number of compounds structurally related to S-adenosylmethionine have been synthesized by Borchardt and collaborators[24] and were tested for inhibitory activity toward S-adenosylmethionine decarboxylases from various sources.[25] None of these compounds was nearly as potent as MGBG, and the only compounds that produced more than 50% inhibition in a standard assay with 0.2 mM S-adeno-

[20] A. E. Pegg and C. Conover, *Biochem. Biophys. Res. Commun.* **69,** 766 (1976).

[21] A. E. Pegg, *FEBS Lett.* **84,** 33 (1977).

[22] M. C. Pankaskie and M. M. Abdel-Monem, *J. Pharm. Sci.* **69,** 1000 (1980).

[23] A. E. Pegg, R. T. Borchardt, and J. K. Coward, *Biochem. J.* **194,** 79 (1981).

[24] R. T. Borchardt, Y. S. Wu, J. A. Huber, and A. F. Wycpalek, *J. Med. Chem.* **19,** 1104 (1976).

[25] A. E. Pegg and G. Jacobs, submitted for publication.

TABLE III
EFFECT OF NUCLEOSIDES ON S-ADENOSYLMETHIONINE DECARBOXYLASES[25]

Nucleoside	Concentration (μM) needed to achieve 50% inhibition of enzyme from			
	Rat prostate	Yeast	*Physarum polycephalum*	*Escherichia coli*
Decarboxylated S-adenosylmethionine	22	32	42	45
5′-Dimethylthioadenosine sulfonium salt	28	27	36	44
S-Methyl-S-adenosyl-L-cysteine	170	430	410	620
S-Adenosyl-4-methylthiobutyric acid	105	360	350	420

sylmethionine were those shown in Table III. Compounds in which the adenosyl moiety was replaced by another base or lacking the sulfonium center were inactive.[25] The strong inhibition of decarboxylation by the product of the reaction with a K_i of 1–5 μM has also been observed by others[26–29] and is a problem in the production of large quantities of decarboxylated *S*-adenosylmethionine by enzymatic methods.

Rat liver *S*-adenosylmethionine decarboxylase was inhibited quite strongly by *S*-5′-deoxyadenosyl-(5′)-2-methylthioethylamine (K_i, 0.9 μM), which was the most active of a series of analogs of decarboxylated *S*-adenosylmethionine tested by Yamanoha and Samejima.[29] Pankaskie and Abdel-Monem[30] reported that *S*-5′-deoxyadenosyl-(5′)-2-methylmethionine (K_i, 18 μM), *S*-5′-deoxyadenosyl-(5′)-1-methyl-3-(methylthio)propylamine (K_i, 12 μM), and *N*-(aminopropyl)-*N*-methyl-5′-amino-5′-deoxyadenosine (K_i, 11 μM) inhibited prostatic *S*-adenosylmethionine decarboxylase. They also report the curious finding that the former two compounds[30] and *S*-adenosylmethionine itself[22] cause a time-dependent inactivation of the enzyme that, they suggest, is due to a decarboxylation-dependent transamination converting the pyruvate prosthetic group to an alanyl residue.[22] There is, at present, no evidence for such inhibition

[26] A. E. Pegg and H. G. Williams-Ashman, *J. Biol. Chem.* **244,** 682 (1969).

[27] J. Jänne, H. G. Williams-Ashman, and A. Schenone, *Biochem. Biophys. Res. Commun.* **43,** 1362 (1971).

[28] H. Pösö, R. Sinervirta, J.-J. Himberg, and J. Jänne, *Acta Chem. Scand., Ser. B* **29,** 932 (1975).

[29] B. Yamanoha and K. Samejima, *Chem. Pharm. Bull.* **28,** 2232 (1980).

[30] M. Pankaskie and M. M. Abdel-Monem, *J. Med. Chem.* **23,** 121 (1980).

occurring *in vivo*, and none of the nucleosides discussed above has yet been shown to influence the synthesis of polyamines within the cell.

More recently a further series of analogs of *S*-adenosylmethionine were prepared and tested for inhibitory properties toward rat liver *S*-adenosylmethionine decarboxylase.[31] It was confirmed that the terminal amino and carboxyl groups were not required for inhibitory binding and, in fact, that the most potent inhibitor tested was 5′-(dimethylsulfonio)-5′-deoxyadenosine, which had a K_i of 2 μM. It was also confirmed that the sulfonium center could be replaced by a nitrogen, and a series of analogs containing nitrogen instead of sulfur were tested and found to be inhibitory.[31] None of these were significantly more potent than the *N*-(aminopropyl)-*N*-methyl-5′-amino-5′-deoxyadenosine already described,[30] and the K_i for this derivative was estimated at 60 μM,[31] which is considerably greater than the earlier report of 11 μM.[30] The recent study also confirmed the time-dependent inactivation of the enzyme by substrate[30] and by all inhibitors containing a terminal amino group,[31] thus supporting the suggestion that this inhibition is active-site directed and involves a transamination of the inhibitor.

[31] M. Kolb, C. Danzin, J. Barth, and N. Claverie, *J. Med. Chem.* **25,** 550 (1982).

[41] Two Enzyme Inhibition Assays for Methylglyoxal Bis(guanylhydrazone)[1]

By P. SEPPÄNEN, L. ALHONEN-HONGISTO, K. KÄPYAHO, and J. JÄNNE

The anticancer drug methylgyoxal bis(guanylhydrazone) (MGBG; methyl-GAG; mitoguazone) is a powerful inhibitor of eukaryotic, putrescine-activated *S*-adenosyl-L-methionine decarboxylase (EC 4.1.1.50) and diamine oxidase (EC 1.4.3.6). Thus both enzymes can be used for the determination of MGBG in biological samples.

Determination of MGBG with the Aid of Adenosylmethionine Decarboxylase

$$S\text{-Adenosyl-L-methionine} \xrightarrow{\text{MGBG}} S\text{-methyladenosylhomocysteamine (decarboxylated adenosylmethionine)} + CO_2$$

[1] P. Seppänen, L. Alhonen-Hongisto, H. Pösö, and J. Jänne, *FEBS Lett.* **111,** 99 (1980).

ISBN 0-12-181994-9

The decarboxylation of adenosylmethionine catalyzed by eukaryotic (putrescine-activated) adenosylmethionine decarboxylase is competitively inhibited by micromolar concentrations of MGBG.[2,3]

Principle

The method represents an enzyme inhibition assay where exogenous adenosylmethionine decarboxylase is inhibited by MGBG present in the sample. By using Dixon's method,[4] the amount of MGBG in the sample can be calculated with the aid of a standard curve.

Preparation of Adenosylmethionine Decarboxylase

A most convenient source of putrescine-activated adenosylmethionine decarboxylase is baker's yeast.[4a] Partial purification of the enzyme[5] is achieved as follows. Baker's yeast (about 25 g) is suspended in an equal volume (w/v) of 25 m*M* Tris-HCl buffer (pH 7.4) containing 1 m*M* dithiothreitol and 0.1 m*M* EDTA. The cell suspension is extruded twice through a chilled (−20°) French press (28,000 psi).[6] The resulting homogenate is centrifuged at 15,000 g_{max} for 15 min. Two-tenths volume of 10% streptomycin sulfate is slowly added to the supernatant fraction, and after occasional stirring for 30 min the suspension is centrifuged at 15,000 g_{max} for 15 min. The resulting supernatant fraction is further fractionated with solid $(NH_4)_2SO_4$ at 0°. The proteins precipitated between 40 and 60% saturation of $(NH_4)_2SO_4$ are collected by centrifugation, dissolved in a small volume of the original buffer, and dialyzed overnight against 50 volumes of the buffer. The enzyme solution can be stored in small portions either frozen or as lyophilized powder. The enzyme preparation is stable for several months.

Assay Method

Preparation of the Sample. The sample can be any biological material: cultured cells, blood leukocytes, tissue homogenates, and extracellular

[2] H. G. Williams-Ashman and S. Schenone, *Biochem. Biophys. Res. Commun.* **46,** 288 (1972).

[3] E. Hölttä, P. Hannonen, J. Pispa, and J. Jänne, *Biochem. J.* **136,** 669 (1973).

[4] M. Dixon, *Biochem. J.* **55,** 170 (1953).

[4a] See also this volume [38] for another preparation of yeast adenosylmethionine decarboxylase.

[5] H. Pösö, R. Sinervirta, and J. Jänne, *Biochem. J.* **151,** 67 (1975).

[6] X-Press, Biox AB, Nacka, Sweden.

fluids, such as blood, plasma, serum, cerebrospinal fluid, and urine. Cell disintegration is most conveniently performed by ultrasonication or with a tissue homogenizer. In some cases, if high levels of endogenous adenosylmethionine decarboxylase activities are anticipated, the sample should be heat-inactivated. If massive precipitation occurs, the actual recovery of added MGBG should be checked. Glass tubes should be avoided when processing and storing the samples, since MGBG sticks to glass surfaces.

Reagents

S-Adenosyl-L-[1-^{14}C]methionine,[7] 3 m*M*. Radioactive adenosylmethionine is diluted with unlabeled adenosylmethionine to give a final specific activity of about 1 Ci/mol.
Putrescine dihydrochloride, 75 m*M* neutralized solution
Potassium phosphate buffer, 1 *M*, pH 7.4
Disodium EDTA, 15 m*M* neutralized solution
Dithiothreitol, 100 m*M*
Adenosylmethionine decarboxylase preparation diluted so that 0.05 ml catalyzes the evolution of about 15,000 cpm of CO_2 in 30 min under standard incubation conditions.
MGBG dihydrochloride[8] (neutralized, in 0.15 *M* NaCl) standards containing 25, 50, 100, and 200 pmol of MGBG in 0.05 ml
Carbon dioxide trapping agent (Soluene[9])
Citric acid, 2 *M*

Procedure. The final incubation mixture contains 0.015 ml of the phosphate buffer, 0.01 ml of dithiothreitol, 0.005 ml of putrescine, 0.01 ml of EDTA, 0.01 ml of adenosylmethionine, 0.05 ml of MGBG standards (or NaCl), and 0.05 ml of the enzyme solution. Incubation is started by the addition of the enzyme and carried out in conical test tubes equipped with rubber stoppers to which center wells containing 0.1 ml of Soluene are attached. The tubes are incubated at 37° for 30 min.

The reaction is stopped by injection of 1 ml of 2 *M* citric acid through the stopper. To release any dissolved CO_2, the incubation is continued for 20 min more. The contents of the center wells are then transferred into toluene-based scintillation fluid and counted for radioactivity.

Calculation of the Results

There is a linear relationship between the concentration of MGBG and the reciprocal of adenosylmethionine decarboxylase activity. Subtract the

[7] A. E. Pegg and H. G. Williams-Ashman, *J. Biol. Chem.* **244,** 682 (1969).
[8] Available from Ega-Chemie, Steinheim/Albuch, Federal Republic of Germany.
[9] Available from Packard Instrument International S.A., Zurich, Switzerland.

blank value (incubated in the absence of the enzyme), take reciprocals ($1 \times cpm^{-1}$) of the adenosylmethionine decarboxylase activities, and plot these values against standard MGBG concentrations (0, 0.5, 1, 2, 3, and 4 μM). With the aid of this standard curve, calculate the concentration of MGBG in the samples.

Determination of MGBG with the Aid of Diamine Oxidase

$$NH_2CH_2CH_2CH_2CH_2NH_2 + O_2 + H_2O \xrightarrow{\text{MGBG}} (NH_2CH_2CH_2CH_2CHO) + H_2O_2 + NH_3$$

$$\downarrow -H_2O$$

(pyrroline ring, N)

Mammalian diamine oxidase is also strongly (noncompetitively) inhibited by micromolar concentrations of MGBG.[3,10]

Principle

This assay is based on the inhibition of exogenous diamine oxidase by MGBG present in the sample.

Preparation of Diamine Oxidase

A convenient source of diamine oxidase is the small intestine of rat. Small intestine is cut off, and the contents are removed. The pieces of intestine are washed with 25 mM potassium phosphate buffer, pH 7.4, containing 10 mM 2-mercaptoethanol and 0.1 mM EDTA. The tissue pieces are homogenized in two volumes of the above buffer and centrifuged for 20 min at 100,000 g_{max}. The resulting supernatant fraction is fractionated with solid $(NH_4)_2SO_4$, and the protein fraction, precipitated between 30 and 50% saturation of $(NH_4)_2SO_4$, is used as the enzyme preparation after overnight dialysis.

Assay Method[11]

Preparation of the Tissue Sample. The tissue samples are prepared as for the adenosylmethionine decarboxylase assay (see above).

[10] A. E. Pegg and S. M. McGill, *Biochem. Pharmacol.* **27,** 1625 (1978).
[11] N. Tryding and B. Willert, *Scand. J. Clin. Lab. Invest.* **22,** 29 (1968).

Reagents

Potassium phosphate buffer, 1 *M*, pH 7.4
Putrescine dihydrochloride, 10 m*M* (neutralized)
[1,4-^{14}C]Putrescine,[12] 0.04 μCi/0.05 ml
Diamine oxidase preparation
MGBG standards containing 25, 50, 100, 200, and 400 pmol/0.05 ml

Procedure. The final incubation mixture contains 0.025 ml of the phosphate buffer, 0.01 ml of putrescine, 0.05 ml of [^{14}C]putrescine, 0.1 ml of the enzyme preparation, 0.05 ml of the MGBG standards, NaCl or sample, and 0.015 ml of water. Incubation is carried out in glass-stoppered test tubes at 37° for 30 min. The reaction is halted by transferring the tubes into an ice bath and adding 2.5 ml of ice-cold water into each tube. The final reaction product (pyrrolidine) is extracted twice with 5 ml of toluene-based scintillation fluid. After each extraction the aqueous phase is frozen in a dry ice–ethanol bath, and the scintillation fluid is poured into scintillation vials.

Calculation of the Results

The standard curve is constructed similarly to that described for adenosylmethionine decarboxylase assay using Dixon's plot. When diamine oxidase is used, the linear standard range is somewhat broader than that of adenosylmethionine decarboxylase.

Clinical Applications

Determination of MGBG in Mononuclear Leukocytes of Patients Receiving the Drug. The narrow therapeutic index of MGBG in the treatment of human malignancies can be partly circumvented by regular determination of the drug levels in target cells. This is especially feasible in case of leukemias. The most reliable picture is obtained when the drug concentrations are measured in the circulating (or bone marrow) mononuclear leukocytes (the buffy coat cells).[13,14] Rapid isolation of mononuclear leukocytes (including leukemic blast cells) can be accomplished as follows. Mix 62 ml of Percoll[15] solution with 38 ml of 0.4 *M* NaCl (results in physiological salt concentration). Dilute heparinized blood with equal vol-

[12] Available from the Radiochemical Centre, Amersham, Bucks, United Kingdom.

[13] P. Seppänen, L. Alhonen-Hongisto, M. Siimes, and J. Jänne, *Int. J. Cancer* **26,** 571 (1980).

[14] M. Siimes, P. Seppänen, L. Alhonen-Hongisto, and J. Jänne, *Int. J. Cancer* **28,** 567 (1981).

[15] Available from Pharmacia Fine Chemicals AB, Uppsala, Sweden.

ume of physiological saline, and layer 2 ml of the diluted blood on top of 3 ml of the Percoll–NaCl solution in a centrifuge tube. Centrifuge at 800 g_{max} for 10 min at room temperature. The leukocyte zone is collected and washed with physiological saline. In case of red cell contamination, the cells are subjected to hypotonic treatment (fourfold dilution with water for 1 min; after this, the salt concentration is adjusted back to 0.15 *M* NaCl). If necessary, the hypotonic treatment is repeated. After centrifugation, the final pellet is suspended in 0.15 *M* NaCl and the cells are disintegrated with ultrasonication (or with any other means). The sonicate is used for the determination of MGBG.

The concentration of MGBG in leukocytes can be expressed as mol $\times 10^{-18}$/cell or as molar concentrations providing that cell volumes have been determined. Therapeutic concentrations in leukemic cells are about 100 to 200 $\times 10^{-18}$ mol/cell or 0.5 to 1 m*M* (if the cell volume is 200–250 $\times 10^{-15}$ liter).

Determination of Plasma MGBG. Plasma MGBG is determined by adding diluted or undiluted plasma directly into the assay mixture. Because of the rapid plasma clearance of MGBG,[13,16,17] monitoring of the plasma levels of the drug is of little value for supervising the treatment, yet the plasma MGBG level is needed for indirect determination of cellular MGBG (see below).

Determination of Whole Blood MGBG. The cellular elements of a heparinized blood sample are disintegrated by ultrasonication or by using Triton X-100 (final concentration 1%, which does not inhibit adenosylmethionine decarboxylase activity). Blood homogenate is then added directly into the assay system.

Indirect Calculation of Leukocytic MGBG Concentration. The MGBG concentration in leukocytes can also be calculated indirectly without a prior separation of the buffy coat cells. The determination is based upon the fact that MGBG does not enter the red cells. Indirect determination is very reliable when most of the leukocytes are blast cells (blast crisis). Both plasma and whole blood MGBG should first be measured. MGBG concentration in plasma (take packed red cell volume into account) is subtracted from whole blood MGBG, thus resulting in the amount of MGBG present in the cellular portion of the blood. The following example illustrates the calculations:

Assume that whole blood MGBG is 13.0 μM, plasma MGBG is 0.5

[16] M. G. Rosenblum, M. J. Keating, B. S. Yap, and T. L. Loo, *Cancer Res.* **41,** 1748 (1981).

[17] K. C. Marsh, J. Liesmann, T. F. Patton, C. J. Fabian, and L. A. Sternson, *Cancer Treat. Rep.* **65,** 253 (1981).

μM, hematocrit is 0.4, and the number of leukocytes is 50×10^9/liter. The cellular MGBG, then, would be

$$\frac{13.0\ \mu\text{mol/liter} - (1 - 0.4) \times 0.5\ \mu\text{mol/liter}}{50 \times 10^9\ \text{cells/liter}} = 254 \times 10^{-18}\ \text{mol/cell}$$

If the cell volume is 250×10^{-15} liter (lymphoblasts), the intracellular MGBG concentration would be 1.0 mM (assuming even distribution).

Determination of Urine MGBG. Determination of the urine MGBG excretion gives some idea of the presence of the drug in body tissues, since the drug is only slowly excreted.[16,17] Urine should always be diluted before the assay, since undiluted urine inhibits adenosylmethionine decarboxylase activity.

Determination of Skin MGBG. The drug concentrations can be determined also in skin biopsies (obtained from patients undergoing topical treatment with MGBG) or any other biopsy material. Skin biopsy sample is processed as follows: homogenize the tissue piece with 20 volumes of 0.2 N perchloric acid. After centrifugation, the supernatant fraction is neutralized and used for the assay of MGBG. Run neutralized perchloric acid blanks in the assay.

Section VI

Putrescine Aminopropyltransferase (Spermidine Synthase) and Spermidine Aminopropyltransferase (Spermine Synthase)

A. Enzyme Assays and Preparations
Articles 42 through 47

B. Enzyme Inhibitors
Articles 48 and 49

[42] Rapid Assays for Putrescine Aminopropyltransferase (Spermidine Synthase) and Spermidine Aminopropyltransferase (Spermine Synthase)[1]

By AARNE RAINA, TERHO ELORANTA, and RAIJA-LEENA PAJULA

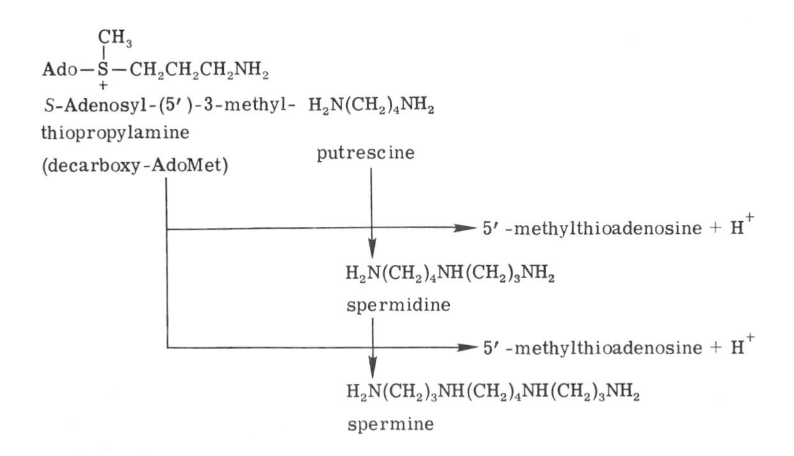

Two methods, based on the use of [^{14}C]decarboxy-AdoMet as a substrate, are described here for the assay of spermidine and spermine synthases. The radioactive product, ^{14}C-labeled polyamine or [*methyl*-^{14}C]methylthioadenosine, is isolated by a simple chromatographic procedure.

Reagents

Potassium phosphate buffer, 1 *M*, pH 7.4
Dithiothreitol, 0.1 *M* (stored frozen)
Putrescine dihydrochloride, 10 m*M* (for spermidine synthase)
Spermidine trihydrochloride, 10 m*M* (for spermine synthase)
Labeled substrates (in 0.001 *M* HCl, stored frozen)
  Method I: [*propylamine*-1-^{14}C]decarboxy-AdoMet, specific activity 5.8 μCi/μmol
  Method II: [*methyl*-^{14}C]decarboxy-AdoMet, specific activity 5–6 μCi/μmol (diluted with unlabeled decarboxy-AdoMet)
Enzyme, 0.005–0.1 unit per assay

[1] See also this volume [43].
[2] A. Raina, R.-L. Pajula, and T. O. Eloranta, *FEBS Lett.* **67,** 252 (1976).

ISBN 0-12-181994-9

The labeled substrates are prepared as follows. Labeled *S*-adenosylmethionine is synthesized from DL-[2-^{14}C]methionine (specific activity 5.8 μCi/μmol) or from L-[*methyl*-^{14}C]methionine (specific activity 57 μCi/μmol) as described by Pegg and Williams-Ashman.[3] Both types of radioactive *S*-adenosylmethionine are now commercially available. Radioactive decarboxy-AdoMet is prepared from *S*-adenosylmethionine using *S*-adenosylmethionine decarboxylase from *Escherichia coli,* purified through step 3 by the method of Wickner *et al.*,[4] as the enzyme. The product is first purified on a Dowex 50-H^+ column, followed by preparative paper electrophoresis using 0.05 *M* citric acid buffer, pH 3.5.[5] Decarboxy-AdoMet and unchanged *S*-adenosylmethionine are eluted separately from the paper and freed of salt by passage through a small Dowex 50, H^+ form, column. *S*-Adenosylmethionine thus obtained can be reused as a substrate in a subsequent decarboxylase reaction. Decarboxy-AdoMet purified by the above procedure is free of any *S*-adenosylmethionine.

Method I[2,6]

Procedure. The assay of spermidine and spermine synthases is conveniently carried out in glass-stoppered centrifuge tubes (volume about 4 ml). The reaction mixture containing 0.1 *M* potassium phosphate, 5 m*M* dithiothreitol, 1 m*M* putrescine (for spermidine synthase) or 1 m*M* spermidine (for spermine synthase), and 20 μ*M* [*propylamine*-1-^{14}C]decarboxy-AdoMet is preincubated for 10 min at 37° before the addition of the enzyme. Enzyme is omitted from the reaction blank. The final incubation volume is 0.1 ml. After incubation at 37° for 10 min, the reaction is stopped by adding 10 μl of 1 *M* potassium hydroxide. The tubes are thoroughly stirred, closed tightly, and incubated for 30 min at 100° to decompose the unreacted decarboxy-AdoMet.[3,5,7] After cooling, 20 μl of 1 *M* perchloric acid are added and the acid supernatant is cleared by low-speed centrifugation. An aliquot of 20 μl of the supernatant is applied on a disk (diameter 24 mm) of phosphocellulose ion-exchange paper (Whatman P-81), and water is added until the disk is thoroughly wet. The numbered disks, up to 50–60 at a time, are washed in a 2-liter beaker four times for 10 min with 10 m*M* Tris-HCl, pH 8.0, containing 100 m*M* NaCl, using

[3] A. E. Pegg and H. G. Williams-Ashman, *J. Biol. Chem.* **244,** 682 (1969).

[4] R. B. Wickner, C. W. Tabor, and H. Tabor, *J. Biol. Chem.* **245,** 2132 (1970). See also this volume [36].

[5] A. Raina and P. Hannonen, *FEBS Lett.* **16,** 1 (1971).

[6] R.-L. Pajula, A. Raina, and T. O. Eloranta, *Eur. J. Biochem.* **101,** 619 (1979).

[7] A. Raina and P. Hannonen, *Acta Chem. Scand.* **24,** 3061 (1970).

mechanical stirring. The disks are then dried and counted for radioactivity in a toluene-based scintillant. Counting efficiency is about 65%.

Comments. The washing procedure described above effectively removes the degradation products derived from decarboxy-AdoMet, while about 90% of the input radioactive spermidine and 99% of spermine is retained by the paper. Although some spermidine (about 10%) is lost during washing, the loss is fairly constant and insignificant when the method is applied for the assay of a large number of samples during enzyme purification. When necessary, this loss can be minimized by washing the disks on a Büchner funnel using 25 m*M* HCl.[2]

Method II

This assay method is a modification of a rapid and sensitive method developed by Hibasami and Pegg.[8] In the original method,[8] the labeled [*methyl*-^{14}C]methylthioadenosine formed from [*methyl*-^{14}C]decarboxy-AdoMet in the propylamine transferase reaction was separated from the substrate by chromatography on a Dowex 50, H^+ form, column. However, the recovery of labeled methylthioadenosine from the Dowex column was incomplete and somewhat variable. Instead, a weak cation-exchange material gave satisfactory results.[9]

Columns. Approximately 0.5 ml of phosphocellulose cation exchanger (Cellex-P, Bio-Rad) is packed in a Pasteur pipette using glass wool as bed support. The column is regenerated with 0.5 *M* HCl and washed with water. The void volume of the column is approximately 0.8 ml.

Procedure. The reaction mixture is the same as that described in Method I except that 20 μM [*methyl*-^{14}C]decarboxy-AdoMet is used as the labeled substrate. The reaction is stopped by the addition of 0.5 ml of 25 m*M* HCl (the pH should drop below 3). Centrifugation of the mixture is needed only if a precipitate appears. An aliquot of 0.5 ml of the mixture is applied to the phosphocellulose column, previously equilibrated with 25 m*M* HCl. The effluent (0.5 ml) is discarded. Radioactive methylthioadenosine and its degradation products are then eluted directly into a scintillation vial with 1.8 ml of 25 m*M* HCl. Ten milliliters of scintillation solution (Aquasol, New England Nuclear) is added to the vial, mixed vigorously, and counted for radioactivity. Counting efficiency is about 82%.

The columns are regenerated with 2 × 2 ml of 0.5 *M* HCl and washed with 2 ml of water. To avoid accumulation of precipitated protein, the

[8] H. Hibasami and A. E. Pegg, *Biochem. J.* **169,** 709 (1978).

[9] T. O. Eloranta, unpublished observation.

columns are occasionally treated with 2 × 2 ml of 0.1 *N* NaOH, washed with water, and regenerated as described above.

Discussion

The two methods described here are sensitive and much more rapid than those previously used[7,10,11] for the assay of propylamine transferase activities of eukaryote tissues. The labeled substrates are fairly expensive and laborious to prepare. On the other hand, only small amounts of decarboxy-AdoMet (about 1–2 nmol per assay mixture) are required for standard assays, as the K_m for decarboxy-AdoMet of spermidine and spermine synthases[6,12] is of the order of 1×10^{-6} *M*. As the endogenous concentration of decarboxy-AdoMet is very low,[13] dialysis of crude tissue extracts usually is not required. The amount of enzyme per assay mixture should be adjusted so that no more than 50% of the decarboxy-AdoMet is consumed during the reaction. The sensitivity of the two methods is about the same, as both [2-^{14}C]methionine and [*methyl*-^{14}C]methionine of high specific activity are now commercially available. The substrate used in Method II is less expensive. The results obtained by these two methods are in good agreement. Method I may be more convenient when analyzing a large number of samples, whereas Method II is especially suitable for a rapid test to determine substrate specificity as regards amine acceptors.[8]

[10] A. E. Pegg and H. G. Williams-Ashman, *Arch. Biochem. Biophys.* **137,** 156 (1970).
[11] J. Jänne, A. Schenone, and H. G. Williams-Ashman, *Biochem. Biophys. Res. Commun.* **42,** 758 (1971).
[12] See this volume [45] and [46].
[13] H. Hibasami, J. L. Hoffman, and A. E. Pegg, *J. Biol. Chem.* **255,** 6675 (1980).

[43] Assay of Aminopropyltransferases[1]

By ANTHONY E. PEGG

Assay Method

Principle. Two separate aminopropyltransferases are present in mammalian cells.[2,3] Spermidine synthase catalyzes the transfer of an aminopropyl group from decarboxylated *S*-adenosylmethionine to putrescine,

[1] See also this volume [42].
[2] P. Hannonen, J. Jänne, and A. Raina, *Biochem. Biophys. Res. Commun.* **46,** 341 (1972).
[3] A. E. Pegg, K. Shuttleworth, and H. Hibasami, *Biochem. J.* **197,** 315 (1981).

ISBN 0-12-181994-9

forming spermidine and 5′-methylthioadenosine. Spermine synthase uses spermidine as the aminopropyl acceptor and produces spermine and 5′-methylthioadenosine. These enzymes are each specific with regard to the amine acceptor[3] and thus differ from the bacterial aminopropyltransferase from *Escherichia coli,* which prefers putrescine, but will use spermidine under some conditions.[4] However, all known aminopropyltransferases produce 5′-methylthioadenosine, which is the basis of the present assay. Decarboxylated *S*-adenosylmethionine labeled in the methyl group is used as a substrate, and the labeled 5′-methylthioadenosine produced in the presence of the appropriate amine acceptor is determined.[5]

Reagents

Sodium phosphate, 0.5 *M*, pH 7.5
Putrescine dihydrochloride, 50 m*M*
Spermidine trihydrochloride, 50 m*M*
Dithiothreitol, 20 m*M*
Decarboxylated *S*-adenosyl-[*methyl*-^{14}C]methionine, 1 m*M* (specific activity 20 μCi/μmol)

Procedure. The assay is carried out by incubation at 37° of centrifuge tubes containing a total volume of 0.2 ml including 20 μl of 0.5 *M* sodium phosphate buffer, 5 μl of 20 m*M* dithiothreitol, 2 μl of 1 m*M* labeled decarboxylated *S*-adenosylmethionine, 10 μl of either 50 m*M* putrescine (for spermidine synthase) or 50 m*M* spermidine (for spermine synthase) and enzyme (0.01–0.2 unit, where 1 unit represents 1 nmol of 5′-methylthioadenosine produced in 30 min) for 10–30 min. Blank tubes contain no putrescine or spermidine, or are not incubated. The reaction is stopped by the addition of 0.2 ml of 4 *M* HCl. After addition of a further 0.6 ml of 2 *M* HCl, the mixture is centrifuged for 10 min at 1600 *g* and the supernatant is applied to a column of Dowex 50W-X4 (H^+ form) made in a Pasteur pipette plugged with glass wool and containing about 1 ml of the resin. The effluent is collected and reapplied to the same column, which is then washed with a further 5 ml of 2 *N* HCl, and the eluate is assayed for radioactivity after the addition of 10 ml of Formula 947 LSC scintillation cocktail (New England Nuclear). Approximately 92% of authentic labeled 5′-methylthioadenosine is eluted in this fashion when added to the samples, and the results can be expressed as picomoles of 5′-methylthioadenosine produced. Results are corrected for the blank value by subtracting the radioactivity present in the 2 *N* HCl eluate when unincubated controls were used. This value (less than 0.1% of the radioactivity

[4] W. H. Bowman, C. W. Tabor, and H. Tabor, *J. Biol. Chem.* **248,** 2480 (1973).
[5] A. E. Pegg and H. Hibasami, *Biochem. J.* **169,** 709 (1978).

added) should not be different from that observed after incubation without the spermidine or putrescine acceptor. If it is different, the presence of an aminopropyl acceptor in the extract used as a source of enzyme should be suspected. An alternative method for the separation of the labeled 5′-methylthioadenosine and decarboxylated *S*-adenosylmethionine uses phosphocellulose instead of Dowex 50. This method, described in detail in this volume,[6] has the advantage that the 5′-methylthioadenosine can be eluted in 0.025 *N* HCl, which gives better counting efficiency.

Alternative Methods

Aminopropyltransferase activities have been assayed by using labeled putrescine or spermidine as acceptor and measuring the corresponding radioactive spermidine and spermine after separation by chromatography or electrophoresis.[4,7,8] An assay that utilizes decarboxylated *S*-adenosylmethionine labeled in the propylamine group and measures the amines produced is also available and is convenient for large numbers of samples.[7] Spectrophotometric and fluorometric assays have also been described.[4,9]

Disadvantages and Advantages of the Present Method

Disadvantages

Availability of Labeled Substrate. Labeled decarboxylated *S*-adenosylmethionine is not commercially available and must be prepared as described below, but *S*-adenosylmethionine labeled in the methyl group at high specific activity is routinely available as a starting material. Unlabeled decarboxylated *S*-adenosylmethionine is also not a commercial product, so the preparation of this nucleoside is needed for all aminopropyltransferase assays.

Extracts to Be Assayed Must Be Free of Amine Acceptors. The method is based on measuring the production of 5′-methylthioadenosine in response to the addition of the appropriate amine acceptor and, therefore, requires that the sample to be assayed does not contain any such amines. Even with crude tissue extracts, this is readily achieved by dialysis for a few hours against 50 m*M* sodium phosphate, pH 7.5, 1 m*M* dithiothreitol

[6] A. Raina, T. O. Eloranta, and R.-L. Pajula, this volume [42].

[7] C. W. Tabor, this series, Vol. 5, p. 761. See also this volume [42].

[8] A. E. Pegg and H. G. Williams-Ashman, *Arch. Biochem. Biophys.* **137,** 156 (1970).

[9] O. Suzuki, T. Matsumoto, M. Oya, Y. Katsumata, and K. Samejima, *Anal. Biochem.* **115,** 72 (1981).

since the sensitivity is such that only a small amount of extract need be added to an assay.

Advantages

Speed and Sensitivity. The assay is rapid and many samples can easily be processed at once using columns in Pasteur pipettes. Separation of 5′-methylthioadenosine from decarboxylated *S*-adenosylmethionine is much easier than separation of the polyamine products from the amine substrate or from decarboxylated *S*-adenosylmethionine. The assay is also very sensitive, since a high specific activity of decarboxylated *S*-adenosylmethionine is used. The sensitivity can be improved even more by using decarboxylated *S*-adenosylmethionine labeled in the methyl group with tritium, which we prepare at a specific activity of 11 Ci/mmol (see below). With this substrate, the assay of samples containing 10^{-5} unit is possible, and samples from cultured cells containing very small amounts of protein can be assayed.

Any Amine Acceptor Can Be Used. The method can be used with amines other than putrescine or spermidine[3,5] to test whether these are substrates for the reaction, since any aminopropyl transfer reaction will generate 5′-methylthioadenosine.

Suitable for Crude Tissue Extracts. Although the method is described as measuring 5′-methylthioadenosine, and this nucleoside may undergo further metabolism in crude extracts, the resulting products are also not retained on Dowex 50 in the presence of 2 *N* HCl and such decomposition does not affect the assay.

Preparation of Decarboxylated *S*-Adenosylmethionine

This is prepared using adenosylmethionine decarboxylase purified from *E. coli* through the step involving DEAE-cellulose chromatography.[10]

Reagents

Tris-HCl buffer, 1 *M*, pH 7.4
$MgCl_2$, 0.5 *M*
Dithiothreitol, 18 m*M*
S-Adenosylmethionine, 4 m*M*
S-Adenosyl-[*methyl*-^{14}C]methionine, 0.3 m*M* (59 mCi/mmol)

[10] R. B. Wickner, C. W. Tabor, and H. Tabor, *J. Biol. Chem.* **245,** 2132 (1970). See also this volume [37] for a more recent purification procedure for *E. coli* adenosylmethionine decarboxylase.

S-Adenosyl-[*methyl*-^{3}H]methionine, 45 μM (11 Ci/mmol)

Enzyme, 250 units/ml in 10 m*M* potassium phosphate, pH 7.5, 1 m*M* dithiothreitol (1 unit catalyzes the production of 1 nmol of decarboxylated *S*-adenosylmethionine in 30 min at 37°)

Procedure. For preparation of unlabeled decarboxylated *S*-adenosylmethionine, a mixture in a test tube of 1 ml of 1 *M* Tris-HCl, pH 7.4, 1 ml of 0.5 *M* $MgCl_2$, 0.55 ml of 18 m*M* dithiothreitol, 1.25 ml of 4 m*M* *S*-adenosylmethionine, and 4 ml of enzyme solution is incubated at 37° for 210 min. The reaction is halted with 2 ml of 50% (w/v) trichloroacetic acid, and the protein pellet is removed after centrifugation at 10,000 *g* for 15 min at 4°. The supernatant is applied to a column (1 cm in diameter × 5 cm) of Dowex 50 W-X2 (H^+ form) that has been prewashed with 50 ml of 6 *N* HCl and then with 200 ml of water. The column is washed with about 50 ml of 1 *N* HCl until the absorbance at 257 is less than 0.05, and then the sulfonium compounds are eluted with 6 *N* HCl. Fractions having absorbance at 257 above 0.2 are combined, evaporated to dryness at 40° under reduced pressure, dissolved in 0.2 ml of 0.1 *N* HCl, and purified either by high voltage electrophoresis[11] or by high-performance liquid chromatography (HPLC).[12] Paper electrophoresis is carried out for 2 hr at 60 V/cm in 0.05 *M* sodium citrate, pH 3.6, on Whatman 3 MM paper. The spot corresponding to decarboxylated *S*-adenosylmethionine (which is well separated from the more slowly migrating spot of *S*-adenosylmethionine itself) is visualized with an ultraviolet lamp, cut out, and eluted with 30 ml of 0.1 *N* HCl overnight at 4°. The eluate is reapplied to a Dowex 50 column as above and washed with 10 ml of 1 *N* HCl; the decarboxylated *S*-adenosylmethionine is eluted with 6 *N* HCl, evaporated to dryness under reduced pressure at 40°, and stored frozen in water adjusted to pH 4 with 0.1 *N* HCl. The concentration is determined from the absorption at 260 nm ($\varepsilon = 15.8$). An alternative procedure for purification that is particularly suitable for the radioactive material is to use HPLC on Whatman Partisil SCX columns as described.[12]

The preparation of labeled decarboxylated *S*-adenosylmethionine is carried out in the same way, but with the following incubation conditions for the enzymatic decarboxylation. For preparation of ^{14}C-labeled material, the tubes contain 1 ml of 1 *M* Tris-HCl, pH 7.4, 1 ml of 0.5 *M* $MgCl_2$, 0.55 ml of 18 m*M* dithiothreitol, 2.5 ml of 0.3 m*M* *S*-adenosyl-[*methyl*-^{14}C]methionine, and 3 ml of enzyme solution and are incubated for 2 hr at 37°. For the preparation of ^{3}H-labeled material of high specific activity, the tubes contain 0.25 ml of 1 *M* Tris-HCl, pH 7.4, 0.25 ml of 0.5 *M*

[11] H. Pösö, P. Hannonen, and J. Jänne, *Acta Chem. Scand., Ser. B* **30,** 807 (1976).

[12] A. E. Pegg and R. A. Bennett, this volume [10].

$MgCl_2$, 0.14 ml of 18 m*M* dithiothreitol, 0.5 ml of 45 μM *S*-adenosyl-[*methyl*-^{3}H]methionine, and 1.1 ml of enzyme solution and are incubated at 37° for 1 hr.

The overall yield of decarboxylated *S*-adenosylmethionine is 20–40% for unlabeled and ^{14}C-labeled material and 10% for the ^{3}H-labeled compound.

[44] Putrescine Aminopropyltransferase (*Escherichia coli*)[1]

By Celia White Tabor and Herbert Tabor

Aminopropyltransferases carry out the biosynthesis of spermidine and spermine by transferring the aminopropyl moiety of decarboxylated adenosylmethionine to putrescine or spermidine, respectively. Methylthioadenosine is formed in these reactions. In fungi and mammalian tissues, two separate enzymes have been found that carry out these two reactions. (For the purification of other aminopropyltransferases, see this volume [45–47].) In *E. coli,* only spermidine is formed, and only one aminopropyltransferase has been found.[1]

Decarboxylated adenosylmethionine + putrescine → spermidine + methylthioadenosine

Assay Methods[2]

Method A. Isotopic Assay

Principle. [^{14}C]Spermidine is formed by the transfer of the propylamine group of decarboxylated adenosylmethionine[3] to [^{14}C]putrescine. The reaction mixture is chromatographed on paper with unlabeled carrier; after staining with ninhydrin, the spermidine area is cut out and counted.

Reagents

Tris-HCl, 1 *M*, pH 8.2

[^{14}C]Putrescine dihydrochloride, 20 *M* (7.5×10^5 cpm/μmol)

Decarboxylated adenosylmethionine, 20 m*M*

Procedure. The incubation mixture contains 5 μl of 1 *M* Tris-HCl, pH 8.2, 5 μl of 20 m*M* [^{14}C]putrescine, 5 μl of 20 m*M* decarboxylated adeno-

[1] W. H. Bowman, C. W. Tabor, and H. Tabor, *J. Biol. Chem.* **248,** 2480 (1973).

[2] For other assay methods for aminopropyltransferases, see this volume [42, 43, 45, 47].

[3] For the preparation of decarboxylated adenosylmethionine, see this volume [11, 43].

ISBN 0-12-181994-9

sylmethionine, and enzyme in a final volume of 50 μl. The reaction tube is incubated for 30–60 min at 37° and then is rapidly frozen to stop the reaction. Aliquots (10 μl) are chromatographed (descending) on Whatman No. 1 paper with 0.1 μmol each of unlabeled putrescine dihydrochloride and spermidine trihydrochloride at 25° for 18 hr, with a solvent containing 1-propanol, concentrated hydrochloric acid, and water (30 : 10 : 10, v/v/v). Papers are dried in air for 6–12 hr, dipped in 2% pyridine in acetone, and dried again at room temperature. The polyamines are detected with 0.2% ninhydrin in 90% 1-propanol, 10% H_2O, and 1% pyridine (v/v/v) (either by spraying or dipping). After drying in air, the paper is heated at 80° for 5 min. The spermidine spot is cut out and counted in a scintillation counter.

Method B. Spectrophotometric Assay

Principle. A coupled assay is carried out with spermidine dehydrogenase from *Serratia marcescens*. The spermidine formed in the assay is oxidized by spermidine dehydrogenase,[4] with potassium ferricyanide as the electron acceptor. This reaction is followed spectrophotometrically.

Reagents

Tris-HCl, 1 *M*, pH 8.2
Putrescine dihydrochloride, 20 m*M*
Decarboxylated adenosylmethionine,[3] 5 m*M*
Spermidine dehydrogenase (600-fold purified)
Potassium ferricyanide, 20 m*M*

Procedure. Twenty microliters of 1 *M* Tris-HCl, pH 8.2, 12 μl of 20 m*M* putrescine, a suitable aliquot of putrescine aminopropyltransferase, 0.1 unit of a partially purified spermidine dehydrogenase (600-fold purified),[4] and 20 μl of 20 m*M* potassium ferricyanide in a final volume of 430 μl are incubated for 5 min at 37° in a 0.5-ml cuvette with a 1-cm light path. Then 20 μl of 5 m*M* decarboxylated adenosylmethionine is added to start the reaction. The decrease in absorbance at 400 nm at 37° resulting from the reduction of potassium ferricyanide is followed in a recording spectrophotometer. (The molar extinction coefficient of potassium ferricyanide is 960 at 400 nm.) The reaction is linear with time for at least 15 min. The activity of the spermidine dehydrogenase is checked after each assay by the addition of 50 nmol of spermidine to the assay mixture, and measurements at 400 nm are continued for several minutes longer.[5]

[4] For the preparation of this enzyme, see this volume [51]. See also this series, Vol. 17B [241].

[5] This method is satisfactory for use with all steps of the purification. However, fractions from steps 1–3 are dialyzed twice for 12–18 hr against 100 volumes of 0.01 *M* Tris-HCl–1 m*M* EDTA, pH 8.2, before assay.

Definition of Unit and Specific Activity. One unit is the amount of enzyme which will form 1 μmol of spermidine in 1 min under the conditions specified for the spectrophotometric assay. This unit is approximately the same as the unit obtained with the isotopic assay. Specific activity is expressed as units per milligram of protein. Protein is estimated by the absorbance at 280 nm and at 260 nm.[6]

Growth of E. coli. A wild-type strain of *E. coli* W is grown overnight in a 300-liter fermentor at 37° in a salts medium[7] containing 0.2% dextrose. The cells are harvested in early stationary phase with a refrigerated Sharples centrifuge and stored at −20°. A loss of 20–25% of the aminopropyltransferase activity is observed in cells that have been stored for a period of 5 months.

Purification Procedure

All steps are carried out at 0–4°.

Step 1. Extraction of Cells. Frozen cells (500 g) are suspended in 2 liters of distilled water, ruptured by homogenization with a Gaulin homogenizer at 6000 psi, and centrifuged for 10 min at 25,000 *g*. The supernatant is collected, and 2 *M* triethanolamine (6 ml) is added to adjust the pH to 7.0.

Step 2. Streptomycin Precipitation. Fresh 12% streptomycin sulfate solution (120 ml) is added dropwise, with stirring, per liter of the supernatant fraction of step 1. The suspension is stirred for 2 hr and then centrifuged for 10 min at 25,000 *g*. About 12 ml of 2 *M* triethanolamine is added to the supernatant solution to bring the pH to 7.5.

Step 3. Ammonium Sulfate Precipitation. Twenty milliliters of 0.5 *M* EDTA, pH 8.3, is added to each 980 ml of the supernatant fraction of step 2 (final EDTA concentration, 1 m*M*). Then 226 g of $(NH_4)_2SO_4$ per liter are added with stirring, and stirring is continued for 2–3 hr, with careful adjusting of the pH to 7.5 by the dropwise addition of concentrated NH_4OH. The precipitate is removed, and 90 g of $(NH_4)_2SO_4$ per liter is added to the supernatant fluid. The suspension is stirred overnight; the precipitate is collected by centrifugation and dissolved in 60 ml of 0.25 *M* Tris-HCl buffer, pH 8.0. The solution is then dialyzed twice for 5–6 hr against 100 volumes of 0.01 *M* Tris-HCl (pH 8.0)–1 m*M* sodium EDTA (buffer A). This preparation is stable on storage at −20° for several weeks.

Step 4. Calcium Phosphate Cellulose Gel Chromatography. Calcium phosphate cellulose gel, prepared by the method of Price and Greenfield[8]

[6] O. Warburg and W. Christian, *Biochem. Z.* **310,** 384 (1941).
[7] H. J. Vogel and D. M. Bonner, *J. Biol. Chem.* **218,** 97 (1956).
[8] V. E. Price and R. E. Greenfield, *J. Biol. Chem.* **209,** 363 (1954).

is packed without pressure into a column 25 cm high × 6.5 cm in diameter. After the column is equilibrated with 1 m*M* potassium phosphate, pH 7.6, the product of step 3 is applied to the column. The column is washed successively with 2 liters each of 5 m*M* potassium phosphate, pH 7.6, and 10 m*M* potassium phosphate, pH 7.6; the enzyme is eluted between 0.8 and 1.8 liters of the 10 m*M* buffer. Fractions with a fourfold or greater purification over step 3 are pooled, and 2 ml of 0.5 *M* sodium EDTA, pH 8.3, is added per liter. The protein is then precipitated by the addition of 400 g of ammonium sulfate per liter of solution. After stirring for 5–6 hr, the precipitate is collected by centrifugation, dissolved in 20 ml of 0.25 *M* Tris-HCl, pH 8.0, and dialyzed as in step 3 against 100 volumes of buffer A containing 0.2 *M* KCl.

Step 5. DEAE-Sephadex Chromatography. Diethylaminoethyl-Sephadex (Pharmacia A-50), equilibrated with buffer A containing 0.2 *M* KCl, is packed into a column (30 × 2.5 cm) and washed with the same buffer. The product of step 4 is applied to the column, and the column is eluted with a 2-liter linear gradient of buffer A containing 0.2 *M* KCl in the mixing flask and 0.6 *M* KCl in the reservoir. The enzyme activity is eluted between 550 and 950 ml. Fractions with a fivefold or greater purification over step 4 are pooled, concentrated to approximately 20 ml by ultrafiltration, and dialyzed against 100 volumes of buffer A containing 0.1 *M* KCl.

Step 6. Hydroxyapatite Chromatography. Hydroxyapatite, prepared by the method of Levin,[9] is gently packed in a column (9 × 1.5 cm) and equilibrated with 1 m*M* potassium phosphate, pH 8.0. The column is charged with the product of step 5 and is then eluted with a 400-ml linear gradient from 1 to 15 m*M* potassium phosphate, pH 8.0, each flask containing 0.1 *M* KCl. The activity is eluted between 150 and 225 ml. Fractions with a twofold or greater purification over step 5 are pooled, and 2 ml of 0.5 *M* sodium EDTA is added per liter. This solution is concentrated by ultrafiltration to about 5 ml and dialyzed against 100 volumes of buffer A containing 0.1 *M* KCl.

Step 7. Preparative Disc Gel Electrophoresis. The product of step 6 is at least 90–95% pure as judged by analytical disc gel electrophoresis. The minor contaminating bands are removed by preparative disc gel electrophoresis at 4° in a Canalco jacketed column PD-2/70. For this purpose, the material from step 6 is concentrated by ultrafiltration to 1.35 mg/ml and brought to 10% sucrose concentration. The enzyme solution, with added Bromphenol Blue, is layered on a column consisting of a 1.5-cm spacer gel and a 3-cm separating gel, pH 8.8. A current of 6 to 8 mA is applied until the indicator dye enters the separating gel. The current is then increased

[9] Ö. Levin, this series, Vol. 5, p. 27.

to 10–12 mA. The column is eluted continuously with 0.375 *M* Tris-HCl, pH 8.8, at a flow rate of approximately 2 ml/min. Fractions of 1 ml are collected. The enzyme is eluted between 20 and 30 ml after elution of the indicator dye. The active fractions are pooled, concentrated by ultrafiltration, and dialyzed as in step 6. This step results in a loss of approximately two-thirds of the activity, but no change in the specific activity. However, the product has only one band on analytical disc gel electrophoresis, coincident with enzyme activity. A representative purification is shown in the table.

Enzyme Stability. The purified enzyme is routinely stored at 0–4° at a concentration of about 0.5 mg/ml in buffer A containing 0.1 *M* KCl. After 6 months under these conditions, the enzyme from step 6 loses about 20% of its activity, and that from step 7 loses about 10%. Rapid inactivation occurs on freezing and thawing. The enzyme is stable in buffer A for at least 30 min at 50°.

Properties

The enzyme is homogeneous by sedimentation equilibrium in the ultracentrifuge and on gel electrophoresis. The molecular weight of the native enzyme is ca 72,000; it contains two equal subunits with molecular weights of 35,000 each. The pH optimum is 10.4. The K_m for putrescine is 1.2×10^{-5} *M*; the V_{max} is 2.0 μmol of spermidine formed per minute per milligram of protein. The K_m for decarboxylated adenosylmethionine is

PURIFICATION OF PUTRESCINE AMINOPROPYLTRANSFERASE[a]

Step	Volume (ml)	Total protein[b] (mg)	Total units	Recovery (%)	Specific activity (units $\times 10^3$ /mg protein)
Crude supernatant	2200	45,500	42.0	100	0.9
Streptomycin supernatant	2340	23,000	36.0	86	1.6
$(NH_4)_2SO_4$ fractionation	152	6,000	47.5	113[c]	7.9
Calcium phosphate cellulose	810	305	27.0	64	89
DEAE-Sephadex	260	20	11.4	27	590
Hydroxyapatite	8.6	4.2	8.3	20	1960
Acrylamide gel electrophoresis	3.5	1.5	2.7	6.5	1830

[a] This is a typical purification from 500 g of frozen cells.
[b] Assayed by the ultraviolet absorbance method.
[c] In other preparations, the recovery of this step was 71–100%.

2.2×10^{-6} *M*. The enzyme is also active with spermidine and cadaverine as the acceptor, but the rates are much slower than with putrescine as the acceptor. No activity is found with 1,3-diaminopropane or monoacetylputrescine. The enzyme is inhibited by *p*-hydroxymercuribenzoate (0.05 m*M*) and *N*-ethylmaleimide (1 m*M* in 0.1 *M* Tris-HCl buffer, pH 8.2). No inhibition occurs with carbonyl-binding reagents. The end products of the reaction, 5′-methylthioadenosine and spermidine, both inhibit the reaction.

[45] Purification of Putrescine Aminopropyltransferase (Spermidine Synthase) from Eukaryotic Tissues[1]

By KEIJIRO SAMEJIMA, AARNE RAINA, BANRI YAMANOHA, and TERHO ELORANTA

Spermidine synthase has been partially purified from several eukaryotic tissues, including rat brain,[2] liver,[3] and ventral prostate.[4] It appears, however, that in no case was the final enzyme preparation more than 10% pure. The recent design of a new affinity chromatographic adsorbent,[5] *S*-adenosyl(5′)-3-thiopropylamine linked to Sepharose (ATPA-Sepharose), has proved to be an important advance in the purification of spermidine synthase. In this chapter we describe purification of spermidine synthase to homogeneity from bovine brain and rat ventral prostate. As the purification procedures for these two tissues are somewhat different, they are treated separately.

Preparation of ATPA-Sepharose[5]

S-Adenosyl(5′)-3-thiopropylamine (ATPA) hydrogen sulfate is prepared by the method of Jamieson.[6] Bromoacetylation of AH-Sepharose (Pharmacia) is carried out by a slightly modified method of Cuatrecasas.[7] To 6 ml of AH-Sepharose 4B is added 0.24 ml of *o*-bromoacetyl-*N*-hy-

[1] See also this volume [47].
[2] A. Raina and P. Hannonen, *FEBS Lett.* **16,** 1 (1971).
[3] P. Hannonen, J. Jänne, and A. Raina, *Biochem. Biophys. Res. Commun.* **46,** 341 (1972).
[4] H. Hibasami, R. T. Borchardt, S. Y. Chen, J. K. Coward, and A. E. Pegg, *Biochem. J.* **187,** 419 (1980).
[5] K. Samejima and B. Yamanoha, *Arch. Biochem. Biophys.* **216,** 213 (1982).
[6] G. A. Jamieson, *J. Org. Chem.* **28,** 2397 (1963).
[7] P. Cuatrecasas, *J. Biol. Chem.* **245,** 3059 (1970).

METHODS IN ENZYMOLOGY, VOL. 94

ISBN 0-12-181994-9

droxysuccinimide solution prepared by mixing 1.2 mmol of *N*-hydroxysuccinimide, 1.0 mmol of bromoacetic acid, and 1.1 mmol of *N,N'*-dicyclohexylcarbodiimide in 8 ml of dioxane. The suspension is gently stirred for 30 min at 4°. Thereafter the Sepharose is washed with 800 ml of 0.1 *M* NaCl and 400 ml of water, and equilibrated with 5% acetic acid. To 5.5 ml of bromoacetamidohexyl-Sepharose 5 ml of ATPA hydrogen sulfate solution (790 mg dissolved in 5% acetic acid) is added, and the mixture is stirred gently for 20 hr at 37°. In the following masking procedures of unreacted bromoacetyl groups, a careful removal of air bubbles is indispensable for obtaining ATPA-Sepharose that gives reproducible results and a high yield in the purification of spermidine synthase. After removal of excess ATPA, 2-mercaptoethanol prepared in 5% acetic acid is added to the suspension to a final concentration of 2 *M*. The suspension in a round-bottom flask is deaerated *in vacuo* for 1 hr at room temperature in a rotary evaporator connected to an aspirator (25 mm Hg), and then kept at room temperature for 3 days. The suspension is washed with 5% acetic acid and water and is equilibrated at 4° with 25 m*M* sodium phosphate buffer, pH 7.2, containing 0.3 m*M* ethylenediaminetetraacetic acid (EDTA) and 0.5 m*M* dithiothreitol (buffer A). After adding 2-ethanolamine hydrochloride to a final concentration of 1 *M*, the suspension is placed *in vacuo* at 4° for 3 hr. The suspension is then allowed to stand for 2 days at 4° and is washed with buffer A before use. ATPA-Sepharose thus obtained contains 1 μmol of ATPA/ml.

Assay Methods

Spermidine Synthase. The assay of spermidine synthase activity is carried out as described in this volume [42], using [*propylamine*-1-^{14}C]decarboxy-AdoMet as the labeled substrate. The activity of the purified enzyme is assayed in the presence of albumin (0.2 mg/ml) in the reaction mixture.

As an alternative procedure, the method of Jänne *et al.*[8] with some modifications[9] can be used. The reaction mixture contains, in a total volume of 0.5 ml, 0.1 *M* sodium phosphate buffer, pH 7.2, 1 m*M* putrescine, 50 μ*M* decarboxy-AdoMet, 5 m*M* dithiothreitol, and enzyme protein. After incubation for 30 min at 37°, spermidine formed is determined by high-performance liquid chromatography.[10]

[8] J. Jänne, A. Schenone, and H. G. Williams-Ashman, *Biochem. Biophys. Res. Commun.* **42,** 758 (1971).

[9] K. Samejima and Y. Nakazawa, *Arch. Biochem. Biophys.* **201,** 241 (1980).

[10] K. Samejima, M. Kawase, S. Sakamoto, M. Okada, and Y. Endo, *Anal. Biochem.* **76,** 392 (1976).

Definition of Unit and Specific Activity. One unit of enzyme activity represents the formation of 1 nmol of spermidine in 1 min under standard assay conditions.

Specific activity is expressed as units per milligram of protein. Protein is measured by the method of Lowry *et al.*[11] either directly or after precipitation with trichloroacetic acid, with crystalline bovine serum albumin as a standard or by ultraviolet absorption.[12]

Purification of Spermidine Synthase from Bovine Brain

Step 1. Extraction of Tissues. Fresh bovine brains (4.2 kg) obtained from a local slaughterhouse and immediately cooled in ice are washed in 0.25 *M* sucrose and homogenized in 125-g portions with an Ultra-Turrax (Janke and Kunkel) homogenizer in two volumes of ice-cold 0.25 *M* sucrose containing 1 m*M* EDTA, 5 m*M* 2-mercaptoethanol, and 0.1 m*M* dithiothreitol. All subsequent operations are performed at 0–4°. The homogenate is centrifuged at 25,000 *g* for 30 min. The supernatant fraction is filtered through glass wool.

Step 2. Ammonium Sulfate Fractionation. The crude supernatant (6270 ml) is fractionated with solid ammonium sulfate (Mann, special enzyme grade). The proteins precipitated between 40 and 60% saturation of ammonium sulfate are collected by centrifugation at 14,000 *g* for 30 min, dissolved in 500 ml of 10 m*M* Tris-HCl, pH 7.5 (measured at room temperature), containing 0.08 *M* NaCl, 1 m*M* 2-mercaptoethanol, and 0.1 m*M* dithiothreitol (buffer B), and dialyzed for 24 hr against 25 volumes of buffer B, changing the buffer twice during dialysis. The dialyzed enzyme preparation is freed of insoluble material by centrifugation at 14,000 *g* for 15 min.

Step 3. DEAE-Cellulose Chromatography. The dialyzed ammonium sulfate fraction (635 ml) is applied to a DEAE-cellulose column (Whatman DE-52, 8.5 × 30 cm) equilibrated with buffer B. The column is washed with 900 ml of buffer B containing 1 m*M* dithiothreitol and connected to a linear gradient of 0.08 to 0.5 *M* sodium chloride in buffer B (total gradient volume is 4000 ml). Spermidine synthase is eluted between 0.11–0.16 *M* sodium chloride. The active fractions (540 ml) are pooled and brought to 0.70 saturation with ammonium sulfate to concentrate the enzyme. The precipitate is divided into two equal portions that are purified separately in the following steps.

Step 4. Chromatography on Hydroxyapatite. This step is included to

[11] O. H. Lowry, N. J. Rosebrough, A. L. Farr, and R. J. Randall, *J. Biol. Chem.* **193,** 265 (1951).

[12] H. M. Kalckar, *J. Biol. Chem.* **167,** 461 (1947).

remove contaminating *S*-adenosylmethionine decarboxylase and spermine synthase activities, which also may have a high affinity for ATPA-Sepharose. These two enzymes are more strongly adsorbed to hydroxyapatite as compared to spermidine synthase.[2,3]

Half of the enzyme preparation obtained in step 3 is dissolved in 17 ml of 0.02 *M* potassium phosphate buffer, pH 7.2, containing 1 m*M* 2-mercaptoethanol and 0.1 m*M* dithiothreitol (buffer C) and dialyzed against 5 liters of buffer C, changing the buffer once during dialysis. After centrifugation at 14,000 *g* for 10 min, the supernatant (27 ml) is applied to a hydroxyapatite column (Hypatite C, Clarkson Chemicals Co.; 2.6 × 17 cm) equilibrated with buffer C. The column is first eluted with buffer C (40 ml) containing 1 m*M* dithiothreitol, followed by a linear gradient of 0.02 to 0.4 *M* potassium phosphate buffer, pH 7.2, containing 1 m*M* 2-mercaptoethanol and 1 m*M* dithiothreitol (total gradient volume is 500 ml). Spermidine synthase is eluted practically unadsorbed. The active fractions (140 ml) are pooled and concentrated by the addition of ammonium sulfate (80% saturation).

Step 5. Affinity Chromatography on ATPA-Sepharose. The precipitate obtained in step 4 is dissolved in 0.02 *M* potassium phosphate buffer, pH 7.2, containing 0.3 *M* NaCl, 0.1 m*M* EDTA, 1 m*M* 2-mercaptoethanol, and 1 m*M* dithiothreitol (buffer D), and passed through a Sephadex G-25 column (1.5 × 17 cm). The desalted enzyme preparation (12.8 ml) is applied at a flow rate of 0.1 ml/min to an ATPA-Sepharose column (1.0 × 2.6 cm) equilibrated with buffer D. The column is washed with 15 ml of buffer D, followed by buffer D containing 0.6 *M* NaCl (15 ml) until no protein is being eluted (monitored by UV absorption). Spermidine synthase is then eluted with 6 ml of buffer D containing 0.25 m*M* decarboxy-AdoMet.[13] The enzyme solution is concentrated by ultrafiltration to 1.4 ml. At this stage the enzyme preparation is nearly homogeneous, but polyacrylamide gel electrophoresis and sodium dodecyl sulfate gel electrophoresis show a faint band of contaminating protein in the high-molecular-weight region.

Step 6. Gel Filtration. A portion of the concentrated enzyme preparation (0.8 ml) is applied to a Sephadex G-150 column (Pharmacia, 1.5 × 45 cm) equilibrated with buffer D and eluted with the same buffer, collecting 2-ml fractions. The active fractions are pooled (9.5 ml), concentrated to 1.1 ml by ultrafiltration, and stored at 0–4°. The enzyme is homogeneous as judged by polyacrylamide gel electrophoresis and sodium dodecyl sulfate gel electrophoresis.

A typical purification is summarized in the table.

[13] Synthesis of decarboxy-AdoMet is described in K. Samejima, Y. Nakazawa, and I. Matsunaga, *Chem. Pharm. Bull.* **26,** 1480 (1978).

PURIFICATION OF SPERMIDINE SYNTHASE FROM BOVINE BRAIN

Step	Total protein (mg)	Specific activity (units/mg)	Total activity (units)	Yield[a] (%)	Purification (fold)
1. Crude extract	50,400	0.10	5050	100	1
2. Ammonium sulfate	10,200	0.26	2670	53	3
3. DEAE-cellulose	1480	1.58	2330	46	16
4. Hydroxyapatite	460	3.60	1650	33	36
5. ATPA-Sepharose	0.92	521	480	9.5	5200
6. Sephadex G-150	0.32	712	230	4.5	7100

[a] The yield at the affinity chromatography step is somewhat variable and can be considerably higher than in this particular experiment depending on the batch of ATPA-Sepharose used.

Purification of Spermidine Synthase from Rat Prostate[5]

Step 1. Extraction of Tissues. Rat ventral prostates are collected and stored frozen at −20°. Frozen prostates (68 g) of Wistar rats are homogenized in 2 volumes of 25 m*M* Tris-HCl buffer, pH 7.2, containing 0.3 m*M* EDTA and 10 m*M* 2-mercaptoethanol. The 10,000 *g* supernatant is passed through glass wool.

Step 2. Ammonium Sulfate Fractionation.[3] The crude supernatant is fractionated with solid ammonium sulfate at 0°. The proteins precipitated between 30 and 60% saturation of ammonium sulfate are collected. The precipitate is dissolved in a small volume of 25 m*M* sodium phosphate, pH 7.2, containing 0.3 m*M* EDTA and 0.5 m*M* dithiothreitol (buffer A) and passed through a Sephadex G-25 column (equilibrated with buffer A) to remove salt.

Step 3. DEAE-Cellulose Chromatography.[3] The desalted protein fraction (1.7 g, 120 ml) is applied to a DEAE-cellulose column (Whatman DE-52, 2.6 × 33 cm) equilibrated with buffer A. The column is eluted with a linear gradient of 0 to 0.4 *M* NaCl prepared in buffer A (total gradient volume 1600 ml). Fractions containing the synthase activity (110 ml) eluted between 0.08 and 0.1 *M* sodium chloride are concentrated and equilibrated with buffer A containing 0.3 *M* NaCl by ultrafiltration in a Centriflo cone (CF 25, Ultrafiltration membrane cones, Amicon) to a final volume of 10 ml.

Step 4. Affinity Chromatography on ATPA-Sepharose. Affinity chromatography is carried out at 4° in a column packed with ATPA-Sepharose (0.5 ml) and equilibrated with buffer A containing 0.3 *M* NaCl. The concentrated enzyme solution (77.4 mg of protein, 10 ml) is applied to the column at a flow rate of 0.1–0.2 ml/min. The column is washed with 10 ml of equilibration buffer. More than 99% of the protein is removed by the washing. The synthase is then eluted with 4 ml of equilibration buffer containing 2.5 m*M* decarboxy-AdoMet.[13] Most of the activity is eluted in the first 2 ml. The effluent is pooled and concentrated by ultrafiltration in a Centriflo cone to a final volume of 0.5 ml (containing 0.4 mg of protein). The synthase is nearly homogeneous in this fraction.

Step 5. Gel Filtration. The concentrated synthase solution is applied to a Sephacryl S-300 column (Superfine from Pharmacia, 1.2 × 100 cm) equilibrated with buffer A containing 0.3 *M* NaCl, and eluted with the same buffer at a flow rate of 3.7 ml/hr. The synthase is eluted in 60–70 ml. The active fraction is stored on ice after concentration to 0.5 ml by ultrafiltration.

The final preparation has a specific activity of 1300 units/mg, representing 4480-fold purification over the crude supernatant fraction. The yield of the enzyme is 39%.

Properties

Stability. Both brain and prostatic spermidine synthases purified through the affinity chromatography step are stable for several months when stored on ice in a buffer containing 0.3 *M* NaCl, 1 m*M* dithiothreitol, and 0.05–0.25 m*M* decarboxy-AdoMet.

Purity. The enzyme preparations after the final step show no evidence of contamination as judged by polyacrylamide gel electrophoresis and sodium dodecyl sulfate gel electrophoresis. No *S*-adenosylmethionine decarboxylase or spermine synthase activities can be detected in the final preparations.

Subunit Molecular Weight and Molecular Size. Polyacrylamide gel electrophoresis in the presence of sodium dodecyl sulfate gives a subunit molecular weight of 35,800 and 37,000 for brain and prostatic enzyme, respectively. Gel filtration on a calibrated column of Sephacryl S-300 reveals a molecular weight of approximately 73,000 for the prostatic enzyme. These data suggest that the native enzyme is composed of two subunits of equal size.

Kinetics. An apparent K_m for decarboxy-AdoMet is about 1 μM for both brain and prostatic spermidine synthases. A K_m of about 0.03 m*M* for putrescine is obtained for the brain enzyme, whereas a somewhat higher value (0.1 m*M*) is obtained for the prostatic spermidine synthase. Both enzymes show a strong substrate inhibition with decarboxy-AdoMet that complicates the study of enzyme kinetics. The enzyme is also weakly inhibited by 5′-methylthioadenosine, one of the reaction products.

Substrate Specificity. Cadaverine, in addition to putrescine, can serve as the propylamine acceptor, although cadaverine gives a reaction rate of only 5–7% of that obtained with putrescine under standard assay conditions (1 m*M* amine). 1,6-Diaminohexane shows some activity (1.5%), whereas 1,3-diaminopropane is totally inactive as a substrate.

[46] Purification of Spermidine Aminopropyltransferase (Spermine Synthase) from Bovine Brain

By AARNE RAINA, RAIJA-LEENA PAJULA, and TERHO ELORANTA

Spermine synthase has been partially purified from rat brain[1] and ventral prostate,[2] tissues that contain the highest activity of this enzyme in the rat.[3] In this chapter we describe the purification of spermine synthase from bovine brain, which also is rich in spermine synthase and is easily obtained in amounts sufficient for large-scale purification. The use of affinity chromatography on spermine-Sepharose considerably facilitates the purification procedure.[4,5]

Assay Methods

Spermine Synthase. The assay of spermine synthase activity is carried out as described elsewhere in this volume [42], using [*propylamine*-1-^{14}C]decarboxy-AdoMet as the labeled substrate.

Definition of Unit and Specific Activity. One unit of enzyme activity represents the formation of 1 nmol of spermine in 1 min under standard assay conditions.

Specific activity is expressed as units per milligram of protein. Protein is measured by the method of Lowry *et al.*[6] either directly or after precipitation with trichloroacetic acid, with crystalline bovine serum albumin as a standard, or by ultraviolet absorption.[7]

[1] P. Hannonen, J. Jänne, and A. Raina, *Biochim. Biophys. Acta* **289,** 225 (1972).

[2] H. Hibasami, R. T. Borchardt, S. Y. Chen, J. K. Coward, and A. E. Pegg, *Biochem. J.* **187,** 419 (1980).

[3] A. Raina, R.-L. Pajula, and T. O. Eloranta, *FEBS Lett.* **67,** 252 (1976).

[4] R.-L. Pajula, A. Raina, and J. Kekoni, *FEBS Lett.* **90,** 153 (1978).

[5] R.-L. Pajula, A Raina, and T. Eloranta, *Eur. J. Biochem.* **101,** 619 (1979).

[6] O. H. Lowry, N. J. Rosebrough, A. L. Farr, and R. J. Randall, *J. Biol. Chem.* **193,** 265 (1951).

[7] H. M. Kalckar, *J. Biol. Chem.* **167,** 461 (1947).

ISBN 0-12-181994-9

Purification Procedure[4,5]

Step 1. Extraction of Tissues. Fresh bovine brains obtained from a local slaughterhouse are immediately cooled in ice. All subsequent operations are performed at 0–4°. The brains (3.6 kg) are freed of membranes, washed in 0.25 *M* sucrose, and homogenized with an Ultra-Turrax homogenizer (Janke and Kunkel) in two volumes of 0.25 *M* sucrose solution containing 0.1 m*M* EDTA, 1 m*M* 2-mercaptoethanol, and 1 m*M* dithiothreitol. The homogenate is centrifuged at 25,000 *g* for 30 min. The supernatant fraction is filtered through glass wool to remove lipids.

Step 2. Ammonium Sulfate Fractionation. The crude supernatant (6430 ml) is fractionated with solid ammonium sulfate (Mann, special enzyme grade). The proteins precipitated between 55 and 75% saturation of ammonium sulfate are collected by centrifugation at 14,000 *g* for 30 min, dissolved in 500 ml of 10 m*M* Tris-HCl (pH 7.5, measured at room temperature) containing 0.08 *M* NaCl and 0.1 m*M* dithiothreitol (buffer A), and dialyzed for 24 hr against 60 volumes of buffer A, changing the buffer twice during dialysis. The dialyzed enzyme preparation is freed of insoluble material by centrifugation at 14,000 *g* for 15 min.

Step 3. DEAE-Cellulose Chromatography. This step is necessary to remove substances, probably nucleic acids, that interfere with the affinity chromatography.[5] The dialyzed ammonium sulfate fraction (730 ml) is applied to a DEAE-cellulose column (Whatman DE-52, 8.5 × 30 cm) equilibrated with buffer A. The column is first washed with 900 ml of buffer A containing 1 m*M* dithiothreitol and connected to a linear gradient of 0.08 to 0.5 *M* NaCl prepared in buffer A (total gradient volume is 4000 ml). Fractions of 15 ml are collected. Spermine synthase activity is eluted between 0.20–0.27 *M* NaCl. The active fractions (795 ml) are combined, 2-mercaptoethanol is added up to 5 m*M*, and the enzyme is concentrated by the addition of ammonium sulfate (80% saturation). The precipitate is dissolved in 50 ml of 50 m*M* Tris-HCl, pH 7.1, containing 0.15 *M* NaCl, 0.1 m*M* EDTA, and 0.1 m*M* dithiothreitol (buffer B), and is dialyzed overnight against 200 volumes of buffer B.

Step 4. First Affinity Chromatography on Spermine-Sepharose. Spermine-Sepharose is prepared by coupling spermine to CH-Sepharose 4B (Pharmacia).[4] The dialyzed enzyme preparation (86 ml) obtained by step 3 is applied at a rate of 1.0 ml/min to a spermine-Sepharose column (1.6 × 11 cm) equilibrated with buffer B. Fractions of 5 ml are collected. The column is washed with 20 ml of buffer B and eluted with 120 ml of buffer B containing 0.3 *M* NaCl until the $A_{280\ nm}$ of the eluate decreases to about 0.1. Spermine synthase activity is eluted with 50 ml of buffer B containing 0.3 *M* NaCl and 1 m*M* spermidine. The active fractions are combined

(24 ml), concentrated by ultrafiltration to a final volume of 3.5 ml, and dialyzed overnight against 2 × 2 liters of buffer B to remove excess salt and spermidine.

Step 5. Second Affinity Chromatography on Spermine-Sepharose. The dialyzed enzyme preparation (4.3 ml) is applied at a flow rate of 0.15 ml/min to a small spermine-Sepharose column (1 × 5 cm) equilibrated with buffer B. The column is first washed with 7 ml of buffer B, followed by 10 ml of buffer B containing 0.3 *M* NaCl. Fractions of 2 ml are collected. Spermine synthase is eluted with 16 ml of buffer B containing 0.3 *M* NaCl and 1 m*M* spermidine. The active fractions are combined (10.2 ml), concentrated by ultrafiltration to a final volume of 2.0 ml, and stored at 0–4°.

A summary of the purification procedure is shown in the table.

Properties

Stability. The final preparation is stable for more than 6 months when stored on ice in a buffer containing 0.3 *M* NaCl and 1 m*M* spermidine.

Purity. The enzyme preparation obtained by the second affinity chromatography on spermine-Sepharose shows no evidence of contamination as judged by polyacrylamide gel electrophoresis and sodium dodecyl sulfate gel electrophoresis. It is free of any *S*-adenosylmethionine decarboxylase and spermidine synthase activities.

Molecular Weight and Subunit Composition. The approximate molecular weight as determined by gel filtration on a calibrated column of Sephadex G-200 or by sedimentation equilibrium method is 88,000–90,000. Polyacrylamide gel electrophoresis in the presence of sodium dodecyl sulfate gives a subunit molecular weight of 45,000, which suggests that the native enzyme is composed of two subunits of equal size.

PURIFICATION OF SPERMINE SYNTHASE FROM BOVINE BRAIN[5]

Step	Total protein (mg)	Specific activity (units/mg)	Total activity (units)	Yield (%)	Purification (fold)
1. Crude extract	53,400	0.07	3520	100	1
2. Ammonium sulfate	8,400	0.22	1850	53	3
3. DEAE-cellulose	1,800	0.65	1160	33	10
4. First spermine-Sepharose	6.8	121	820	23	1830
5. Second spermine-Sepharose	2.1	396	830	24	6000

Kinetics. An apparent K_m value for decarboxy-AdoMet of about 0.6 μM is obtained using the integrated Michaelis–Menten equation.[5] Contrary to the results obtained with the purified spermidine synthase,[8] no evidence of substrate inhibition is observed with decarboxy-AdoMet up to a concentration of 0.4 mM. The apparent K_m value for spermidine is about 60 μM.

The reaction rate is linear with enzyme concentration up to 1.5 μg/ml, but declines with time when a highly purified enzyme is used. This is obviously due to the accumulation of 5′-methylthioadenosine, a product of the spermine synthase reaction, which is a potent inhibitor of the enzyme with an apparent K_i of about 0.3 μM.[5] A linear reaction rate with time is observed if 5′-methylthioadenosine phosphorylase purified from human placenta[9] is added to the reaction mixture.[10]

Spermine is a weak inhibitor of purified spermine synthase.[5]

Other Inhibitors. The purified spermine synthase is weakly inhibited by putrescine. This inhibition appears to be competitive with respect to spermidine. The enzyme is also inhibited by *N*-ethylmaleimide and *p*-chloromercuribenzoate in the absence of sulfhydryl compounds.

Substrate Specificity. The enzyme shows a strict specificity for spermidine as the propylamine acceptor. *S*-Adenosylmethionine cannot replace decarboxy-AdoMet in the synthesis of spermine.

[8] See Samejima *et al.*, this volume [45].
[9] G. Cacciapuoti, A. Oliva, and V. Zappia, *Int. J. Biochem.* **9**, 35 (1978).
[10] R.-L. Pajula, to be published.

[47] Putrescine Aminopropyltransferase (Spermidine Synthase) of Chinese Cabbage

By Ram K. Sindhu and Seymour S. Cohen

Spermidine synthase catalyzes the transfer of the propylamine moiety from *S*-adenosyl-(5′)-3-methylthiopropylamine (decarboxylated AdoMet) to putrescine to form spermidine, with a stoichiometric release of 5′-methylthioadenosine and a proton. The enzyme was first discovered in *Escherichia coli*[1] and was later purified and characterized from this source.[2] Spermidine synthase has been partially purified from several rat

[1] H. Tabor, S. M. Rosenthal, and C. W. Tabor, *J. Biol. Chem.* **233**, 907 (1958).
[2] W. H. Bowman, C. W. Tabor, and H. Tabor, *J. Biol. Chem.* **248**, 2480 (1973).

ISBN 0-12-181994-9

tissues.[3–5] The purification of this enzyme to homogeneity from bovine brain and rat ventral prostate has finally been accomplished with the aid of affinity chromatography.[6] The work on spermidine synthase in plants, however, is scanty. Using the assay method to be described below, we have shown the presence of this enzyme in soluble extracts of spinach and Chinese cabbage. Also protoplasts prepared from healthy or virus-infected cabbage leaves contain the enzyme, and about a quarter of the activity resides in the low-speed sedimentable fraction containing the chloroplasts.[7,8] A small fraction of the enzyme activity can be found associated with chloroplasts after purification of the organelles on Ludox gradients[9] as described by Price *et al.*[10] In this section we shall describe the assay method using a labeled diamine acceptor, as well as the partial purification and some properties of spermidine synthase from healthy Chinese cabbage leaves. A similar assay, using labeled spermidine as aminopropyl acceptor, has been used in the detection of the spermine synthase of Chinese cabbage.

Choice of Assay Method with Crude Plant Extracts. Four types of assay methods have been devised for spermidine synthase in crude extracts. The first uses labeled putrescine as one of the substrates and depends upon the isolation of a purified labeled product.[1,11] The second involves the use of a more difficultly prepared, appropriately labeled decarboxylated AdoMet.[12] The third depends on the recovery of the labeled product, 5′-methylthioadenosine, from a methyl-labeled decarboxylated AdoMet.[13] Fourth, the formation of spermidine has been determined by a coupled enzymatic assay.[2,14] In crude systems the last two methods are defeated by the presence of additional degradative enzymes. The first procedure is facilitated by the availability of commercially available iso-

[3] A. Raina and P. Hannonen, *FEBS Lett.* **16,** 1 (1971).

[4] P. Hannonen, J. Jänne, and A. Raina, *Biochem. Biophys. Res. Commun.* **46,** 341 (1972).

[5] H. Hibasami, R. T. Borchardt, S. Y. Chen, J. K. Coward, and A. E. Pegg, *Biochem. J.* **187,** 419 (1980).

[6] K. Samejima, A. Raina, and T. O. Eloranta, this volume [45].

[7] S. S. Cohen, R. Balint, and R. K. Sindhu, *Plant Physiol.* **68,** 1150 (1981).

[8] S. S. Cohen, R. Balint, R. K. Sindhu, and D. Marcu, *Med. Biol.* **59,** 394 (1981).

[9] R. K. Sindhu, K. J. McCarthy, R. Balint, and S. S. Cohen, unpublished observation.

[10] C. A. Price, M. Bartolf, W. Ortiz, and E. M. Reardon, *in* "Plant Organelles: Methodological Surveys in Biochemistry" (E. Reid, ed.), Vol. 9, p. 25. Ellis Horwood, Ltd., Chichester, 1979.

[11] J. Jänne, A. Schenone, and H. G. Williams-Ashman, *Biochem. Biophys. Res. Commun.* **42,** 758 (1971).

[12] A. Raina, R.-L. Pajula, and T. O. Eloranta, *FEBS Lett.* **67,** 252 (1976).

[13] H. Hibasami and A. E. Pegg, *Biochem. J.* **169,** 709 (1978).

[14] O. Suzuki, T. Matsumoto, M. Oya, Y. Katsumata, and K. Samejima, *Anal. Biochem.* **115,** 72 (1981).

topic putrescine. Furthermore, decarboxylated AdoMet can now be synthesized chemically.[15] In estimating formation of the product, the newly synthesized spermidine can be purified as the distinctive fluorescent dansyl derivative and its radioactivity estimated by scintillation counting. For these reasons we have adopted the first procedure in determining the enzyme activity in crude plant extracts. In doing so we have improved the purification of the amine product and the isolation of the dansyl derivative.

Assays of enzyme activities should be determined from constant initial rates in ranges of enzyme concentration providing a linear dependence of activity on concentration. With crude plant extracts or supernatant fluids from which chloroplasts were removed, reaction rates decreased rapidly with increasing extract concentrations. The effect was due mainly to the dilution of radioactive putrescine by putrescine in the extract. When the specific radioactivity of the putrescine reisolated from the reaction mixture was used to calculate spermidine produced, there was an initial proportionality of rate to extract concentration and a second phase in which the rate deviated from linearity by 20–25%. No inhibitory material leading to inactivation of the enzyme or the destruction of decarboxylated AdoMet has been detected in the extracts.

Assay Method

Principle. The assay is based on the transfer of the propylamine moiety of decarboxylated AdoMet to labeled putrescine, ion-exchange separation of the labeled spermidine, followed by conversion to the dansyl derivative and purification by thin-layer chromatography (TLC). The radioactivity of dansylated amine is determined.

Reagents

Gly-NaOH buffer, 1 *M*, pH 8.8
Decarboxylated AdoMet[16]
[^{14}C]Putrescine dihydrochloride (102 mCi/mmol)
Enzyme extract[17]: 0.001–0.03 unit

[15] K. Samejima, Y. Nakazawa, and I. Matsunaga, *Chem. Pharm. Bull.* **26,** 1480 (1978).

[16] Decarboxylated AdoMet is prepared either from [1-^{14}C]AdoMet or [*methyl*-^{3}H]AdoMet with the aid of *E. coli* enzyme (for details, see Cohen *et al.*[7]; also this volume [11,43]). The synthetic materials supplied to us by Dr. K. Samejima of Tokyo and Dr. V. Zappia of Naples are similarly active with spermidine synthase of *E. coli* and of Chinese cabbage.

[17] For the preparation of the enzyme extract from protoplasts, see Cohen *et al.*[7] When the leaves are used, a 50% extract (w/v) is prepared by grinding the tissue in 1 *M* Gly-NaOH, pH 8.8, in a chilled pestle–mortar at 0°.

Procedure. The reaction is carried out routinely in 12 × 75 mm screw-capped tubes. The reaction mixture contains 150 m*M* Gly-NaOH buffer, pH 8.8; 37 μ*M* [^{14}C]putrescine dihydrochloride (1.2 μCi); 25 μ*M* decarboxylated AdoMet; appropriate amounts of enzyme extract and H_2O in a total volume of 0.325 ml. Comparable control incubation mixtures are set up routinely that lack either decarboxylated AdoMet or the enzyme extract. The tubes are incubated at 37° for 1 hr, and the reaction is stopped by adding 1 ml of cold 5% perchloric acid. After centrifugation the precipitates are washed twice with 1 ml each of 3% perchloric acid. To the combined supernatants are added 0.03 μmol of [^{12}C]spermidine trihydrochloride to each tube as carrier, and the solutions are placed on 1 × 5 cm columns of Dowex 50W (H^+ form, 200–400 mesh, 8% cross-linked), which have been prepared as described by Inoue and Mizutani.[18] The columns are washed with 40 ml of 0.1 *M* sodium phosphate buffer, pH 8.0, containing 0.7 *M* NaCl; 10 ml of 1 *N* HCl, and 40 ml of 2.3 *N* HCl. The flow rate is about 1–2 ml/min. The latter wash removed more than 95% of the substrate, i.e., [^{14}C]putrescine.[19] About 96% of the spermidine is retained on the column and is then eluted with 30 ml of 6 *N* HCl.[20] Labeled spermidine is then dansylated[21] (after being dried under reduced pressure at 45° and taken up in 30 m*M* HCl) and extracted into 0.5 ml of benzene (J. T. Baker); 50 μl of the benzene extract is then applied to a preactivated (110°, 1 hr) silica gel G plate (LK 6D, 250 μm, Whatman, Inc.). The plate is developed once in chloroform–triethylamine (80 : 16, v/v). This solvent is useful because the R_f of dansyl spermidine in this solvent is 1.7 times greater than that of dansyl putrescine. This separation further reduces the contamination by residual labeled dansyl putrescine. The spots comigrating with authentic dansyl spermidine are scraped and counted in 1 ml of H_2O and 10 ml of Aquasol-2 (New England Nuclear). Recovery of the triamine from the plate is about 80%.

The initial ion-exchange separation of substrate spermidine and product spermine and the purification of dansyl spermine by TLC with a solvent in which this derivative moves ahead of the residual derivatized spermidine has also been used in assaying spermine synthase. These procedures are particularly important for the detection and estimation of spermine synthase, which is present in cabbage extracts at approximately one-tenth of the activity of spermidine synthase. Washing the Dowex column with 3.3 *N* HCl removes 99% of the spermidine, whereas about

[18] H. Inoue and A. Mizutani, *Anal. Biochem.* **56,** 408 (1973).

[19] W. A. Gahl, A. M. Vale, and H. C. Pitot, *Biochem. J.* **187,** 197 (1980).

[20] These columns can be reused after washing with H_2O.

[21] S. S. Cohen, S. Morgan, and E. Streibel, *Proc. Natl. Acad. Sci. U.S.A.* **64,** 669 (1969).

93% of spermine remains bound to the Dowex-50W (H^+) column and can be eluted with 6 *N* HCl.

As an alternative procedure for studies with the partially purified enzyme, the fluorometric assay of Suzuki *et al.*[14] using oat seedling polyamine oxidase may be used in a coupled system. In the latter case the volume of the reaction mixture is reduced to one half. The final concentration of sodium phosphate buffer, putrescine dihydrochloride, and decarboxylated AdoMet are 33 m*M*, 200 μM, and 25 μM, respectively.

Definition of Enzyme Unit and Specific Activity. One enzyme unit is defined as the amount of enzyme required to form 1 nmol of spermidine per minute under the standard assay conditions.

Specific activity is expressed as units per milligram of protein. Protein is determined by the method of Lowry *et al.*,[22] with crystalline bovine serum albumin as the standard. The protein in a crude extract or in the supernatant fraction is precipitated with cold trichloroacetic acid. The precipitate is separated by centrifugation, extracted twice with cold acetone–methanol (7 : 2, v/v) to free it of interfering chlorophyll, and then extracted with 1 *N* NaOH. Chlorophyll-free fractions are estimated directly after suitable dilutions.

Purification Procedure

The enzyme is prepared from healthy Chinese cabbage (*Brassica pekinesis,* var. Pak Choy) leaves. The plants are grown in a controlled environment chamber set for 18-hr days at 28° and 20,000 lux, using incandescent fluorescent lighting, and 6-hr dark periods at 22°. When the plants are about 4 weeks old, actively growing leaves (<5 cm long) are taken, the midrib is removed with a razor blade, and the deribbed leaves are weighed. The leaves are then washed with H_2O and blotted dry. All the purification steps are carried out at 0–4°, unless stated otherwise.

Step 1. Eighty-three grams of Chinese cabbage leaves are chilled, cut, and homogenized in a chilled Waring blender for 5 min with 160 ml of 1 *M* Gly-NaOH buffer, pH 8.8. Care is taken that the temperature does not increase during homogenizing. The crude extract is filtered through eight layers of cheesecloth (186 ml) and centrifuged at 12,000 *g* for 30 min, eliminating most of the chlorophyll at this step.

Step 2. The supernatant (176 ml) is brought to pH 5.1 with 0.1 *M* acetic

[22] O. H. Lowry, N. J. Rosebrough, A. L. Farr, and R. J. Randall, *J. Biol. Chem.* **193,** 265 (1951).

acid. A freshly prepared solution of streptomycin sulfate (652.5 mg in 16 ml of cold H_2O) is added slowly with stirring. This is allowed to stand for 20 min and then centrifuged at 12,000 *g* for 10 min. The chlorophyll-free supernatant is brought back to pH 8.8 with 1 *M* NaOH.

This solution (352 ml) is fractionated by addition of solid ammonium sulfate (Reagent grade, J. T. Baker). $(NH_4)_2SO_4$ (69 g) is added slowly with stirring. After standing for 1 hr, the precipitate is removed by centrifuging at 12,000 *g* for 30 min. To the supernatant, more $(NH_4)_2SO_4$ (75.3 g) is then added slowly with stirring. After standing for 1 hr, the precipitate is collected by centrifugation at 12,000 *g* for 30 min; after draining, it is dissolved in 40 ml of 25 m*M* sodium phosphate buffer, pH 7.2.

To this fraction, 40 ml of chilled acetone (stored at −20°) is added slowly with constant stirring. A heavy precipitate forms, which is discarded after centrifugation at 10,000 *g* for 10 min. To the clear supernatant 120 ml of acetone (−20°) is added with stirring. This precipitate, containing the bulk of the spermidine synthase, is collected by centrifugation at 10,000 *g* for 10 min, drained, and dissolved in 25 m*M* sodium phosphate buffer, pH 7.2, containing 0.1 *M* KCl. This solution is freed of any insoluble material by centrifugation at 12,000 *g* for 20 min and stored frozen at −20° in 1-ml aliquots.

Step 3. A portion of the enzyme preparation after acetone precipitation (2.5 ml), is applied to a Sephadex G-100 column (Pharmacia, 1.5 × 86 cm), equilibrated with 25 m*M* sodium phosphate buffer, pH 7.2, containing 0.1 *M* KCl, eluted with the same buffer in 1.2-ml fractions. All the enzyme activity is recovered in a single, fairly sharp peak. Active fractions are pooled (20.4 ml) and concentrated to 2 ml by filtration through Centriflo CF-25 cones (Amicon). This preparation is stored frozen at −20° in small aliquots. Using this procedure it is possible to purify the enzyme 160-fold with a recovery of 71%.

A typical purification is summarized in the table.

PURIFICATION OF SPERMIDINE SYNTHASE FROM CHINESE CABBAGE LEAVES

Step	Total activity (units)	Total protein (mg)	Specific activity (units/mg protein)	Degree of purification (fold)	Recovery (%)
Crude extract	14.3	1790	0.008	1	100
Supernatant (before streptomycin)	15.9	1350	0.012	1.5	111
Acetone fractionation	26.1	40	0.65	82	180
Sephadex G-100	10.2	8.0	1.28	160	71

Properties

The plant enzyme loses 50–75% of its activity when dialyzed after $(NH_4)_2SO_4$ fractionation. Introduction of 1 mM dithiothreitol and/or 1 mM EDTA in the dialysis medium does not prevent this loss in the enzyme activity. Addition of a supernatant from a heat-inactivated enzyme preparation to a dialyzed inactivated enzyme preparation does not restore the activity.

Enzyme precipitated by acetone does not adhere to columns of either spermidine- or spermine-Sepharose prepared and used as described by Pajula *et al.*[23] The plant enzyme is retained on *S*-adenosyl-(5′)-3-thiopropylamine (ATPA)-Sepharose column,[24] but elution with as high as 3.5 mM decarboxylated AdoMet does not elute more than 12% of the enzyme activity. Such a relatively inactive eluted enzyme preparation is not further activated by various column fractions. This loss in the activity of enzyme eluted from ATPA-Sepharose is not due to the presence of a high concentration of decarboxylated AdoMet. The enzyme activity is assayed after removing excess decarboxylated AdoMet by filtration through Centriflo CF-25 cones (Amicon).

Stability. When protoplasts from Chinese cabbage leaves are disrupted in 1 M triethanolamine sulfate, pH 8.2, and stored at $-15°$ (3.1×10^6 cells/ml), more than 90% of the enzyme activity is lost after 1 week.[7] However, after acetone precipitation the enzyme is quite stable when stored frozen at $-20°$, and no decrease in the activity is observed for at least 1 month. After Sephadex chromatography, however, there is about 30% loss in activity after storage for 1 month at $-20°$.

Molecular Weight. The molecular weight of the plant enzyme was determined by using a calibrated column of Sephadex G-100 as described by Andrews.[25] The apparent molecular weight was estimated to be 81,000.

pH Optimum. The optimum pH for the cabbage enzyme is 8.8 in Gly-NaOH buffer. No enzyme activity is detected at pH 6.0 in phosphate or at 10.0 in Gly-NaOH buffer. By contrast, the bacterial[2] and the mammalian[26] spermidine synthases have been shown to have pH optima at 10.4 and 10, respectively. At a given pH, the activity of the plant enzyme in Tris buffer is generally lower than that in phosphate or Gly-NaOH buffers. At pH 8.0 and 8.8, the enzyme activities in Tris buffer are about 70% and 88% compared to those found in phosphate and Gly-NaOH buffers, respectively, at the same pH.

[23] R.-L. Pajula, A. Raina, and J. Kekoni, *FEBS Lett.* **90,** 153 (1978).

[24] ATPA-Sepharose was kindly supplied by Dr. K. Samejima, Tokyo Biochemical Research Institute, Tokyo 171, Japan. See also this volume [45].

[25] P. Andrews, *Biochem. J.* **91,** 222 (1964).

[26] K. Samejima and B. Yamanoha, *Arch. Biochem. Biophys.* **216,** 213 (1982). We thank Dr. K. Samejima for sending us a preprint of his manuscript.

[48] Aminopropyltransferase Substrates and Inhibitors

By JAMES K. COWARD, GARY L. ANDERSON, and KUO-CHANG TANG

The aminopropyltransferases spermidine synthase (EC 2.5.1.16) and spermine synthase catalyze the transfer of an aminopropyl group from *S*-adenosyl-3-methylthio-1-propylamine (decarboxylated *S*-adenosylmethionine, dcSAM) to putrescine and spermidine, respectively. The second product of the reaction in both cases is the nucleoside 5′-deoxy-5′-methylthioadenosine (MTA). We have developed methods for the syntheses of dcSAM and MTA, in addition to related adenosine 5′-thioethers and sulfonium compounds, for use in studying the mechanism of these enzymes and their inhibition *in vitro* and *in vivo*.[1–3] These synthetic mate-

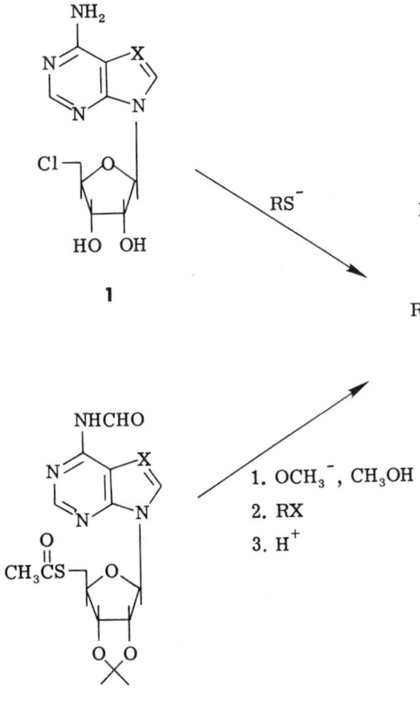

NH_2, N, X, N, N, Cl, O, HO, OH

1

RS^-

NH_2, N, X, N, N, RS, O, HO, OH

3

(1)

NHCHO, N, X, N, N, O, CH_3CS, O, O, O

2

1. OCH_3^-, CH_3OH
2. RX
3. H^+

[1] J. K. Coward, N. C. Motola, and J. D. Moyer, *J. Med. Chem.* **20,** 500 (1977).
[2] G. L. Anderson, D. L. Bussolotti, and J. K. Coward, *J. Med. Chem.* **24,** 1271 (1981).
[3] K.-C. Tang, R. Mariuzza, and J. K. Coward, *J. Med. Chem.* **24,** 1277 (1981).

ISBN 0-12-181994-9

rials have been used for preliminary kinetic studies on crude spermidine synthase from rat prostate,[1] for studies on the inhibition of partially purified polyamine biosynthetic enzymes,[4,5] and for studies of polyamine biosynthesis in cultured mammalian cells.[6,7] The procedures described below are based on the general reactions shown in Eq. (1).

Synthetic Procedures

*5'-Deoxy-5'-chloroadenosine (**1**, X = N)*

To 100 ml of hexamethylphosphoramide (HMPA) at 0–4° (ice bath) under N_2 was added 12.5 ml (20.48 g, 172 mmol) of freshly distilled thionyl chloride. Adenosine (10.68 g, 40 mmol) was then added in one portion. The resulting solution was stirred for 2 hr at ambient temperature and then added to 600 ml of ice water. The pH of the aqueous HMPA reaction mixture was adjusted to ca 9 with concentrated NH_4OH, and then it was heated on a steam bath to give a homogeneous solution. This hot solution was filtered, then cooled to yield the desired product, **1** (X = N) (11.2 g, 98%). The material obtained in this manner was homogeneous by all chromatographic and spectral criteria. Recrystallization could be accomplished from H_2O, if desired; mp 160° (decomp.) with softening from 80°.

*5'-Deoxy-5'-chlorotubercidin (**1**, X = CH)*

This compound was prepared in a similar manner to that described for **1** (X = N), using 1.02 g (3.85 mmol) of tubercidin[8] in 9 ml of HMPA containing 0.9 ml (1.47 g, 12.4 mmol) thionyl chloride. After stirring at ambient temperature for 2 hr, the reaction solution was added to 60 ml of ice water, the pH was adjusted to ca. 9, and solution was obtained by heating on a steam bath. Chloroform (25 ml) was added to the cooled, filtered solution with stirring, and a crystalline product (**1**, X = CH) was precipitated from the solution on standing overnight in the refrigerator (yield, 961 mg; 85%). Recrystallization from H_2O afforded 541 mg (47%) of a white crystalline material, mp 250° (decomp.) with softening from 150°.

[4] H. Hibasami, R. T. Borchardt, S. Y. Chen, J. K. Coward, and A. E. Pegg, *Biochem. J.* **187,** 419 (1980).

[5] K.-C. Tang, A. E. Pegg, and J. K. Coward, *Biochem. Biophys. Res. Commun.* **96,** 1371 (1980).

[6] A. E. Pegg, R. T. Borchardt, and J. K. Coward, *Biochem. J.* **194,** 79 (1981).

[7] A. E. Pegg, K.-C. Tang, and J. K. Coward, *Biochemistry* **21,** 5082 (1982).

[8] Tubercidin was generously provided by Dr. George Whitfield, The Upjohn Company, Kalamazoo, Michigan.

*5'-Deoxy-5'-methylthioadenosine (**3**, X = N, R = CH_3)*

A solution of aqueous $NaSCH_3$ was prepared by slowly adding CH_3SH to 12 ml of H_2O containing 3.85 g (96 mmol) NaOH until a weight gain of 4.47 g (93 mmol) was observed; 10 ml of 2 *N* NaOH was then added to give a final solution containing 3 mmol of $NaSCH_3$ per gram. To 1.2 g (3.6 mmol) of this aqueous $NaSCH_3$ solution was added 910 mg (3.2 mmol) of **1** (X = N). After heating on a steam bath for 15 min, the reaction solution was set at ambient temperature for 2 days, during which time a white precipitate formed. The solid product was collected by filtration and washed several times with water to give 706 mg (80%) of **3** (X = N, R = CH_3) with mp 210°. Recrystallization from water afforded the product as fine white needles, mp 211–212° (lit[9]: mp 211–212°).

*5'-Deoxy-5'-methylthiotubercidin (**3**, X = CH, R = CH_3)*

The aqueous solution of $NaSCH_3$ (1 g, 3 mmol) described above for the preparation of **3** (X = N, R = CH_3) was added to 285 mg (1 mmol) of **1** (X = CH), and the reaction mixture was heated on a steam bath. Within 5 min a light yellow solution was observed, and within 15 min a white precipitate formed. The solid product was then collected by filtration, washed several times with water to give 267 mg (90%) of **3** (X = N, R = CH_3) with mp 177°. Recrystallization from water gave white needles, mp 177° (lit[1]: mp 172°).

*5'-Deoxy-5'-(γ-amino)propylthioadenosine (**3**, X = N, R = —$(CH_2)_3NH_2$)*

The compound, also known as decarboxylated *S*-adenosylhomocysteine, can be prepared from either **1** (route A) or **2** (route B).

Route A. To a three-necked round-bottom flask containing liquid NH_3 (25 ml) under a stream of dry N_2 was added 262 mg (1.4 mmol) of 3-(benzylthio)-1-propylamine,[10] followed by sodium metal (ca 150 mg) until the opaque blue color persisted for at least 15 min. Then compound **1** (285 mg, 1 mmol) was added in one portion as the color changed from blue to a dull gold. After allowing the reaction mixture to stir at −70° for at least 6 hr, NH_3 was allowed to evaporate slowly overnight as the temperature rose to ambient. The residue was dissolved in water (25 ml), and the solution was neutralized to pH 6.0 with 12 *N* HCl. After filtration, the solution was lyophilized, the crude residue was dissolved in ca 5 ml of H_2O, and purified by preparative thin-layer chromatography (TLC) on two silica gel plates (Brinkmann No. 5766) in butanol–acetic acid–water (12 : 3 : 5). After three developments, extraction of the product (R_f = 0.3–

[9] K. Kikugawa, K. Iizuka, Y. Higuchi, H. Hirayama, and M. Ichino, *J. Med. Chem.* **15,** 387 (1972).

[10] G. A. Jamieson, *J. Org. Chem.* **28,** 2397 (1963).

0.4) with CH_3OH and evaporation of the solvent *in vacuo* gave 142 mg (37%) of a yellow, gummy residue, homogeneous by TLC ($R_f = 0.62$) in 5% Na_2HPO_4 on cellulose. This material was converted to the hydrogen sulfate salt by dissolving in 1 *N* H_2SO_4, filtering, and adding EtOH to precipitate the salt on cooling. The salt was collected on a filter, and additional salt was obtained by further additions of EtOH to the filtrate. The collected salts were combined and recrystallized from H_2O-EtOH to give 55 mg of **3** (X = N, R = —$(CH_2)_3NH_2$) as the HSO_4^- salt, mp 173–174° decomp. (lit[10]: mp 173–175° decomp.).

Route B. The nucleoside **2**[11] (2.0 g, 5 mmol) was dissolved in 200 ml of methanol containing 324 mg (6 mmol) of sodium methoxide. After stirring for 10 min at ambient temperature, the yellow solution was treated with 1.6 g (6 mmol) of 3-bromopropylphthalimide and allowed to stir overnight at ambient temperature. The solvent was then removed *in vacuo,* and the oily residue was partitioned between $CHCl_3$ and water. The dried organic extracts were concentrated *in vacuo,* and the desired product was separated from *sym*-adenosine-5′-disulfide by preparative TLC on silica gel in EtOAc: yield, 980 mg (38%) of a viscous straw-colored oil; $R_f = 0.44$ on silica gel in EtOAc. The phthalimidopropylthioadenosine compound (980 mg, 1.6 mmol) was dissolved in 200 ml of EtOH, and 1.5 ml (25 mmol) of 85% hydrazine was added. The resulting solution was heated at reflux for 18 hr, after which the solvent was removed *in vacuo* to give an oily residue. This residue was dissolved in 20 ml of water, acetic acid was added to pH 4, and the turbid solution was clarified by filtration through a Millipore filter. The filtrate was concentrated *in vacuo,* the pH was adjusted to 11, and the solution was extracted with $CHCl_3$. The dried organic extracts were evaporated *in vacuo* to give an oily residue: net weight, 480 mg (82%); $R_f = 0.50$ on silica gel in butanol–acetic acid–water (12 : 3 : 5). This oil was converted to the hydrogen sulfate salt of **3** (X = N, R = —$(CH_2)_3NH_2$) as described for (route A): net weight, 412 mg (21% overall from **2**); mp 173–174° decomp.

S-Adenosyl-1,8-diamino-3-thiooctane

$$\left(\mathbf{3},\ X = N,\ R = -CH\begin{matrix} \diagup (CH_2)_5NH_2 \\ \diagdown (CH_2)_2NH_2 \end{matrix}\right)$$

A solution of 4.7 g (17.4 mmol) of 1,8-diazido-3-(*S*-acetyl)thiooctane (prepared from 1,8-dichloro-3-octanol, described in detail below) in 60 ml

[11] T. Nielson, T.-Y. Shen, and W. V. Ruyla, French Patent 1,589,694; *Chem. Abstr.* **74,** p126013c (1971). The 5′-thioacetyl derivative, **2**, was prepared from the corresponding 5′-tosylate using KSAc in DMSO at ambient temperature (yield=80%), rather than in refluxing 2-butanone (yield=25%) as described in this patent.

of DMSO was deoxygenated with a stream of N_2 for $1\frac{1}{2}$ hr. At this time, 3.3 g (11.57 mmol) of **1** (X = N) was added, followed by 14 ml of 4 *N* NaOH. The resulting solution was stirred at ambient temperature for 36 hr, at which time analysis by TLC showed >95% conversion to product. The reaction solution was then added to 1200 ml of H_2O to give a milky white mixture, from which the desired diazidooctyl thioadenosine could be isolated as a solid by repeated freezing and thawing (yield, 4.8 g; 87%). This product was pure by all chromatographic and spectral criteria, but could be recrystallized from CH_3OH-H_2O if desired.

The diazido derivative just described (2.388 g, 5 mmol) and 4.28 g (16.3 mmol) of triphenylphosphine were dissolved in 24 ml of anhydrous pyridine, and the resulting solution was stirred at ambient temperature for 3 hr. At this time, 5 ml of 15 *M* NH_4OH was added, and stirring continued overnight. Pyridine and excess NH_4OH were then removed under high vacuum at room temperature, the resulting residue was triturated with H_2O, and the combined aqueous layers were then washed with benzene (3×) and ether (3×). The resulting aqueous solution was then concentrated to dryness by lyophilization to give 1.32 g (62%) of analytically pure **3**

$$\left(X = N,\ R = \text{—CH} \begin{matrix} \diagup (CH_2)_5NH_2 \\ \\ \diagdown (CH_2)_2NH_2 \end{matrix} \right)$$

1,8-Diazido-3-[S-acetyl]thiooctane

6-Chlorohexanoyl Chloride. 2.26 g (16.6 mmol) of zinc chloride was added to 77.3 g (667 mmol) of ε-caprolactone in an ice bath to give a reddish solution, to which 94.0 g (790 mmol) of thionyl chloride was added dropwise. After addition, the color of the reaction mixture was a dark brown, which gradually became lighter in color as the reaction mixture was heated at 50–60° overnight. Nuclear magnetic resonance (NMR) spectra indicated that all the starting lactone had been consumed by this time. After the removal of the excess of thionyl chloride under reduced pressure, the crude product was vacuum distilled (bp 70–72°/1.5 torr) to give 57.49 g (51%) of a clear, colorless liquid.

1,8-Dichloro-3-octanone. A well-stirred, ice-cooled solution of 30.35 g (179.6 mmol) of 6-chlorohexanoyl chloride in 300 ml of dry CCl_4 was degassed with nitrogen for $\frac{1}{2}$ hr. $AlCl_3$ (26.34 g, 179.6 mmol) was added in portions, then ethylene was bubbled in at a rate so that no excess ethylene escaped from the reaction vessel. Ethylene was allowed to bubble through the reaction mixture at 0° for 2 hr and at ambient temperature overnight. The reaction mixture was poured into 800 ml of ice water and extracted with chloroform (2 × 400 ml). The combined chloroform extracts were washed with saturated aqueous $NaHCO_3$ (500 ml), H_2O (500 ml), and saturated NaCl (500 ml), and dried over $MgSO_4$. After the removal of the solvent, 15.2 g (43%) of crude 1,8-dichlorooctanone was obtained, which was used without further purification.

1,8-Dichloro-3-octanol. To a well-stirred solution of 15.08 g (77 mmol) of 1,8-dichloro-3-octanone in 15 ml of 95% ethanol, cooled in an ice bath, was added a solution of 1.97 g (52 mmol) of sodium borohydride in 7 ml of H_2O; 20 ml of concentrated NH_4OH was added, and the resulting solution was stirred at ambient temperature for 1 hr after the ice bath was removed. The reaction mixture was poured into 350 ml of H_2O and extracted with $CHCl_3$ (2 × 350 ml). The combined extracts were washed with 5% HCl solution (300 ml) and dried over $MgSO_4$. After removal of the solvent under reduced pressure, the residue was vacuum distilled (bp 95–97°/0.3 torr) to give 10.49 g (69%) of pure 1,8-dichloro-3-octanol.

1,8-Diazido-3-hydroxyoctane. A mixture of 5.5 g (27.6 mmol) of 1,8-dichloro-3-octanol, 5.5 g (84.57 mmol) of sodium azide, and a catalytical amount of anhydrous lithium iodide in 25 ml of dry DMF was heated to 60° ± 5° (oil bath) for 1 day and then stirred at ambient temperature for another day. The solvent was then removed to near dryness under vacuum at a temperature of ca 30°. The residue was then partitioned between 250 ml each of $CHCl_3$ and H_2O. The organic layer was separated and washed with H_2O (2 × 250 ml) and saturated aqueous NaCl (250 ml), and dried over $MgSO_4$. After removal of the solvent, the crude product was distilled (bp 105–106°/0.015 torr) to give 5.25 g (89.7%) of pure 1,8-diazido-3-hydroxyoctane.

1,8-Diazido-3-tosyloxyoctane. 1,8-Diazido-3-hydroxyoctane (4.24 g; 20 mmol) in 40 ml of dry pyridine was cooled in an ice bath, and 16.9 g (88.89 mmol) of recrystallized tosyl chloride was added in portions, and the resulting solution was stirred at 4° for 1 day. A white precipitate, presumably pyridinium hydrochloride, was formed from a clear pink solution after several hours. The reaction mixture was poured into 750 ml of ice water and extracted with $CHCl_3$ (3 × 250 ml). The combined $CHCl_3$ extracts were washed with H_2O (2 × 250 ml), cold 5% H_2SO_4 (2 × 250 ml), saturated aqueous $NaHCO_3$ (250 ml) and saturated aqueous NaCl (250 ml), and dried over $MgSO_4$. After removal of the solvent and drying under high vacuum, 9.92 g (~100%) of the crude desired product was obtained and used without any further purification.

1,8-Diazido-3-(S-acetyl)thiooctane. Potassium thioacetate (5.15 g; 45 mmol), previously triturated with 2-butanone, was added to a solution of 13.9 g (~28.9 mmol) of crude 1,8-diazido-3-tosyloxyoctane in 130 ml of dry DMSO, and the resulting solution was stirred at ambient temperature for 1½ days. The reaction mixture was then concentrated under high vacuum, and the residue was partitioned between 500 ml each of H_2O and $CHCl_3$. The organic layer was washed with H_2O (4 × 500 ml) and saturated aqueous NaCl (2 × 500 ml), and dried over $MgSO_4$. After removal of the solvent and drying under high vacuum, the residual crude product was purified by distillation; bp 165–170°/1.2 torr; yield 6.34 g (78.6%), a slightly yellow-brown liquid. A second distillation (bp 130–132°/0.15 torr) gave 5.78 g (72%) of the pure 1,8-diazido-3-(*S*-acetyl)thiooctane as a yellow liquid.

*Synthesis of Radiolabeled **1** and **3***

The procedures described above can be adapted for the synthesis of radiolabeled adenosine 5′-thioethers.

*[8-^{14}C]5′-Deoxy-5′-chloroadenosine (**1**, X = N).* [8-^{14}C]Adenosine (0.125 mg, 54.7 mCi/mmol) was diluted with nonradioisotopic adenosine (2.375 mg) to give a final specific activity of 2.7 mCi/mmol. HMPA (25 μl) and thionyl chloride (2.5 μl) were added to [8-^{14}C]adenosine (2.5 mg) contained in a 0.5-ml glass culture tube, the mixture was capped and vortexed, and the resulting yellow solution was left to stand at room temperature for 12 hr. Distilled water (100 μl) was added to quench the reaction; the mixture was vortexed and applied to a 1.0 ml Bio-Rad AG 50W-X8 resin column (100–200 mesh, H^+ form) that had been prewashed with distilled water (40 ml). The column was washed with distilled water (30 ml) and then 50% NH_4OH solution, and the ammoniacal liquors were collected until no significant radioactivity (<500 cpm/ml) could be detected in the eluates. Prior to lyophilization, a small aliquot of the ammoniacal fraction was subjected to liquid chromatographic analysis on a Partisil-10 ODS column loaded with 4.5% octadecyl silica by using methanol–water (20 : 80, v/v) as the eluting solvent. This chromatographic system effects a satisfactory separation of adenosine and 5′-chloroadenosine and is of preparative value. Examination of the radiochromatogram of the ammoniacal eluate showed that no significant amount of [8-^{14}C]adenosine was present and that 5′-chloro[8-^{14}C]adenosine was formed in almost quantitative yield (24.4 μCi, 97%). 5′-Deoxy-5′-chloro[2-^{3}H]adenosine (**1**, X = N) may be prepared by a similar procedure.[12]

*5′-Deoxy-5′-[$^{14}CH_3$]methylthioadenosine (**3**, X = N, R = $^{14}CH_3$).* $^{14}CH_3SH$ (0.96 mg, 20 μmol, specific activity, 12.5 μCi μmol^{-1}) was converted to the sodium salt by quickly adding 0.2 ml of 2 *N* NaOH to the cooled (−70°) gas in a freshly opened ampoule. The resulting solution was

[12] N. Kamatani, W. A. Nelson-Rees, and D. A. Carson, *Proc. Natl. Acad. Sci. U.S.A.* **78**, 1219 (1981).

added to a flask containing 56 mg (0.2 mmol) of 5′-chloro-5′-deoxyadenosine (**1**). The ampoule was then rinsed with an additional 0.2 ml of 2 *N* NaOH, which was added to the reaction flask, followed by 0.075 ml of a 20% (by weight) aqueous solution of nonradioactive $NaSCH_3$. The reaction solution was heated at 80° for 1 hr, then cooled, acidified to pH 6 with acetic acid, and diluted to ca 10 ml with H_2O. The resulting yellow solution was applied to a Dowex AG 50-X8 (H^+ form) column (1 × 7 cm), and the column was washed with water (250 ml) to elute a small amount of UV-absorbing material (λ_{max} 260 nm). Then the eluent was changed to 1.0 *N* NH_4OH, and fractions of 2.8 ml were collected. The desired product was eluted in fractions 10–60; however, fractions 10–19 were shown to contain a radioactive impurity, and therefore fractions 20–60 were pooled and lyophilized to give a white, fluffy solid: $\lambda_{max}^{H_2O}$ = 260 nm; R_f = 0.76 on silica gel in butanol–acetic acid–water (12 : 3 : 5); yield, 18.5 μmol (9.2%); specific activity, 0.24 μCi μmol^{-1}.

5′-Deoxy-5′-[$^{14}CH_3$]methylthiotubercidin (**3**, *X = CH, R = $^{14}CH_3$*). This compound was prepared in the same way as described above for radioactive compound **3** (X = N, R = ^{14}CH), except that the desired product was obtained as a solid directly from the reaction solution, as described for nonradioactive compound. Additional product could be obtained from the reaction filtrate by preparative TLC on cellulose in butanol–acetic acid–water (12 : 3 : 5). The material with R_f = 0.7 was extracted from the cellulose with 0.1 *N* HCl and the aqueous extract was lyophilized to give a white, fluffy solid: $\lambda_{max}^{0.1\,N\,H^+}$ = 272 nm; total yield, 72 μmol (3.6%); specific activity, 0.33 μCi μmol^{-1}.

(±) *S-Adenosyl-3-methylthio-1-propylamine (Decarboxylated S-Adenosylmethionine)*

The HSO_4^- salt of **3** (X = N, R = —$(CH_2)_3NH_2$) (55 mg, 125 μmol) was methylated with CH_3I in ca 2.5 ml of HCOOH : CH_3COOH (1 : 1) for 48 hr at ambient temperature protected from the light.[10] The yellow-orange reaction solution was then diluted with 5 ml of H_2O and extracted with ether (3 × 5 ml) to remove excess CH_3I. The resulting colorless aqueous solution was lyophilized to give a fluffy solid residue; $\lambda_{max}^{pH\,3}$ = 260 nm. Thin-layer chromatography was on cellulose in 5% aqueous Na_2HPO_4: R_f = 0.73 vs R_f = 0.62 for starting material, **3**[X = N, R = —$(CH_2)_3NH_2$]. On standing in 0.1 *N* NaOH for 10 min at ambient temperature, λ_{max} = 268 nm, indicative of base-catalyzed fragmentation of the adenosine sulfonium salt to yield adenine.[13]

[13] L. W. Parks and F. Schlenk, *J. Biol Chem.* **230,** 295 (1958); J. Baddiley, W. Frank, N. A. Hughes, and J. Wieczorkowski, *J. Chem. Soc.* p. 1999 (1962); R. T. Borchardt, *J. Am. Chem. Soc.* **101,** 458 (1979).

This general procedure has been used to prepare sulfonium salts of many types of nucleoside 5′-thioethers (**3**).[1-3]

Comments

The procedures described herein for the synthesis of compound **3** from compound **1** are readily carried out with minimal prior experience. The synthesis of various stable precursors of RS^- [Eq. (1)] is somewhat more difficult, and handling of amino thiolate anions in basic conditions requires, as much as possible, exclusion of oxygen from the reactive mixture. Similarly, use of **2** as a stable precursor of 5′-deoxy-5′-mercaptoadenosine is plagued by the inherent instability of the free thiol. Rapid oxidative dimerization to the symmetrical disulfide occurs, even under apparently O_2-free conditions. Thus, the yields of **3** from **2** are generally lower than the yields of **3** from **1**. In general, we have found that if the precursor to RS^- [Eq. (1)] can be obtained in a pure state, e.g., distillable thiol acetate[3] or recrystallizable thiol hydrochloride,[2] the conversion of **1** → **3** goes in reproducibly high yields.

[49] Inhibition of Aminopropyltransferases

By ANTHONY E. PEGG

Many compounds have been tested for inhibitory activity toward the mammalian aminopropyltransferases spermidine synthase and spermine synthase.[1-8] The assay of these enzymes is in this volume,[9] and the synthesis of the inhibitors having the most potential for further studies is

[1] H. Hibasami, R. T. Borchardt, S. Y. Chen, J. K. Coward, and A. E. Pegg, *Biochem. J.* **187,** 419 (1980).
[2] H. Hibasami and A. E. Pegg, *Biochem. Biophys. Res. Commun.* **81,** 1398 (1978).
[3] R.-L. Pajula and A. Raina, *FEBS Lett.* **99,** 343 (1979).
[4] R.-L. Pajula, A. Raina, and T. O. Eloranta, *Eur. J. Biochem.* **101,** 619 (1979).
[5] K.-C. Tang, A. E. Pegg, and J. K. Coward, *Biochem. Biophys. Res. Commun.* **96,** 1371 (1980).
[6] A. E. Pegg and J. K. Coward, *Adv. Polyamine Res.* **3,** 153 (1981).
[7] K. Samejima and Y. Nakazawa, *Arch. Biochem. Biophys.* **201,** 241 (1980).
[8] M. C. Pankaskie, M. M. Abdel-Monem, A. Raina, T. Wang, and J. E. Foker, *J. Med. Chem.* **24,** 549 (1981).
[9] See [42] and [43].

ISBN 0-12-181994-9

described by Coward *et al.*[10] The properties of aminopropyltransferase inhibitors are summarized here. In most cases, results are given as the concentration needed to produce 50% inhibition of the enzyme activity in a standard assay system in 50 mM sodium phosphate, pH 7.5, containing 2.5 mM putrescine or spermidine and 20 μM decarboxylated *S*-adenosylmethionine. The results are given in this form because of the complex kinetics of the aminopropyltransferase reactions which are strongly inhibited by excess decarboxylated *S*-adenosylmethionine,[1,11,12] and because more detailed studies, including measurements of K_i, have been carried out for only a very few of these inhibitors.

Polyamine Inhibitors

None of the many di- and polyamines tested were potent inhibitors of the aminopropyltransferases, and even at 10 mM concentrations they produce little inhibition in the standard assays containing 2.5 mM of the amine acceptors.[2,12] When the substrate concentration was reduced to 0.25 mM, inhibition of rat prostate spermidine synthase was produced by all α,ω-diamines having 3–12 carbon atoms, but not by polyamines. The most potent inhibitor was cadaverine, which has a K_i of 600 μM compared to the K_m for putrescine of 15 μM. (Cadaverine is also a substrate for the enzyme.) The next most potent inhibitor was 1,3-diaminopropane, which gave 50% inhibition when present at 6 mM.[2,12] Spermine synthase from rat brain was inhibited by putrescine[4] (K_i of 1.7 mM compared to K_m for spermidine of 60 μM). Rat prostatic spermine synthase was inhibited by 1,3-diaminopropane (50% inhibition at 1.2 mM), 1,8-diaminooctane (50% inhibition at 3 mM). The spermidine analogs, 1,9-diamino-4-azanonane, 1,9-diamino-5-azanonane, and 3,3′-diaminodipropylamine, were also weak inhibitors of spermine production, but are all weak substrates for the enzyme.[2,12] It appears that mammalian aminopropyltransferases are very specific with regard to the amine acceptor and unphysiological di- and polyamines have little influence on the reaction. However, it should be noted that very large doses of 1,3-diaminopropane have been used to repress ornithine decarboxylase activity in mammalian cells,[13] and at these doses interference with aminopropyltransferase activity is quite possible.

[10] J. K. Coward, G. L. Anderson, and K.-C. Tang, this volume [48].

[11] J. K. Coward, N. C. Motola, and J. D. Moyer, *J. Med. Chem.* **20,** 500 (1977).

[12] A. E. Pegg, K. Shuttleworth, and H. Hibasami, *Biochem. J.* **197,** 315 (1981).

[13] O. Heby and J. Jänne, *in* "Polyamines in Biology and Medicine" (L. J. Marton and D. R. Morris, eds.), p. 243. Dekker, New York, 1981.

Hibasami and colleagues[14] observed that cyclohexylamine and dicyclohexylamine were strong inhibitors of rat prostatic spermidine synthase. Inhibition was apparently competitive with putrescine (K_i for dicyclohexylamine of 0.2 μM compared to a K_m for putrescine of 29 μM). Spermine synthase was not affected. A decline in hepatic spermidine content was produced by giving 200 mg/kg doses of dicyclohexylamine to rats after partial hepatectomy showing that inhibition occurs *in vivo*.

5′-Methylthioadenosine and Related Compounds

5′-Methylthioadenosine is a very potent inhibitor of spermine synthase from bovine brain[3,4] (K_i of 0.3 μM) and from rat prostate[1] (50% inhibition by 10 μM). Rat prostatic spermidine synthase is also powerfully inhibited by this nucleoside[1] (50% inhibition at 30 μM). 5′-Ethylthioadenosine also inhibited these enzymes at similar concentrations.[1] The 7-deaza analog of 5′-methylthioadenosine, 5′-methylthiotubercidin, was a good inhibitor, but it was slightly less potent, giving 50% inhibition of spermidine synthase at 45 μM and of spermine synthase at 15 μM.[1] However, this derivative is not a substrate for 5′-methylthioadenosine phosphorylase,[11] a very active and widely distributed enzyme that maintains 5′-methylthioadenosine at very low concentrations in mammalian cells.[6] Therefore, the 5′-methylthiotubercidin is much more stable in the cell and can be used to inhibit the aminopropyltransferases *in vivo*.[6,15]

Other Nucleosides Related to *S*-Adenosylmethionine

Many other nucleosides have been tested for inhibition of aminopropyltransferase activities. The most potent compound reported by Samejima and Nakazawa[7] was *S*-5′-deoxyadenosyl-(5′)-2-methylthioethylamine, which inhibited spermidine synthase from rat prostate by 80% when present at 1 m*M*. *S*-5′-Deoxyadenosyl-(5′)-1-methyl-3-(methylthio) propylamine inhibited bovine brain spermine synthase by 77% at 1 m*M*.[8] The most potent inhibitor of spermidine synthase among a series of 50 compounds tested by Hibasami *et al.*[14] was *S*-adenosyl-L-homocysteine sulfone, which produced 50% inhibition at 20 μM; *S*-adenosyl-3-thiopropylamine sulfone gave 50% inhibition at 50 μM, and *S*-adenosyl-4-methylthiobutyric acid inhibited 50% at 40 μM. In the same study, the following compounds were the most inhibitory to prostatic spermine synthase; *S*-adenosylmethionine (50% inhibition at 30 μM), *S*-adenosyl-4-methylthiobutyric acid (40 μM), 5-[5′-deoxy-5′-(*C*)-4′,5′-didehydroadenosyl-L-

[14] H. Hibasami, M. Tanaka, J. Nagai, and K. Ikeda, *FEBS Lett.* **116,** 99 (1980).
[15] A. E. Pegg, R. T. Borchardt, and J. K. Coward, *Biochem. J.* **194,** 79 (1981).

ornithine (compound A9154C) (35 μM) and *S*-adenosyl-4-thiobutyrate methyl ester (75 μM). Some of the compounds described above have been tested for the ability to influence polyamine synthesis *in vivo,* but none were found to be specific inhibitors of polyamine accumulation, although several had complex inhibitory actions on cell growth.[8,15,16]

Transition-State Inhibitors

By far the most potent and specific inhibitors of spermidine synthase are the putative transition-state analogs synthesized by Coward and colleagues.[5,6,10] The most active compound is *S*-adenosyl-1,8-diamino-3-thiooctane, which produced 50% inhibition at 0.4 μM in the standard assay system for prostatic spermidine synthase. Spermine synthase was inhibited by only 15% with 100 μM, showing the excellent discrimination of this inhibitor. The corresponding methyl sulfonium salt, *S*-adenosyl-1,8-diamino-3-methylthiooctane, was less potent than the thioether, but was still a specific inhibitor of spermidine synthase producing 50% inhibition at 20 μM. The greater potency of the thioether, which lacks the charge on the sulfur, resembles the very strong inhibition of methyltransferases by *S*-adenosylhomocysteine and may indicate that charge dispersal facilitates binding. Irrespective of mechanism, this feature is fortunate since the thioether is chemically more stable than the methylsulfonium salt and is readily taken up by mammalian cells. Exposure of SV3T3, 3T3, or HTC cells to this inhibitor results in a dose-dependent reduction of intracellular spermidine.[17] A 50% reduction was achieved by exposure of SV3T3 cells to 1 μM concentrations of the drug and a 90% reduction by 50 μM.[17]

Although a full kinetic analysis of the inhibition of spermidine synthase by *S*-adenosyl-1,8-diamino-3-thiooctane has not yet been carried out, the effects of varying either of the substrate concentrations are consistent with its acting as a transition-state analog.[6] At physiological concentrations of putrescine and decarboxylated *S*-adenosylmethionine, 50% inhibition of spermidine synthase activity was produced by 20 nM concentrations of the drug.[6]

[16] A. Raina, K. Tuomi, and R.-L. Pajula, *Adv. Polyamine Res.* **3,** 163 (1981).
[17] A. E. Pegg, K.-C. Tang, and J. K. Coward, *Biochemistry* **21,** 5082 (1982).

Section VII

Amine Oxidases and Dehydrogenases

[50] Purification of Putrescine Oxidase from *Micrococcus rubens* by Affinity Chromatography[1,2]

By MASATO OKADA, SEIICHI KAWASHIMA, and KAZUTOMO IMAHORI

$$\text{Putrescine} + O_2 + H_2O \rightarrow \Delta'\text{-pyrroline} + H_2O_2 + NH_3$$

Assay Method

Principle. Putrescine oxidase activity is measured, using putrescine as the substrate, by the formation of hydrogen peroxide which can be determined spectrophotometrically by oxidation of *o*-dianisidine in the presence of peroxidase.

Reagents

Sodium borate buffer, 0.1 *M*, pH 8.0
Putrescine dihydrochloride, 10 m*M* aqueous solution
o-Dianisidine dihydrochloride, 1% aqueous solution (filtered)
Peroxidase (from horseradish), 60 units/ml

Procedure. A mixture is prepared in a final volume of 30 ml containing 25 ml of buffer, 3.0 ml of putrescine solution, 1.0 ml of *o*-dianisidine solution, and 1.0 ml of peroxidase solution. An aliquot of this mixture is placed in a 1-ml cuvette with a 1-cm light path. Enzyme is added (0.001–0.1 unit) and the increase in the absorbance at 460 nm is followed in a Gilford recording spectrophotometer at 25°. The molar extinction coefficient of oxidized *o*-dianisidine is 11,300 at 460 nm.[3]

Definition of Unit and Specific Activity. One enzyme unit is defined as the amount that oxidizes 1.0 μmol of putrescine in 1 min at 25°. Specific activity is expressed as the number of units per milligram of protein. The protein concentrations are estimated from the absorbance at 280 nm, assuming $E^{0.1\%}_{280\ \mathrm{nm}} = 1.0$. For the pure enzyme, the value of $E^{0.1\%}_{280\ \mathrm{nm}} = 1.23$[2] is used.

[1] M. Okada, S. Kawashima, and K. Imahori, *J. Biochem.* (*Tokyo*) **85,** 1225 (1979).

[2] This enzyme has been purified previously by a different procedure by R. J. DeSa [*J. Biol. Chem.* **247,** 5527 (1972)] and by H. Yamada (this series, Vol. 17B [237]) and was shown to contain 1 mol of FAD moiety per mole of enzyme.

[3] *Worthington Enzyme Manual,* p. 19 (1972).

ISBN 0-12-181994-9

Purification Procedure

Preparation of Affinity Adsorbent. Sepharose 4B (Pharmacia) is activated by the cyanogen bromide method of March *et al.*[4] Forty milliliters of a suspension of washed Sepharose, consisting of equal volumes of gel and water, is added to 40 ml of 2 *M* sodium carbonate and stirred by magnetic stirrer. Four grams of cyanogen bromide dissolved in 2 ml of acetonitrile is added at once to the suspension. After 2 min, the suspension is poured onto a glass-filter funnel, filtered and washed with 400 ml each of ice-cold 0.1 *M* sodium bicarbonate (pH 9.5) and ice-cold water under weak vacuum. After the last wash, the activated Sepharose is suspended in 80 ml of water containing 30.2 g of 1,12-diaminododecane previously adjusted to pH 9.0 with 6 *M* HCl. The mixture is swirled at 4° for 24 hr. After the coupling reaction, the Sepharose is washed successively with 500 ml each of water, 0.1 *M* acetic acid, water, 0.1 *M* sodium bicarbonate (pH 9.0), water, 2.0 *M* NaCl, and water.

Preparation of Crude Extract. The wet cells (560 g) of *Micrococcus rubens*[5] (IAM 1315; Institute of Applied Microbiology, University of Tokyo, Japan), which had been grown at 30° with aeration (equivalent volume of air to volume of medium per minute) in a medium containing 0.4% peptone, 0.2% yeast extract, 0.1% NaCl, 0.02% putrescine, and 0.02% *N*-(3-aminopropyl)-1,3-diaminopropane, are suspended in 1500 ml of 10 m*M* phosphate buffer (pH 7.2) and are passed twice through a Dyno-Mill homogenizer (Model KDL, Willy A. Bachofen Manufacturing Engineers, Basel 5/Switzerland) at 3000 rpm at a flow rate of 2 liters/hr. The homogenate is centrifuged at 15,500 rpm for 1.5 hr.

Ammonium Sulfate Fractionation. The crude extract is precipitated by 40–70% saturation with ammonium sulfate. The precipitate is dissolved in 200 ml of 5 m*M* phosphate buffer (pH 7.2) and dialyzed against the same buffer overnight.

DEAE-Cellulose Fractionation. The dialyzed supernatant is placed on a DEAE-cellulose (Whatman DE-52) column (5 × 50 cm) equilibrated with 10 m*M* phosphate buffer (pH 7.2). The column is washed with 500 ml of the same buffer containing 0.1 *M* NaCl, then the enzyme is eluted with a linear gradient of NaCl from 0.1 to 0.4 *M* in a total volume of 3000 ml. The active fractions, which are eluted with the buffer containing about 0.25 *M* NaCl, are pooled and precipitated with 70% saturation of ammonium sulfate. The precipitate is collected by centrifugation (10,000 rpm, 30 min) and dissolved in 150 ml of 10 m*M* phosphate buffer (pH 7.2), then the enzyme solution is dialyzed against the same buffer.

[4] S. C. March, I. Parikh, and P. Cuatrecasas, *Anal. Biochem.* **60,** 149 (1974).
[5] This strain is now named *Micrococcus roseus.*

PURIFICATION OF PUTRESCINE OXIDASE

Step	Total protein (mg)	Total activity (units)	Specific activity (units/mg)	Yield (%)
Crude extract	223,000	3650	0.016	100
Ammonium sulfate	63,000	3500	0.056	96
DEAE-cellulose	3,200	2700	0.84	74
Affinity column	77	2700	35	74

Affinity Chromatography. The dialyzed enzyme is applied to a 1,12-diaminododecane-Sepharose column (2 × 14 cm) equilibrated with 10 m*M* phosphate buffer (pH 7.2). The column is washed with 500 ml of the same buffer, then proteins are eluted with a linear gradient of NaCl (0 to 1.0 *M*) in a total volume of 2000 ml. Subsequently, putrescine oxidase is eluted with buffer containing 1.0 *M* NaCl and 5 m*M* putrescine. The active fractions are pooled and dialyzed against 10 m*M* phosphate buffer (pH 7.2) saturated with ammonium sulfate. The precipitate is collected by centrifugation and dissolved in 10 m*M* phosphate buffer (pH 7.2). The solution is then dialyzed against the same buffer. Putrescine oxidase obtained by this procedure is electrophoretically homogeneous. A typical purification is summarized in the table.

[51] Purification of Spermidine Dehydrogenase from *Serratia marcescens* by Affinity Chromatography[1]

By MASATO OKADA, SEIICHI KAWASHIMA, and KAZUTOMO IMAHORI

$$\text{Spermidine} \rightarrow \text{1,3-diaminopropane} + \Delta'\text{-pyrroline}$$

Assay Method[2]

Principle. Spermidine dehydrogenase activity is assayed in terms of the consumption of ferricyanide by following the absorbance at 400 nm. Two moles of ferricyanide are reduced as the electron acceptor when 1 mol of spermidine is oxidized by spermidine dehydrogenase.

[1] M. Okada, S. Kawashima, and K. Imahori, *J. Biochem.* (*Tokyo*) **85,** 1225 (1979).

[2] This enzyme was previously purified by a different procedure by C. W. Tabor and P. D. Kellogg (*J. Biol. Chem.* **245,** 5424 (1970); see also this series, Vol. 17B [241]). The enzyme was shown to contain 1 mol of FAD and 1 mol of protoporphyrin IX per mole of enzyme.

ISBN 0-12-181994-9

Reagents

Potassium phosphate buffer, 1.0 *M*, pH 7.2
Potassium ferricyanide, 0.1 *M* aqueous solution
Spermidine trihydrochloride, 50 m*M* aqueous solution

Procedure. A mixture is prepared in a final volume of 50 ml containing 5 ml of the buffer, 0.5 ml of potassium ferricyanide solution, and 0.5 ml of spermidine solution. A half milliliter of the mixture is placed in a 1-ml cuvette with a 1-cm light path. Enzyme is added (0.01–1 unit) and the decrease in absorbance at 400 nm is followed in a Gilford recording spectrophotometer at 25°. The molar extinction coefficient of potassium ferricyanide is 960 at 400 nm. Two moles of potassium ferricyanide are used for the oxidation of 1 mol of spermidine.

Definition of Unit and Specific Activity. One enzyme unit is defined as the amount that oxidizes 1 μmol of spermidine in 1 min at 25°. Specific activity is expressed as the number of units per milligram of protein. The protein concentrations are estimated from the absorbance at 280 nm, assuming $E_{280\text{ nm}}^{0.1\%} = 1.0$. For the pure enzyme, the value of $E_{280\text{ nm}}^{0.1\%} = 1.60$ is used.[2]

Purification Procedure

Preparation of Affinity Adsorbent. Sepharose 4B (Pharmacia) is activated with cyanogen bromide. Forty milliliters of a suspension of Sepharose, consisting of equal volumes of gel and water, is added to 40 ml of 2 *M* sodium carbonate and stirred by magnetic stirrer. To this suspension, 4.0 g of cyanogen bromide dissolved in 2 ml of acetonitrile is added at once and stirred vigorously for 2 min. The activation is terminated by pouring the suspension onto a glass-filter funnel. After the filtration, the activated gel is washed with 400 ml each of ice-cold 0.1 *M* sodium bicarbonate (pH 9.5) and ice-cold water. After the last wash, the gel is suspended in 80 ml of cold 0.1 *M* sodium bicarbonate (pH 9.0) and mixed with 80 ml of water containing 21.7 g of 1,8-diaminooctane, the pH of which having been previously adjusted to 9.0 with 6 *N* HCl. The mixture is swirled at 4° for 24 hr. Subsequently, the coupling gel is washed consecutively with 500 ml each of water, 0.1 *M* acetic acid, water, 0.1 *M* sodium bicarbonate (pH 9.0), water, 2.0 *M* NaCl, and water.

Preparation of Crude Extract. Serratia marcescens (IAM 1067; Institute of Applied Microbiology, University of Tokyo, Japan) is grown at 30° with aeration (equivalent volume of air to volume of medium per minute) in a medium (pH 7.0) containing 0.2% sodium citrate, 0.1% K_2HPO_4, 0.1% NaCl, 0.02% $MgSO_4$, 0.0007% $FeSO_4$, 0.0015% $MnSO_4$, 0.02% pu-

PURIFICATION OF SPERMIDINE DEHYDROGENASE

Step	Total protein (mg)	Total activity (units)	Specific activity (units/mg)	Yield (%)
Cell extract	165,000	980	0.006	100
DEAE-cellulose	1,300	460	0.370	47
Affinity column	0.91	400	440	41

trescine, and 0.02% *N*-(3-aminopropyl)-1,3-diaminopropane. The wet cells (300 g) are suspended in 950 ml of 5 m*M* sodium phosphate buffer and ground with glass beads by passing through a Dyno-Mill homogenizer (Model KDL, Willy A. Bachofen Manufacturing Engineers, Basel 5/Switzerland) at 3000 rpm at a flow rate of 2 liters/hr. The homogenate is centrifuged at 15,500 rpm for 1.5 hr.

DEAE-Cellulose Fractionation. The crude extract is precipitated by 80% saturation with ammonium sulfate. The precipitate is dissolved in 400 ml of 5 m*M* phosphate buffer (pH 7.2) and dialyzed against the same buffer. The dialyzed solution is placed onto a DEAE-cellulose column (5 × 50 cm) equilibrated with 5 m*M* phosphate buffer (pH 7.2), and the column is washed with 1500 ml of the same buffer. The proteins are eluted with a linear gradient of NaCl from 0 to 0.25 *M* in a total volume of 3000 ml. The active fractions, which are eluted at about 0.1 *M* NaCl, are pooled and precipitated by 80% saturation with ammonium sulfate. The precipitate is dissolved in 100 ml of 5 m*M* phosphate buffer (pH 7.2) and dialyzed against the same buffer.

Affinity Chromatography. The dialyzed enzyme is applied to a 1,8-diaminooctane–Sepharose column (2 × 14 cm) equilibrated with 5 m*M* phosphate buffer (pH 7.2). The column is washed with 500 ml of the same buffer, then proteins are eluted with a linear gradient of NaCl (0 to 1.0 *M*) in a total volume of 1000 ml. Subsequently, spermidine dehydrogenase is eluted with the buffer containing 1.0 *M* NaCl and 5 m*M* *N*-(3-aminopropyl)-1,3-diaminopropane trihydrochloride. The active fractions are pooled and precipitated by 80% saturation with ammonium sulfate. The precipitate is collected by centrifugation and dialyzed against 5 m*M* phosphate buffer. The enzyme is electrophoretically homogeneous and is purified by about 1200-fold.

A typical purification is summarized in the table.

[52] Polyamine Oxidase[1] (Rat Liver)[2]

By ERKKI HÖLTTÄ

$$\text{Spermine} + O_2 + H_2O \rightarrow \text{spermidine} + \text{3-aminopropanal} + H_2O_2 \quad (1)$$

$$N^1,N^{12}\text{-Diacetylspermine} + O_2 + H_2O \rightarrow N^1\text{-acetylspermidine} + \text{3-acetamidopropanal} + H_2O_2 \quad (2)$$

Assay Methods

Radiochemical Assay [Eq. (1)]

Principle. Enzyme activity is measured by determining labeled spermidine formed from [^{14}C]spermine. The reaction product and substrate are separated by paper electrophoresis.

Reagents

Glycine–NaOH buffer, 1 *M*, pH 9.5, at 20°
Dithiothreitol, 0.25 *M*
Benzaldehyde, 0.25 *M*
Spermine tetrahydrochloride, 20 m*M*, neutralized to pH 7 with NaOH
[*tetramethylene*-1,4-^{14}C]Spermine, 100 mCi/mmol
Trichloroacetic acid, 50% (w/v), containing 1 m*M* spermidine

Procedure. The reaction mixture contains 0.1 *M* glycine–NaOH buffer, 5 m*M* dithiothreitol, 5 m*M* benzaldehyde (when added), 0.4–1 m*M* [^{14}C]spermine (0.1–1 μCi), and enzyme in a final volume of 0.25 ml. The higher substrate concentrations are used for assays of crude cell extracts. The reaction is initiated by addition of the enzyme. A blank sample lacking enzyme is satisfactory for most purposes. The incubation is carried out in air at 37° for 30–60 min with constant shaking. The reaction is stopped with 50 μl of 50% trichloroacetic acid containing spermidine as a marker. After centrifugation (2000 *g* for 5 min), an aliquot (20–30 μl) of the supernatant is subjected to paper electrophoresis on paper strips (3 × 39 cm; Whatman No. 1). Citric acid buffer (0.1 *M*, pH 3.6) is used for electrophoresis (300 V, 2 hr). After the run, the paper strips are dried at 100° for 5 min, then impregnated with 1% ninhydrin solution (in

[1] E. Höltta, *Biochemistry* **16,** 91 (1977).
[2] See also this volume [53] by T. A. Smith for an analogous reaction in oat seedlings that converts spermidine and spermine to diaminopropane.

ISBN 0-12-181994-9

ethanol) and dried again. The colored spermidine fraction (migrating faster than spermine) is then cut off and counted in toluene-based scintillation fluid in a liquid scintillation spectrometer.

Fluorometric Assay[3] *[Eq. (2)]*

Principle. Dansylation technique is used for the measurement of N^1-acetylspermidine formed from N^1,N^{12}-diacetylspermine.[4] The dansyl derivatives are separated on thin-layer chromatography.

For the assay procedure, see Seiler *et al.*[3]

Comments

Alternative assays in which the formation of 3-aminopropanal[1] or hydrogen peroxide[1,4] are measured have been described, but they are not as reliable as the above assays for use with crude cell extracts. Polyamine oxidase activity has also been measured by determining the conversion of [^{14}C]spermidine to labeled putrescine and subsequently to Δ^1-pyrroline by a coupled diamine oxidase reaction.[5]

Definition of Unit and Specific Activity. For the radiochemical assay, one unit of enzyme activity is defined as the amount of enzyme catalyzing the formation of 1 nmol of spermidine per minute from spermine. Specific activity is expressed as the number of units per milligram of protein, as determined by the method of Lowry *et al.*[6]

Purification Procedure[1]

Polyamine oxidase activity is found both in peroxisomes (highest specific activity) and cytosol (highest proportion of the total activity). The cytosolic activity is, at least partly, derived from broken peroxisomes (which are very fragile) during subcellular fractionation. Purification of the enzyme from the peroxisomal fraction of rat liver is described below.

All steps are carried out at 0–4°.

Step 1. Preparation of Crude Extract. The livers from 20 rats (155 g of liver) are homogenized with 4 volumes of 0.25 *M* sucrose–10 m*M* Tris buffer, pH 7.4, using a Potter–Elvehjem homogenizer. The homogenate is

[3] N. Seiler, F. N. Bolkenius, B. Knödgen, and P. S. Mamont, *Biochim. Biophys. Acta* **615,** 480 (1980). See also this series, Vol. 17B [256] and this volume [2].

[4] F. N. Bolkenius and N. Seiler, *Int. J. Biochem.* **13,** 287 (1981).

[5] G. Quash, T. Keolouangkhot, L. Gazzolo, H. Ripoll, and S. Saez, *Biochem. J.* **177,** 275 (1979).

[6] O. H. Lowry, N. J. Rosebrough, A. L. Farr, and R. J. Randall, *J. Biol. Chem.* **193,** 265 (1951).

centrifuged in a Sorvall RC-2 B centrifuge at 600 *g* for 10 min. The supernatant fraction obtained is then centrifuged at 15,000 *g* for 10 min, and the resulting pellet is used for further purification. The pellet is suspended in 50 ml of 10 m*M* Tris–0.1 m*M* EDTA–0.1 m*M* dithiothreitol buffer, pH 7.8 (buffer A), and made 0.1% with respect to Triton X-100. The suspension is centrifuged at 25,000 *g* for 30 min to obtain a soluble fraction which is then dialyzed overnight against two 3-liter volumes of buffer A.

Step 2. DEAE-Cellulose Chromatography. Whatman DE-52 column (3 × 40 cm) is packed and equilibrated with buffer A. The crude extract of step 1 (120 ml) is applied to the column which is then washed with 150 ml of buffer A. A constant flow rate (about 60 ml/hr) is maintained with a peristaltic pump. The enzyme is eluted with a linear gradient of 0.1 to 0.4 *M* NaCl in buffer A. The total gradient volume is 1000 ml. Fractions of 10 ml are collected. The enzyme is eluted at about 0.25 *M* NaCl. The most active fractions (80 ml) are pooled for the next step.

Step 3. Hydroxyapatite Chromatography. A Clarkson Hypatite C column (3 × 12 cm) is equilibrated with 10 m*M* potassium phosphate buffer, pH 7.8, containing 0.1 m*M* dithiothreitol. The enzyme preparation of step 2 is adsorbed to the column and then eluted with a linear gradient of 0.01–0.3 *M* potassium phosphate buffer, pH 7.8, containing 0.1 m*M* dithiothreitol (total gradient volume is 500 ml). A flow rate of about 30 ml/hr is used and fractions of 10 ml are collected. The enzyme starts to emerge after 200 ml of the fluid has passed through the column. The fractions showing the highest activity (70 ml) are combined and concentrated in an Amicon ultrafiltration cell (using a PM-10 membrane) to a volume of 5 ml.

Step 4. Gel Filtration Chromatography. The concentrated enzyme of step 3 is applied through a plunger-type flow adaptor to a Sephadex G-100 column (Pharmacia, 2.6 × 62 cm) equilibrated with 10 m*M* Tris–0.1 m*M* EDTA–0.1 m*M* dithiothreitol–50 m*M* NaCl buffer, pH 7.8 (buffer B). An upward-flow technique is used. A constant flow rate of buffer (15 ml/hr) is maintained by use of a Mariotte flask. Fractions of 5 ml are collected. The enzyme is eluted at 1.4 times the void volume (V_0 is 120 ml, as determined with Blue Dextran 2000). The fractions with the highest activity (25 ml) are pooled and concentrated by ultrafiltration to a small volume (about 1.5 ml).

Step 5. Polyacrylamide Gel Electrophoresis. On a small scale (without a commercial preparative electrophoresis apparatus) the concentrated enzyme of step 4 can be purified on 7.5% polyacrylamide gels (1 × 10 cm) prepared by the method of Davis[7] (omitting the spacer gel). The stability

[7] B. J. Davis, *Ann. N. Y. Acad. Sci.* **121,** 404 (1964).

PURIFICATION OF POLYAMINE OXIDASE FROM RAT LIVER[a]

Step	Total protein (mg)	Total activity (units)	Specific activity (units/mg)	Yield (%)	Purification (fold)
Crude extract	6380	220	0.034	100	1
DEAE-cellulose	275	138	0.50	63	15
Hydroxyapatite	42	112	2.66	51	77
Sephadex G-100	6.2	102	16.7	47	480
Gel electrophoresis	0.40	57	141	26	4090

[a] The enzyme activity was assayed (at pH 9.2) with spermine as the substrate in the absence of benzaldehyde. In the presence of benzaldehyde the activities are about 10-fold higher.

of the enzyme is slightly better on pH 7.5 gels[8] than on pH 8.9 gels.[7] To the electrode buffers is added 0.1 m*M* dithiothreitol. Preelectrophoresis for 1–2 hr is carried out to remove ammonium persulfate. The enzyme sample of step 4 is brought to a 10% sucrose concentration by adding solid sucrose. About 2.3 mg of protein is layered on the gel surface, and a drop of Bromophenol Blue marker (0.05%) is added. A current of 2 mA per gel column is used. The gels are run in a 0–4° cold room. The run is stopped when the indicator dye has reached the gel bottom. The gel is then sliced into 0.25-cm pieces which are broken up and eluted with 0.5 ml of buffer B for 24 hr. The two most active fractions from each gel are combined. A typical purification is summarized in the table.

Properties

Stability. The purified enzyme is quite stable at 0–4° for at least 2 months. The enzyme can also be stored at −20° or at −70° without loss of activity.

Purity. The enzyme preparations of the final step show no evidence of contamination as judged by analytical polyacrylamide gel electrophoresis at pH 7.5 and 8.9. The enzyme activity is localized in the position of the single protein band.

Molecular Weight. The average molecular weight of the enzyme is 60,000, as determined by sucrose density gradient centrifugation,[9] gel filtration,[9,10] and sodium dodecyl sulfate gel electrophoresis.[11] The sedimentation constant is 4.5 S.

[8] H. R. Maurer, "Disk-Elektrophorese," p. 42. de Gruyter, Berlin, 1968.

[9] G. K. Ackers and R. L. Steere, *Methods Virol.* **2,** 325 (1967).

[10] P. Andrews, *Biochem. J.* **96,** 595 (1965).

[11] K. Weber and M. Osborn, *J. Biol. Chem.* **244,** 4406 (1969).

pH Optimum. The maximum enzyme activity is obtained at pH 10 of glycine–NaOH buffer (pH measured at 20°).

Substrate Specificity[12] *and Kinetic Constants*. Polyamine oxidase catalyzes the oxidation of spermine and spermidine and their *N*-acetyl derivatives (acetylated at the propylamino moiety) at the secondary nitrogen giving rise to a shorter polyamine or its acetyl derivative and 3-aminopropanal or acetamidopropanal. The best substrate for the enzyme appears to be N^1-acetylspermine (K_m 0.6 μM), followed by N^1,N^{12}-diacetylspermine (K_m = 5 μM) and N^1-acetylspermidine (K_m = 14 μM) as determined for the partially purified enzyme from rat liver cytosol.[4] The K_m value for spermine is 15 μM (5 μM in the presence of benzaldehyde), and for spermidine it is 50 μM (15 μM with benzaldehyde) with the purified enzyme above. The closely related amines N^1,N^8-diacetylspermidine, N^8-acetylspermidine, N^1-acetyl-1,3-diaminopropane, N^1-acetylputrescine,[4] putrescine, cadaverine, and diaminopropane, and monoamines,[1] do not act as substrates.

Inhibitors. N^8-acetylspermidine (K_i = 11 μM) and *N*-(3-aminopropyl)-1,3-diaminopropane (K_i = 60 μM) are noncompetitive inhibitors with respect to N^1-acetylspermidine as the substrate.[4] 3-Aminopropanal is also a potent inhibitor, whereas cadaverine, putrescine, diaminopropane,[1] and *N*-acetylputrescine[4] are only weak inhibitors.

The enzyme is inhibited by sulfhydryl reagents (Hg^{2+}, *p*-hydroxymercuribenzoate, *N*-ethylmaleimide, and iodoacetamide). Preincubation (30 min) with the carbonyl reagents (phenylhydrazine, hydroxylamine, semicarbazide, and $NaBH_4$) resulted in a loss of enzyme activity. Quinacrine (0.1 m*M*) caused an 80% inhibition of the activity. Iron chelators decreased the enzyme activity, whereas copper chelators and 5 m*M* EDTA, 10 m*M* NaN_3, and 50 m*M* NaF were without effect.

Cofactors. The enzyme has a tightly bound FAD as a prosthetic group. There is some evidence that iron (Fe^{2+}) may be a metal cofactor.

Stimulation by Aldehydes. The oxidation rates of spermine and spermidine in particular are stimulated by various aldehydes, such as benzaldehyde, in relatively high concentrations (K_a = 0.2 m*M*). Pyridoxal (but not pyridoxal phosphate) shows 30% of the stimulatory activity of benzaldehyde. The aldehydes most probably form Schiff bases with the primary amino groups of spermine and spermidine (thus reducing their positive charge), as the oxidation of N^1-acetylspermidine is not affected by benzaldehyde. Acetylation thus serves as a physiological means to increase the degradation of polyamines. Yet it is also possible that the aldehydes trap the inhibitory reaction products, which cannot otherwise be eliminated *in*

[12] For the preparation of the acetylated derivative of putrescine and spermidine, see also this volume [72].

vitro. With crude peroxisomal extracts the oxidation of spermidine is not as strictly dependent on the presence of benzaldehyde as with the purified enzyme.

[53] Polyamine Oxidase (Oat Seedlings)

By TERENCE A. SMITH

$NH_2CH_2CH_2CH_2NHCH_2CH_2CH_2CH_2NH_2 + O_2$
Spermidine

$\longrightarrow NH_2CH_2CH_2CH_2NH_2 + \Delta'\text{-pyrroline} + H_2O_2$

Diaminopropane Δ′-Pyrroline

$NH_2CH_2CH_2CH_2NHCH_2CH_2CH_2CH_2NHCH_2CH_2CH_2NH_2 + O_2$
Spermine

$\longrightarrow NH_2CH_2CH_2CH_2NH_2 + NH_2CH_2CH_2CH_2N(CH{=}CH{-}CH_2{-}CH_2) + H_2O_2$

Diaminopropane 1-(3-Aminopropyl) pyrroline[1]

Amine oxidases and amine dehydrogenases are found in various microorganisms.[2–4] In the higher plants, amine oxidases that effect the oxidation of polyamines by the above mechanisms have been found only in the Gramineae (the grasses)[5–13]; they have not yet been detected in animals.[14]

[1] The structure is shown as a derivative of Δ^2-pyrroline, although the position of the double bond is not yet established.

[2] R. H. Weaver and E. J. Herbst, *J. Biol. Chem.* **231,** 647 (1958).

[3] K. Isobe, Y. Tani, and H. Yamada, *Agric. Biol. Chem.* **45,** 727 (1981).

[4] C. W. Tabor and P. D. Kellogg, *J. Biol. Chem.* **245,** 5424 (1970).

[5] T. A. Smith, *Biochem. Biophys. Res. Commun.* **41,** 1452 (1970).

[6] T. A. Smith, *Phytochemistry* **11,** 899 (1972).

[7] T. A. Smith, *Phytochemistry* **13,** 1075 (1974).

[8] T. A. Smith and D. A. Bickley, *Phytochemistry* **13,** 2437 (1974).

[9] T. A. Smith, *Phytochemistry* **15,** 633 (1976).

[10] T. A. Smith, *Phytochemistry* **16,** 1647 (1977).

[11] Y. Suzuki and E. Hirasawa, *Phytochemistry* **12,** 2863 (1973).

[12] E. Hirasawa and Y. Suzuki, *Phytochemistry* **14,** 99 (1975).

[13] Y. Suzuki and H. Yanagisawa, *Plant Cell Physiol.* **21,** 1085 (1980).

[14] See, however, this volume [52] for a polyamine oxidase in animal tissues that converts spermidine and acetylspermidine to putrescine.

METHODS IN ENZYMOLOGY, VOL. 94

ISBN 0-12-181994-9

Preliminary work with oats, barley, and maize had indicated that the enzymes are associated with a particulate fraction,[6,9,11] though a more recent study, supporting an earlier observation,[8] demonstrates that this is an artifact of extraction and that this enzyme normally occurs in the cell wall.[15] The enzyme is particularly active in vascular tissue.[5,15]

Assay Methods

There are several methods for the determination of this enzyme. Perhaps the least ambiguous depends on the estimation of oxygen consumption using a Clark electrode.[8] However, measurement of the hydrogen peroxide by a coupled peroxidase system[7] is more sensitive, owing to the greater baseline stability.[8]

Reagents

Tris buffer, 0.1 *M*, pH 6.5 (correct at 30°), air equilibrated
Guaiacol, 25 m*M*
Peroxidase, Sigma, type II (200 purpurogallin units/mg), 1 mg/ml
Spermidine, 25 m*M*

Procedure. Buffer (2 ml), guaiacol (0.1 ml), peroxidase (0.1 ml), and the oat leaf polyamine oxidase preparation (0.2 ml) are placed in optically matched cylindrical spectrophotometer cells (1.2 × 12.5 cm). After preincubation for 2 min in a water bath at 30° the tube is transferred to a spectrophotometer cuvette holder, also thermostatted at 30°. Spermidine (0.1 ml) is added, and the increase in absorbance at 470 nm is determined. The spectrophotometer output is fed into a logarithmic recorder (full-scale deflection = 0.2 absorbance unit). Activity is determined on extrapolation of the curve to zero time.

Activity declines rapidly during the assay, especially with spermine as substrate.[10] Spermidine is therefore to be preferred in the assay. The use of fluorogenic peroxidase substrates[16] should allow a considerable increase in sensitivity.

Purification Procedure[9]

Material. Oats (*Avena sativa* L., cv 'Black Supreme') are grown in the dark for 3 weeks at 22° in sand culture in a medium containing 2 m*M* K_2SO_4, 3 m*M* $MgSO_4$, 8 m*M* $CaCl_2$, 4 m*M* NaH_2PO_4, and 12 m*M* $NaNO_3$ together with Fe (as FeEDTA) 5.6 ppm, Mn 0.6 ppm, Cu 0.06 ppm, Zn 0.07 ppm, Mo 0.05 ppm, and B 0.7 ppm.

[15] R. Kaur-Sawhney, H. E. Flores, and A. W. Galston, *Plant Physiol.* **68,** 494 (1981).
[16] K. Zaitsu and Y. Ohkura, *Anal. Biochem.* **109,** 109 (1980).

TABLE I
PURIFICATION OF POLYAMINE OXIDASE FROM OAT LEAVES[9]

Step	Volume (ml)	Total protein (mg)	Total activity (nkat)[b]	Specific activity (nkat/mg protein)	Purification (fold)	Yield (%)
Aqueous extraction[a]	550	800	5600	6.8	1	(100)
pH 4 precipitation	22	34	4740	140	20	85
Acetone precipitation	2.4	2.9	4580	1620	240	82
Gel filtration	66	0.22	1340	6020	890	24

[a] The extract was prepared from 150 g of 3-week-old dark-grown shoots.
[b] The enzyme was assayed by the peroxidase–guaiacol method at pH 6 with spermine as substrate.

Extraction. The shoots are blended in 4 volumes of cold water, the macerate is squeezed through muslin, and the residue is reextracted with 2 volumes of cold water.

pH 4 Precipitation. The combined substrates are adjusted to pH 4 with saturated citric acid. The precipitate is collected by centrifuging at 2500 *g* for 15 min and extracted in 0.1 *M* pH 6 citrate buffer containing 1 *M* NaCl. On recentrifuging at 3000 *g* for 15 min the precipitate is discarded.

Acetone Precipitation. The supernatant is cooled to 0°, and 1 volume of acetone at −15° is added. The precipitate collected by centrifugation is extracted with 1 *M* NaCl in 0.1 *M* pH 6 citrate buffer; on recentrifugation, the precipitate is discarded.

Gel Filtration. The supernatant is applied to a column of Sephadex G-100 (3.5 × 87 cm) equilibrated with 1 *M* NaCl in 0.1 *M* pH 6 citrate buffer. Fractions (8 ml each at 24 ml/hr) eluted in the same buffer are collected; the active band is pooled and concentrated by Millipore filtration (UM-10 filter).

On polyacrylamide gel electrophoresis the final product migrates as a single band at pH 4.3 toward the anode, and enzyme activity is coincident with the major protein band.[10]

A typical purification is summarized in Table I.

Properties of the Oat Leaf Enzyme

Activators and Inhibitors. Oxidation of the polyamines is strongly inhibited by $NH_2(CH_2)_{10}NH(CH_2)_3NH_2$, a spermidine homolog.[10] The fungicide, guazatine, $[NH_2C(=NH)NH(CH_2)_8]_2NH$ (KenoGard AB, Stockholm, Sweden), is also a powerful inhibitor (K_i ca 10^{-8} *M*). The

TABLE II
COMPARISON OF THE PROPERTIES OF OAT, BARLEY, AND MAIZE SEEDLING POLYAMINE OXIDASES

Property	Oats[9,10]	Barley[7,9]	Maize[11-13]
Molecular weight	80,000	85,000	65,000
Relative activity, spermine vs spermidine[a]	1	14	0.7
pH Optimum for spermidine	6.5	8.0[b]	6.3
pH Optimum for spermine	6.5	4.8	5.5
Cofactors	ND[c]	ND	FAD
Maximum specific activity (nkat/mg protein)	6020	—	476

[a] Activity at the pH optimum for the respective substrate.
[b] The enzyme is unstable at this pH.
[c] ND, Not detected.

enzyme is activated by the chlorides of Li^+, Na^+, K^+, and Rb^+ at 1 M in order of increasing effectiveness.[17] The K_m values for spermidine and spermine, respectively, are 8×10^{-6} M and 2×10^{-6} M in air-saturated buffer. The K_m values for oxygen are 1.83×10^{-4} M and 0.85×10^{-4} M with spermidine and spermine, respectively, as substrates.[10] For the barley enzyme the K_m for oxygen is relatively high, and oxygen in air-saturated buffer is limiting.[8] Properties of the polyamine oxidases from oat, barley, and maize seedlings are compared in Table II.

Stability. In intact oat leaves, loss of activity at −15° is 5% per week; for the purified enzyme, loss of activity at −15° is 10% per week.[10]

[17] T. A. Smith and P. E. Gay, unpublished (1981).

[54] Purification of Bovine Plasma Amine Oxidase

By B. MONDOVÌ, P. TURINI, O. BEFANI, and S. SABATINI

Bovine plasma amine oxidase is a Cu^{2+}-containing enzyme that catalyzes the oxidation of primary amines[1] as follows:

$$R{-}CH_2{-}NH_3^+ + O_2 + H_2O \rightarrow RCHO + H_2O_2 + NH_4^+$$

[1] E. A. Zeller, *Helv. Chim. Acta* **21,** 880 (1938).

ISBN 0-12-181994-9

We have summarized a simple purification procedure[2] according to Turini *et al.*[3] which yields the highest specific activity reported in the literature.[4-8]

Assays

The enzyme activity is assayed according to Tabor *et al.*[9] One enzyme unit catalyzes the formation of 1 μmol of benzaldehyde per minute. Specific activity is expressed in units/milligram protein. The protein concentration is determined by a biuret method[10] and using the absorbance at 280 nm. $E_{1\,\text{cm}\,280}^{1\%} = 13$ was calculated on the basis of dry-weight determinations.

Purification Procedure

All operations are carried out at cold room temperature.

Beef blood (5 liters) is mixed with 500 ml of 3.8% sodium citrate and centrifuged at 1500 *g*; the plasma is collected.

Ammonium sulfate (209 g/liter) is added to the plasma; the mixture is stirred overnight and centrifuged at 13,700 *g* for 20 min; 129 g of ammonium sulfate per liter are added to the supernatant, and the mixture is again stirred overnight. The precipitate is collected after centrifugation at 13,700 *g* for 20 min, dissolved in 10 m*M* potassium phosphate at pH 8, and dialyzed for 20 hr against two changes of 3 m*M* potassium phosphate at pH 8. The dialyzed enzyme, after centrifugation at 20,000 *g* for 20 min, is applied to a column (2 × 40 cm) of AH-Sepharose 4B (Pharmacia) equilibrated with 10 m*M* potassium phosphate at pH 8 and washed with 500 ml of the same buffer. Three hundred milliliters of potassium phosphate at pH 8 are placed in each of five chambers of a Varigrad (Technicon Chromatography Corp., New York) at concentrations of 0.01, 0.05, 0.1, 0.5, and 1 *M*. The gradient formed by the Varigrad is used to elute the enzyme from the column. The fractions containing a specific activity greater than

[2] See this series, Vol. 17B [232] and Vol. 2 [64] for another preparation of this enzyme.

[3] P. Turini, S. Sabatini, O. Befani, F. Chimenti, C. Casanova, P. L. Riccio, and B. Mondovì, *Anal. Biochem.* **125,** 294 (1982).

[4] H. Yamada and K. T. Yasunobu, *J. Biol. Chem.* **237,** 1511 (1962).

[5] K. T. Yasunobu, H. Ishizaki, and N. Minamiura, *Mol. Cell. Biochem.* **13,** 3 (1976).

[6] M. C. Summers, R. Markovic, and J. P. Klinman, *Biochemistry* **18,** 1969 (1979).

[7] M. Ishizaki and K. T. Yasunobu, *Biochim. Biophys. Acta* **611,** 27 (1980).

[8] A. Svenson and P. A. Hynning, *Prep. Biochem.* **11**(1), 99 (1981).

[9] C. W. Tabor, H. Tabor, and S. M. Rosenthal, *J. Biol. Chem.* **208,** 645 (1954).

[10] J. Goa, *Scand. J. Clin. Invest.* **5,** 218 (1953).

TABLE I
PURIFICATION OF BOVINE PLASMA AMINE OXIDASE[a]

Step	Volume	Total protein (mg)	Total activity (units)	Specific activity (units/mg)	Recovery (%)	Purification (fold)
1. Plasma	2000	164,000[b]	41	0.00025	100	1
2. First ammonium sulfate fractionation	1900	71,300[b]	35.6	0.0005	86.8	2
3. Second ammonium sulfate fractionation	300	15,100[b]	30.2	0.0020	73.5	8
4. AH-Sepharose 4B	260	104[c]	11.4	0.11	28	440
5. Concanavalin A–Sepharose[d]	80	30[c]	9.9	0.33	24.1	1320

[a] The starting material used was 5 liters of beef blood.
[b] Protein concentrations are determined by the biuret method.[10]
[c] Protein concentrations are determined by absorbance at 280 nm assuming $E_{1\,\mathrm{cm}\,280}^{1\%} = 13$.
[d] Data of steps 1–4 correspond to data reported previously.[3] Data of step 5 are the average of 1[illegible] different samples recently purified.

0.1 are pooled, concentrated to about 10 ml by vacuum dialysis,[11] and dialyzed against 0.1 *M* potassium phosphate at pH 7.2.

The enzyme solution collected from four AH-Sepharose 4B columns is placed on top of a column (1 × 15 cm) of concanavalin A–Sepharose (Pharmacia), previously equilibrated with 0.1 *M* potassium phosphate at pH 7.2, and washed with 500 ml of the same buffer.

The enzyme is eluted with 0.3 *M* methyl α-D-glucopyranoside in 0.1 *M* potassium phosphate at pH 7.2.

Three activity peaks are separated, but the yield is very variable in different enzyme preparations. The three separated fractions show similar patterns in disc gel electrophoresis and in the ultracentrifuge, and the carbohydrate content of each is in the 12–15% range. Active fractions are collected and concentrated by vacuum dialysis.

Table I summarizes the purification procedure.

Properties

The concentrated enzyme solution is pink. It can be stored at −20° for 3 months without appreciable loss of activity or can be lyophilized in 0.01 *M* potassium phosphate buffer at pH 7.2 with a loss of about 10%. The

[11] E. A. Kabat and M. M. Mayer, "Experimental Immunochemistry," p. 729. Thomas, Springfield, Illinois, 1961.

purified enzyme has a specific activity of 0.23–0.35 and gives a single electrophoretic band on polyacrylamide disc gel and SDS electrophoresis. Two minor components migrating toward the cathode (less than 5%) are present in some preparations; both have amine oxidase activity.

Bovine plasma amine oxidase has a visible spectrum with a maximum at 480 nm. The enzyme, which has an M_r of about 180,000, appears to contain two copper atoms and one carbonyl group per molecule and to be formed of two subunits not covalently bound; each polypeptide chain should contain a disulfide bridge.[3]

TABLE II
RELATIVE RATES OF OXIDATION OF VARIOUS SUBSTRATES BY PURIFIED BOVINE PLASMA AMINE OXIDASE[a]

Substrate	Relative rate (%)
Spermine	100
Spermidine	100
Benzylamine	40
Butylamine	40
Kynuramine	20
Phenylethylamine	20
Tyramine	15
Tryptamine[b]	0
Serotonin[c]	0
Octopamine	0
Histamine[d]	0
Norepinephrine	0
1,4-Diaminobutane	0
1,5-Diaminopentane	0

[a] The enzyme activity is studied by polarographic determination of oxygen uptake in a YSI oxygraph Model 53 equipped with a Clark electrode. The reaction mixture contains 1.7 ml of 0.1 *M* phosphate buffer, pH 7.2, 0.52 mg of purified enzyme, and 1 μmol of substrate in a total volume of 2.0 ml.

[b] Acts as a noncompetitive inhibitor, showing a K_i = 0.1 m*M*.[e]

[c] Acts as a noncompetitive inhibitor, showing a K_i = 0.3 m*M*.[e]

[d] Acts as a noncompetitive inhibitor, showing a K_i = 6.6 m*M*.[e]

[e] B. Mondovì, P. Guerrieri, M. T. Costa, and S. Sabatini, *Adv. Polyamine Res.* **3,** 75 (1981).

This enzyme actively oxidizes spermine with the formation of terminal aldehydes on each three-carbon moiety of spermine. Spermidine is also a good substrate, but oxidation occurs only on the three-carbon moiety with the formation of a monoaldehyde.[12] Putrescine and cadaverine are not oxidized by this enzyme, although other monoamines and diamines serve as substrates (see Table II). Enzymes with similar substrate specificities have not been found in other mammalian tissues, except for the blood of several ruminants. However, Seiler[13] has found evidence for the terminal oxidation of spermine in *in vivo* experiments in rats after the parenteral administration of high doses of spermine.

[12] C. W. Tabor, H. Tabor, and U. Bachrach, *J. Biol. Chem.* **239,** 2194 (1964).
[13] N. Seiler and M. J. Al-Therib, *Biochem. J.* **144,** 29 (1974).

Section VIII

Spermidine Acetylation and Deacetylation

[55] Spermidine N^1-Acetyltransferase

By FULVIO DELLA RAGIONE and ANTHONY E. PEGG

An inducible spermidine acetylase activity was first observed in rat liver cytosol extracts after treatment with carbon tetrachloride.[1] More recent studies have shown that this activity is induced in response to a variety of substances including growth hormone,[2] thioacetamide,[2] dialkylnitrosamines,[3] folic acid,[4] and spermidine.[5] The enzyme is distinct from a previously described polyamine acetylase that is located in chromatin and also acts on histones.[6,7] The latter enzyme forms 45% N^8-acetylspermidine,[7] but the inducible enzyme forms exclusively N^1-acetylspermidine and is, therefore, correctly described as spermidine N^1-acetyltransferase.[8,9]

Assay Method

Principle. Activity is measured by the incorporation of radioactivity from labeled acetyl-CoA into monoacetylspermidine that is retained on cellulose phosphate paper disks.[7]

Reagents

Tris-HCl buffer, 1.0 *M*, pH 7.8
[1-^{14}C]Acetyl-CoA, 4 μCi/ml (49 mCi/mmol)
Spermidine trihydrochloride, 30 m*M*

Procedure. The reaction is carried out in conical centrifuge tubes that contain a total volume of 0.1 ml including 0.01 ml of 30 m*M* spermidine, 0.01 ml of 1 *M* Tris-HCl, pH 7.8, 0.01 ml of [1-^{14}C]acetyl-CoA (4 μCi/ml), and enzyme solution containing 10^{-4} to 10^{-2} unit. After incubation at 30° for 10 min, the amount of labeled acetylspermidine formed is determined as described by Libby.[7] The reaction is terminated by addition of 0.02 ml of 1 *M* hydroxylamine hydrochloride, and the tubes are heated at 100° for

[1] I. Matsui and A. E. Pegg, *Biochem. Biophys. Res Commun.* **92,** 1009 (1980).
[2] I. Matsui and A. E. Pegg, *Biochim. Biophys. Acta* **633,** 87 (1980).
[3] I. Matsui and A. E. Pegg, *Cancer Res.* **42,** 2990 (1982).
[4] I. Matsui and A. E. Pegg, *FEBS Lett.* **139,** 205 (1982).
[5] I. Matsui, H. Pösö, and A. E. Pegg, *Biochim. Biophys. Acta* **719,** 199 (1982).
[6] J. Blankenship and T. Walle, *Arch. Biochem. Biophys.* **179,** 235 (1977).
[7] P. R. Libby, *J. Biol. Chem.* **252,** 233 (1978); see also this volume [56].
[8] I. Matsui, L. Wiegand, and A. E. Pegg, *J. Biol. Chem.* **256,** 2454 (1981).
[9] F. Della Ragione and A. E. Pegg, *Biochemistry* **24,** 6152 (1982).

ISBN 0-12-181994-9

3 min. After centrifugation at 3000 *g* for 5 min to remove any protein precipitate, aliquots of 0.05 ml of the supernatant are applied to cellulose phosphate paper disks (2.3 cm in diameter, Whatman P81). The disks are washed five times with 1-ml portions of distilled water and three times with 1-ml portions of ethanol on a sintered-glass funnel. The disks are dried in air, placed in 10 ml of a toluene-based scintillation fluid, and assayed for radioactivity. A blank value is obtained by omitting spermidine or by omitting the enzyme solution.

Definition of Unit. A unit is defined as the amount of enzyme that catalyzes the production of 1 nmol of monoacetylspermidine per minute at 30°.

Purification Procedure

Step 1. Induction of Enzyme and Preparation of Extracts. Enzyme activity is induced in 50 female Sprague–Dawley rats (about 200 g body weight) by intraperitoneal treatment with 1.5 ml/kg of carbon tetrachloride 6 hr before death. (This treatment enhances the enzyme activity at least 50-fold.) The livers are removed and homogenized at 4° in 2 volumes of 0.25 *M* sucrose, 50 m*M* Tris-HCl, pH 7.5, 25 m*M* KCl, and 5 m*M* $MgCl_2$. All further steps are carried out at 0–4°. The homogenate is centrifuged at 25,000 *g* for 30 min, and the supernatant is removed and centrifuged at 105,000 *g* for 1 hr.

Step 2. Fractionation with Ammonium Sulfate. Proteins precipitating between 20 and 50% saturation with ammonium sulfate are collected, dissolved in 50 m*M* Tris-HCl, pH 7.5, and dialyzed for 4 hr against 100 volumes of this buffer.

Step 3. DEAE-Cellulose Chromatography. The dialyzed sample is clarified by centrifugation at 25,000 *g* for 20 min and applied to a column (5 × 30 cm) of DEAE-cellulose (Whatman DE-52) previously equilibrated with 50 m*M* Tris-HCl, pH 7.5. The sample is applied at a flow rate of 60 ml/hr. The column is then washed with 150 ml of equilibration buffer and eluted with a linear gradient (2 liters total volume) of from 0.1 to 0.6 *M* NaCl in this buffer at a flow rate of 80 ml/hr. Fractions of 20 ml are collected and assayed, and those containing the enzyme (which elutes at about 0.25 *M* NaCl) are pooled, concentrated by ultrafiltration, and diluted to a salt concentration of 0.1 *M* NaCl with 50 m*M* Tris-HCl, pH 7.5. (Attempts to precipitate the enzyme with ammonium sulfate and/or to dialyze the solution result in a large loss of activity due to extreme instability at this stage.)

Step 4. Affinity Chromatography on Norspermidine–Sepharose.[9] *sym*-Norspermidine (3,3′-diaminodipropylamine, available commercially from Eastman Organic Chemicals, Rochester, New York) is linked to Sepharose 4B as follows. Ten grams of 6-aminohexanoic acid–Sepharose 4B (Pharmacia Fine Chemicals, Uppsala, Sweden) is swollen overnight in 1 liter of 0.5 *M* NaCl and washed with four changes of 1 liter of 0.5 *M* NaCl and then with water to remove the salt. The washed gel is mixed with 18 mmol of norspermidine (using a solution adjusted to pH 5.5 with 2 *N* HCl), and the total volume is made up to 100 ml with water. The mixture is stirred slowly while 2 g of 1-ethyl-3-(3-dimethylamino-propyl)carbodiimide dissolved in water is added dropwise. The pH is maintained at 5.5, and the mixture is shaken gently at room temperature for 24 hr. The gel is washed with 8 liters of 0.5 *M* NaCl and stored in 50 m*M* Tris-HCl, pH 7.5, containing 1 *M* NaCl and 0.02% sodium azide at 4° until used.

A column of norspermidine-Sepharose (1.5 × 10 cm) is prepared and equilibrated with 50 m*M* Tris-HCl. The enzyme solution from step 3 is applied to this column at a flow rate of 40 ml/hr. The column is then washed with 0.5 *M* NaCl in 50 m*M* Tris-HCl, pH 7.5, until the absorbance of the eluate at 280 nm is about 0.05. The eluting solution is then changed to 0.8 *M* NaCl in 50 m*M* Tris-HCl, pH 7.5, and washing is continued (usually for about 60 ml) until the absorbance of the eluate is less than 0.01 at 280 nm. Spermidine N^1-acetyltransferase is then eluted with 5 m*M* spermidine in 0.8 *M* NaCl and 50 m*M* Tris-HCl, pH 7.5. Fractions containing activity are concentrated to approximately 1 ml by ultrafiltration and freed from NaCl by repeatedly diluting to 5 ml with 50 m*M* Tris-HCl, pH 7.5, containing 10 m*M* spermidine. The spermidine is needed to stabilize the enzyme activity.

The material eluted from the norspermidine-Sepharose column can be further purified to homogeneity in two ways. Either a second passage through this absorbent or affinity chromatography on Cibacron Blue agarose can be used. For the former, a column of norspermidine–Sepharose (1 × 2 cm) is equilibrated with 50 m*M* Tris-HCl, pH 7.5. The enzyme solution from step 4 is applied, and the column is washed with 0.5 *M* NaCl and 0.8 *M* NaCl in 50 m*M* Tris-HCl, pH 7.5, as described above except that a flow rate of 20 ml/hr is used. The column is washed until the absorbance at 280 nm at the end of washing with 0.8 *M* NaCl is less than 0.005. The enzyme is then eluted by 5 m*M* spermidine in 0.8 *M* NaCl, 50 m*M* Tris-HCl, pH 7.5.

The alternative procedure is to apply the sample to a column (0.8 × 4 cm) of Cibacron Blue agarose (AffiGel Blue, Bio-Rad Laboratories, Richmond, California) equilibrated with 50 m*M* Tris-HCl, pH 7.5, 1 m*M* spermidine. The column is loaded and eluted at a rate of 18 ml/hr. It is washed with the equilibration buffer until the absorbance at 280 nm is reduced to

PURIFICATION OF SPERMIDINE N^1-ACETYLTRANSFERASE FROM RAT

Fraction	Total protein (mg)	Total units	Specific activity (units/mg)	Purification (fold)	Yield (%)
1. 105,000 *g* supernatant	28,864	2078	0.07	1	100
2. Ammonium sulfate	12,332	1726	0.14	2	83
3. DEAE-cellulose	532	1500	2.82	39	72
4. Chromatography on norspermidine–Sepharose	0.66	860	1303	18,037	41
5a. Second chromatography on norspermidine–Sepharose	0.052	419	8069	112,069	20
or					
5b. Chromatography on Cibacron Blue agarose	0.048	389	8102	112,527	19

zero (about 30 ml), and the enzyme is then eluted by the addition of 1 m*M* coenzyme A to the buffer. Three-milliliter fractions are collected, and the activity in each is assayed after removal of coenzyme A (which is inhibitory) by repeated concentration and dilution with 50 m*M* Tris-HCl, pH 7.5, 10 m*M* spermidine. The activity commences to elute in about the second fraction, and about 40 ml is needed to recover all the activity. These active fractions are concentrated and the purified enzyme stored in 0.2 ml of 50 m*M* Tris-HCl, pH 7.5, 10 m*M* spermidine.

A summary of the purification is shown in the table. It can be seen that both methods for the final step give a similar specific activity. Using the two steps sequentially in either order did not increase the specific activity. The final preparation gave a single band on analysis by polyacrylamide gel electrophoresis under native or denaturing conditions, and in the former case the enzyme activity corresponded exactly to the band.[9] These observations provide convincing proof of the homogeneity of the final preparation.

Properties

The enzyme has an apparent molecular weight of 114,000 and is made up of two subunits of molecular weight 60,000.[9] It is very sensitive to heat denaturation and to incubation in dilute solutions in the absence of other proteins. For these reasons assays are best carried out at 30°; when the purified enzyme is assayed it is essential to add a stabilizing protein (e.g., 0.5 mg of bovine serum albumin per milliliter) to ensure a linear rate of reaction.

The purified enzyme is very unstable in the absence of spermidine but can be stored frozen at −20° in the presence of 10 m*M* spermidine with a loss of activity of 30% in a week. The enzyme will not catalyze the acetylation of histones, *sym*-homospermidine, or putrescine, but it will also acetylate spermine, monoacetylspermine, *sym*-norspermidine, *sym*-norspermine, and, at a much slower rate, 1,3-diaminopropane.[8,9] This suggests that the enzyme requires a substrate with the structure $H_2N(CH_2)_3NHR$ and acetylates the primary amino group, which is consistent with the identification of N^1-acetylspermidine as the product of its reaction with spermidine. The enzyme has no known cofactors.

[56] Purification of Two Spermidine *N*-Acetyltransferases (Histone *N*-Acetyltransferases)[1] from Calf Liver Nuclei

By PAUL R. LIBBY

Spermidine + acetyl-CoA →
N^1-acetylspermidine (55%) + N^8-acetylspermidine (45%) + CoA

Assay Method

Principle. Acetylspermidine is formed from [1-^{14}C]acetyl-CoA, and bound to cellulose phosphate paper. The paper is washed, dried, and counted.[2]

Reagents and Materials

Bicine, 0.5 *M*, pH 8.8
Spermidine · 3HCl (3 m*M*)
[1-^{14}C]Acetyl-CoA, 0.04 m*M*
Enzyme
Hydroxylamine (1 *M*) · HCl
Cellulose phosphate paper (Whatman P-81), 2.4 cm

Procedure. The enzyme is incubated with 10 μl of spermidine, 10 μl of acetyl-CoA, and 20 μl of buffer in a final volume of 100 μl and incubated

[1] EC 2.3.1.48. See also this volume [55] for purification of spermidine *N*′-acetyltransferase.
[2] P. R. Libby, *J. Biol. Chem.* **253,** 233 (1978).

ISBN 0-12-181994-9

at 37° for 5 min. The reaction is stopped by chilling and the addition of 20 μl of hydroxylamine · HCl. The tubes are placed in a boiling water bath for 3 min. If necessary, protein is removed by centrifugation and 50 μl of the solution is spotted on a cellulose phosphate disk, which is then immersed in a beaker of tap water. The water in the beaker is changed once, and then each disk is exhaustively washed on a sintered-glass filter. Finally, the disks are washed with methanol, dried, and counted in a liquid scintillation spectrometer.

Unit. One unit of enzyme is the amount that forms 1 pmol of acetylspermidine per minute. Protein concentrations are determined by the fluorescamine reaction.[3]

Purification of Enzyme

Reagents. All buffers contained 0.1% (9 m*M*) thioglycerol.

Buffer A: 0.25 *M* sucrose, 5 m*M* potassium phosphate, pH 7.4, 3 m*M* magnesium chloride

Buffer B: 400 m*M* potassium chloride in 5 m*M* potassium phosphate, pH 7.4

Buffer C: 20% saturated ammonium sulfate, pH 7.4, in 5 m*M* potassium phosphate, pH 7.4

Buffer D: 20% saturated ammonium sulfate, in 0.5 *M* potassium phosphate, pH 8.05

Buffer E: 5 m*M* potassium phosphate, pH 7.4

Buffer F: 250 m*M* potassium phosphate, pH 7.4

Buffer G: 320 m*M* potassium chloride in 5 m*M* potassium phosphate, pH 7.4

Saturated ammonium sulfate: a solution of ammonium sulfate, saturated at room temperature, is titrated to pH 7.4 with ammonium hydroxide and chilled.

Purification Procedures

All procedures are carried out at 2–4° unless otherwise noted. Calf liver is obtained fresh from a local slaughterhouse and transported to the laboratory on ice. The liver is minced, and 200-g portions are blended with 500 ml of buffer A in a Waring blender for 60 sec. The pellet from a 1000 *g* centrifugation is washed twice with buffer A and stored at −70°.

[3] P. Bohlen, S. Stein, W. Dairman, and S. Udenfriend, *Arch. Biochem. Biophys.* **155,** 213 (1973).

Nuclei from 1400 g of calf liver are thawed and suspended in buffer B (1300 ml). The nuclei are blended for 30 sec and centrifuged for 10 min at 10,000 *g*. The pellet is reextracted with an additional liter of buffer B. The two extracts are combined.

One-tenth volume (240 ml) of alumina C_γ gel suspension (Calbiochem) is added to the nuclear extract, and the suspension is stirred for 2 hr. The gel is centrifuged at 1000 *g* and washed four times with buffer B. The enzyme is extracted from the gel with two 250-ml washes with buffer C and one 250-ml wash with buffer D. The extracts are pooled and successively brought to 35 and 60% saturation with solid ammonium sulfate. The 35 to 60% precipitate is dialyzed against buffer E until free of ammonium sulfate.

The enzyme is loaded onto a column of DEAE-cellulose (2.5 × 23 cm) equilibrated with buffer E and washed with 50 ml of buffer E. The enzyme is eluted by a convex gradient of potassium chloride in buffer E established with 250 ml of buffer E in the constant-volume mixing chamber, and buffer B as the limit. Active fractions are eluted at about 0.18 *M* KCl, pooled, brought successively to 45 and 62% saturation with ammonium sulfate, and dialyzed against buffer E until free of ammonium sulfate.

The enzyme is applied to a column of hydroxyapatite (2.5 × 3.2 cm) in buffer E. The enzyme is eluted with a linear gradient (100 ml of buffer E and 100 ml of buffer F). Two peaks of enzyme activity, A and B, are eluted at about 0.11 and 0.18 *M* phosphate, respectively. Active fractions of each peak are pooled separately and each brought to 70% saturation with ammonium sulfate; the precipitates are dialyzed against buffer E.

Each enzyme is separately applied to a column of Ultrogel AcA-34 (2.6 × 83 cm) and eluted with buffer E at a flow rate of 16 ml/hr; 3.4-ml fractions are collected. Enzyme A eluted with the peak activity at tube 62; enzyme B eluted with the peak activity at tube 58. Active fractions are pooled and concentrated on an Amicon XM 100-A membrane.

Enzymes A and B are separately applied to a column of DEAE-Sephadex A-25 (1.5 × 14 cm) equilibrated with buffer E. A linear gradient is used to elute the enzymes with 100 ml of buffer E and 100 ml of buffer G. Enzyme A is eluted with the peak activity occurring at 0.17 *M* KCl; enzyme B, with the peak activity occurring at 0.2 *M* KCl. Active fractions of each enzyme are pooled, dialyzed against 50% glycerol in buffer E, and stored at −20°.

The table summarizes a typical purification of the two enzymes. Enzyme A has been purified 4580-fold from the nuclear pellet with a yield of 5.5%. Enzyme B has been purified 5100-fold from the nuclear pellet with a yield of 1.6%.

PURIFICATION OF CALF LIVER NUCLEAR SPERMIDINE *N*-ACETYLTRANSFERASE

Fraction	Volume (ml)	Protein (mg)	Spermidine acetyl-transferase (units)	Recovery (%)	Specific activity (units/mg)	Purifi-cation (fold)
Nuclei	1640	59,000	16,500	100	0.28	—
Extract	2350	45,000	15,600	94.7	0.35	—
Alumina eluate	760	7,900	11,400	69.0	1.44	5.2
Ammonium sulfate	96	2,040	15,100	91.3	7.42	26.5
DEAE pool	264	520	10,200	61.7	19.6	70
Acetyltransferase A						
Hydroxyapatite	40	38.4	2,340	14.2	61.0	218
Sephadex pool	16.8	5.38	1,170	7.1	218	779
Sephadex A-25 pool	2.75	0.72	920	5.5	1280	4580
Acetyltransferase B						
Hydroxyapatite	39	14.7	2,050	12.4	140	499
Sephadex pool	5.4	1.4	690	4.2	490	1770
Sephadex A-25 pool	2.2	0.18	260	1.6	1430	5100

Properties of the Enzymes

Substrate Specificity. As has been discussed,[2] the spermidine *N*-acetyltransferase activities of calf liver nuclei can be attributed to the histone *N*-acetyltransferase enzymes of the nucleus. This conclusion arose from parallel purification of the two activities during a 5000-fold purification, identical heat-denaturation kinetics, and inhibition of one enzyme activity (i.e., spermidine acetyltransferase) in the presence of the other acceptor (i.e., histone). Spermidine is the favored polyamine substrate for both enzymes, with spermine and norspermidine also efficiently acetylated. Diamines are relatively poorly acetylated.

Characteristics of the Enzymes. pH optima for both enzymes are pH 8.8. Both enzymes are strongly inhibited by low levels of *p*-chloromercuribenzoate (50% inhibition at 2 and 8 μM for enzymes A and B, respectively).[4] EDTA was inhibitory only at high concentrations (30–50 mM).

Kinetics. Michaelis constants for spermidine are 1.9×10^{-4} M and 1.6×10^{-4} M for enzymes A and B, respectively. The Michaelis constants for acetyl-CoA are 9.5×10^{-6} M and 8.5×10^{-6} M, respectively.

Stability. Enzyme B is significantly more stable at 47° than is enzyme A. Both enzymes, however, may be stored for several months at −20° in 50% glycerol without significant loss of activity. Freezing and thawing of partially purified enzymes is deleterious.

[4] These inhibitions by *p*-chloromercuribenzoate were seen in the presence of 18 μM thioglycerol.

[57] Acetylspermidine Deacetylase[1] (Rat Liver)

By PAUL R. LIBBY

N^1-Acetylspermidine + H_2O → acetic acid + spermidine

Assay

Principle. Labeled acetic acid derived from N^1-acetylspermidine is extracted into ethyl acetate and counted.[2]

Reagents

Sodium borate buffer, 0.2 *M*, pH 10.4
Dithiothreitol, 0.1 *M*
N^1-[*acetyl*-^{3}H]Acetylspermidine,[3] 0.1 m*M*
Enzyme
HCl, 1 *M*, containing 1% acetic acid
Ethyl acetate

Procedure. Enzyme is incubated with 10 μl of acetylspermidine, 10 μl of dithiothreitol, and 20 μl of buffer in a final volume of 100 μl for 5 min in a 37° water bath. The reaction is stopped by chilling and the addition of 100 μl of HCl–acetic acid. Ethyl acetate (1.5 ml) is added, and the tubes are vigorously vortexed. After brief centrifugation to separate phases, 0.5 ml of the ethyl acetate layer is counted in a liquid scintillation spectrometer using a commercial aqueous counting solution.

Unit. One unit of enzyme activity is defined as the amount that releases 1 pmol of acetic acid from N^1-acetylspermidine per minute. Protein is determined by the fluorescamine reaction.[4]

Purification of Enzyme

Reagents. All buffers contained 0.1% (9 m*M*) thioglycerol.

Buffer A: 100 m*M* sodium borate, pH 9.2
Buffer B: 10 m*M* sodium borate, pH 9.2

[1] EC 3.5.1.
[2] P. Libby, *Arch. Biochem. Biophys.* **188,** 360 (1978).
[3] H. Tabor, C. W. Tabor, and L. de Meis, this series, Vol. 17B, p. 829. See also this volume [72] for a later preparation of N'-acetylspermidine.
[4] P. Bohlen, S. Stein, W. Dairman, and S. Udenfriend, *Arch. Biochem. Biophys.* **155,** 213 (1973).

ISBN 0-12-181994-9

Buffer C: 400 mM potassium chloride in 10 mM sodium borate, pH 9.2

Buffer D: 200 mM potassium phosphate, pH 9.2, in 10 mM sodium borate, pH 9.2

Saturated ammonium sulfate: a solution of ammonium sulfate, saturated at room temperature, is titrated to pH 9.2 with concentrated ammonium hydroxide and chilled to 2–4°. This solution does not have thioglycerol added to it.

Purification Procedures

All procedures are carried out at 2–4°. In one typical preparation, 34 g rat liver were minced and blended with 100 ml of buffer A in a Waring blender for 30 sec and centrifuged at 7000 *g* for 15 min. The supernatant was successively brought to 25 and 50% saturation by the dropwise addition, with stirring, of the saturated ammonium sulfate solution, the 25–50% fraction was collected by centrifugation, taken up in a minimal amount of buffer B, and dialyzed free of ammonium sulfate.

The enzyme was then applied to a DEAE-cellulose column (2.5 × 38.2 cm) equilibrated with buffer B and washed free of nonadsorbed protein. A linear gradient was then established with 175 ml buffer B and 175 ml buffer C. Enzyme was eluted at about 0.08 M KCl. The most active fractions were pooled, brought successively to 33 and 67% ammonium sulfate with the saturated solution of ammonium sulfate, and centrifuged; the 33 to 67% fraction was taken up in buffer B and dialyzed free of ammonium sulfate.

The enzyme was then applied to a column of Ultrogel AcA-34 (2.5 × 42.2 cm; LKB Instrument Co.) equilibrated with buffer B and eluted at a flow rate of 12 ml/hr. Fractions of 2.8 ml were collected. The most active fractions (33 to 41) were pooled.

The enzyme was then applied to a column of hydroxyapatite (1.5 ×

PURIFICATION OF ACETYLSPERMIDINE DEACETYLASE

Fraction	Protein (mg)	Deacetylase (units)	Recovery (%)	Specific activity (units/mg)	Purification (fold)
Extract	3170	59,500	—	19	—
Ammonium sulfate	2000	35,300	59	18	—
DEAE pool	146	12,200	21	84	4.4
AcA-34 pool	26	7,340	12	280	15
Hydroxyapatite	11	6,230	10.5	540	29

13.8 cm) in buffer B. A linear gradient was established with 150 ml of buffer B and 150 ml of buffer D. The enzyme was eluted with the peak of activity at about 0.015 *M* phosphate. The active fractions were pooled and dialyzed against 50% glycerol in buffer B to remove phosphate and concentrate the enzyme for storage. The results of this purification are summarized in the table. The enzyme was apparently purified 29-fold with an apparent yield of 11%. We have previously discussed the possibility that an appreciable portion of the activity in the original homogenate represents oxidative degradation of acetylspermidine.[2]

Properties

Stability and Storage. In the presence of glycerol and thioglycerol, the enzyme retained full activity at −20° for a period of up to 6 months. Under assay conditions (i.e., pH 10.4, 10 m*M* dithiothreitol) the enzyme retains activity for 30 min, although at a diminishing rate.

Activators and Inhibitors. The optimum pH in the presence of sodium borate buffer is apparently 10.4, but no higher pH values were tested. If sodium bicarbonate–carbonate buffers are used, the pH optimum is 10.0 and a broader profile is seen. Dithiothreitol activates the enzyme 64% at 1 m*M* and 95% at 10 m*M*, *p*-chloromercuribenzoate inhibits 73%.[5] EDTA, 150 m*M*, is not inhibitory. The enzyme is inhibited by the polyamines, spermidine, spermine, and putrescine.

Kinetic Properties. The apparent K_m values for N^1-acetylspermidine and N^1-acetylspermine are 3 μM and 16 μM, respectively. Insufficient N^8-acetylspermidine was available to measure the K_m, but limited data suggested that it was similar to the N^1 derivative. Free polyamines showed competitive inhibition with N^1-acetylspermidine, the K_i values of putrescine, spermidine, and spermine being 250, 55, and 36 μM, respectively.

Specificity. The enzyme hydrolyzes N^1- and N^8-acetylspermidine, and N^1-acetylspermine. It does not hydrolyze N^1- or N^8-diacetylspermidine, N^1-acetylputrescine,[6] or *N*-acetylhistones. On the basis of competition studies, it apparently does not hydrolyze *N*-acetylmethionine, *N*-acetylglucosamine, *N*-acetylgalactosamine, *N*-acetylmannosamine, or *N*-acetylneuraminic acid, since none of these compounds affected the deacetylation of acetylspermidine when present at 1 m*M* concentrations. (These studies were carried out at both pH 10.4 and pH 8.5.)

[5] The enzyme is probably highly sensitive to *p*-chloromercuribenzoate, since the final concentration of thioglycerol in the reaction mixture (from the enzyme) was 0.09 m*M*.

[6] Since an appreciable amount of N^1-acetylputrescine partitions into the ethyl acetate layer, the standard assay is unsuitable for routine measurement of acetylputrescine hydrolysis.

Section IX

Other Enzymes Involved in Polyamine Synthesis and Metabolism

[58] Putrescine Synthase from *Lathyrus sativus* (Grass Pea) Seedlings

By K. S. SRIVENUGOPAL and P. R. ADIGA

Agmatine iminohydrolase (EC 3.5.3.12, agmatine deiminase) (reaction 2a) and an activity degrading *N*-carbamoylputrescine (NCP) to putrescine, CO_2, and NH_3 (NCP amidohydrolase) have been detected in some plant extracts[1-3] and were implicated earlier as catalyzing sequential reactions in putrescine biosynthesis in plants.[4] However, we have now found that the enzymes agmatine iminohydrolase, putrescine transcarbamylase (EC 2.1.3.6, putrescine carbamoyltransferase), ornithine transcarbamylase (EC 2.1.3.3, ornithine carbamoyltransferase), and carbamate kinase (EC 2.7.2.2) copurify from the extracts of *Lathyrus sativus* seedlings, the relative specific activities of the enzymes remaining constant. These activities reside in a single polypeptide, and the trivial name putrescine synthase is used to designate this multifunctional protein.[5] The significance of the association of these activities and the operation of an "agmatine cycle" involving intact transfer of carbamoyl group from NCP to ornithine to form citrulline have been discussed.[5]

Overall Reactions

$$\text{Agmatine} + \text{ornithine} + H_2O + P_i \rightarrow \text{putrescine} + \text{citrulline} + NH_3 + P_i \quad (1a)$$

$$\text{Agmatine} + \text{ADP} + P_i + H_2O \rightarrow \text{putrescine} + \text{ATP} + 2NH_3 + CO_2 \quad (1b)$$

Constituent Activities of Putrescine Synthase

$$\text{Agmatine} + H_2O \xrightarrow[\text{(agmatine iminohydrolase)}]{} \text{NCP} + NH_3 \quad (2a)$$

$$\text{NCP} + P_i \underset{\text{(putrescine transcarbamylase)}}{\rightleftharpoons} \text{putrescine} + \text{carbamoyl phosphate} \quad (2b)$$

$$\text{Carbamoyl phosphate} + \text{ornithine} \underset{\text{(ornithine transcarbamylase)}}{\rightleftharpoons} \text{citrulline} + P_i \quad (2c)$$

$$\text{Carbamoyl phosphate} + \text{ADP} + H_2O \underset{\text{(carbamate kinase)}}{\rightleftharpoons} \text{ATP} + CO_2 + NH_3 \quad (2d)$$

[1] T. A. Smith, *Phytochemistry* **8,** 2111 (1969).
[2] T. A. Smith, *Phytochemistry* **4,** 599 (1965).
[3] R. K. Sindhu and H. V. Desai, *Phytochemistry* **19,** 317 (1980).
[4] T. A. Smith and J. L. Garraway, *Phytochemistry* **3,** 23 (1964).
[5] K. S. Srivenugopal and P. R. Adiga, *J. Biol. Chem.* **256,** 9532 (1981).

ISBN 0-12-181994-9

Assay Method[6]

Principle. Putrescine synthase is assayed by measuring citrulline formed with agmatine or NCP and ornithine as substrates (overall reaction 1a). The product is separated from the amine substrates on Dowex 50 (H^+) resin and quantitated by a color reaction.[7]

Reagents

Tris-HCl buffer, 0.25 *M*, pH 8.8
Dithiothreitol, 0.1 *M*
Ornithine · HCl, 0.1 *M*
NH_4OH, 2 *M*
Agmatine · HCl[8] or NCP · HCl, 0.1 *M*
Na_2HPO_4, 0.1 *M*
$MgSO_4$, 0.25 *M*
Perchloric acid, 20%
Bovine serum albumin, 1 mg/ml

Chromatographic Column. Small glass columns (0.7 × 10 cm) plugged with glass wool at the bottom are filled with a Dowex 50 suspension (H^+ form, 8% cross-linked, 200–400 mesh). With an appropriate manifold, 8–10 columns can be used simultaneously.

Procedure. A boiled enzyme blank is necessary for assay at all purification steps.[9] Assay mixtures (0.5 ml) contain 25 μmol of Tris-HCl (0.1 ml), 1 μmol of Na_2HPO_4 (0.01 ml), 1 μmol of dithiothreitol (0.01 ml), 5 μmol of $MgSO_4$ (0.02 ml), 2.5 μmol of ornithine–HCl (0.025 ml), 2.5 μmol of agmatine, HCl or NCP · HCl (0.025 ml), 10 μg of bovine serum albumin (0.01 ml), and various amounts of enzyme. The assay mixtures are incubated at 37°, and the reaction is arrested by the addition of 0.1 ml of $HClO_4$. After removal of the denatured proteins by brief centrifugation, the samples are applied to the Dowex column.[9] The column is washed with distilled water (5 ml), then the amino acid fraction containing citrulline is selectively eluted with 2.5 ml of 2 *M* NH_4OH. The amine substrates, which interfere with the color reaction by producing excess of

[6] Details of individual assay procedures for all the constituent activities of putrescine synthase have been given elsewhere.[5]

[7] L. M. Prescott and M. E. Jones, *Anal. Biochem.* **32,** 408 (1969).

[8] Agmatine · SO_4 (Sigma) required purification on a Dowex 50 (H^+) column using an HCl gradient (0 to 4 *M*) to remove putrescine contamination. NCP · HCl was prepared by the procedure described elsewhere.[11]

[9] Approximately 1% of the amine substrates was eluted with the citrulline from the Dowex 50 column. Hence, the boiled enzyme controls also produced a significant amount of color in the colorimetric procedure used for citrulline determination.

chromogen, are preferentially retained on Dowex resin during this step.[10] An aliquot of the NH_4OH eluate is subjected to the color reaction by the method of Prescott and Jones.[7] The difference between the color obtained with active enzyme and with boiled enzyme represents the citrulline formed, and the amount of ureidoamino acid is determined by reference to a standard curve.[11]

Definition of Unit and Specific Activity. One unit is defined as the amount of enzyme that catalyzes the formation of 1 μmol of citrulline in 60 min at 37°. Specific activity is expressed as units per milligram of protein.

Purification Procedure[12]

Growth of Seedlings. Lathyrus sativus seeds, obtained from the Plant Breeding Section of Indian Agricultural Research Institute, New Delhi, are washed, soaked, and grown in the dark on sterile cotton pads to get 5-day-old etiolated seedlings as described earlier.[13] All the steps are carried out at 4°, and centrifugation is at 25,000 *g* for 15 min.

Step 1. Crude Extract. The fresh seedlings are washed with distilled water and homogenized in a chilled Waring blender with one volume of 0.05 *M* imidazole-HCl buffer (pH 8.0) containing 5 m*M* 2-mercapto-ethanol. The homogenate is passed through four layers of cheesecloth to remove fibrous material, and centrifuged.

Step 2. $MnCl_2$ Treatment. The crude extract is adjusted to 7.5 m*M* $McCl_2$ concentration by the addition of 1 *M* stock solution. After stirring for 30 min, the precipitated nucleoproteins are removed by centrifugation.

Step 3. $(NH_4)_2SO_4$ Fractionation. The supernatant from step 2 is brought to pH 7.0 by the addition of 2 *M* NH_4OH. Precooled solid $(NH_4)_2SO_4$ is added in small amounts with stirring to achieve 45% saturation. After 30 min of stirring, the precipitate is centrifuged and discarded. The resulting supernatant is adjusted to 85% saturation in $(NH_4)_2SO_4$ and stirred for 1 hr. The precipitate obtained after centrifugation is dissolved in 10 m*M* imidazole buffer (pH 7.5) containing 2 m*M* 2-mercaptoethanol and dialyzed against two changes of 2 liters of the same buffer.

[10] Fresh Dowex 50 preparations worked well. Often a material interacting with reagents of color reaction was found to elute with NH_4OH from used, but regenerated, Dowex resin.

[11] K. S. Srivenugopal and P. R. Adiga, *Anal. Biochem.* **104,** 440 (1980). See also this volume [59].

[12] An alternative purification procedure for putrescine synthase using organomercurial-Sepharose and DEAE-Sephadex chromatography has also been reported.[5]

[13] M. R. Suresh, S. Ramakrishna, and P. R. Adiga, *Phytochemistry* **15,** 483 (1976).

Step 4. Affinity Chromatography on Putrescine–CH Sepharose

PREPARATION OF PUTRESCINE-CH SEPHAROSE. The activation and washing procedures described by March *et al.*[14] are employed to obtain CH-Sepharose by allowing 6-aminohexanoic acid (Sigma) to react with CNBr-activated Sepharose. The free carboxyl groups are coupled to putrescine by carbodiimide condensation[15] at pH 4.8 with two additions of 1,3-(dimethylaminopropyl)carbodiimide.

AFFINITY CHROMATOGRAPHY. The dialyzed $(NH_4)_2SO_4$ fraction (step 3) is adsorbed on the affinity column (5 ml bed volume) equilibrated with 50 m*M* imidazole buffer at a flow rate of 10 ml/hr. After extensive washing with this buffer to remove all the unadsorbed proteins (until a base line $A_{280\ nm}$ is obtained), the elution of the bound enzyme is carried out with the addition of 2 m*M* putrescine · 2HCl in the washing buffer.[16] It is advisable to collect the fractions directly into test tubes containing 10 μg of bovine serum albumin, which stabilizes the enzyme. Since the substrate elution was found to yield a pure enzyme, the fractions with high $A_{280\ nm}$ are pooled and dialyzed against 10 m*M* imidazole buffer to remove the diamine.

A typical purification is summarized in the table.

Properties

The enzyme was homogeneous on polyacrylamide gel electrophoresis at pH 4.0 and 8.3. It exhibited a single protein band (stainable with Coomassie Blue) corresponding to M_r 55,000 on sodium dodecyl sulfate–polyacrylamide gels and catalyzed overall reaction 1a stoichiometrically. The constituent activities of putrescine synthase were found to copurify with similar specific activity ratios during the four-step purification procedure. Mg^{2+} and P_i were essential for the overall reactions Mn^{2+} was found to activate agmatine iminohydrolase fourfold.

pH Optima. pH 7.5 was found to be optimal for most of the constituent activities, but putrescine synthase was most active at pH 8.8 in the overall reaction.

Inhibitors. Putrescine synthase was inhibited by sulfhydryl blocking agents such as *p*-chloromercuribenzoate and *N*-ethylmaleimide. Arcain, the diguanido analog of agmatine, caused a 30% inhibition of the agmatine

[14] S. C. March, I. Parikh, and P. Cuatrecasas, *Anal. Biochem.* **60,** 149 (1974).

[15] P. Cuatrecasas, *J. Biol. Chem.* **245,** 3059 (1970).

[16] Prior to substrate-specific elution, washing the column with 0.2 *M* KCl in the washing buffer did result in elution of some nonspecific proteins. However, this step was not necessary, since elution with putrescine led to desorption of only putrescine synthase.

PURIFICATION OF PUTRESCINE SYNTHASE FROM *Lathyrus sativus* SEEDLINGS

Fraction	Protein (mg)	Total enzyme units[a]: Agmatine + ornithine → citrulline	NCP + ornithine → citrulline	Purification (fold)	Recovery (%)
Crude extract	4680	270 (0.06)	330 (0.07)	1.0	100
$MnCl_2$ supernatant	3450	214 (0.06)	270 (0.08)	1.0	80
$(NH_4)_2SO_4$, 45–85% saturation	1050	180 (0.17)	230 (0.22)	2.8	67
Putrescine-CH Sepharose affinity step	5	60 (11.9)	70 (14.0)	200	23

[a] Numbers in parentheses indicate specific activities.

iminohydrolase reaction. Inorganic phosphate was found to inhibit the arsenolysis of NCP.

Substrate Specificity. Arginine, creatine, and *N,N'*-dicarbamoylputrescine did not serve as substrates.

Stability. The purified enzyme was highly unstable; it lost all component activities within 48 hr after purification when stored at 4°. Prolonged dialysis and freezing–thawing also led to considerable losses in the activity. The presence of bovine serum albumin during the assay and storage stabilized the enzyme considerably.

[59] Putrescine Carbamoyltransferase[1] (*Streptococcus faecalis*)

By VICTOR STALON

$$NH_2CO_2PO_3^{2-} + NH_3^+(CH_2)_4NH_3^+ \rightleftharpoons NH_2C(=O)NH(CH_2)_4NH_3^+ + HOPO_3^{2-} + H^+$$

Carbamoyl-phosphate — Putrescine — Carbamoyl-putrescine

[1] Putrescine carbamoyltransferase (EC 2.1.3.6). See also this volume [58].

ISBN 0-12-181994-9

Assay Method[2]

The enzyme can be measured in the reverse direction of its physiological function by carbamoylputrescine synthesis from carbamoylphosphate and putrescine.

Reagents

Tris-HCl buffer, 1 *M*, pH 7.0
Putrescine, 0.1 *M*, pH 7.0
Dilithium carbamoylphosphate, 20 m*M* in water, prepared just before use to avoid decomposition[3]

Assay Mixture. The total incubation volume is 2.0 ml, which contains 100 μmol of Tris buffer, 20 μmol of carbamoylphosphate, and a suitable aliquot of extract, generally 1–5 μl of cell extract containing 5–10 mg of protein per milliliter. A vessel containing the assay mixture but no enzyme is necessary to correct for a small nonenzymatic synthesis of carbamoylputrescine during the 10-min incubation at 37°. The reaction is started by the addition of carbamoylphosphate solution after the vessel has been at 37° for a 5-min preincubation and is stopped after an additional 10 min by the addition of 2 ml of 1 *N* HCl. The entire 4 ml is assayed for carbamoylputrescine by the method of Archibald.[4]

Definition of Unit and Specific Activity. One unit of putrescine carbamoyltransferase is defined as the amount that catalyzes the synthesis of 1 μmol of carbamoylputrescine per hour. Specific activity is expressed as units per milligram of protein. Protein is determined by the method of Lowry *et al.*[5] in the early steps of the purification. The Kalckar spectrophotometric method is used for the last purification steps.[6]

Purification Procedure[7]

Cell Growth. The enzyme is isolated from *Streptococcus faecalis* (American Type Culture Collection No. 11,700). The organism is grown in a medium containing, per liter, 10 g of yeast extract, 5 g of tryptone, 5 g of NaCl, 1.4 g of KH_2PO_4, 14.3 g of Na_2HPO_4, 1.7 g of K_2SO_4, 0.5 mg of $MgSO_4 \cdot 4H_2O$, 0.5 mg of $FeCl_3$. Agmatine is sterilized separately by filtration and added to a final concentration of 25 m*M* just before inoculation.

[2] R. J. Roon and H. A. Barker, *J. Bacteriol.* **109,** 44 (1972).
[3] C. M. Allen, Jr. and M. E. Jones, *Biochemistry* **3,** 1238 (1964).
[4] R. M. Archibald, *J. Biol. Chem.* **156,** 121 (1944).
[5] O. Lowry, N. J. Rosebrough, A. L. Farr, and R. J. Randall, *J. Biol. Chem.* **193,** 265 (1951).
[6] H. M. Kalckar, *J. Biol. Chem.* **167,** 461 (1947).
[7] B. Wargnier, N. Lauwerr, and V. Stalon, *Eur. J. Biochem.* **101,** 143 (1979).

Induction of putrescine carbamoyltransferase occurs when other energy sources of the medium (carbohydrates and arginine) are exhausted and when growth depends on agmatine. Maximum activity is observed when cells have entered the stationary phase of growth, and under anaerobic conditions. Aeration invariably results in lower putrescine carbamoyltransferase formation.

Thirty liters of medium are made up in a glass carboy. The medium is inoculated to a density of 10^7 cells/ml with a culture of *Streptococcus faecalis* that had been grown overnight at 37°. The concentration of cells is determined by optical density. Induction is followed by the pH level. The inoculated medium has a pH between 6.8 and 7.0. As the cells grow and utilize the carbohydrates of the medium, the pH drops to 6.2. The pH rises to 8.0 or slightly above when ammonia and putrescine are produced from agmatine. When the cells enter the stationary phase of growth, they are harvested by centrifugation in a Sharples centrifuge, and the pellet is washed with a solution of 0.9% NaCl, and stored at −20°. Unless otherwise indicated, further steps are performed in the cold at 4°.

Step 1. Extract. The bacterial mass obtained after centrifugation is suspended in twice its volume of 100 m*M* potassium phosphate buffer (pH 6.0) supplemented with 10 m*M* putrescine and 1 m*M* dithiothreitol. This suspension is placed in a 10-kc Raytheon sonic oscillator at maximum power for 30 min. The cell extract is centrifuged in the Sorvall at 20,000 *g* for 30 min; the cell pellet is resuspended in the extraction buffer and disrupted again. This operation is repeated one time. The supernatants of each extraction are pooled and used for fractionation.

Step 2. Heat Denaturation. The cell extract is heated at 65° for 10 min with gentle stirring in a water bath. After cooling in an ice bath, the coagulated proteins are separated by centrifugation for 30 min at 20,000 *g*. The precipitate is washed with 50 m*M* potassium phosphate buffer (pH 7.5), supplemented with 0.5 m*M* dithiothreitol, and centrifuged again; the two supernatants are pooled.

Step 3. Ammonium Sulfate Precipitation. Solid ammonium sulfate is added slowly to the heat-denatured supernatant, with constant stirring, to 40% saturation. After 30 min, the precipitate is removed by centrifugation at 20,000 *g* for 20 min. The supernatant is brought to 70% saturation with solid ammonium sulfate. The precipitate is collected by centrifugation, suspended in 50 m*M* potassium phosphate buffer (pH 7.6), and thoroughly dialyzed against 300 m*M* potassium phosphate (pH 7.6) to remove the ammonium sulfate.

Step 5. Sephadex G-200 Chromatography. A column (2.5 × 45 cm) of Sephadex G-200 is equilibrated with 50 m*M* potassium phosphate buffer (pH 7.6). The enzyme solution is applied to the column, and the column is

PURIFICATION OF PUTRESCINE CARBAMOYLTRANSFERASE FROM *Streptococcus faecalis* ATCC 11,700

Step	Total protein (mg)	Total enzyme activity (units)	Specific enzyme activity (units/mg)	Yield (%)
Extract	9790	10,400,000	1,070	100
Heat (65°)	3020	8,400,000	3,000	81
Ammonium sulfate	2280	8,150,000	3,550	78
DEAE-Sephadex chromatography	358	7,400,000	20,600	71
Sephadex G-200 chromatography	182	5,400,000	27,600	52

developed with the equilibrating buffer; 1.5-ml fractions are collected. The most active fractions of putrescine carbamoyltransferase are pooled, then concentrated by ultrafiltration.

A representative purification of putrescine carbamoyltransferase is shown in the table.

When subject to acrylamide gel electrophoresis, the final preparation still contains at least two protein bands; the major band yields 95% of the putrescine carbamoyltransferase. The second (minor) band appearing in the gels is ornithine carbamoyltransferase. The remaining ornithine carbamoyltransferase can be removed either by a second chromatography on Sephadex G-200 superfine or by chromatography on hydroxyapatite. Hydroxyapatite is equilibrated with 1 m*M* potassium phosphate buffer (pH 6.8). A column (1.6 × 20 cm) is poured and enzyme (14 mg) from step 5, dialyzed against the same 1 m*M* phosphate, is applied. The column is washed with the same buffer and eluted with a 300-ml linear gradient made of 150 ml of 1 m*M* potassium phosphate buffer (pH 6.8) in the mixing chamber, and 150 ml of the same buffer (100 m*M*). The most active fractions are pooled. This step does not increase significantly the specific activity of the enzyme, but it removes the contaminating traces of ornithine carbamoyltransferase.

Stability of Enzyme Preparation. The enzyme (2.5 mg/ml) has been stored for 3 years with no loss of activity at −20° in 50 m*M* potassium phosphate buffer (pH 7.5).

Properties

Physical Properties. The enzyme is composed of three identical subunits with a molecular weight of 40,000 as judged by treatment of the enzyme with the bifunctional reagent glutaraldehyde, followed by polyacrylamide gel electrophoresis in the presence of sodium dodecyl sulfate.

Kinetic Properties. The observed pH optimum for carbamoylputrescine synthesis depends on the concentration of substrates, putrescine, and carbamoylphosphate. A fairly broad pH optimum ranging from 7 to 9 is observed when the concentration of putrescine and carbamoylphosphate is 10 m*M*. When the putrescine concentration is lowered to 0.5 m*M*, the enzyme shows a bell-shaped curve whose pH optimum of activity is 9.0, whereas lowering both carbamoylphosphate and putrescine concentrations to 0.5 m*M* gives rise to a curve whose pH optimum is 7.8. The K_m values at pH 7.0 for putrescine and carbamoylphosphate are 2×10^{-2} m*M* and 0.17 m*M*, respectively. The arsenolytic cleavage of carbamoylputrescine gives K_m values of 10 m*M* for carbamoylputrescine and 4 m*M* for arsenate.

Substrate Specificity. Putrescine carbamoyltransferase is able to carbamoylate a number of diamines. K_m and V_{max} (relative to putrescine) are cadaverine, 7.7 m*M* (14.4); ornithine, 13 m*M* (7.4); spermidine, 19.7 m*M* (2.9); spermine, 0.35 m*M* (0.05): 1,6-diaminohexane, 6.6 m*M* (0.04); 1,3-diaminopropane, 2.9 m*M* (0.015). A few other metabolites, lysine, 2,4-diaminobutyrate, β-alanine, and ethylenediamine are also carbamoylated by putrescine carbamoyltransferase, but at a lower rate than 1,3-diaminopropane.

Inhibitors. The synthesis of carbamoylputrescine is inhibited by the product, orthophosphate (K_i = 4 m*M*), and by arsenate (K_i = 12 m*M*) and pyrophosphate (K_i = 0.06 m*M*). These inhibitions are competitive with respect to carbamoylphosphate. Putrescine analogs, such as alternative substrates listed above and glutamate, γ-aminobutyrate, norvaline, arginine, guanidobutyrate, and oxalurate, inhibit the reaction.

Presence of Putrescine Carbamoyltransferase in Other Bacteria. *Streptococcus faecalis* ATCC 14508 and *Pediococcus acidilacti* ATCC 8042 have been found to contain an induced putrescine carbamoyltransferase.

Acknowledgment

The author wishes to acknowledge permission to reproduce the table from *European Journal of Biochemistry*.

[60] Agmatine Coumaroyltransferase (Barley Seedlings)

By COLIN R. BIRD and TERENCE A. SMITH

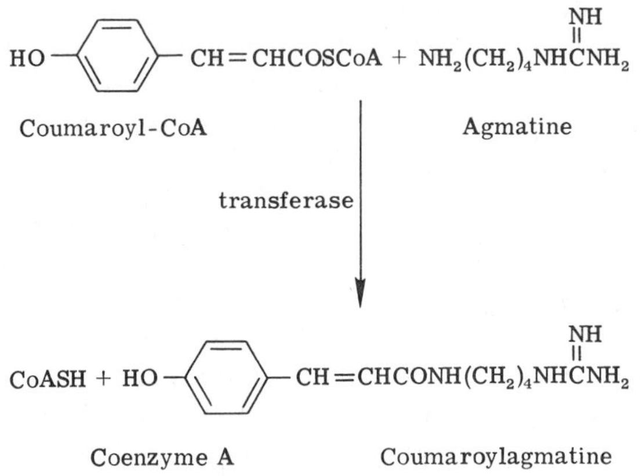

$$\text{HO}-\langle\bigcirc\rangle-\text{CH}=\text{CHCOSCoA} + \text{NH}_2(\text{CH}_2)_4\text{NHC}(=\text{NH})\text{NH}_2$$

Coumaroyl-CoA Agmatine

transferase ↓

$$\text{CoASH} + \text{HO}-\langle\bigcirc\rangle-\text{CH}=\text{CHCONH}(\text{CH}_2)_4\text{NHC}(=\text{NH})\text{NH}_2$$

Coenzyme A Coumaroylagmatine

Coumaroylagmatine and its antifungal dimers known as the hordatines, which are present in the shoots of barley seedlings, have been characterized by Stoessl,[1,2] and their distribution has been studied by Smith and Best.[3] Other similar cinnamic acid–amine conjugates are found throughout the plant kingdom.[4,5]

Following suggestions by Zenk and Gross,[6] coumaroyl coenzyme A and agmatine were found to be the substrates for an enzyme (EC 2.3.1. -; *p*-coumaroyl-CoA agmatine *N*-*p*-coumaroyltransferase) that occurs in barley seedlings, catalyzing the formation of coumaroylagmatine.[7]

Synthesis of Coumaroyl-CoA[8]

*Synthesis of N-Hydroxysuccinimide Ester of Coumaric Acid. trans-p-*Coumaric acid (15 mmol) is dissolved with heating in ethyl acetate. *N*-

[1] A. Stoessl, *Phytochemistry* **4,** 973 (1965).
[2] A. Stoessl and C. H. Unwin, *Can. J. Bot.* **48,** 465 (1970).
[3] T. A. Smith and G. R. Best, *Phytochemistry* **17,** 1093 (1978).
[4] T. A. Smith, *Prog. Phytochem.* **4,** 27 (1972).
[5] T. A. Smith, *in* "The Biochemistry of Plants" (E. E. Conn, ed.), Vol. 7, p. 249. Academic Press, New York, 1981.
[6] M. H. Zenk and G. G. Gross, *Recent Adv. Phytochem.* **4,** 87 (1972); **12,** 139 (1978).
[7] C. R. Bird and T. A. Smith, *Phytochemistry* **20,** 2345 (1981).
[8] J. Stöckigt and M. H. Zenk, *Z. Naturforsch., B: Anorg. Chem., Org. Chem.* **30,** 352 (1975).

METHODS IN ENZYMOLOGY, VOL. 94

ISBN 0-12-181994-9

Hydroxysuccinimide (15 mmol) is then added and the solution cooled to 30°. Dicyclohexylcarbodiimide (17 mmol) is added and the mixture is left for 24 hr at room temperature. The dicyclohexylurea is then removed by filtration, and the filtrate is extracted with 1 *M* sodium bicarbonate. The ethyl acetate phase is dried over sodium sulfate. After evaporation of the solvent, the ester is purified by thin-layer chromatography on silica gel G with chloroform–methanol (20 : 1) as solvent. The upper yellow band at R_f 0.46[9] is eluted with chloroform, which is then evaporated.

Thiol Ester Exchange. Nitrogen is slowly bubbled through an aqueous solution of CoA (10 μmol in 2 ml) and solid sodium bicarbonate (100 μmol) is added, followed by the *N*-hydroxysuccinimide ester of coumaric acid (50 μmol). Acetone is used to form a single phase. After storage at 4° for 24 hr, the acetone is evaporated under nitrogen at room temperature. The aqueous phase is desalted on Dowex 50-X8 (H^+ form) resin. After extraction with ethyl acetate, the aqueous phase is freeze dried and further purified by thin-layer chromatography on cellulose Whatman CC41 with butanol–acetic acid–water (4 : 1 : 5; upper phase). In this solvent coumaroyl-CoA has an R_f of 0.5. The coumaroyl-CoA is eluted with water and freeze dried.

Assay Method

Principle. The coumaroyl-CoA has λ_{max} at 333 nm and coumaroylagmatine has λ_{max} at 292 and 304 nm. The depletion of coumaroyl-CoA during the reaction can be measured by the reduction of the absorbance at 333 nm.

Reagents

p-Coumaroyl coenzyme A, 1 m*M*
Agmatine sulfate, 10 m*M*
Tris buffer, 0.1 *M*, pH 7.5, containing 1 m*M* EDTA and 25 m*M* 2-mercaptoethanol

Procedure. The incubation mixture contains 10 μl of coumaroyl-CoA, enzyme, and Tris buffer to a final volume of 1 ml (in a semimicro quartz cuvette). A reference cuvette containing the same proportion of enzyme and Tris buffer is also prepared. The two cuvettes are equilibrated in the spectrophotometer at 30° for 3 min. The reaction is initiated by the addition of 20 μl of agmatine to the assay mixture, and the depletion of *p*-coumaroyl-CoA is followed by the decline in absorbance at 333 nm on a recording spectrophotometer.

[9] J. Negrel, unpublished observation, 1981.

Purification of the Enzyme

All work is done at 0–4°. Two buffers are used, either 0.1 *M* Tris-HCl, pH 8.5 (buffer A), or 0.1 *M* Tris-HCl, pH 7.5 (buffer B). Both buffers contain 1 m*M* EDTA and 25 m*M* 2-mercaptoethanol. The purification is summarized in the table.

Preparation of Barley Extracts. Shoots (20 g) from barley seedlings grown for 3 days in the dark at 22° are blended with 60 ml of buffer A. The extract is squeezed through four layers of muslin and centrifuged at 40,000 *g* for 30 min; the precipitate is discarded.

Ammonium Sulfate Precipitation. Ammonium sulfate is dissolved in the supernatant to 20% saturation, and the mixture is stirred for 30 min. After centrifugation at 40,000 *g* for 30 min, the precipitate is discarded. Ammonium sulfate is then dissolved in the supernatant to 60% saturation, and stirred for 30 min. After centrifugation at 40,000 *g* for 30 min, the precipitate is redissolved in 4 ml of buffer B.

Gel Filtration. The enzyme is applied to a BioGel A-0.5m column (3.5 × 55 cm) that has been preequilibrated in buffer B. A flow rate of 20 ml per hour is used, and 4-ml fractions are collected. The active fractions (V_e 330 ml) are combined.

Affinity Chromatography. The enzyme is pumped onto a preequilibrated CH-Sepharose 4B-agmatine column (1.5 × 15.5 cm) that had been prepared according to the manufacturers' instructions. The inert proteins are eluted in 160 ml of buffer B, and the enzyme is eluted with buffer B, containing 3.5 m*M* agmatine, in 5-ml fractions.

Concentration. The combined active fractions are concentrated five-fold by ultrafiltration using an Amicon UM-10 membrane. The agmatine is removed by dialysis against 3 × 200 volumes of buffer B.

PURIFICATION OF AGMATINE COUMAROYLTRANSFERASE FROM BARLEY SEEDLINGS

Step	Volume (ml)	Total activity (nkat)	Total protein (mg)	Specific activity (nkat/mg protein)	Yield (%)	Purification (fold)
Crude extract (20 g fresh weight)	62	14	161	0.087	100	1
$(NH_4)_2SO_4$ (20–60%)	48	5.9	66	0.089	42	1
Gel filtration	42	5.7	23	0.25	41	3
Affinity chromatography	48	1.5	0.22	6.55	10.5	75

Properties

Specificity. The enzyme apparently is active only with agmatine as the amine donor; no activity is detected with putrescine, spermidine, spermine, cadaverine, arginine, or homoarginine as substrates. Coumaroyl-CoA, feruloyl-CoA, and caffeoyl-CoA[10] will all act as substrates.

Kinetics. At pH 7.5 and 30° the apparent K_m for coumaroyl-CoA at 20 μM agmatine is 0.74 μM. The apparent K_m for agmatine at 10 μM coumaroyl-CoA is 2.8 μM in crude extracts and 12 μM in purified extracts.

Molecular Weight. Gel filtration on BioGel A-0.5m indicates that the molecular weight is 40,000.

Distribution. There is no activity in the seed; maximum activity occurs 3–4 days after germination when grown at 22° in the dark. Five days after germination, no activity can be detected.

Stability. The purified enzyme after concentration can be stored at −40° for at least 1 month with negligible loss of activity.

pH Optimum. Activity is maximal at pH 7.5, and half-maximal activity occurs at pH 8 and pH 7.

[10] The authors are indebted to J. Negrel for the synthesis of feruloyl- and caffeoyl-CoA.

[61] γ-Glutamylamine Cyclotransferase[1] (Rabbit Kidney)

By MARY LYNN FINK and J. E. FOLK

O
‖
C — NHR
|
CH_2
|
CH_2
|
H — C — NH_2
|
CO_2H
(L)

⟶ H_2NR +

H_2C — CH_2
/ \
C CH
O⫽ \ N / \ CO_2H
H
(L)

γ-Glutamylamine cyclotransferase, an enzyme found in extracts of animal tissues and cells, bacteria, and plants, catalyzes the conversion of L-γ-glutamylamines to free amines and 5-oxo-L-proline. Among its sub-

[1] M. L. Fink, S. I. Chung, and J. E. Folk, *Proc. Natl. Acad. Sci. U.S.A.* **77,** 4564 (1980); M. L. Fink and J. E. Folk, *Mol. Cell. Biochem.* **38,** 59 (1981); *Adv. Polyamine Res.* **3,** 187 (1981).

ISBN 0-12-181994-9

strates are ε-(L-γ-glutamyl)-L-lysine, and L-γ-glutamyl derivatives of alkylamines, histamine, putrescine, and the polyamines.

Assay Methods

Principle. The assay procedure described here in detail is based on measuring the release of lysine from ε-(L-γ-glutamyl)-L-lysine. In practice, a variety of other L-γ-glutamylamines may be substituted for ε-(L-γ-glutamyl)-L-lysine. Enzyme activity is assayed by measurement of the release of free amine.

Preparation of ε-(L-γ-Glutamyl)-L-lysine. Benzyloxycarbonyl-L-glutamic acid α-benzyl ester (Vega Biochemicals) is coupled to *N*-α-benzyloxycarbonyl-L-lysine benzyl ester benzenesulfonate (Bachem) by the use of the mixed anhydride procedure.[2] The compound, α-*N*-benzyloxycarbonyl-ε-*N*-(benzyloxycarbonyl-L-γ-glutamyl)-L-lysine dibenzyl ester, is isolated from the reaction mixture and purified as described by Kornguth *et al.*[3] Protecting groups are removed from the intermediate in one step by catalytic hydrogenation. The final product, ε-(L-γ-glutamyl)-L-lysine, is recrystallized from water–ethanol.

Reagents

ε-(L-γ-Glutamyl)-L-lysine, 0.02 *M*, in water
Sodium phosphate buffer, 0.2 *M*, pH 7.5
Trichloroacetic acid, 10%

Procedure. Substrate solution (0.02 ml), buffer (0.02 ml), and enzyme solution (0.02 ml), are mixed and incubated at 37° for 15–120 min. Aliquots (0.03 ml) of the reaction mixture are removed and treated with 10% trichloroacetic acid (0.03 ml) to terminate the reaction. Precipitated protein is removed by centrifugation. Lysine is determined quantitatively, as described,[1,4] with the use of a Dionex D-400 analyzer equipped with a column (0.4 × 8 cm) of DC-6A resin (Dionex). Elution is carried out with a single buffer (sodium citrate buffer, 0.26 *M* in Na^+, pH 4.8, containing 0.55% ethanol) at a flow rate of 43.5 ml/hr, and detection is performed on stream with *o*-phthalaldehyde (flow rate 21 ml/hr). Retention times for ε-(L-γ-glutamyl)-L-lysine and lysine are 145 and 555 sec, respectively. Endogenous levels of lysine present in enzyme samples are determined in enzyme controls run without substrate. γ-Glutamylamines slowly un-

[2] J. R. Vaughan, Jr. and R. L. Osato, *J. Am. Chem. Soc.* **74,** 676 (1952).
[3] M. L. Kornguth, A. Neidle, and H. Waelsch, *Biochemistry* **2,** 740 (1963).
[4] J. E. Folk, M. H. Park, S. I. Chung, J. Schrode, E. P. Lester, and H. L. Cooper, *J. Biol. Chem.* **255,** 3695 (1980).

dergo spontaneous cyclization with the release of free amines. Therefore, for each set of determinations, a substrate control is run.

Alternative Methods. L-γ-Glutamylamines, such as L-γ-glutamylputrescine (available from Vega Biochemicals) or L-γ-glutamyl-*n*-butylamine,[5] may be used as substrates in the assay described above. The latter compound is especially useful as a substrate in assaying for enzyme activity in crude samples. The amine product *n*-butylamine is not present in extracts of tissues and can be easily separated from endogenous amines by ion-exchange chromatography.

L-γ-Glutamylmonodansylcadaverine, N^5-{5-[5-(dimethylamino)-1-naphthalenesulfonylamido]pentyl}-L-glutamine, is a useful substrate for testing for, and locating, fractions of enzymic activity during enzyme purification. Separation of fluorescent product from substrate is accomplished by thin-layer chromatography on polyamide sheets[6] using dilute aqueous pyridine acetate buffer, pH 5.65.

Enzyme Purification

Step 1. Preparation of Tissue Extracts. Freshly excised or frozen kidneys from New Zealand White rabbits are diced and homogenized for 3–5 min in 2 volumes of 0.25 *M* sucrose with a Polytron homogenizer (Brinkmann). The homogenate is centrifuged for 1 hr at 100,000 *g* and the supernatant is filtered through cheesecloth. This step and subsequent ammonium sulfate fractionations are carried out at 4°. Chromatography steps are conducted at room temperature.

Step 2. Chromatography on DEAE-Cellulose. The supernatant obtained from 20 g of kidney tissue is applied to a column (3 × 11 cm) of DEAE-cellulose. The column is washed with 200 ml of 5 m*M* sodium phosphate buffer at pH 7.5 (buffer A), and the enzyme is eluted at a flow rate of 5 ml/min with a 300-ml linear gradient established between buffer A and 100 m*M* sodium phosphate buffer at pH 7.5. The fractions containing activity (eluted at about 50 m*M* buffer) are combined.

Step 3. First Ammonium Sulfate Fractionation and Gel Filtration. Solid $(NH_4)_2SO_4$ (31 g/100 ml) is added, and the precipitate obtained upon centrifugation for 25 min at 20,000 *g* is discarded. Solid $(NH_4)_2SO_4$ (22 g/100 ml) is added to the supernatant, and the precipitate collected by centrifugation is dissolved in a minimal volume of buffer A. This solution is chromatographed on a column (1.5 × 45 cm) of Sephadex G-100 in

[5] This compound is prepared by a procedure similar to the one described in this chapter for ε-(L-γ-glutamyl)-L-lysine.

[6] L. Lorand and L. K. Campbell, *Anal. Biochem.* **44,** 207 (1971).

PURIFICATION OF γ-GLUTAMYLAMINE CYCLOTRANSFERASE[a]

Step	Volume (ml)	Protein (mg/ml)	Specific activity (units/mg)	Yield (%)
1. Homogenate	52	59	0.13	100
Supernatant	35	26	0.25	57
2. DEAE-chromatography	60	6	0.37	33
3. First ammonium sulfate fractionation and gel filtration	11	0.25	10	7
4. Second ammonium sulfate fractionation and gel filtration	8	0.11	32	7

[a] Results were obtained starting with 20 g of rabbit kidney.

buffer A at a flow rate of 12 ml/hr. The peak of activity emerges at 52 ml, near the end of the descending shoulder of a large protein peak. Only those fractions of specific activity >9 units per milligram of protein are combined and used in the next step.

Step 4. Second Ammonium Sulfate Fractionation and Gel Filtration. Solid $(NH_4)_2SO_4$ (66 g/100 ml) is added, and the precipitate obtained is dissolved in a minimal volume of buffer A. This solution is chromatographed on a column (1.25 × 40 cm) of Sephadex G-50 in buffer A at a flow rate of 12 ml/hr. The peak of activity emerges at 23 ml. Fractions containing activity are combined.

A representative purification of γ-glutamylamine cyclotransferase is given in the table.

Definition of Units and Specific Activity. A unit is defined as the amount of enzyme that produces 1 μmol of lysine in 1 hr in the assay described. Specific activity is defined as units per milligram of protein. Protein is determined by the method of Lowry *et al.*[7] using bovine serum albumin as standard.

Properties

Enzyme prepared as described above shows no noticeable loss in activity over a period of several weeks when stored at −20°. γ-Glutamylamine cyclotransferase exhibits a broad pH optimum between 6.8 and 8.5. It gradually loses activity below pH 6.8 and above pH 8.5.

[7] O. H. Lowry, N. J. Rosebrough, A. L. Farr, and R. Randall, *J. Biol. Chem.* **193,** 265 (1951).

Substrate Specificity. The enzyme is inactive toward L-γ-glutamyl-α-amino acids, with the exception of L-γ-glutamylglycine, for which it displays weak activity. Thus, the specificity of the enzyme is distinct from that of γ-glutamylamino acid cyclotransferase, which acts only on γ-glutamylamino acids.[8] In addition to ε-(L-γ-glutamyl)-L-lysine, a wide variety of other L-γ-glutamylamines are substrates. Included are L-γ-glutamyl derivatives of methylamine, ethylamine, isopropylamine, isobutylamine, *n*-butylamine, histamine, α-*N*-acetyllysine methyl ester, and β-alanine. Both mono- and bis-γ-glutamyl derivatives of putrescine, spermidine, and spermine are substrates.

In contrast to the broad specificity of the enzyme toward the amine portion of substrates, the L-γ-glutamyl moiety cannot be replaced by L-β-aspartyl, D-γ-glutamyl, β-aminoglutaryl, glutaryl, acetyl, or formyl groups.

Identification of L-γ-Glutamylamines Using γ-Glutamylamine Cyclotransferase

Current methods for identifying L-γ-glutamylamines are based on their chromatographic properties and on the finding of equimolar amounts of glutamic acid and amine in their acid hydrolyzates. As additional proof of structure, we routinely employ γ-glutamylamine cyclotransferase in a manner complementary to the above methods. The enzyme requirement for the L-γ-glutamyl moiety allows identification of amine derivatives that act as substrates for the enzyme as L-γ-glutamylamines.

The amine derivative is separated from free amine by ion-exchange chromatography. Fractions containing the amine derivative are collected, adjusted to pH 7.5, and, without further purification, treated with enzyme. Detection of free amine with concomitant disappearance of amine derivative upon rechromatography provides evidence that the derivative is an L-γ-glutamylamine. The amide bond in γ-glutamylamines is resistant to hydrolysis by proteolytic enzymes. Thus, amines bound to protein through γ-glutamyl linkages can be analyzed as described above after isolation by ion-exchange chromatography from exhaustive proteolytic digests of protein.

[8] M. Orlowski, P. G. Richman, and A. Meister, *Biochemistry* **8,** 1048 (1969); M. Orlowski and A. Meister, *J. Biol. Chem.* **248,** 2836 (1973).

Section X

Metabolism of 5′-Methylthioadenosine and 5-Methylthioribose

[62] Purification and Properties of 5′-Methylthioadenosine Phosphorylase from *Caldariella acidophila*[1]

By VINCENZO ZAPPIA, MARIA CARTENÌ-FARINA, GIOVANNA ROMEO, MARIO DE ROSA, and AGATA GAMBACORTA

Methylthioadenosine + $P_i \rightleftharpoons$ methylthioribose 1-phosphate + adenine

5′-Methylthioadenosine nucleosidase represents the main enzyme related to the degradation of 5′-methylthioadenosine (MTA), a product of *S*-adenosylmethionine (AdoMet) metabolism. The thioether is enzymatically formed from AdoMet by at least four independent pathways.[2–5] Two classes of MTA nucleosidases have been reported in the literature: a hydrolytic nucleosidase, described in several prokaryotes,[2,6] that cleaves 5′-methylthioadenosine into adenine and 5-methylthioribose (MTR), and a phosphorolytic nucleosidase, purified from various mammalian tissues,[3,4] that cleaves MTA into adenine and methylthioribose 1-phosphate (MTR-1-P).

The physiological significance of the cleavage of MTA is probably related to the removal of the thioether, which in turn exerts a significant inhibition on several methyl transfer reactions[7,8] and an antiproliferative effect on stimulated human lymphocytes[5] and virally transformed mouse fibroblasts.[9] The reaction also plays a primary role in the salvage and recycling of adenine and methionine[10] (Scheme 1).

This article reports the purification and properties of 5′-methylthioadenosine phosphorylase from the soluble fraction of *Caldariella acidophila,* an extreme thermophilic bacterium living optimally at 87°.[11] The

[1] M. Cartenì-Farina, A. Oliva, G. Romeo, G. Napolitano, M. De Rosa, A. Gambacorta, and V. Zappia, *Eur. J. Biochem.* **101,** 317 (1979).

[2] A. J. Ferro, A. Barrett, and S. K. Shapiro, *Biochim. Biophys. Acta* **438,** 487 (1976).

[3] G. Cacciapuoti, A. Oliva, and V. Zappia, *Int. J. Biochem.* **9,** 35 (1978).

[4] V. Zappia, A. Oliva, G. Cacciapuoti, P. Galletti, G. Mignucci, and M. Cartenì-Farina, *Biochem. J.* **175,** 1043 (1978).

[5] A. J. Ferro, *in* "Transmethylation" (E. Usdin, R. T. Borchardt, and C. R. Creveling, eds.), p. 117. Elsevier, Amsterdam, 1979.

[6] J. A. Duerre, *J. Biol. Chem.* **237,** 3737 (1962).

[7] V. Zappia, C. R. Zydeck-Cwick, and F. Schlenk, *J. Biol. Chem.* **244,** 4499 (1969).

[8] V. Zappia, M. Cartenì-Farina, G. Cacciapuoti, A. Oliva, and A. Gambacorta, *in* "Natural Sulfur Compounds: Novel Biochemical and Structural Aspects" (D. Cavallini, G. E. Gaull, and V. Zappia, eds.), p. 133. Plenum, New York, 1980.

[9] A. E. Pegg, R. T. Borchardt, and J. K. Coward, *Biochem. J.* **194,** 79 (1980).

[10] H. G. Williams-Ashman, J. Seidenfeld, and P. Galletti, *Biochem. Pharmacol.* **31,** 277 (1982).

[11] M. De Rosa, A. Gambacorta, and J. D. Bu'Lock, *J. Gen. Microbiol.* **86,** 156 (1975).

ISBN 0-12-181994-9

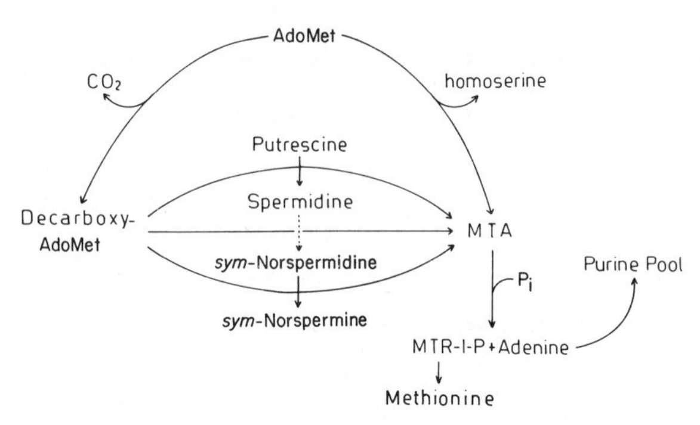

SCHEME 1. Pathway of polyamine biosynthesis in *Caldariella acidophila*. Decarboxy-AdoMet = *S*-adenosyl-(5′)-3-methylthiopropylamine; CH_3S-Ado = 5′-methylthioadenosine (MTA); CH_3S-Rib-1-P = 5-methylthioribose 1-phosphate (MTR-1-P). From Carteni-Farina *et al.*[1]

enzyme shows a high degree of thermostability and thermophilicity, its temperature optimum being at 95°. For its phosphorolytic mechanism, the enzyme from *C. acidophila* resembles MTA phosphorylase purified from mammalian tissues[3,4] rather than the enzyme from other prokaryotes.[2,6] In this respect it is important to note that many archaebacteria have developed biochemical strategies that are more closely related to the eukaryotic cells than to the eubacteria.[12]

Assay Method

Principle. The routine assay for enzymatic activity is performed at 70° for 10 min by determining the rate of release of 5-[*methyl*-^{14}C]methylthioribose from 5′-[*methyl*-^{14}C]methylthioadenosine[3]. With these assay conditions the chemical hydrolysis of substrate is less than 5% and is thus negligible.

Reagents

Phosphate buffer, 0.8 *M*, pH 7.2, at 70° ($\Delta pH/\Delta t$ is about 0.003)
5′-Methylthioadenosine, 6 m*M* (specific activity: 10^5 counts min^{-1} μmol^{-1})
Trichloroacetic acid, 0.3 *M*
Scintillation fluid: Instagel (Packard)

Procedure. The assay medium contains 80 μmol of potassium phosphate buffer, pH 7.2 (100 μl), 0.3 μmol of 5′-[*methyl*-^{14}C]methylthioadenosine (50 μl), and an appropriate amount of the enzyme in a total

[12] C. R. Woese and R. Gupta, *Nature (London)* **289,** 95 (1981).

volume of 0.4 ml. The incubation is carried out for 10 min at 70° in 1-ml sealed glass vials in order to reduce evaporation. The reaction is stopped by the addition of 0.6 ml of 0.3 *M* trichloroacetic acid, and the precipitate is removed by centrifugation (1000 *g*). The supernatant is then applied to a Dowex 50, H^+ column (0.3 × 4 cm) equilibrated with 0.2 *M* trichloroacetic acid; MTR-1-P is eluted with 1 ml of 0.2 *M* trichloroacetic acid, and MTA is quantitatively retained by the resin.

Definition of Enzyme Activity. A unit of enzyme is defined as the amount that catalyzes the conversion of 1 μmol of substrate per minute.

Purification Procedure

All purification steps are performed at room temperature.

Growth of Caldariella acidophila. *Caldariella acidophila,* strain MT-4, has been isolated from an acidic hot spring in Agnano, Naples.[11] The bacteria are grown at 87° in 25-liter batches in a fermentor with slow mechanical agitation and an aeration flux of 2.5–3 liters/min. The culture medium contains, per liter, 1 g of yeast extract, 1 g of casamino acids, 3.1 g of KH_2PO_4, 2.5 g of $(NH_4)_2SO_4$, 0.2 g of $MgSO_4 \cdot 7H_2O$, and 0.25 g of $CaCl_2 \cdot H_2O$. The pH is adjusted to 3.0 with 0.1 *M* H_2SO_4. The culture vessel is inoculated by adding 3 liters of 12-hr broth culture. The doubling time of the microorganism in these conditions is 7.5 hr. The cells are then harvested in the late-exponential phase (25 hr of incubation) by continuous-flow centrifugation in Alfa-Laval Model LKB 102 B-25 separator.

The yield from a typical 25-liter fermentation is approximately 30 g of wet cells.

Bacteria are washed three times with isotonic saline solution and stored at −20° without any loss of enzymatic activity.

Preparation of Extract. Twenty grams of frozen bacteria are washed with 25 ml of 0.01 *M* Tris-HCl buffer, pH 7.5, and collected by centrifugation. The pellet is ground with 35 g of glass beads and 35 ml of buffer Tris-HCl (0.01 *M*, pH 7.5) in the stainless steel chamber of a Sorvall Omni-Mixer for 5 min at full speed; 0.5 ml of a freshly prepared solution of DNase (1 mg/ml) is added to the homogenate, and the mixture is incubated for 30 min at 30° with occasional stirring. The beads are removed by centrifugation at 650 *g* for 10 min and centrifuged again at 1500 *g* for 10 min. The two supernatants are combined and centrifuged at 35,000 *g* for 45 min.

pH Precipitation. The high-speed supernatant 50 ml is adjusted to pH 4.8 with 0.2 *M* HCl. The mixture is stirred for 15 min and the precipitate obtained by centrifugation at 16,000 *g* for 10 min was discarded. After neutralization the supernatant (40 ml), containing the MTA phosphorylase activity, is dialyzed overnight against several changes of 5 mM potassium phosphate buffer, pH 8.

DEAE-Cellulose Chromatography. The dialyzed enzyme (43 ml) is loaded into DEAE-cellulose column (3 × 10 cm) equilibrated with 5 mM potassium phosphate buffer (pH 8). After washing with 150 ml of the same buffer, the elution is performed with a linear gradient (400 ml) of KCl from 0 to 0.3 M in the same buffer. Active fractions (70 ml) are combined and dialyzed overnight against 20 volumes of 20 mM potassium phosphate buffer (pH 7.2).

Hydroxyapatite Chromatography. The dialyzed enzyme solution (74 ml) is applied to a hydroxyapatite column (3 × 7 cm) previously equilibrated with 20 mM potassium phosphate buffer (pH 7.2). After washing with 150 ml of the same buffer, the enzyme is eluted with a linear gradient of 500 ml of potassium phosphate buffer at pH 7.2 (20 to 100 mM).

5′-Methylthioadenosine phosphorylase activity is associated with a single peak eluted in presence of 70 mM phosphate. Fractions with higher activity (60 ml) are combined, concentrated under vacuum, and dialyzed overnight against 25 mM potassium phosphate buffer (pH 7.2).

Sephadex G-150 Gel Filtration. The concentrated enzyme (5 ml) is loaded onto a Sephadex G-150 column (2.5 × 100 cm) equilibrated with 25 mM potassium phosphate buffer (pH 7.2) and eluted as sharp single peak (40 ml) with the same buffer at a rate of 0.5 ml/min.

Preparative Isoelectric Focusing. The isoelectric focusing experiments are performed with an LKB 8102 column, using carrier ampholytes (Ampholine), pH range 4–6, purchased from the same source. Column assembly and the preparation of the sucrose density gradient are according to LKB instructions sheet, positioning the anode at the bottom. The enzyme sample from Sephadex G-150, concentrated under vacuum (15 ml), is layered in the middle of the column by a syringe filled with a Teflon capillary tube. The electrofocusing experiment is performed for 120 hr at 2 ± 0.5° by applying a maximum power of 6 W. Two-milliliter fractions are then collected from the bottom of the column; the enzyme activity and the pH value of each fraction are determined. The pH is measured at 25° with a Radiometer pH meter Model TTT-1, equipped with the scale expander 630 T. The electrofocusing pattern is obtained by plotting enzyme activity against elution volume. The active fractions are pooled (6 ml) and extensively dialyzed against phosphate buffer (0.25 M, pH 7.2); the specific activity is determined. It should be pointed out that in this step the recovery of the enzyme activity is almost complete.

A typical purification is summarized in the table.

The recovery of enzyme activity varies in different enzyme preparations from 30 to 35%, and the final preparation has a specific activity of 145–150 units per milligram of protein.

PURIFICATION OF 5′-METHYLTHIOADENOSINE PHOSPHORYLASE

Step	Total activity (units)[a]	Specific activity (units/mg)	Purification (fold)	Yield (%)
Crude extract	2570	0.0081	1.0	100
Precipitation, pH 4.7	2500	0.0113	1.4	95
DEAE-cellulose	1850	0.053	6.5	72
Hydroxyapatite	1130	0.190	23.1	44
Sephadex G-150	870	2.07	253	34
Isoelectric focusing	830	2.46	302	32

[a] Units = micromoles of 5′-methylthioadenosine converted per minute.

The homogeneity of the purified enzyme is evaluated by disk electrophoresis on a polyacrylamide gel according to the procedure of Davis.[13] The enzyme activity is associated with the only detectable band at 2 cm from the start (the front 5 cm).

Properties

Stability. The reaction rate is markedly increased by temperature elevation, exhibiting a maximum at 95°, as reported in Fig. 1. At 86° the reaction rate is one-half of the maximal value, and no activity is detect-

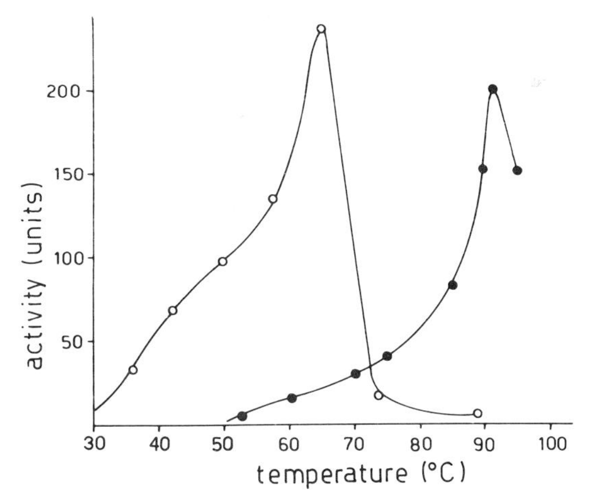

FIG. 1. Effect of temperature on activity of 5′-methylthioadenosine phosphorylase from human placenta[3] (○) and *Caldariella acidophila* (●). The assay was performed as indicated under Materials and Methods. From Carteni-Farina *et al.*[1]

[13] B. J. Davis, *Ann. N.Y. Acad. Sci.* **121,** 404 (1964).

	R^1	R^2
(I)	$-NH_2$	$-S-CH_3$
(II)	$-OH$	$-S-CH_3$
(III)	$-NH_2$	$-\overset{+}{S}(CH_3)_2$
(IV)	$-NH_2$	$-S-CH_2-CH_2-CH_2-CH_3$
(V)	$-NH_2$	$-S-CH_2-CH(CH_3)_2$
(VI)	$-OH$	$-S-CH_2-CH(CH_3)_2$
(VII)	$-NH_2$	$-S-CH_2-CH_2-CH(NH_2)-COOH$
(VIII)	$-NH_2$	$-S-CH_2-CH_2OH$
(IX)	$-NH_2$	$-OH$

Substrate analogs	Concentration (m*M*)	Enzyme activity (units/mg)	Relative activity (%)
5′-Methylthioadenosine (I)	0.75	135	100.0
5′-Methylthioinosine (II)	0.75	12	8.8
5′-Dimethylthioadenosine sulfonium salt (III)	1.00	n.d.	–
5′-*n*-Butylthioadenosine (IV)	0.75	126	93.3
5′-Isobutylthioadenosine (V)	0.75	131	97.0
5′-Isobutylthioinosine (VI)	1.00	11	8.1
Adenosylhomocysteine (VII)	1.00	n.d.	–
5′-Thioethanoladenosine (VIII)	1.00	n.d.	–
Adenosine (IX)	1.00	n.d.	–

FIG. 2. Substrate specificity of 5′-methylthioadenosine phosphorylase. Results are mean values of three separate experiments. n.d., not detected. Modified from Carteni-Farina *et al.*[1]

able at 40°. Only a low decrease of enzyme activity is observable at the maximum assayed temperature of 100°.

This remarkable thermophilicity is similar to that observed for many enzymatic proteins from thermophilic organisms.[14] In fact, the optimum temperature for the enzymatic activity is about 10° higher than the optimal growth temperature of the microorganism. In the same figure is reported the thermal profile of MTA phosphorylase purified from human placenta,[3] which shows an optimum of activity at 67°. The enzyme from *C. acidophila* also shows an elevated thermostability: no loss of activity is observable even after preincubation for 1 hr at 100°.

The purified enzyme shows no loss in activity after storage for several months at −20°.

The thermostability of the MTA phosphorylase is uncommon even if

[14] H. Zuber, "Enzymes and Proteins from Thermophilic Microorganisms." Birkaueser, Basel, 1976.

compared to other enzymes from thermophiles in that the enzyme is stable also at low protein concentration and the highest level of purity.[15]

Specificity and Kinetics. The specificity of the enzyme is rather strict: among the structural analogs assayed, reported in Fig. 2, only 5′-isobutylthioadenosine and 5′-*n*-butylthioadenosine equal the activity of the natural substrates; the modification of the purine moiety results in a resistance to enzymatic hydrolysis: only about 8% of activity is retained in 5′-methylthioinosine and 5′-isobutylthioinosine. Replacement of the bivalent sulfur of the thioether by a charged sulfonium group results in a loss of activity. The positively charged group probably prevents the catalytic interaction with the enzyme. The replacement of the methyl group by polar substituents such as —OH, $—CH_2CH_2OH$, or $—CH_2CH_2CH(NH_2)COOH$ results in a loss of enzymatic activity. The compounds which were inactive as substrates failed to show any inhibitory activity when assayed as inhibitors.

The mechanism of the reaction is sequential according to the kinetic data.[1]

The apparent K_m values are 0.095 m*M* for 5′-methylthioadenosine and 6.1 m*M* for phosphate.

[15] M. R. Heinrich, "Extreme Environments: Mechanisms of Microbial Adaptation," p. 147. Academic Press, New York, 1976.

[63] 5-Methylthioribose Kinase (*Enterobacter aerogenes*)

By ADOLPH J. FERRO and KEVIN S. MARCHITTO

$$\text{MTR} + \text{ATP} \xrightarrow{Mg^{2+}} \text{MTR-1-P} + \text{ADP}$$

5-Methylthioribose (MTR) is synthesized from 5′-methylthioadenosine (MTA) in most prokaryotes[1,2] and plants[3] via a nucleosidic cleavage that also liberates free adenine. MTR kinase catalyzes the subsequent ATP-dependent phosphorylation of MTR to form 5-methylthioribose 1-phosphate (MTR-1-P) and ADP.[4] MTR kinase activity has been found in extracts prepared from *Enterobacter aerogenes*[4] and from several plant

[1] J. A. Duerre, *J. Biol. Chem.* **237,** 3737 (1962).
[2] A. J. Ferro, A. Barrett, and S. K. Shapiro, *Biochim. Biophys. Acta* **438,** 487 (1976).
[3] A. B. Guranowski, P. K. Chiang, and G. L. Cantoni, *Eur. J. Biochem.* **114,** 293 (1981).
[4] A. J. Ferro, A. Barrett, and S. K. Shapiro, *J. Biol. Chem.* **253,** 6021 (1978).

ISBN 0-12-181994-9

tissues.[4a] MTR-1-P is also synthesized in animals by the action of MTA phosphorylase, which directly cleaves MTA to MTR-1-P and adenine.[5–8] Regardless of its route of synthesis, MTR-1-P appears to be an intermediate in the recycling of MTA back into methionine.[4,9–11]

Assay Method

Principle. The radiochemical assay measures the conversion of MTR to MTR-1-P by chromatographic separation of the reactant and product on Dowex 1-X8 columns and is based on the procedure used for isolating ribose 1-phosphate.[12]

Procedure. The reaction mixture contains imidazole-HCl (50 m*M*, pH 7.3), ATP (1 m*M*), $MgSO_4$ (5 m*M*), dithiothreitol (10 m*M*), and [$^{14}CH_3$]methylthioribose in a final volume of 0.2 ml. After a preincubation at 37° for 10 min, an appropriate amount of enzyme is added. After a 20-min incubation at 37°, the reaction is terminated by the addition of 3 volumes of ethanol, and precipitated protein is removed by centrifugation. A 0.2-ml aliquot of the supernatant fluid is applied to a Dowex 1-X8 formate column (0.7 × 2 cm) and eluted with 12 ml of 0.01 *N* sodium formate, pH 5.0, followed by 12 ml of 0.75 *N* sodium formate, pH 5.0. The 0.01 *N* eluate contains [$^{14}CH_3$]MTR, and the 0.75 *N* eluate contains the product [$^{14}CH_3$]MTR-1-P. Recovery of radioactivity applied to the column is 96–100%. The amount of radioactivity in a 3-ml aliquot of the 0.75 *N* eluate is determined in a liquid scintillation spectrometer with 0.4% 2,5-diphenyloxazole (PPO) in toluene–Triton X-100 (2 : 1).

Definition of Unit and Specific Activity. One unit is the amount of enzyme that catalyzes the formation of 1 μmol of MTR-1-P per minute at 37°. Specific activity is expressed as units per milligram of protein. Protein is determined by the method of Lowry *et al.*[13]

Preparation of [$^{14}CH_3$]MTR. [$^{14}CH_3$]MTR is prepared from

[4a] M. M. Kushad, D. G. Richardson, and A. J. Ferro, *Biochem. Biophys. Res. Commun.* **108,** 167 (1982).

[5] A. E. Pegg and H. G. Williams-Ashman, *Biochem. J.* **115,** 241 (1969).

[6] D. L. Garbers, *Biochim. Biophys. Acta* **523,** 82 (1978).

[7] G. Cacciapuoti, A. Oliva, and V. Zappia, *Int. J. Biochem.* **9,** 35 (1978).

[8] A. J. Ferro, A. A. Vandenbark, and K. Marchitto, *Biochim. Biophys. Acta* **588,** 294 (1979).

[9] S. K. Shapiro and F. Schlenk, *Biochim. Biophys. Acta* **633,** 176 (1980).

[10] P. S. Backlund, Jr. and R. A. Smith, *J. Biol. Chem.* **256,** 1533 (1981).

[11] S. K. Shapiro and A. Barrett, *Biochem. Biophys. Res. Commun.* **102,** 302 (1981).

[12] P. E. Plesner and H. Klenow, this series, Vol. 3, p. 181.

[13] O. H. Lowry, N. J. Rosebrough, A. L. Farr, and R. J. Randall, *J. Biol. Chem.* **193,** 265 (1951).

[$^{14}CH_3$]MTA obtained by the acid hydrolysis of *S*-adenosyl-L-[$^{14}CH_3$]methionine.[14–16]

Purification Procedure[4]

Unless otherwise stated, all the following purification steps are carried out at 4° and all buffers contain 5 m*M* 2-mercaptoethanol.

Growth of Cells. Enterobacter aerogenes (ATCC 8724) is grown in a mineral salts–glucose medium with aeration at 37° to mid-log phase (18.2 g wet weight of cells). The cells are harvested by centrifugation and stored at −20°.

Preparation of Cell Extracts. The cells are suspended in 3 volumes (milliliters/wet weight) of 0.05 *M* potassium phosphate buffer, pH 7.3. The suspended cells are passed through a French pressure cell twice at 20,000 psi and cellular debris is removed by centrifugation at 30,000 *g* for 45 min.

Ammonium Sulfate Fractionation. To 65 ml of crude extract (31.4 mg of protein per milliliter) a saturated solution of ammonium sulfate, pH 7.5, is added slowly to a final concentration of 50% saturation. The mixture is stirred for an additional 20 min and then centrifuged at 30,000 *g* for 40 min. The supernatant fluid is discarded and the precipitate is dissolved in 0.05 *M* potassium phosphate buffer, pH 7.3.

Sephadex G-200 Gel Filtration. The protein solution from the preceding step is applied in two aliquots to a Sephadex G-200 column (2.7 × 85 cm) set up for ascending flow and equilibrated with 0.05 *M* potassium phosphate buffer, pH 7.3. Proteins are eluted from the column with the same buffer, and the fractions containing MTR kinase activity are pooled. The kinase activity is well separated from proteins eluting at the void volume.

DEAE-Cellulose Chromatography. The combined Sephadex G-200 fractions (230 mg of protein) are adsorbed on a DEAE-cellulose column (1.5 × 20 cm) that had been equilibrated with 0.05 *M* potassium phosphate buffer, pH 7.3. The column is eluted with a 500-ml linear gradient of 0.05 to 0.5 *M* potassium phosphate buffer, pH 7.3. The active fractions, which are eluted by approximately 0.2 *M* potassium phosphate, are pooled and concentrated by the addition of solid ammonium sulfate to 60% saturation.

A summary of the purification procedure is presented in the table.

[14] F. Schlenk, C. R. Zydek-Cwick, and J. L. Dainko, *Biochim. Biophys. Acta* **320,** 357 (1973).

[15] F. Schlenk, C. R. Zydek-Cwick, and N. K. Hutson, *Arch. Biochem. Biophys.* **142,** 144 (1971).

[16] See also this volume [64].

PURIFICATION OF 5-METHYLTHIORIBOSE FROM *Enterobacter aerogenes*

Fraction	Volume (ml)	Total protein (mg)	Total units	Specific activity (mU/mg)	Yield (%)
1. Extract supernatant	65	2040	1.50	0.73	100
2. Ammonium sulfate	28	1230	2.24	1.83	150
3. Sephadex G-200	110	230	0.90	3.90	60
4. DEAE-cellulose	31	21	0.60	28	40

Properties[4]

Stability. At a protein concentration of at least 5 mg/ml the enzyme can be stored either at 4° or −20°. Stability at −20° is enhanced by the addition of glycerol at a final concentration of 20%.

pH Optimum. The pH optimum of the enzyme is approximately 7.3. Enzyme activity is the highest in either potassium phosphate or imidazole-HCl buffer.

Requirements for Cations and Sulfhydryl Reducing Agents. The enzyme has absolute requirements for both a divalent cation and sulfhydryl reagent for activity. These requirements are optimally met with 5 mM Mg^{2+} and 20 mM dithiothreitol, respectively. Mg^{2+} can be partially (20%) replaced by 10 mM Mn^{2+}, but not by Ca^{2+}, whereas dithiothreitol may be partially (70%) replaced by 20 mM 2-mercaptoethanol and to a lesser extent by 20 mM glutathione (44%).

Specificity for Phosphate Donor. The nucleotide specificity is very narrow. GTP and UTP are not active as substrates, and CTP results in only 20% of the activity seen with ATP.

Specificity for Phosphate Acceptor. In addition to MTR, the following 5-modified ribose compounds also are substrates: 5-ethylthioribose, 5-propylthioribose, 5-isopropylthioribose, 5-butylthioribose, and 5-isobutylthioribose.

Kinetic Constants. The K_m value for MTR is 8.1×10^{-6} M, and the K_m for ATP is 7.4×10^{-5} M.

Regulatory Properties. None of the following sugars or sugar phosphates have any effect as inhibitors of the reaction: ribose, glucose, fructose, 2-deoxyribose 1-phosphate, ribose 1-phosphate, ribose 5-phosphate, glucose 1-phosphate, glucose 6-phosphate, fructose 1-phosphate, and fructose 6-phosphate. The presence of ADP in the reaction at equimolar concentration with ATP results in 52% inhibition of the rate.

[64] 5′-Methylthioadenosine Nucleosidase (*Lupinus luteus* Seeds)

By ANDRZEJ B. GURANOWSKI, PETER K. CHIANG, and GIULIO L. CANTONI

5′-Methylthioadenosine + H_2O → 5′-methylthioribose + adenine

Assay Method

Principle. Nucleosidase activity can be determined by measurement of radioactive adenine formation from 5′-methylthiol[*adenine*-U-^{14}C]-adenosine.

Preparation of Radioactive 5′-Methylthioadenosine. 5′-Methylthio[*adenine*-U-^{14}C]adenosine was prepared as follows. To the reaction mixture containing 50 m*M* HEPES[1] buffer, pH 7.6, 100 m*M* KCl, 5 m*M* $MgCl_2$, 10 m*M* L-methionine, and 72 μ*M* [*adenine*-U-^{14}C]adenosine triphosphate, excess AdoMet[1] synthetase from yeast was added. (A purification procedure of the yeast AdoMet synthetase is described below.) After a 3-hr incubation at 37°, the mixture was heated at 100° to degrade the labeled AdoMet to MeSAdo,[1] which was next purified by chromatography on cellulose plates developed in distilled water. The labeled MeSAdo was stored with 10 m*M* dithiothreitol. In the chromatographic system used, MeSAdo migrated as a sharp band ($R_f = 0.45$), and was readily separated from unconverted ATP and methionine, which migrated with the front, as well as from undegraded AdoMet ($R_f = 0.03$–0.08) and traces of adenine ($R_f = 0.28$). 5′-[*methyl*-^{14}C]Methylthioadenosine was prepared in a similar way, and in this case the incubation mixture contained 5 m*M* ATP and 400 μ*M* L-[*methyl*-^{14}C]methionine.

Reagents and Chemicals

Buffers A and B: 5% glycerol in 10 or 30 m*M* potassium phosphate (pH 6.8), respectively

Buffer C: 50 m*M* Bicine[1]–KOH (pH 8.3) and 5% glycerol

Thin-layer chromatographic plates and aniline phthalate spray (E. Merck, Darmstadt, Federal Republic of Germany)

[1] Abbreviations: MeSAdo, 5′-deoxy-5′-methylthioadenosine; AdoHcy, *S*-adenosylhomocysteine; AdoMet, *S*-adenosylmethionine; HEPES, 4-(2-hydroxyethyl)-1-piperazine-ethanesulfonic acid; Bicine, *N,N*-bis(2-hydroxyethyl)glycine; DEAE, diethylaminoethyl; EDTA, ethylenediaminetetraacetic acid.

METHODS IN ENZYMOLOGY, VOL. 94

ISBN 0-12-181994-9

AdoHcy[1]-Sepharose prepared by the procedure of Kim *et al.*[2]
Hydroxyapatite (BioGel HTP; Bio Rad Laboratories)

Enzymes. AdoMet synthetase II was purified by a modification of the procedure described by Chiang and Cantoni.[3] The enzyme extract was processed essentially as described up to step 5, which was the first DEAE-cellulose[1] column, and then the fractions containing the highest specific activity of AdoMet synthetase II were concentrated to 5 ml by Aquacide III. The concentrated enzyme solution was applied onto a 400-ml column (2.6 × 86 cm) of Sephadex G-100, equilibrated with 20 m*M* potassium phosphate (pH 6.8), 5 m*M* dithiothreitol, 0.5 m*M* EDTA.[1] The peak fractions from the Sephadex G-100 column were next adsorbed onto a 40-ml column (2.5 × 7.5 cm) of hydroxyapatite that had been equilibrated with 20 m*M* potassium phosphate (pH 6.8), 5 m*M* dithiothreitol, and 0.5 m*M* EDTA. The column was washed with the same buffer, then AdoMet synthetase II was eluted with a 500-ml linear gradient of 20 to 180 m*M* potassium phosphate (pH 6.8), containing 5 m*M* dithiothreitol and 0.5 m*M* EDTA. The fractions containing the highest enzyme activity were eluted at about 6.5 mmho. They were concentrated to 4 ml by Aquacide III, and applied onto a Sephadex G-150 column (2.6 × 60 cm) that had been equilibrated with 50 m*M* Tris (pH 7.8), 0.5 m*M* EDTA, 40 m*M* KCl, and 5 m*M* dithiothreitol. AdoMet synthetase was chromatographed with the same buffer. The specific activity of the AdoMet synthetase II purified at this stage was about 30 units/mg, each unit being as defined previously,[3] and the enzyme was homogeneous by polyacrylamide gel electrophoresis.

Procedure. The standard incubation mixture contained, in a final volume of 50 μl: 50 m*M* Bicine–KOH (pH 8.3), 100 μ*M* radioactive MeSAdo, and the nucleosidase. Incubation at 37° was for 2–30 min depending on enzyme activity, and the reaction was stopped by heating at 100° for 1 min. Aliquots of the reaction mixture were applied to a cellulose paper, and chromatograms were developed with water. The spots containing radioactive adenine were cut out and immersed in 5 ml of aquasol; the radioactivity was determined.

Definition of Unit. One unit of MeSAdo nucleosidase catalyzes the cleavage of 1 μmol of substrate in 1 min at 37°.

Purification

The ionic strength of the extracting fluid strongly affects the solubility of lupin seed proteins.[4] Since MeSAdo nucleosidase can be extracted in good yields with low-ionic-strength buffers, it can be separated from the

[2] S. Kim, S. Nochumson, W. Chin, and W. K. Paik, *Anal. Biochem.* **84,** 415 (1978).
[3] P. K. Chiang and G. L. Cantoni, *J. Biol. Chem.* **252,** 4506 (1977).
[4] H. Jakubowski and J. Pawelkiewicz, *Acta Biochim. Pol.* **21,** 271 (1974).

bulk of other seed proteins that are insoluble under these conditions. After selective extraction, the four conventional steps described here yield a preparation about 20% pure with an 1860-fold purification. The last step utilizes adsorption on AdoHcy-Sepharose and desorption with MeSAdo. The purified MeSAdo nucleoside is a homogeneous preparation with a molecular weight of 62,000, representing an overall purification of 8900-fold and a yield of 12%. The enzyme consists of two subunits, each with a molecular weight of 31,000. The purified enzyme has a specific activity of 27 units per milligram of protein.

The meal of yellow lupin seeds (670 g) was extracted with 2700 ml of buffer A for 30 min. The extract was next passed through cheesecloth and centrifuged at 30,000 *g* for 20 min. The MeSAdo nucleosidase was precipitated with ammonium sulfate (27–47% saturation) from the supernatant. The precipitate was dissolved in buffer B; after dialysis against this buffer and centrifugation, the supernatant was applied onto a DEAE-Sephacel column (5 × 33 cm) equilibrated with the same buffer. The column was washed with buffer B,[1] and a 3.5-1 linear gradient of 30 to 250 m*M* potassium phosphate (pH 6.8) and 5% glycerol was applied. The nucleosidase activity emerged at about 200 m*M* potassium phosphate. The enzyme was precipitated with ammonium sulfate (60% saturation), dissolved in a small volume of buffer B, and chromatographed on a Sephadex G-200 column (2.5 × 80 cm). It eluted at $V_e/V_o = 1.8$ and was concentrated by ammonium sulfate precipitation. The enzyme was dissolved in buffer A, dialyzed, and applied onto a hydroxyapatite column (1 × 10 cm) equilibrated with the same buffer. The column was washed with 30 ml of buffer A, and a 140-ml linear gradient of 10 to 80 m*M* potassium phosphate (pH 6.8) and 5% glycerol was applied. The active enzyme fractions, which eluted between 15 and 25 m*M* potassium phosphate, were pooled, concentrated with Aquacide I, dialyzed against buffer C,[1] and then applied onto an AdoHcy-Sepharose column (1 × 7 cm) equilibrated with the same buffer. The column was first washed with 30 ml of buffer C, and next with 20 ml of buffer C containing 200 m*M* KCl. Finally, MeSAdo nucleosidase was eluted from the affinity column with buffer C containing 200 m*M* KCl and 10 μM MeSAdo. The active fractions were pooled, concentrated with Aquacide I, dialyzed against buffer C, and stored frozen at −20°. All the operations were carried out at about 4°.

Properties

Michaelis Constants. The K_m value for MeSAdo estimated for lupin MeSAdo nucleosidase was 0.41 μM, and is about the same as those of

MeSAdo phosphorylase from rat liver[5] and MeSAdo nucleosidase from *Escherichia coli*,[6] 0.47 and 0.31 μM, respectively. The MeSAdo K_m values are 10.3 and 15 μM for MeSAdo nucleosidase from *Vinca rosea*[7] and tomato,[8] respectively. The MeSAdo K_m values reported for MeSAdo phosphorylase from rat prostate and human placenta are 0.3 mM[9] and 43.5 μM,[10] respectively.

Substrate Specificity. A number of thioether analogs of MeSAdo modified in the purine base, sugar, or aliphatic residues were checked as potential substrates for the lupin MeSAdo nucleosidase.[11] The enzyme liberates the free base from most of them, and tolerates structural changes both in the nucleoside moiety and aliphatic residue of MeSAdo. The nucleosidase discriminates, however, between the analogs with and without an α-amino group in their aliphatic residues, since AdoHcy and *S*-adenosylcysteine are not degraded, in contrast to *S*-adenosylpropionic acid. Under our experimental conditions, MeSAdo and its adenosyl analogs are better substrates than their deaminated (inosyl) counterparts. The V_{max} for MeSAdo is 50 times higher than that for 5′-methylthioinosine, the K_m for the latter being 55 μM. 5′-Deoxyadenosine, a highly active substrate for the MeSAdo-cleaving enzyme of Sarcoma 180 cells,[12] is a very poor substrate for the lupin MeSAdo nucleosidase.

Inhibition Studies. The following compounds were checked as potential effectors of the reaction catalyzed by the lupin MeSAdo nucleosidase: ribose, 5′-methylthioribose, adenine, 3-deazaadenine, hypoxanthine, cytosine, uracil, adenosine, 3-deazaadenosine, AMP, ATP, AdoHcy, putrescine, spermidine, spermine, 9-*erythro*-(2-hydroxyl-3-nonyl)adenine, sinefungin, decoyinine, (*S*)-9-(2,3-dihydroxypropyl)adenine and adenosine 5′-carboxamide, methionine, homocystein, dithiothreitol, and *p*-hydroxymercuribenzoate. Only some of them can exert an inhibitory effect. Adenine, a product of the nucleosidase activity, is a potent competitive inhibitor (K_i = 11 μM). In contrast, 5′-methylthioribose, the other product of the reaction, is a poor inhibitor (K_i = 1060 μM). The enzyme can be inhibited by 3-deazaadenine (K_i = 19 μM), and 9-*erythro*-

[5] A. J. Ferro, N. C. Wrobel, and J. A. Nicolette, *Biochim. Biophys. Acta* **570,** 65 (1979).

[6] A. J. Ferro, A. Barrett, and S. K. Shapiro, *Biochim. Biophys. Acta* **438,** 487 (1976).

[7] C. Baxter and C. J. Coscia, *Biochem. Biophys. Res. Commun.* **54,** 147 (1973).

[8] Y. Yu, D. O. Adams, and S. F. Yang, *Arch. Biochem. Biophys.* **198,** 280 (1979).

[9] A. E. Pegg and H. G. Williams-Ashman, *Biochem. J.* **115,** 241 (1969).

[10] M. Cartenì-Farina, F. Della Ragione, G. Ragosta, A. Oliva, and V. Zappia, *FEBS Lett.* **104,** 266 (1979). See also this volume [60].

[11] A. B. Guranowski, P. K. Chiang, and G. L. Cantoni, *Eur. J. Biochem.* **114,** 293 (1981).

[12] T. M. Savarese, G. W. Crabtree, and R. E. Parks, Jr., *Biochem. Pharmacol.* **28,** 2227 (1979).

(2-hydroxyl 3-nonyl)adenine (K_i = 37 μM), the latter being known chiefly as an inhibitor of adenosine deaminase.[13] AdoHcy is a less effective inhibitor (K_i = 135 μM). Sinefungin,[14,15] decoyinine,[16] (*S*)-9-(2,3-dihydroxypropyl)adenine (a known inhibitor of other enzymatic reactions[17,18]), and adenosine 5′-carboxamide (known as an effector of coronary dilatory action[19]) can also inhibit the lupin MeSAdo nucleosidase. Their corresponding K_i values are 122, 133, 638, and 482 μM, respectively. Neither dithiothreitol nor *p*-hydroxymercuribenzoate can affect the reaction of lupin MeSAdo nucleosidase, and this suggests that sulfhydryl groups are not essential for the enzyme activity. The other compounds have no apparent effect on lupin MeSAdo nucleosidase, and it also has been reported that adenosine, ATP, AdoHcy, and putrescine do not inhibit MeSAdo phosphorylase of *Drosophila melanogaster*.[20]

[13] H. J. Schaeffer and C. F. Schwender, *J. Med. Chem.* **17,** 6 (1974).

[14] R. W. Fuller, *in* "Transmethylation" (E. Usdin, R. T. Borchardt, and C. R. Creveling, eds.), p. 251. Elsevier/North-Holland, Amsterdam, 1979.

[15] R. T. Borchardt and C. S. G. Pugh, *in* "Transmethylation" (E. Usdin, R. T. Borchardt, and C. R. Creveling, eds.), p. 197. Elsevier/North-Holland, Amsterdam, 1979.

[16] R. J. Suhadolnik, "Nucleoside Antibiotics." Wiley-Interscience, New York, 1970.

[17] H. J. Schaeffer, D. Vogel, and R. Vince, *J. Med. Chem.* **8,** 502 (1965).

[18] E. DeClercq, J. Descamps, P. DeSomer, and A. Holy, *Science* **200,** 563 (1978).

[19] G. Raberger, W. Schutz, and O. Kraupp, *Arch. Int. Pharmacodyn. Ther.* **230,** 140 (1977).

[20] L. Shugart, M. Tancer, and J. Moore, *Int. J. Biochem.* **10,** 901 (1979).

Section XI

Methods for the Study of Polyamines in Lymphocytes and Mammary Gland

[65] Methods for the Study of the Physiological Effects of Inhibitors of Polyamine Biosynthesis in Mitogen-Activated Lymphocytes

By CHRISTINE E. SEYFRIED and DAVID R. MORRIS

Inhibitors of ornithine decarboxylase and *S*-adenosylmethionine decarboxylase have achieved considerable importance in studies of the cell biology of the polyamines.[1] These inhibitors also may have important clinical applications.[2–4] Mitogen-activated small lymphocytes are a convenient system for evaluating the influence of these antimetabolites on cellular physiology. The major advantage of these cells is that they can be synchronously triggered to undergo a defined sequence of events leading ultimately to DNA synthesis and cell division.[5,6] The induction of ornithine decarboxylase[7,8] and of *S*-adenosylmethionine decarboxylase[7–9] are prominent among the prereplicative events in these cells and lead to rather large elevations of the cellular levels of putrescine, spermidine, and spermine.[10] Inhibitors of these two enzymes have been shown to be effective in blocking polyamine accumulation in intact lymphocytes and to inhibit DNA replication in the mitogen-activated cells (reviewed by Heby and Jänne[1] and Morris and Harada[11]). There is no influence of these inhibitors on prereplicative events, including RNA synthesis, protein synthesis, or time of entry of the cells into S phase.

[1] O. Heby and J. Jänne, *in* "Polyamines in Biology and Medicine" (D. R. Morris and L. J. Marton, eds.), p. 243. Dekker, New York, 1981.

[2] C. W. Porter, C. Dave, and E. Mihich, *in* "Polyamines in Biology and Medicine" (D. R. Morris and L. J. Marton, eds.), p. 407. Dekker, New York, 1981.

[3] J. Koch-Weser, P. S. Schechter, P. Bey, C. Danzin, J. R. Fozard, M. J. Jung, P. S. Mamont, N. J. Prakash, N. Seiler, and A. Sjoerdsma, *in* "Polyamines in Biology and Medicine" (D. R. Morris and L. J. Marton, eds.), p. 437. Dekker, New York, 1981.

[4] M. Siimes, P. Seppänen, L. Alhonen-Hongisto, and J. Jänne, *Int. J. Cancer* (in press).

[5] N. R. Ling and J. E. Kay, "Lymphocyte Stimulation." North-Holland Publ., Amsterdam, 1975.

[6] D. A. Hume and M. J. Weidemann, "Mitogenic Lymphocyte Transformation." Elsevier/North-Holland Biochemical Press, Amsterdam, 1980.

[7] J. E. Kay and A. Cook, *FEBS Lett.* **16**, 9 (1971).

[8] J. E. Kay and V. J. Lindsay, *Exp. Cell Res.* **77**, 428 (1973).

[9] R. H. Fillingame and D. R. Morris, *Biochem. Biophys. Res. Commun.* **52**, 1020 (1973).

[10] R. H. Fillingame and D. R. Morris, *Biochemistry* **12**, 4479 (1973).

[11] D. R. Morris and J. J. Harada, *in* "Polyamines in Biomedical Research" (J. M. Gaugas, ed.), p. 1. Wiley, New York, 1980.

ISBN 0-12-181994-9

The purpose of this chapter is to summarize procedures for preparing, culturing, and monitoring physiological responses of small lymphocytes. The preparation of cells from two sources is described. Where large quantities of cells are required, bovine lymph node cells are purified by passage over a glass bead column. When smaller quantities will suffice, cells are prepared from mouse mesenteric lymph nodes.

I. Preparation and Culture of Lymphocytes

A. Bovine Cells

Materials and Reagents. Standard tissue culture glass-washing procedure is used. Solutions are sterilized by filtration, and glassware and other equipment are autoclaved.

Culture medium: RPMI 1640[12,13] supplemented with penicillin (100 units/ml), streptomycin (50 units/ml), and 1-mercaptoethanol (5×10^{-5} *M*)

Newborn calf serum. Add to RPMI 1640 for a final concentration of 10% (v/v).

Mycostatin (Nystatin, E. R. Squibb and Sons, Inc.): A 500,000-unit vial is reconstituted with 10 ml of sterile distilled water. When required, add to culture medium to a final concentration of 50 units/ml. The stock solution is stored at −20°.

Aerosporin (Polymixin B sulfate, Burroughs Wellcome Co.): Reconstitute 500,000 units with 5 ml of sterile water. When required, add to culture medium to a final concentration of 100 units/ml. Store the stock solution at 4°.

Polycarbonate bottle (290 ml), sterile, containing 100 ml of culture medium supplemented with Mycostatin and Aerosporin.

Forceps, 5 inch, with blunt ends; 2 pairs, sterile

Scissors, iris, straight with sharp points; 1 pair, sterile

Scissors, 5½ inch, 1 blunt and 1 pointed end; 2 pairs, sterile

Petri dishes, 100 × 20 mm; 3, sterile

Beaker, 150 ml; sterile

Beaker, 400 ml; 3, sterile

Stainless steel filter pans with screen (Cellector tissue sieve, Bellco Glassware); 3 fitted with 20, 40, and 60 mesh screens, sterile

Spatula, spoon end; sterile

Erlenmeyer flask, 1 liter; sterile

Scalpel handle, Bard-Parker, size No. 4

Scalpel blades, Bard-Parker, size No. 21, treated to remove toxic

[12] Moore, G. E., R. E. Gerner, and H. A. Franklin, *J. Am. Med. Assoc.* **199,** 519 (1967).
[13] Morton, H. C., *In Vitro* **9,** 468 (1970).

antirust compound: blades are soaked in carbon tetrachloride and scrubbed with cotton swabs. They are then soaked and scrubbed in 95% ethanol and in deionized water. After air drying, the blades are stored in 95% ethanol and are flamed before use.

Column feed bottles, sterile. Two 500-ml bottles with efflux tubes of neoprene at the bottom closed by clamps. The efflux tubes are connected by a glass Y tube that is fitted to a No. 10 vented stopper. The two bottles are closed on top by No. 4 vented stoppers.

Glass beds, 1 mm (Propper Manufacturing Co.); 2 pounds in a beaker covered with water, autoclaved.

Glass bead column; sterile. A glass column (35 × 5.5 cm), fitted on top to the No. 10 stopper of the column feed bottles. The stopcock at the bottom of the column has two outlets, one fitted with a No. 9 vented stopper to empty into a sterile 1-liter Erlenmeyer flask and one connected to 6 inches of Tygon tubing for draining excess medium. The column is sterilized with a glass tube (1.5 × 40 cm) inside supported by a nylon mesh network at the base of the column.

Beaker, 1 liter; nonsterile

Powder funnel; sterile

Polycarbonate centrifuge bottle (290 ml, Sorvall); sterile

Conical centrifuge tubes (50 ml, Falcon 2070); sterile

Prescription bottles, 32-ounce (Sani-glas, Brockway Glass Co., Inc.); washed and sterilized

Glass scintillation vials (20 ml, Wheaton Glass Co.); washed, capped loosely with 2.4 cm (i.d.) lids and sterilized

Cheesecloth funnel, long stem, 4.5-cm top diameter glass funnel containing four layers of cheesecloth and connected to a 50-ml graduated cylinder with a No. 4 rubber stopper.

Buffered saline, 10× solution: For 100 ml, add 8.0 g of NaCl, 0.15 g of KH_2PO_4 and 0.29 g of $Na_2HPO_4 \cdot 7H_2O$ to 90 ml of distilled water. Adjust the pH to 7.0, and bring the volume to 100 ml. Sterilize by autoclaving, and store at 4°.

Concanavalin A (Con A) solution. Dilute the 10× buffered saline 10-fold with distilled H_2O and cool to 4° on ice. Weigh an appropriate amount of Con A and add to the cold saline solution in a beaker. Mix very slowly on a stirring motor using a pipette to wet the Con A and wash down the sides of the beaker. Filter the solution through a 0.22-μm disposable filter. Dilute a sample of the Con A solution 1 : 10 in distilled water and read the $A_{280\text{ nm}}$. Calculate the concentration of Con A in solution assuming that 1 mg/ml gives $A_{280\text{ nm}} = 1.3$. The Con A solution must be kept cold and used within 4 hr.

Procedures. Unless otherwise specified, these procedures are carried out in a laminar-flow hood, using standard sterile technique.

1. The glass bead column is prepared by pouring ca 1 inch of glass beads into the column through the glass tube, using the powder funnel. This procedure prevents the nylon net support from moving when the remaining beads are poured. A sterile spatula is used to move the beads from the beaker, and additional sterile water can be added if necessary. Remove the glass tubing and funnel and continue to pour the beads into the column to a height 2 inches below the top. Wash the column with 500 ml of culture medium, containing newborn calf serum, Mycostatin, and Aerosporin, collecting the effluent in a nonsterile beaker. After washing, place the column in a 37° incubator until needed.
2. A maximum of four lymph nodes are obtained at a slaughterhouse from freshly killed steers. The suprapharyngeal lymph nodes are placed immediately into a sterile petri dish, from which they are transferred with forceps to the polycarbonate bottle containing 100 ml of culture medium, Mycostatin, and Aerosporin. If there is too much fat surrounding a node for it to fit through the neck of the bottle, trim the excess with a pair of sterile scissors. The capsule surrounding the node should be kept intact. The bottle is placed on ice until used.
3. In a tissue culture hood, transfer the nodes to a sterile petri dish. Trim all excess fat and connective tissue from the capsule and hilus of the nodes. Transfer the cleaned nodes to a new petri dish, and add approximately 10 ml of medium containing newborn calf serum, Mycostatin, and Aerosporin. Working with one-quarter of a node at a time, slide the scalpel immediately under the cortex, cutting it away from the medulla. As the medulla is prepared, it is transferred to a petri dish and covered with 10 ml of medium. The cortex should be lightly scraped with the scalpel to remove any free material, which is added to the petri dish. Chop the medulla into small pieces with scissors and transfer to a 150-ml beaker along with the associated medium. Using the sharp, pointed iris scissors, disperse the cells by cutting briskly for 15 min. It is important that no large pieces of tissue remain at the end of this step. The tissue is poured through the stainless steel filter pans in succession from 20 mesh to 60 mesh. The cells are collected in 400-ml beakers beneath the pans. Most of the cells will pass through the screen with the medium, but large pieces retained should be pressed gently with the spatula and rinsed with medium. Cells should also be scraped from underneath each pan with the spatula.
4. Transfer the cell suspension (no more than 150 ml) to one of the column feed bottles. Fill the other bottle with 500 ml of culture medium. The feed bottles are placed above the glass bead column and the No. 10

stopper is attached sterilely. A sterile 1-liter Erlenmeyer flask is placed on the sterile outlet of the glass bead column. Allow the cells to flow onto the column, collecting the effluent in a nonsterile beaker. Wash the cells onto the column with a small amount of medium and incubate for 40 min at 37°.

5. The cells are eluted from the column with culture medium at a flow rate of approximately 10 ml/min. The eluate is collected via the sterile outlet of the column into the sterile, 1-liter Erlenmeyer flask.

6. When elution of the column is complete, quickly remove the flask from the base of the column and cover it with sterile aluminum foil. In a tissue culture hood, divide the cell suspension between two sterile polycarbonate bottles and centrifuge at 300 *g* for 15 min at room temperature.

7. Wash the cells by resuspending them in 10 ml of medium and transfer to a sterile 50-ml conical centrifuge tube. Rinse the bottles with an additional 10 ml of medium, and combine with the cell suspensions. Centrifuge for 15 min at 300 *g;* repeat this washing procedure twice more.

8. After washing the cells, resuspend in approximately 50 ml of medium and pour the suspension through the cheesecloth-funnel assembly into the graduated cylinder, noting the final volume.

9. The cells are then counted in a hemacytometer and placed in 100-ml aliquots of culture medium at a final concentration of 2 to 3×10^6 cells per milliliter in 32-ounce prescription bottles. The bottles are placed, flat side down, in a humid incubator containing an atmosphere of 5% CO_2 at 37°. The yield is generally 2×10^{10} cells per lymph node.

10. The cells are incubated overnight and stimulated on the following day with Con A at a final concentration of 18 μg/ml.

B. Murine Cells

Materials and Reagents. These are as for bovine cells, unless otherwise indicated.

- Dissection board and pins
- Plastic petri dishes, 60 × 15 mm; one for every 3 mice (Falcon 3002 or similar)
- Forceps, 5 inch, straight with blunt ends; 2 pairs, sterile
- Scissors, iris, straight with sharp points; 2 pairs, sterile
- Wash bottle containing 95% ethanol
- Stainless steel screen, 60 mesh, 2 × 2 inches; sterile
- Spatula, spoon end; sterile
- Centrifuge tubes, 15 ml with screw cap; sterile (Corning 325310 or similar)
- Beaker, 2 liter
- Forceps, watchmakers with extra fine point; 1 pair, sterile
- Cheesecloth
- Dry ice

Scissors, 5½-inch with 1 sharp and 1 blunt point; 1 pair, sterile
Trypan blue, 0.4% suspension in saline, sterile (GIBCO No. 630-5250 or similar)

Procedures

1. To kill mice by CO_2 inhalation, cover dry ice in the bottom of a beaker with cheesecloth and add a small amount of water. Place one mouse at a time in the beaker, cover the top with foil, and wait 5 min. Pin the mouse to the corkboard with limbs extended. Using the squirt bottle of ethanol, wash the abdominal area. With blunt-ended scissors and forceps, cut through the skin layer and retract it from the body. Using a pair of sterile scissors and forceps, open the peritoneal cavity and retract as for the skin. Lift the gut to locate the mesenteric lymph node, which is single and elongated and is located next to the ascending colon. After removal, place in a petri dish containing 5 ml of culture medium supplemented with Mycostatin.

2. In a tissue culture hood, trim all adipose and connective tissue using the watchmaker forceps and iris scissors, and transfer the cleaned nodes to a fresh dish containing culture medium without Mycostatin. Place the stainless steel screen over the bottom half of a petri dish and place the nodes on the screen, covering them with 1 or 2 drops of medium (surface tension should prevent its passing through). Mince the nodes with the scissors for about 1 min. Gently press the released cells through the screen with the spatula, rinsing the minced tissue with 1 ml of medium every 30 sec.

3. Transfer the cell suspension to a sterile conical tube, and rinse the dish once with 1–2 ml of media. Allow any large clumps of tissue to settle for 2 min, and transfer the suspension to a new centrifuge tube. Centrifuge the cells at 300 *g* for 10 min at room temperature. Wash the pellet three times in 10 ml of medium by resuspension and centrifugation.

4. Determine the number of viable cells using a hemacytometer and trypan blue. The value generally is ca 5×10^7 cells per mouse. Adjust the cell density to 3×10^6 viable cells per milliliter, and culture overnight. The cells are recounted the next morning and harvested by centrifugation; the density is readjusted to 2 to 3×10^6 viable cells per milliliter. Con A is added at a final concentration of 5 μg/ml.

II. Measurements of Macromolecular Synthesis

A. Protein and RNA Synthesis

Protein and RNA synthesis are most conveniently measured by the use of radioactive precursors, although chemical measurements of accumulation are also successfully used.[10] These parameters are usually measured during the prereplicative phase, prior to entry of the cells into S phase of the cell cycle at 24 hr after activation. This is done to minimize secondary effects on protein and RNA synthesis as a result of influences of inhibitors on later stages of the cell cycle. [^{3}H]Adenine is used to monitor RNA synthesis, since this obviates the large transport component involved in the rate of nucleoside incorporation.[10,14]

Materials and Reagents

Conical centrifuge tubes, 15 ml, glass
L-[4,5-^{3}H]Leucine, ca 5 Ci/mmol
[2-^{3}H]Adenine, ca 10 Ci/mmol
Trichloroacetic acid, 5% (w/v)
Potassium hydroxide, 0.3 *N*
Phosphoric acid, 0.6 *N*
Perchloric acid, 1.3 *N*
Scintillation vials, 20 ml
Scintillation fluid: xylene and Triton X-114 (3 : 1), 3 g of 2,5-diphenyloxazole per liter

Procedures

1. At approximately 20 hr after activation with mitogen, label the cells for 2 hr with [^{3}H]leucine or [^{3}H]adenine (both at 5 μCi/ml).
2. Transfer the cultures to conical centrifuge tubes in an ice-water bath. Harvest the cells by centrifugation and wash with trichloroacetic acid as described for cells labeled with [^{3}H]thymidine in the next section. The washed precipitates are suspended in 0.5 ml of 0.3 *N* KOH incubated at 37° for 1 hr.
3. For the cells labeled with [^{3}H]leucine, the dissolved pellet is transferred to a scintillation vial containing 0.5 ml of 0.6 *N* phosphoric acid. Add 10 ml of scintillation fluid and count.
4. For the cells labeled with [^{3}H]adenine, add 0.5 ml of 1.3 *N* perchloric acid to the KOH hydrolyzate, and allow the tubes to stand on ice for 1 hr. Centrifuge the samples at 1000 *g* for 15 min and transfer the supernatant solutions to scintillation vials. Add 10 ml of scintillation fluid and count.

B. DNA Synthesis

[^{3}H]Thymidine Incorporation in Whole Cells

Materials

Trichloroacetic acid, 5%
Potassium hydroxide, 0.3 *N*

[14] R. H. Fillingame, Ph.D. Thesis, University of Washington, Seattle, 1973.

Perchloric acid, 1.3 *N*
Conical centrifuge tubes, 15 ml, glass
Diphenylamine
Acetaldehyde, 16 mg/ml in H_2O. Store frozen.
Glacial acetic acid
Sulfuric acid, concentrated
[*methyl*-^{3}H]Thymidine, ca 6.7 Ci/mmol
Disposable glass culture tubes, 15 × 85 mm
Scintillation fluid: xylene and Triton X-114 (3 : 1), 3 g of 2,5-diphenyloxazole per liter
Standard DNA solution: Dissolve 1.2 mg of calf thymus DNA in 5 ml of 0.005 *N* NaOH and stir overnight at room temperature. Dilute 1 : 10 in standard saline citrate (0.15 *M* NaCl–0.015 *M* sodium citrate) and read the $A_{260\ nm}$. DNA at 1 mg/ml in standard saline citrate will have an $A_{260\ nm}$ of 20. Adjust the concentration of the stock solution to 200 μg of DNA per milliliter with 0.005 *N* NaOH and add an equal volume of 10% trichloroacetic acid (w/v). Hydrolyze for 20 min at 90°. Store this solution (100 μg of DNA per milliliter final concentration) frozen.

Procedures

1. Three-milliliter cultures are labeled with [^{3}H]thymidine (5 μCi/ml) for 2 hr. After labeling, the cultures are rapidly cooled by transferring them to conical centrifuge tubes in an ice-water bath; they are centrifuged at 300 *g* for 10 min at 4°. All subsequent procedures are carried out at 4° unless otherwise specified.

2. The supernatant solutions are discarded into a radioactive waste receptacle, and the pellets are resuspended in 3 ml of cold 5% trichloroacetic acid. The tubes are centrifuged for 10 min at 1000 *g*, and the pellets obtained are washed 3 more times in 3 ml of 5% trichloroacetic acid by resuspension and centrifugation.

3. The precipitates are dissolved in 0.5 ml of 0.3 *N* KOH and incubated for 1 hr at 37°. The tubes are cooled on ice, and 0.5 ml of 1.3 *N* perchloric acid is added to each sample. Stir this mixture well with a sealed Pasteur pipette and leave standing on ice for 1 hr. Centrifuge for 10 min at 1000 *g*. Resuspend the pellets in 1.0 ml of 5% trichloroacetic acid and hydrolyze for 1 hr at 90°. After 1 hr at 4°, centrifuge the samples for 10 min at 1000 *g*.

4. For counting, transfer 0.2-ml samples to scintillation vials containing 1 ml of water and 10 ml of scintillation fluid.

5. Transfer 0.5 ml of supernatant solutions to a culture tube (15 × 85 mm) for assay by the diphenylamine reaction[15] as follows: Dissolve 1.5 g

[15] K. Burton, *Biochem. J.* **62,** 315 (1956).

of diphenylamine in 100 ml of glacial acetic acid. Add 1.5 ml of concentrated sulfuric acid. This solution should be prepared shortly before addition to the samples. Combine 20 ml of the diphenylamine–acetic acid–sulfuric acid mixture with 0.1 ml of acetaldehyde (16 mg/ml). Add 1.0 ml of the reagent mixture to each 0.5-ml sample, mix immediately, and cover. Include as a standard curve 0–50 μg of calf thymus DNA by combining the appropriate amount of the standard DNA solution and 5% trichloroacetic acid for a final volume of 0.5 ml. Leave the samples at room temperature for 18 ± 2 hr and read the $A_{600\text{ nm}}$.

6. Express the incorporation data as counts per minute per microgram of DNA.

[^{3}H]Thymidine Autoradiography

If a decrease in the rate of [^{3}H]thymidine incorporation into DNA is observed, it could arise from an inhibition of DNA replication per se or from a decreased number of cells in the S phase of the cycle. These alternatives are distinguished by determining the fraction of the cell population in S phase by autoradiography.

Materials

[*methyl*-^{3}H]Thymidine, ca 6.7 Ci/mmol
Conical centrifuge tubes, 15 ml, glass or plastic
Hypotonic sucrose solution. Dissolve 34.2 g of sucrose, 102.9 mg of $CaCl_2$, and 17.5 mg of NaCl in 1 liter of distilled H_2O. Sterilize by filtration and store frozen.
Carnoy's fixative. Combine glacial acetic acid and absolute methanol (1 : 3). This should be made fresh daily.
Vortex mixer with rheostat
Glass microscope slides, 3 × 1 inch, with one side of one end frosted
Pasteur pipettes
NT-2B Nuclear track emulsion (Eastman Kodak), warmed to 37° in an incubator for approximately 3 hr before use.
Lighttight slide boxes, each containing a small amount of desiccant wrapped in cheesecloth.
Test tube rack, three-tiered with spaces at least 1 inch on the diagonal to hold slides for drying after coating with emulsion. Place a layer of paper towels on top of the middle level to support the slides and absorb any excess emulsion.
Glass beaker, 30 ml, containing 15 ml of distilled H_2O
Hot air dryer
Staining dishes, 4 (LabTek 4456 or equivalent)
Slide-staining holder (LabTek 4465 or equivalent)

Dektol developer (Kodak 146-4726) diluted 1 : 1 with distilled H_2O
Fixer (Kodak 197-1746)
Stop solution: Dilute 34 ml of glacial acetic acid with 1.1 liters of distilled H_2O. Store in a brown glass bottle at room temperature.
Aluminum foil for wrapping slide boxes for storage
Timers, interval; 2
Spatula with spoon end
Giemsa blood stain

Preparation of Cells

1. Lymphocyte cultures are labeled for 2 hr with 5 μCi of [*methyl*-^{3}H]thymidine per milliliter. Each culture should be at least 1 ml and preferably 3 ml, so that [^{3}H]thymidine incorporation into DNA can be assessed in parallel on a 2-ml sample as described above in Section II,B ([^{3}H]thymidine incorporation in whole cells) procedures.

2. Transfer 1-ml samples of each culture to conical centrifuge tubes in an ice-water bath. Harvest the cells by centrifugation at 300 *g* for 1.5 min. Aspirate all but 0.2 ml of supernatant solution, and resuspend the pellet with gentle vortexing. Add 5 ml of hypotonic sucrose solution, and incubate for 5 min at room temperature. Centrifuge the suspension for 1.5 min at 300 *g* at room temperature. Remove all but 0.2 ml of the supernatant solution and resuspend as before.

3. Slowly add 5.0 ml of Carnoy's fixative over a period of 30 sec. Leave for 10 min at room temperature. Centrifuge for 1.5 min at 300 *g* at room temperature. Aspirate all but 0.2 ml of the supernatant solution, and resuspend the pellet in 5.0 ml of Carnoy's fixative as before. After 10 min, centrifuge for 1.5 min at 300 *g*.

4. Aspirate all but 0.2 ml of the supernatant solution, gently resuspend the cells, and, using a Pasteur pipette, drop the cells onto three glass slides labeled appropriately with pencil on the frosted side. The drop should be placed about one-fourth of the length of the slide from one end. Allow the slides to dry at room temperature.

Exposure of Slides

1. A 45° water bath, NT-2B emulsion, spatula, aluminum foil, 30-ml beaker with water, test tube rack, air dryer, slide boxes, and the slides to be coated are placed in a photographic dark room. All work is done with only a safety light on.

2. Emulsion (15 ml) is added to the beaker of water, and the mixture stirred with a spatula. The beaker is then placed in the 45° bath to prevent the solution from solidifying.

3. Dip the lower half of each slide (containing the sample) in the warmed emulsion and place it in the test tube rack to dry. Include at least four blank slides to use for testing the correct development time. The use of an air dryer at this step will lessen the drying time to approximately 10 min, compared to 25 min without one. Do not use heat to dry the slides, as it will tend to raise the background seen after development.

4. Place the slides in the lighttight boxes containing desiccant. Duplicate sets of slides should be placed in separate boxes to allow for different times of exposure. The slide boxes are then wrapped in three layers of aluminum foil and stored at 4°, away from any ambient radiation. Four days of exposure are usually sufficient.

Development of Slides

1. Fill the four staining dishes with *(a)* diluted Dektol, *(b)* water; *(c)* stop solution, and *(d)* fixer respectively. All reagents should be at room temperature.

2. In the dark room, remove the slides from the slide boxes and place in the slide-taining holder. First, develop a test slide to ensure the right exposure times. Develop for 2–3 min in the Dektol, dip once into the stop solution, once in the water, and 10 min in the fixer. At this point the test slide can be washed briefly in water and held up to a light to ensure that the emulsion is clear. If not, increase the development time in Dektol. Wash the developed slides in running water for 20 min and air dry.

3. Stain the dry slides in Giemsa blood stain for 30 min followed by rinsing in running water for 10 min.

4. The fraction of cells labeled is scored under a microscope at 400×.

LABELING OF THYMIDINE NUCLEOTIDE POOLS

A change in the rate of [^{3}H]thymidine incorporation into DNA could be due to an alteration in the rate of DNA replication or to a change in the uptake of thymidine or in its conversion to dTTP. The rate of labeling of the thymidine nucleotide pools is measured as follows.[16]

Materials and Reagents

Phosphate-buffered saline: In 950 ml of distilled H_2O, combine 0.10 g of $CaCl_2$, 0.20 g of KCl, 0.20 g of KH_2PO_4, 0.10 g of $MgCl_2 \cdot 6H_2O$, 8.0 g of NaCl, and 2.16 g of $Na_2HPO_4 \cdot 7H_2O$. Adjust the pH to 7.4, bring volume to 1.0 liter, sterilize by filtration, and store at 4°.

Hypoxanthine, 0.29 m*M*: Dissolve 0.4 mg hypoxanthine in 10 ml of distilled water. Store at 4°.

[16] C. E. Seyfried and D. R. Morris, *Cancer Res.* **39,** 4861 (1979).

Amethopterin: Stock solution should be made immediately before use by dissolving 1.5 mg of amethopterin in 0.95 ml of H_2O containing 30 μl of NaOH (5 mg/ml).

[^{3}H]Thymidine stock solution (1.83 mCi/ml; 3.27 mCi/μmol): Lyophilize 2.1 mCi of [*methyl*-^{3}H]thymidine (ca 6.1 Ci/mmol). Resuspend in 1.05 ml of distilled H_2O and 0.1 ml of 3 m*M* cold thymidine

Glass conical centrifuge tubes, 15 ml

Trichloroacetic acid solution, 6% (w/v) in H_2O

[2-^{14}C]Thymidine solution: 0.01 μCi in 100 μl of distilled H_2O (ca 50 mCi/mmol)

Ether, anhydrous

Cellulose PEI plates on plastic backing sheets (J. T. Baker Chemicals, No. 4473). The plates are prewashed by running 5 cm in 10% NaCl and then in distilled water to the top. They are air dried, run to the top a second time in distilled water, air dried, and stored at $-20°$ until use.

LiCl, 3 *M*: Dissolve 63.59 g of LiCl in 500 ml of distilled H_2O.

Acetic acid, 1 *M*: Combine 57.1 ml of glacial acetic acid and 942.9 ml of distilled H_2O.

Rectangular thin-layer chromatographic developing tanks

Standard marker solution for thin-layer chromatography: Combine equal volumes of the following solutions made up in distilled water: thymidine (5 mg/ml), TMP (5 mg/ml), TDP (5 mg/ml), and TTP (5 mg/ml). These solutions are stored frozen.

NH_4OH, 2 *M*: Combine 13.5 ml of ammonium hydroxide and 86.5 ml of distilled H_2O.

Scintillation fluid: As in Section II,A,1.

Procedures

1. Three-milliliter cultures of lymphocytes are stimulated with Con A. Hypoxanthine, 30 μl of 0.29 m*M* stock solution, is added to each culture as a purine source 33 hr after stimulation. Amethopterin, 10 μl of 3.3 m*M* stock solution, is added at 47.5 hr to block endogenous thymidine synthesis, and 100 μl of the [^{3}H]thymidine stock solution is added at 48 hr after Con A addition. The cultures are labeled with [^{3}H]thymidine for 0.5 hr at 37°; the vials are then transferred to an ice-water bath for rapid cooling. All subsequent steps are carried out at 0–4°.

2. Transfer the cell suspension to a conical centrifuge tube and centrifuge at 300 *g* for 10 min at 4°. Aspirate the supernatant solution into radioactive waste, and resuspend the pellet in 5 ml of cold phosphate-buffered saline. Centrifuge at 300 *g* for 10 min, aspirate the supernatant solution, and wash the pellet again in 5 ml of phosphate-buffered saline.

3. Suspend the washed cell pellet in 1.0 ml of 6% TCA and incubate for 1 hr in an ice-water bath. Collect the precipitate by centrifugation at 1000 *g* for 10 min, and transfer the supernatant solution to a new conical centrifuge tube. The pellet may be analyzed for radioactivity in DNA as described under thymidine incorporation in Section II,B.

4. [2-^{14}C]Thymidine, 100 μl of stock solution, is added to each supernatant solution to monitor recovery. A control consisting of 1.0 ml of 6% trichloroacetic acid should also be included in this step. Extract the acid from each sample by adding 3 ml of anhydrous ether and vortexing extensively. The ether is then drawn off with a Pasteur pipette, and the extraction is repeated twice more. The ether-extracted supernatant is frozen in an acetone–dry ice bath and lyophilized to dryness.

5. Resuspend each of the lyophilized samples in 150 μl of the standard marker solution. Count 20 μl of each sample in 250 μl of 2 *M* NH_4OH, 1.5 ml of H_2O, and 10 ml of scintillation fluid. At the same time, count 20 μl of the [^{14}C]thymidine stock in the same way.

6. Spot samples on the PEI plates at 2.5-cm intervals and at least 1 cm from the edges. Spot each sample on two plates, since two solvent systems are required for complete analysis of the thymidine nucleotide pools. The samples can be easily applied using a 10-μl micropipette, filling it to volume, and placing the pipette just close enough to the plate to allow capillary action to empty the contents on the desired spot. Repeat again to load a total of 20 μl of each sample. It helps to have a cool blower directed at the plate while the sample is loaded to increase the rate of evaporation and, consequently, the rate at which the sample is drawn onto the plate. Spot 10 μl of the standard marker solution on each plate as a reference for identifying the components.

7. To prepare the solvent system for separating TDP and TTP, combine 2.5 parts 3 *M* LiCl and 7.5 parts 1 *M* acetic acid. Place this mixture in the bottom of a chromatographic tank, cover, and allow the atmosphere to equilibrate for about 30 min. Quickly place the plate or plates in the tank, cover, and run the plates 15 cm. The same procedure is used to separate thymidine, TMP, and TDP except that 1 part 3 *M* LiCl is combined with 9 parts 1 *M* HAc to prepare the solvent.

8. After running, the plates are air dried and the spots are detected visually under an ultraviolet light (254 nm) and lightly circled with a No. 1 pencil. The spots are cut out with scissors and placed in scintillation vials; the radioactivity is eluted by adding 250 μl of 2 *M* NH_4OH. After elution, 1.5 ml of H_2O and 10 ml of scintillation fluid are added to each vial, and the samples are counted. Areas of each plate between spots should be counted in the same way to assess background radioactivity.

9. A 10-μl sample of the [^{3}H]thymidine stock solution (diluted 1 : 100

in distilled water) is counted in 250 μl of 2 *M* NH_4OH, 1.5 ml of H_2O, and 10 ml of scintillation fluid, and the specific radioactivity is calculated. The radioactivity in the thymidine nucleotides is corrected for recovery of [^{14}C]thymidine and the specific radioactivity of the [^{3}H]thymidine stock solution; the results are expressed as picomoles per 10^6 cells.

DNA Synthesis in Isolated Nuclei

DNA synthesis in isolated nuclei reflects the physiological state of the cells from which the nuclei are obtained and is independent of the *in vivo* labeling kinetics of the thymidine nucleotide pools.[17]

Materials

- Conical centrifuge tubes, 15 ml, glass or plastic
- Buffer A: 10 m*M* Tris-HCl (pH 7.8 at 22°), 4 m*M* $MgCl_2$, 1 m*M* EDTA, 2 m*M* dithiothreitol
- Dounce homogenizer with loosely fitting pestle
- Reaction mixture: 100 m*M* HEPES (pH 7.7), 18 m*M* $MgCl_2$, 300 m*M* NaCl, 15 m*M* ATP, 0.3 m*M* dATP, 0.3 m*M* dCTP, 0.3 m*M* dGTP, 30 μ*M* [^{3}H]dTTP, 0.3 Ci/mmol
- Trichloroacetic acid, 20%
- Glass fiber filters
- Vacuum filtering apparatus (for example, Millipore Corp. XX10 025 00)
- Trichloroacetic acid, 5%, containing 5 m*M* $Na_4P_2O_7$
- Ethanol, 95%
- Perchloric acid, 0.4 *N*, containing 5 m*M* $Na_4P_2O_7$
- Glass scintillation vials, 20 ml
- Scintillation fluid: xylene and Triton X-114 (3 : 1), 3 g of 2,5-diphenyloxazole per liter

Procedures. All procedures are carried out at 4° or on ice.

1. Lymphocyte cultures (100 ml) are harvested by centrifugation at 300 *g* for 15 min. One culture is sufficient for three reaction mixtures. The cell pellets are resuspended in 5 ml of phosphate-buffered saline (see labeling of pools, Section II,B) and centrifuged at 250 *g* for 10 min.
2. Resuspend the cells in buffer A at 1 × 10^8 cells/ml and transfer to the homogenizer. The cells are swollen for 5 min and then disrupted by three strokes of the pestle.
3. Transfer the suspension to a centrifuge tube and isolate a crude nuclear pellet by centrifugation at 250 *g* for 10 min. Resuspend the nuclear pellet in 1 ml of buffer A per 2.5 × 10^7 cells.

[17] J. C. Knutson and D. R. Morris, *Biochim. Biophys. Acta* **520,** 291 (1978).

4. Warm 2.0 ml of the reaction mixture to 37° in a water bath. DNA synthesis is initiated by the addition of 4.0 ml of the nuclear suspension to the reaction mixture. Samples (0.6 ml) are removed at 2-min intervals over a 20-min period and placed in test tubes in an ice bath. After removal, the samples are quenched by the addition of 1 ml of 20% trichloroacetic acid and stored frozen overnight.

5. Collect the precipitates by filtering onto glass fiber disks, washing with 100 ml of 5% trichloroacetic acid + $Na_4P_2O_7$, and drying with 3 ml of 95% EtOH. Transfer the disks to a scintillation vial, add 1 ml 0.4 *N* perchloric acid + $Na_4P_2O_7$, and hydrolyze the DNA by incubating for 20 min at 90°. Add 10 ml of scintillation fluid and count.

III. Polyamine Analysis

The procedures used for polyamine analysis are those described by Seiler in this volume [1–3]. Because of interfering impurities in the lymphocyte samples, the extracts must be prepurified on Dowex 50 columns prior to dansylation and thin-layer chromatography; this sample preparation is described in detail here. A procedure has been reported for thin-layer chromatographic analysis of polyamines in lymphocyte extracts that does not require prior purification,[18] but we have not tried it in our laboratory.

Materials and Reagents

Perchloric acid, 0.2 *N*

[^{3}H]Spermidine, ca 15 Ci/mmol. NOTE: The purity of the [^{3}H]spermidine should be assessed by dansylation and thin-layer chromatography.

Dowex AG 50W-X8, 200–400 mesh hydrogen form, stored in 6 *N* HCl

HCl, 2.5 *N* and 6.0 *N*

Tubes, 15-ml conical, glass or plastic

Rotary evaporator

Dry block heating module for evaporating samples

Acetone and H_2O (1 : 1)

Ethyl acetate

Fluorometer

Procedures

1. Lymphocyte cultures, 5 ml each, are harvested by centrifugation at 300 *g*, and the cell pellets are washed twice with phosphate-buffered

[18] E. Höltta, J. Jänne, and T. Hovi, *Biochem. J.* **178,** 109 (1979).

saline (see labeling of pools, Section II,B). Resuspend the cells in 0.5 ml of 0.2 *N* perchloric acid at 4° and incubate for 1 hr.

2. Centrifuge the acid precipitates at 1000 *g* and transfer 470 μl of the supernatants to new test tubes. Add 1 μCi of [^{3}H]spermidine to assess recovery. Remove 10 μl of each sample for counting.

3. Apply the remaining supernatant solutions to 2.5-ml Dowex columns that have been washed with distilled water to neutrality. Wash the samples onto the columns with 4 ml of 2.5 *N* HCl. Wash the columns with 2.5 *N* HCl for 40 min at 1.2 ml/min. Elute the polyamines from the columns with 6 *N* HCl at 1.2 ml/min for 40 min.

4. Evaporate the eluates to dryness, and wash the sides of the flask once with 0.5 ml of distilled H_2O and twice with 0.5 ml of acetone–H_2O (1 : 1), transferring each wash to a test tube. Evaporate to dryness and redissolve in 0.5 ml of 0.2 *N* perchloric acid.

5. Dansylate the samples and separate by thin-layer chromatography as described by Seiler (this volume, [2]).

6. The spots containing putrescine, spermidine, and spermine are eluted into 2 ml of ethyl acetate. The fluorescence is measured at 510 nm, exciting at 365 nm. Samples of the ethyl acetate eluates are counted to assess recovery.

IV. Specificity of Inhibitor Action

In evaluating the action of a drug on any cell system, one must assess the specificity of its action. In the case of the inhibitors of polyamine biosynthesis, are the physiological effects of the inhibitor due to polyamine depletion or to other pharmacological actions of the drug? With the lymphocyte system, we have developed a set of criteria for drug specificity that should be met, where possible, for any new drugs tested. These criteria have been fulfilled in the lymphocyte system for methylglyoxal bis(guanylhydrazone), α-methylornithine, and methylglyoxal bis(guanylhydrazone) in combination and for difluoromethylornithine.[16,19,20] The criteria are as follows:

1. The dose–response curves for inhibition of polyamine biosynthesis and the measured physiological effect (commonly [^{3}H]thymidine incorporation) should be similar.

2. Addition of exogenous polyamines to the inhibitor-treated cells should reverse the physiological defect. There are two complicating fac-

[19] R. H. Fillingame, C. M. Jorstad, and D. R. Morris, *Proc. Natl. Acad. Sci. U.S.A.* **72,** 4024 (1975).

[20] D. R. Morris, C. M. Jorstad, and C. E. Seyfried, *Cancer Res.* **37,** 3169 (1977).

tors in this experiment. As has been shown with methylglyoxal bis(guanylhydrazone) and spermidine,[2] there may be competition for entry into the cells between the drug and the added polyamine. A second complicating factor is that newborn calf serum, commonly used in culturing the cells, contains high levels of an amine oxidase active against spermidine and spermine[21]; the oxidized products of these polyamines are highly toxic to animal cells.[22-24] Therefore, prior to addition of exogenous polyamines, the cells must be placed either in medium containing horse serum,[19] which is low in the amine oxidase[24,25] or in a serum-free medium.[20]

3. Delaying the addition of the drug, to allow limited polyamine accumulation prior to inhibition, should show a relationship between polyamine levels and the physiological effect.

4. The physiological response of the cells to structural variants of the drug in question should be related to their effectiveness in inhibiting polyamine biosynthesis. This has been tested in the lymphocyte system for analogs of methylglyoxal bis(guanylhydrazone).[14] In the case of the ornithine analogs, relatively high concentrations are sometimes required to inhibit polyamine biosynthesis *in vivo*. An important control, that has been suggested,[26] is to test DL- or L-ornithine at the concentration of the ornithine analog used. In the case of 9L rat brain tumor cells, 50 m*M* ornithine produced the same effect as 50 m*M* α-methylornithine on cell proliferation, with no influence on intracellular polyamine content.[26]

[21] J. G. Hirsch, *J. Exp. Med.* **97,** 345 (1953).

[22] R. A. Alarcon, G. E. Foley, and E. J. Modest, *Arch. Biochem. Biophys.* **94,** 540 (1961).

[23] U. Bachrach, S. Abzug, and A. Bekierkunst, *Biochim. Biophys. Acta* **134,** 174 (1967).

[24] M. L. Higgins, M. C. Tillman, J. P. Rupp, and F. R. Leach, *J. Cell. Physiol.* **74,** 149 (1969).

[25] H. Blaschko, *Adv. Comp. Physiol. Biochem.* **1,** 67 (1962).

[26] J. Seidenfeld and L. J. Marton, *in* "Polyamines in Biology and Medicine" (D. R. Morris and L. J. Marton, eds.), p. 311. Dekker, New York, 1981.

[66] Use of Mammary Gland Tissue for the Study of Polyamine Metabolism and Function

By TAKAMI OKA and JOHN W. PERRY

The mammary gland provides a useful and unique model for the study of polyamine metabolism and function in the hormonal induction of cell growth and differentiation. The growth and development of the mammary gland are virtually arrested in the adult, nonpregnant state. During preg-

ISBN 0-12-181994-9

nancy mammary epithelial cells undergo extensive proliferation to form a network of lobuloalveolar cells that synthesize the characteristic milk components during the period of lactation. The concentrations of polyamines in mammary cells begin to rise during pregnancy when cell proliferation occurs and reach maximal levels during the lactation period, when milk proteins are formed.[1,2] The activities of several enzymes involved in the biosynthesis of polyamines also increase markedly during these periods.[1,2]

The morphological and biochemical changes that occur during pregnancy and lactation can be induced in culture by cultivating mouse mammary tissue explants in a chemically defined synthetic medium containing appropriate combinations of hormones.[3] When mammary tissue explants are cultured in the presence of insulin, the mammary epithelium undergoes proliferation. The addition of glycocorticoid with insulin promotes development of cellular organelles such as rough endoplasmic reticulum. Under the influence of insulin, cortisol, and prolactin, mammary epithelial cells express their differentiative function by production of the milk proteins casein and α-lactalbumin. Because this organ culture system provides a well-controlled environment and is free of certain problems that are inherent in experiments performed *in vivo,* it has proved to be useful for delineating the complex regulatory mechanism of the development of the mammary gland. This chapter describes the various experimental procedures employed in the study of polyamine biosynthesis and function in the hormone-induced development of the mammary gland in culture.

Organ Culture of the Mammary Gland

The animals were killed by cervical dislocation, and the abdominal mammary glands were removed bilaterally. Tissue explants weighing approximately 1 mg each were prepared by cutting tissue at 25° in Medium 199 (Hanks' salts) and cultured as described previously.[4] All procedures were done under sterile conditions. Insulin (crystalline pork zinc insulin, Eli Lilly) and bovine (or ovine) prolactin (NIH) were added to the medium at a final concentration of 5 μg/ml. Cortisol was used at a concentration of

[1] T. Oka, T. Sakai, D. W. Lundgren, and J. W. Perry, *in* "Hormones, Receptors and Breast Cancer" (W. L. McGuire, ed), p. 301. Raven, New York, 1978.

[2] D. H. Russell and T. McVicker, *Biochem. J.* **130,** 71 (1972).

[3] Y. J. Topper and T. Oka, *in* "Lactation" (G. L. Larson and V. R. Smith, eds.), p. 327. Academic Press, New York, 1974.

[4] Y. J. Topper, T. Oka, and B. K. Vonderhaar, this series, Vol. 39, p. 443.

0.03 μM in the study of α-lactalbumin production and 3 μM for the induction of casein synthesis.[5]

Assay of the Activity of Enzymes Involved in Polyamine Biosynthesis

The activity of ornithine decarboxylase was determined by measuring the liberation of $^{14}CO_2$ from DL-[1-^{14}C]ornithine. Tissue explants were blotted, weighed, and homogenized at 4° in a small glass homogenizer (Kontes, 1 ml capacity) with 0.3–0.4 ml of a solution containing 50 mM Tris-HCl, pH 7.5, 4 mM EDTA, 5 mM dithiothreitol, and 40 μM pyridoxal phosphate. The homogenate was centrifuged at 105,000 g for 60 min at 4°, and the supernatant fraction was removed for the enzyme assay. This fraction, which could be kept frozen at −20° for 2–3 days without any detectable loss of activity, contained virtually all the enzyme activity present in the homogenate. The amount of tissue explants used for the assay varied from 10 to 15 mg in the case of tissue from midpregnant mice to about 50 mg of virgin mouse tissue because of the variation in the basal level of the enzyme activity at different physiological states of the animals. Because enzyme activity in the mammary tissue can be altered by other factors such as osmolarity of the culture medium,[6] temperature during the preparation of explants, and the circadian rhythm of the animals, care must be taken to use consistent procedures in culture experiments. The enzyme assay was carried out in 25-ml Erlenmeyer flasks that contained 2 ml of the following reaction mixture: 5 mM dithiothreitol, 40 μM pyridoxal phosphate, 4 mM EDTA, 50 mM Tris-HCl (pH 7.5), 1 μM DL-[1-^{14}C]ornithine (specific activity 8 mCi/mmol), and 0.1 to 0.2 ml of the enzyme solution. The DL-[1-^{14}C]ornithine obtained from commercial sources was dissolved in 1 ml of 0.01 N HCl and evaporated to dryness in a rotary evaporator before use. This procedure eliminated the $^{14}CO_2$, which is apparently a contaminant in the original commercial preparation, and reduced the blank value in the assay below 80 cpm. A blank contained an equal amount of the reaction mixture without the enzyme solution. After addition of the reaction mixture and the enzyme, the flasks were closed with a rubber cap from which a glass well was suspended and shaken in a water bath at 37° for 60 min. Incubations were terminated by injecting 0.2 ml of Hyamine solution into the center glass well and 0.5 ml of 5 N H_2SO_4 into the incubation medium. Details of the procedure for trapping and counting $^{14}CO_2$ are given elsewhere.[7] Enzyme activity is

[5] M. Ono and T. Oka, *Cell* **19,** 473 (1980).
[6] J. W. Perry and T. Oka, *Biochim. Biophys. Acta* **629,** 24 (1980).
[7] T. Oka and J. W. Perry, *J. Biol. Chem.* **251,** 1738 (1976).

expressed as picomoles of $^{14}CO_2$ formed per hour per milligram of tissue explants.

The activity of *S*-adenosyl-L-methionine decarboxylase was assayed by measuring $^{14}CO_2$ release from carboxyl-labeled substrate. The preparation of enzyme extract from explants was essentially identical to that described for ornithine decarboxylase, except that about 100 mg of tissue explants were homogenized in 0.5 ml of a solution containing 25 m*M* sodium phosphate buffer, pH 7.6, 1 m*M* dithiothreitol, 0.1 m*M* EDTA, and 0.5 m*M* putrescine. The enzyme reaction was carried out in a total volume of 2 ml consisting of 0.1–0.2 ml of enzyme solution and the homogenizing buffer, which contained, per milliliter, 0.05 μCi of adenosyl[*carboxy*-^{14}C]methionine (specific activity 55 mCi/mmol). Other details of the procedure were essentially the same as those described for ornithine decarboxylase. Enzyme activity was expressed as picomoles of $^{14}CO_2$ formed per hour per milligram of tissue explants. The enzyme activity in mammary epithelium of cultured tissue is stimulated by the synergistic activity of insulin and glucocorticoid.[8] The enzyme from lactating mouse mammary gland has been purified to apparent homogeneity.[9] An antibody to the pure enzyme was used in measuring levels of the enzyme in the extract of mammary tissue.[9]

The activity of spermidine synthase was assayed by measuring the production of spermidine. Enzyme extract from mammary explants was prepared by homogenizing 50–100 mg of explants at 4° in 0.5 ml of 25 m*M* phosphate buffer, pH 7.2, containing 1 m*M* dithiothreitol and 0.1 m*M* EDTA. The homogenate was centrifuged for 60 min at 105,000 *g* at 4°, and the resultant supernatant solution was used for the enzyme assay. The reaction was carried out at 37° for 60 min in test tubes that contained 25 m*M* sodium phosphate buffer (pH 7.6), 0.1 m*M* EDTA, 1 m*M* dithiothreitol, 0.1 m*M* decarboxylated adenosylmethionine, 1 m*M* [1,4-^{14}C]putrescine (specific activity, 55 mCi/mmol), and 0.1 ml of the enzyme solution in a volume of 0.15 ml as described previously.[10] A blank contained an equal amount of the reaction mixture without enzyme. Decarboxylated adenosylmethionine was synthesized enzymatically by the method of Tabor[11] with *E. coli S*-adenosylmethionine decarboxylase purified through step 4 by the method of Wickner *et al.*[12] The activity was

[8] T. Oka and J. W. Perry, *J. Biol. Chem.* **249,** 7674 (1974).

[9] T. Sakai, C. Hori, K. Kano, and T. Oka, *Biochemistry* **18,** 5541 (1979).

[10] J. Jänne, A. Schenone, and H. G. Williams-Ashman, *Biochem. Biophys. Res. Commun.* **42,** 758 (1971).

[11] C. W. Tabor, this series, Vol. 5, p. 754.

[12] R. B. Wickner, C. W. Tabor, and H. Tabor, *J. Biol. Chem.* **245,** 2132 (1970).

expressed as picomoles of spermidine formed per hour per milligram of tissue explant.[13] The increase in enzyme activity in cultured explants is dependent on the presence of insulin and glucocorticoid.[13,14]

The activity of arginase in mammary explants was determined by measuring the formation of urea from arginine as described earlier.[15] About 20–30 mg of explants were homogenized in 1 ml of distilled water, and the homogenate was treated at 55° for 5 min in the presence of 0.01 *M* $MnCl_2$. The assay was initiated by the addition of 1 ml of 0.25 *M* L-arginine (pH 9.7) and 0.01 ml of 0.1 *M* $MnCl_2$. After incubation at 37° for 15 min, the reaction was stopped by the addition of 2 ml of 6% (w/v) trichloroacetic acid, and the protein precipitate was removed by centrifugation. The amount of urea formed was determined colorimetrically[15] using a standard curve constructed with known concentrations of urea. Activity was expressed as micrograms of urea formed per 10 min per milligram wet weight of tissue.[16] The synergistic action of insulin and prolactin enhances the enzyme activity in cultured tissue.[16]

In the mammary gland of virgin and pregnant mice, mammary epithelial cells are surrounded by a considerable number of fat cells. To study the effects of hormones on enzyme activity in the epithelium of mammary explants, it is important to distinguish the response of epithelial cells from that of fat cells. This can be achieved by determining enzyme activity both in whole explants and in the isolated epithelial cells derived from the mammary explants of the parallel culture. The differential response of epithelial and fat cells to hormones was observed in the study of several enzymes involved in polyamine biosynthesis in cultured mammary tissue.[7,8,13]

Determination of Polyamine Content

The content of putrescine, spermidine, and spermine in mammary explants was quantitated with ninhydrin using an amino acid analyzer.[17] About 50–100 mg of explants were homogenized in 1 ml of 3% perchloric acid. After centrifugation, the residue was washed by another 1 ml of 3% perchloric acid, and the combined supernatant fractions were neutralized with 4 *N* KOH. The precipitate was removed by centrifugation, and the

[13] T. Oka, J. W. Perry, and K. Kano, *Biochem. Biophys. Res. Commun.* **79,** 979 (1977).

[14] T. Sakai, D. W. Lundgren, and T. Oka, *J. Cell. Physiol.* **95,** 259 (1978).

[15] R. T. Schimke, *J. Biol. Chem.* **39,** 3809 (1964).

[16] T. Oka and J. W. Perry, *Nature* (*London*) **250,** 660 (1974).

[17] D. W. Lundgren, P. M. Farrell, and P. A. di Sant'Agnese, *Proc. Soc. Exp. Biol. Med.* **152,** 81 (1976).

supernatant solution was lyophilized. The dried residue was dissolved in a potassium chloride–potassium citrate buffer (2.34 M K^+), pH 5.64, and assayed for the amount of polyamine.[14] This method is more sensitive than the high-voltage electrophoretic method used in earlier studies.[8] In cultured mammary tissue from virgin mice, insulin increases the content of putrescine and spermidine about threefold over the initial level during a 3-day culture, whereas the content of spermine remains relatively unchanged.[14] In culture of mammary tissue from midpregnant mice, the synergistic action of insulin, cortisol, and prolactin increases the concentration of spermidine in mammary epithelium about threefold in 2 days.[8] The combination of insulin and prolactin or the combination of insulin and cortisol enhances the polyamine level to a small extent.[8] The concentration of spermine in mammary epithelium increased approximately 30% during 2 days, but no appreciable difference was found among various combinations of the three hormones.[8]

Uptake of Polyamines

Mouse mammary gland has been shown to possess a transport system for spermidine, spermine, and putrescine.[18] In order to measure the uptake of polyamine, 10–20 mg of explants were incubated in Medium 199 with appropriately labeled polyamines. The amount of radioactive polyamine added to the culture was 1–2 μCi/ml and 0.01 μCi/ml for ^{3}H-labeled and for ^{14}C-labeled polyamine, respectively. Tritium- or ^{14}C-labeled polyamines (specific activity 100–1000 mCi/mmol and 5–20 mCi/mmol, respectively) were obtained from commercial sources, checked for purity, and, whenever necessary, purified by high-voltage electrophoresis prior to use.[8] After addition of labeled polyamine, explants were harvested at various times, weighed, and placed for washing on a Whatman GS/C disk filter paper that had been immersed for at least 15 min in Medium 199 containing 40 mM corresponding unlabeled polyamine to prevent adsorption of the residual external radioactive polyamines. Explants were washed on a Millipore filter apparatus with 15 ml of Medium 199 and 4 ml of phosphate-buffered 0.15 M NaCl (pH 7.2), using suction. Each washing solution also contained the appropriate unlabeled polyamine at 40 mM. This washing procedure removed approximately 10% of the radioactivity associated with explants, which presumably represented nonspecific adsorption of the polyamines to the cells. The washed explants on the filters were transferred into a scintillation vial, digested with 1 ml of tissue solubilizer, and assayed for radioactivity

[18] K. Kano and T. Oka, *J. Biol. Chem.* **251,** 2795 (1976).

in a toluene-based scintillation fluid with a liquid scintillation spectrometer. Insulin stimulates the uptake of polyamines by enhancing V_{max} for polyamine influx and preventing efflux of polyamine.[18] Prolactin, in the presence of insulin, elicits a greater increase in V_{max} for polyamine uptake.[18]

Assessment of the Function of Polyamines in the Mammary Gland

Hormonal stimulation of the mammary gland in culture causes marked enhancement of polyamine accumulation prior to acceleration of DNA synthesis[14] and milk-protein synthesis.[8] Since the mammary epithelium in virgin mice is essentially nonproliferative, but can be induced to undergo DNA synthesis in culture with insulin, this system is suited for examining the role of polyamines in the proliferation of mammary epithelium. On the other hand, the mammary cells in pregnant mice are in an active phase of proliferation and can be readily stimulated to undergo differentiation. Thus organ culture of pregnant mouse mammary tissue provides a system to study the function of polyamines in mammary differentiation.

The biological importance of polyamine accumulation in the development of the mammary gland can be assessed by employing various culture conditions in which the intracellular level of polyamines is varied by alterations in hormone combination,[8] osmolarity,[6] amino acid composition of culture medium,[16] use of several inhibitors of ornithine decarboxylase,[19] and *S*-adenosylmethionine decarboxylase,[8] and finally, by use of mammary explants derived from virgin and pregnant mice containing initially a low and a high level of polyamines.[14,19] It is also possible to change the intracellular concentration of polyamines by exogenously adding them to culture.[8,14] Combinations of these experimental approaches have yielded some useful information concerning the role of polyamines in the development of the mammary gland.

In cultured mammary explants from virgin mice, both putrescine and spermidine appear to be important for hormonal induction of DNA synthesis in mammary epithelium.[14,19] Furthermore, several lines of evidence suggest that spermidine (0.04 m*M*) can simulate the action of glucocorticoid on induction of α-lactalbumin synthesis in cultured tissue from midpregnant mice.[8] It is noteworthy that stimulation of synthesis of milk protein in rabbit mammary tissue can occur in the presence of only insulin and prolactin and does not require the action of glucocorticoid and spermidine.[20] On the other hand, synthesis of milk protein in rat mammary

[19] H. Inoue and T. Oka, *J. Biol. Chem.* **255,** 3308 (1980).
[20] L. M. Houdebine, E. Devinoy, and C. Delouis, *Biochimie* **60,** 735 (1978).

gland is dependent on both glucocorticoid and spermidine, but spermidine, in place of cortisol, is unable to act synergistically with insulin and prolactin.[21] The reason for the observed species differences remains unknown. It should be emphasized that some caution must be exercised in interpreting these experimental results obtained by use of enzyme inhibitors and supraphysiological conditions because of other possible side effects.

[21] F. F. Bolander and Y. J. Topper, *Biochem. Biophys. Res. Commun.* **90,** 1131 (1979).

[67] Ornithine Decarboxylase Assay Permitting Early Determination of Histocompatibility in the Mixed Lymphocyte Reaction

By ACHILLES A. DEMETRIOU, CELIA WHITE TABOR, and HERBERT TABOR

The rejection of transplanted organs is a consequence of genetic disparity between donor and recipient. The human major histocompatibility complex (MHC), termed HLA, is localized on chromosome six and it includes three loci (*A, B, C*) whose products can be recognized by complement-dependent lymphocytotoxicity techniques. Another important component is the *HLA-D* locus, whose products are recognizable by the mixed lymphocyte culture (MLC) test. In the MLC test, lymphocytes from a donor and a recipient are mixed *in vitro,* and antigenic determinants coded by the *HLA-D* region, which are present on the cell surface, provoke a mixed lymphocyte reaction of proliferation and blast transformation when genetically dissimilar lymphocytes interact. At present, the most widely used assay of mixed lymphocyte reactivity measures the incorporation of tritiated thymidine into DNA and requires 4–5 days.[1] This long waiting period has hampered the application of this technique to the prospective selection of recipients for cadaver organ transplantation. Changes in the activity of ornithine decarboxylase, at 18 hr after mixing in the MLC, can be used as an early, sensitive indicator of the degree of histocompatibility.[2]

[1] F. H. Bach and K. Hirschhorn, *Science* **142,** 813 (1964).

[2] A. A. Demetriou, M. W. Flye, C. W. Tabor, and H. Tabor, *Transplantation* **27,** 190 (1979).

METHODS IN ENZYMOLOGY, VOL. 94

ISBN 0-12-181994-9

Assay Procedure

Mixed Lymphocyte Cultures[3]

Using sterile techniques, lymphocytes are obtained from about 50 ml of heparinized blood by the Ficoll–Hypaque sedimentation method of Boyum.[4] The cells are suspended in RPMI 1640 (Gibco Laboratories) supplemented with 10% fetal pig serum or 10% fetal calf serum, 100 units of penicillin per milliliter, and 100 μg of streptomycin per milliliter, at a concentration of 5×10^6 cells/ml. One-tenth milliliter of a lymphocyte suspension from each of two different individuals is mixed in a series of 9 wells of a Microtiter plate (Dynatech Laboratories, Inc.). Control wells contain 0.2 ml of a lymphocyte suspension from a single individual. In the "one-way" MLC reaction, the 0.1 ml of cells from one individual are irradiated with 2000 R from a Gammator M cesium source (Isomedix, Inc.) and mixed with 0.1 ml of untreated lymphocytes from the second individual. The plates are incubated at 37°, 100% humidity, with 5% CO_2 and 95% air; the usual incubation time is 18 hr.

Ornithine Decarboxylase Assay

Principle. Ornithine decarboxylase activity is determined by the release of $^{14}CO_2$ from [1-^{14}C]ornithine.[5] The reaction is carried out in scintillation vials, and $^{14}CO_2$ is trapped in a Hyamine-soaked filter paper.[6]

Reagents

- Buffer A: 50 m*M* Tris-HCl, pH 7.1, 50 μM pyridoxal phosphate, 5 m*M* dithiothreitol, and 0.1 m*M* EDTA
- L-[1-^{14}C]Ornithine, specific activity 45 mCi/mmol (New England Nuclear).

Procedure. The lymphocytes in the wells of the Microtiter plates are suspended and transferred to centrifuge tubes; the contents of three replicate wells are pooled in a single tube. The cells are collected by sedimentation, washed twice with 0.15 *M* NaCl (0.5 ml), and disrupted by freezing and thawing in 0.5 ml of buffer A. The assay mixture consists of 0.2 ml of extract and 0.5 μCi (10 μl) of L-[1-^{14}C]ornithine. Immediately after the labeled ornithine is added, the scintillation vials are tightly capped. Each vial cap contains a 1.8-cm square of Whatman No. 3 filter

[3] See also this volume [65] for the preparation of lymphocytes.
[4] A. Boyum, *Scand. J. Clin. Lab. Invest.* **21,** 77 (1968).
[5] J. Jänne and H. G. Williams-Ashman, *J. Biol. Chem.* **246,** 1725 (1971).
[6] R. B. Wickner, C. W. Tabor, and H. Tabor, *J. Biol. Chem.* **245,** 2132 (1970).

paper soaked with 20 μl of Hyamine hydroxide wedged on its inner surface. The vials are incubated at 37° for 60 min. The reaction is terminated by adding 0.5 ml of 5 *N* H_2SO_4, and incubation is continued for 60 min at 37° with gentle shaking. The filter papers containing the radioactivity are then counted in scintillation vials containing 5 ml of Aquascint-II (ICN Pharmaceuticals, Inc.). All assays are done in triplicate after pooling cell suspensions from three wells; the results vary by less than 10%.

Interpretation. A 10-fold increase in ornithine decarboxylase activity is seen in mixed lymphocyte cultures from outbred animals or genetically unrelated individuals at 18 hr after lymphocyte mixing. No such increase is seen when cells from genetically inbred animals, major histocompatibility complex-compatible animals, or identical twins are mixed.[2] It is possible to use changes in ornithine decarboxylase activity as an early assay of histocompatibility in mixed lymphocyte cultures. Further studies are needed, however, to determine whether the enzyme assay can detect small differences in the major histocompatibility complex and thus have clinical usefulness.

Section XII

Analogs and Derivatives

[68] Novel Polyamines in *Thermus thermophilus:* Isolation, Identification, and Chemical Synthesis

By TAIRO OSHIMA

An extreme thermophile, *Thermus thermophilus,*[1] is a treasure-house for discovery of new naturally occurring polyamines, and a variety of novel polyamines[2] have been isolated from the thermophilic cell. The major polyamines in the bacterial cell grown under the optimal conditions are thermine[3] and thermospermine.[4] In addition, the extreme thermophile contained norspermidine, spermidine, and *sym*-homospermidine as minor components.[5] Two pentaamines have been detected in the cells grown at higher temperatures.[6,7] Some strongly basic, unidentified compounds are also present in the cells (see Fig. 1). Some of these novel polyamines are found also in other organisms.

Polyamine Analysis

The polyamine composition could be analyzed by gas chromatography, thin-layer chromatography, ion-exchange chromatography, or high-performance liquid chromatography (see Bachrach[8] for a review). In the author's laboratory, polyamines in cells and tissues are assayed by a simplified method using an automatic amino acid analyzer.[4] A short column (8 mm wide × 7 cm high) packed with CK-10S cation exchange resin (Mitsubishi Chemical Co., Tokyo; Durrum DC-1A, Aminex A-6, or

[1] T. Oshima and K. Imahori, *Int. J. Syst. Bacteriol.* **24,** 102 (1974).

[2] Trivial names, systematic names, and chemical structures of the usual and novel polyamines described in this chapter are norspermidine(=caldine): 1,7-diamino-4-azaheptane; $NH_2(CH_2)_3NH(CH_2)_3NH_2$; spermidine: 1,8-diamino-4-azaoctane, $NH_2(CH_2)_3$-$NH(CH_2)_4NH_2$; *sym*-homospermidine: 1,9-diamino-5-azanonane, $NH_2(CH_2)_4NH(CH_2)_4$-NH_2; thermine(=norspermine): 1,11-diamino-4,8-diazaundecane, $NH_2(CH_2)_3NH(CH_2)_3$-$NH(CH_2)_3NH_2$; thermospermine: 1,12-diamino-4,8-diazadodecane, $NH_2(CH_2)_3$-$NH(CH_2)_3NH(CH_2)_4NH_2$; spermine: 1,12-diamino-4,9-diazadodecane, $NH_2(CH)_2)_3$-$NH(CH_2)_4NH(CH_2)_3NH_2$; caldopentamine: 1,15-diamino-4,8,12-triazapentadecane, $NH_2(CH_2)_3NH(CH_2)_3NH(CH_2)_3NH(CH_2)_3NH_2$; and homocaldopentamine: 1,16-diamino-4,8,12-triazahexadecane, $NH_2(CH_2)_3NH(CH_2)_3NH(CH_2)_3NH(CH_2)_4NH_2$.

[3] T. Oshima, *Biochem. Biophys. Res. Commun.* **63,** 1093 (1975).

[4] T. Oshima, *J. Biol. Chem.* **254,** 8720 (1979).

[5] T. Oshima and M. Baba, *Biochem. Biophys. Res. Commun.* **103,** 156 (1981).

[6] T. Oshima, *J. Biol. Chem.* **257,** 9913 (1982).

[7] T. Oshima and S. Kawahata, *J. Biochem.* (*Tokyo*) **93,** in press.

[8] U. Bachrach, *Adv. Polyamine Res.* **2,** 5 (1978).

ISBN 0-12-181994-9

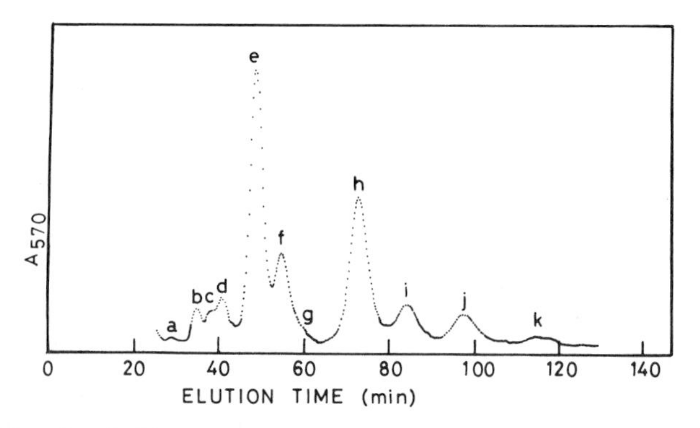

FIG. 1. Polyamines in *Thermus thermophilus* cells grown at 80°. The cells were harvested at the mid-log phase. Polyamines were extracted in 5 ml of 5% trichloroacetic acid from 1 g of wet cells. The extract (0.2 ml) was diluted with 0.8 ml of H_2O and applied on Hitachi 034 automatic amino acid analyzer (full scale = 0.08 absorbance unit) equipped with a CK-10S column (0.8 × 7 cm) and developed with the potassium citrate-HCl buffer containing 2.01 *M* KCl and 5% ethanol (v/v) as described in the text. Compound a = putrescine, b = norspermidine, c = spermidine, d = *sym*-homospermidine, e = thermine, f = thermospermine, g = unknown, h = caldopentamine, i = homocaldopentamine, and j and k = unknown.

other equivalent products can be used) was connected to an analyzer. The column was eluted with a single buffer at a flow rate of 0.5 ml/min, 65°. The standard buffer was prepared by dissolving 38.1 g of $CH_3COONa \cdot 3H_2O$, 7.2 g of CH_3COOH, 146.1 g of NaCl, and 0.4 g of Brij 35 in about 900 ml of H_2O. 2-Propanol (50 ml) was added to the solution, and the final volume was adjusted to 1.000 ml by adding H_2O. Usually no pH adjustment was done. In most cases, the column was used continuously without alkali washing, and the samples were injected every 150–180 min using an automatic sampler. Polyamines were detected with ninhydrin reaction. To determine the factor for the color development, a standard solution containing 50 nmol/ml each of the authentic polyamines was applied to the analyzer.

When pentaamine analysis was concerned, an alternative buffer was often used.[6] The buffer was prepared by dissolving 32.4 g of $K_3C_6H_5O_7 \cdot H_2O$ (potassium citrate), 65.0 ml of 1 *N* HCl, 150 g of KCl, 50 ml of ethanol, and 0.4 g of Brij 35 in H_2O; the final volume was 1000 ml.

The cell suspension (1 g wet cells in 2 ml of H_2O or suitable buffer) or tissue homogenate was mixed with an equal volume of 10% trichloroacetic acid followed by centrifugation. The precipitate was washed with 5% trichloroacetic acid. The supernatant and the wash were combined, and an aliquot was applied to the analytical column after appropriate dilution with H_2O. In case of *T. thermophilus* cells at the mid-log phase,

0.2 ml of trichloroacetic acid extract (ca 5 ml per gram of wet cells) was diluted with 0.8 ml of H_2O, and the diluted solution was applied on the analytical column.

For conjugated polyamine analysis, a part of the trichloroacetic acid extract was mixed with an equal volume of 12 *N* HCl, and heated at 105° for 16 hr. The hydrolyzate was evaporated under reduced pressure at 40° and dissolved in an appropriate volume of H_2O. In cases of animal and plant tissues, it was often necessary to remove free amino acids prior to the analysis. For this purpose, an aliquot of the extract or the HCl hydrolyzate (containing up to 5 μmol of total polyamine) was applied on a short column of Dowex 50W-X4, H^+ form (2.7 mm wide × 5.0 cm long) after being diluted with H_2O. The column was washed with 1 ml of H_2O, 10 ml of sodium phosphate buffer[9] (dissolve 1.7 g of $Na_2HPO_4 \cdot 12H_2O$, 0.04 g of $NaH_2PO_4 \cdot 2H_2O$, and 2.0 g of NaCl in final 50 ml of H_2O), and 2 ml of 1 *N* HCl. The polyamines were eluted with 8 ml of 6 *N* HCl. The eluate was evaporated under reduced pressure at 40° and then dissolved in an appropriate volume of H_2O.

Figure 1 shows an example of polyamine composition of the extreme thermophile, *T. thermophilus*.

Notes on the Analytical Method. In the routine assay described above, elution times of homospermidine (1,9-diamino-4-azanonane) and thermine are close to each other. Those of agmatine and spermine (or thermospermine) are also close. It is sometimes necessary to confirm the absence of homospermidine and agmatine by using a buffer containing lower concentration of salt. Spermine and thermospermine cannot be separated by the present method. To confirm the presence of thermospermine or spermine, it is necessary to isolate the tetraamine to be studied and to record its infrared (IR) spectrum.

Isolation and Identification of Novel Polyamines

Culture. The polyamine composition of *T. thermophilus* strain HB8 (ATCC 27634) depends on the growth conditions. In general, the younger cells contain more polyamines than the older cells. Triamines are rich in the cells grown at lower temperatures (55–60°). Thermine and pentaamines are rich in cells grown at high temperature extreme (80–83°). A synthetic medium[5] used for the growth contained 20 g of sucrose, 20 g of monosodium glutamate, 2 g of NaCl, 0.25 g of KH_2PO_4, 0.5 g of K_2HPO_4, 0.5 g of $(NH_4)_2SO_4$, 0.125 g of $MgCl_2 \cdot 6H_2O$, 0.025 g of $CaCl_2 \cdot 2H_2O$, 6 mg of $FeSO_4 \cdot 7H_2O$, 0.8 mg of $CoCl_2 \cdot 6H_2O$, 20 μg of $NiCl_2 \cdot 6H_2O$,

[9] H. Inoue and A. Mizutani, *Anal. Biochem.* **56,** 408 (1973).

1.2 mg of $Na_2MoO_4 \cdot 2H_2O$, 0.1 mg of $VOSO_4$, 0.5 mg of $MnCl_2 \cdot 4H_2O$, 60 μg of $ZnSO_4 \cdot 7H_2O$, 15 μg of $CuSO_4 \cdot 5H_2O$, 100 μg of biotin, and 1 mg of thiamin per liter. The final pH was adjusted to 7.0–7.2. The addition of biotin and trace metals was essential for the growth.

Isolation of Caldopentamine.[6] The thermophile cells grown in the synthetic medium at 80° and harvested at the mid-log phase (250 g wet weight) were suspended in 500 ml of H_2O. To the suspension 500 ml of 10% trichloroacetic acid were added, and the mixture was centrifuged at 5000 rpm. The precipitate was resuspended in 1 liter of 5% trichloroacetic acid, and the suspension was centrifuged. The supernatants were combined (total volume = 1.96 liters, containing 225 μmol of caldopentamine) and diluted to 5 liters with H_2O. The diluted extract was applied on a Dowex 50W-X4, H^+ form (2 cm wide × 22 cm long). The column was washed with 100 ml of H_2O, 100 ml of phosphate buffer (3.4 g of $Na_2HPO_4 \cdot 12H_2O$, 0.08 g of $NaH_2PO_4 \cdot 2H_2O$, and 4.0 g of NaCl in a final volume of 100 ml), 200 ml of H_2O, and finally 200 ml of 2 *N* HCl. The polyamines were eluted with 6 *N* HCl. Fractions (5 ml/tube) were collected. An aliquot (4 μl) of each fraction was diluted with 1 ml of H_2O and applied on the analytical column; chromatography was carried out as described in the preceding section. Fractions containing caldopentamine were combined and evaporated under reduced pressure at 40°. The residue was dissolved in about 10 ml of H_2O, and charged on a CK-10S column (9 mm in diameter × 40 cm long). The column was eluted with 0.1 *M* potassium citrate-HCl buffer, pH 6.0, containing 1.74 *M* KCl (final K^+ concentration = 2.04 *M*). If necessary, repeat the chromatography. Fractions that contained caldopentamine, but were not contaminated with other polyamines, were combined (total volume 250 ml) and diluted to 1 liter. The diluted solution was passed through a column of Dowex 50W-X4 (6 mm wide × 7 cm long). The column was washed with 5 ml of H_2O and 2 ml of 2 *N* HCl and then eluted with 6 *N* HCl. Fractions (20 drops/tube) containing caldopentamine were combined and evaporated under reduced pressure at 40°. The residue was dissolved in the smallest volume of hot 50% ethanol. The crystals of the hydrochloride salt formed by cooling the solution were collected by filtration and dried at 60°. The yield was 22 mg.

Other novel polyamines (thermine, thermospermine, and homocaldopentamine) were isolated in a similar manner. Triamines were also isolated with a slight modification.[5] The CK-10S column was eluted with 0.28 *M* sodium acetate buffer, pH 5, containing 2.22 *M* NaCl.

Identification. Chemical structure of an unusual polyamine can be predicted from its ^1H NMR, ^{13}C NMR, and mass spectra. The table summarizes spectral data and the possible assignments. Copies of IR spectra

(KBr disc) of the novel polyamines described in this chapter will be available from the author upon request.

Chemical Synthesis

Novel polyamines can be synthesized by adding an aminopropyl or aminobutyl group to a terminal amino group of a polyamine. Ganem *et al.*[10] reported an elegant method for the synthesis of thermospermine. Okada *et al.*[11] also reported the chemical synthesis of polyamines and related compounds. Yanagawa *et al.*[12] reported the synthesis and biochemical properties of a polyamine analog.

The following methods may not be the optimal ones from the point of view of an organic chemist, but they are suitable for enzymologists, biochemists, and biologists, since the methods do not require the sophisticated equipment or techniques that are often used by organic chemists but that are unfamiliar to biochemists and biologists. Isolation of an intermediate is not involved in the following methods. Starting materials can be easily purchased from commercial companies. All solvents used are familiar ones for biochemists.

Caldopentamine.[6] An aminopropyl group can be added to a polyamine by the following reaction.

$$R-NH_2 + CH_2=CH-CN \longrightarrow R-NHCH_2CH_2CN \xrightarrow{+H_2} R-NHCH_2CH_2CH_2NH_2$$

Dissolve thermine free base [*N,N'*-bis(3-aminopropyl)-1,3-propanediamine, 188 mg, Tokyo Kasei Co., Tokyo; available also from Eastman Kodak Company] in 5 ml of ethanol. Add 53 mg of acrylonitrile and keep the solution at room temperature for 16 hr. Adjust the mixture to pH 6.0 with acetic acid after the addition of 5 ml of 2% $CoCl_2$ solution (in H_2O). Add slowly 5 ml of 10% $NaBH_4$ solution. During the addition, the mixture should be kept at pH 6 by adding acetic acid. Reflux the resulting black solution for 1 hr. Caldopentamine formed is separated from unreacted thermine and other products by using a CK-10S column chromatography as described in the section Isolation and Identification of Novel Polyamines. The reaction mixture can be applied on a CK-10S column (9 mm wide × 40 cm long) after diluting with 3 times the volume of H_2O. Crystals of the hydrochloride salt were obtained from a hot 50% ethanol solution. The overall yield was 100 mg.

[10] K. Chantrapromma, J. S. McManis, and B. Ganem, *Tetrahedron Lett.* **21,** 2475 (1980).
[11] M. Okada, S. Kawashima, and K. Imahori, *J. Biochem.* (*Tokyo*) **85,** 1235 (1979).
[12] H. Yanagawa, Y. Ogawa, and F. Egami, *J. Biochem.* (*Tokyo*) **80,** 891 (1976).

^{1}H-NMR, ^{13}C-NMR, AND MASS SPECTRA OF POLYAMINES

Trivial name	Chemical structure, mass spectrum[a]	^{1}H NMR[b] δ (ppm)	^{1}H NMR[b] Assignment	^{13}C NMR[b] δ (ppm)	^{13}C NMR[b] Assignment
sym-Homospermidine	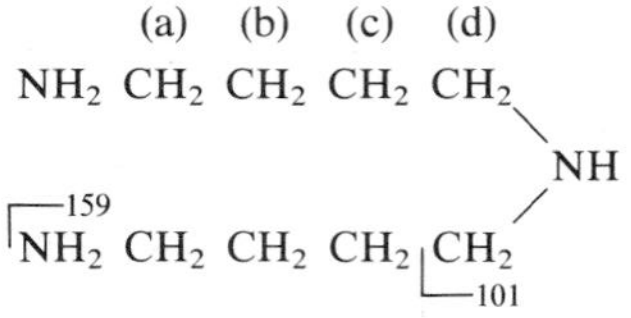 (a) (b) (c) (d) NH_2 CH_2 CH_2 CH_2 CH_2 \ NH	1.8 (1)	(b)+(c)	22.4 (1) 23.6 (1)	(b),(c)
	159 ┌ NH_2 CH_2 CH_2 CH_2 | CH_2 / └ 101	3.1 (1)	(a)+(d)	38.4 (1) 46.5 (1)	(a) (d)
Thermine	(a) (b) (c) (d) NH_2 CH_2 CH_2 CH_2 NH CH_2 \ (e) CH_2	2.2 (1)	2×(b)+(e)	23.4 (1) 24.5 (2)	(e) (b)
	188 ┌ NH_2 CH_2 | CH_2 ┌144 CH_2 NH CH_2 / └ 158	3.2 (2)	2×(a)+2×(c) +2×(d)	37.5 (2) 45.4 (2) 45.5 (2)	(a) (c) (d)

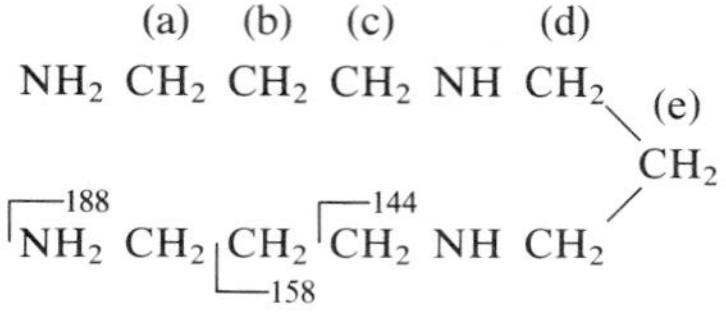

Thermospermine	(a) (b) (c) (d) $NH_2\ CH_2\ CH_2\ CH_2\ NH\ CH_2$— (e) (f) (g) (h) (i) $CH_2\ CH_2\ NH\ CH_2\ CH_2\ CH_2$— (j) $CH_2\ NH_2$ (fragments: 158, 144, 172, 202)	1.8 (1) 2.1 (1) 3.2(3)	(h)+(i) (b)+(e) (a)+(c)+(d) +(f)+(g)+(j)	23.5 (1) 23.6 (1) 24.6 (1) 24.8 (1) 37.4 (1) 39.6 (1) 45.3 (1) 45.5 (1) 45.6 (1) 47.9 (1)	(b),(e), (h),(i) [23.5–24.8] (a),(j) [37.4–39.6] (c),(d), (f),(g) [45.3–47.9]
Spermine	(a) (b) (c) (d) (e) $NH_2\ CH_2\ CH_2\ CH_2\ NH\ CH_2\ CH_2$ | $NH_2\ CH_2\ CH_2\ CH_2\ NH\ CH_2\ CH_2$ (fragments: 172, 202, 158)	1.8 (1) 2.1 (1) 3.2 (3)	(e) (b) (a)+(c)+(d)	23.6 (1) 24.6 (1) 37.4 (1) 45.4 (1) 47.8 (1)	(e) (b) (a) (c) (d)
Caldopentamine	(a) (b) (c) (d) (e) (f) $NH_2\ CH_2\ CH_2\ CH_2\ NH\ CH_2\ CH_2\ CH_2$ NH $NH_2\ CH_2\ CH_2\ CH_2\ NH\ CH_2\ CH_2\ CH_2$ (fragments: 245, 201, 215)	2.1 (1) 3.2 (2)	(b)+(e) (a)+(c)+(d) +(f)	26.8 (1) 28.0 (1) 38.8 (1) 46.4 (1) 46.8 (2)	(e) (b) (a) (c) (d)+(f)

[a] Figures indicate *m/e* values of major peaks in the spectrum and their assignments.

[b] Polyamine hydrochloride salt was dissolved in D_2O. The pH adjustment was not done. Figures in parentheses indicate the relative intensity.

Thermine can also be synthesized in a similar manner using norspermidine as the starting material.

Thermospermine. An aminobutyl group can be added to an amino group of a polyamine by the following reaction.[11]

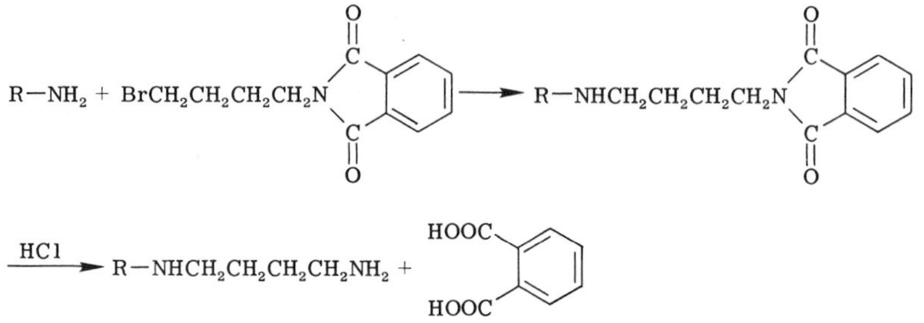

$$R{-}NH_2 + BrCH_2CH_2CH_2CH_2N(CO)_2C_6H_4 \longrightarrow R{-}NHCH_2CH_2CH_2CH_2N(CO)_2C_6H_4$$

$$\xrightarrow{HCl} R{-}NHCH_2CH_2CH_2CH_2NH_2 + C_6H_4(COOH)_2$$

Place 130 mg of norspermidine (Tokyo Kasei or Eastman Kodak) and 5 ml of 2-propanol in a Pyrex test tube 18 mm in diameter with a screw cap. Add 280 mg of *N*-(4-bromobutyl)phthalimide (Aldrich Chemical Co., Milwaukee, Wisconsin) to the solution, cap the tube tightly, and incubate the test tube in a heating block (105°). After 16 hr, add 5 ml of 12 *N* HCl and place again in the heating block (105°) for 8 hr. Discard the upper layer by using a Pasteur pipette. Dilute the aqueous (lower) layer with 100 ml of H_2O, and then apply on a CK-10S column. Develop the column with 0.1 *M* potassium citrate-HCl buffer, pH 6.0, containing 1.74 *M* KCl. A by-product, $NH_2(CH_2)_4N((CH_2)_3NH_2)_2$, can be separated from thermospermine. Salts in the pooled fraction can be removed by using a short column of Dowex 50W-X4 as described above. Thermospermine hydrochloride salt can be precipitated by adding an excess amount of methanol–ethanol (1 : 1) mixture after dissolving the salt in the minutest quantity of H_2O. The yield was about 100 mg (28%).

sym-Homospermidine (yield 22%) and homocaldopentamine (yield 5%) were synthesized in a similar manner from putrescine and thermine, respectively.

The purity of the synthesized polyamine can be checked by using analytical chromatography as described in the preceding section. Impurity can also be detected with a ^{13}C NMR or IR spectrum. If the preparation is contaminated with the by-product(s) and/or the starting material, repeat the ion-exchange chromatography with a CK-10S column.

Distribution in Other Organisms

The novel polyamines are also present in other extreme thermophiles. Polyamine composition of *Thermus aquaticus* is similar to that of *T. ther-*

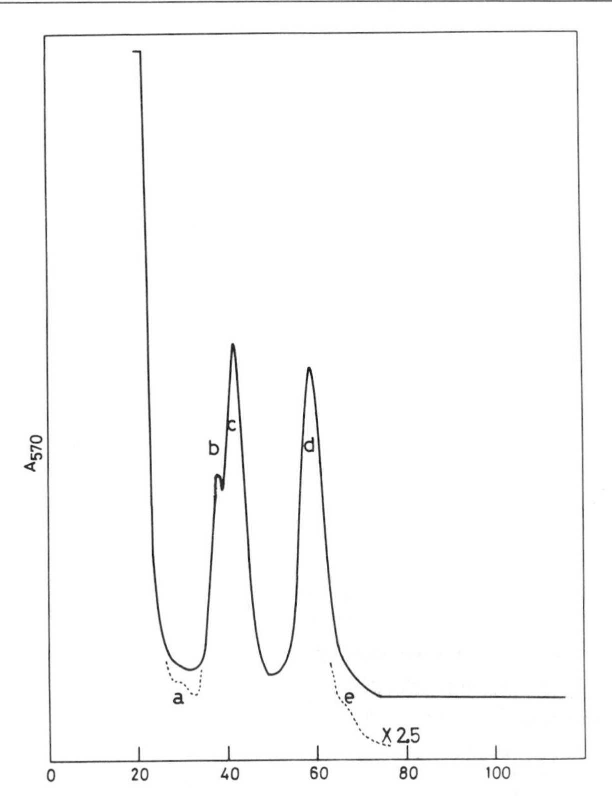

FIG. 2. Polyamines in an Antarctic shrimp, *Euphausia superba*. The shrimp (10.3 g wet weight) was homogenized in 10 ml of H_2O. Trichloroacetic acid (10%, 10 ml) was added. Polyamines were reextracted from the pellet with 20 ml of 5% trichloroacetic acid. The supernatants were combined and charged on a Dowex 50W-X4 column. Polyamines were eluted with 6 *N* HCl after the column had been washed with the phosphate buffer described in the text to remove most of free amino acids in the extract; the HCl was evaporated, and the amines were finally dissolved in 2 ml of H_2O, as described in the text. An aliquot (0.4 ml) was diluted with 0.6 ml of H_2O and analyzed using a CK-10S column (0.8 × 7 cm) developed with the standard buffer described in the text. Compounds a, b, c, d, and e were putrescine, norspermidine, spermidine, thermine, and spermine and/or thermospermine, respectively.

mophilus. Extreme acidothermophiles, *Caldariella acidophila*[13] and *Sulfolobus acidocaldarius,*[14] contained norspermidine, thermine, and thermospermine and/or spermine, but no *sym*-homospermidine. On the other hand, moderate thermophiles, *Bacillus stearothermophilus* and *Bacillus acidocaldarius,* contained no novel polyamines.

[13] M. DeRosa, S. DeRosa, A. Gambacorta, M. Cartenì-Farina, and V. Zappia, *Biochem. Biophys. Res. Commun.* **69,** 253 (1976).

[14] T. Oshima, unpublished data.

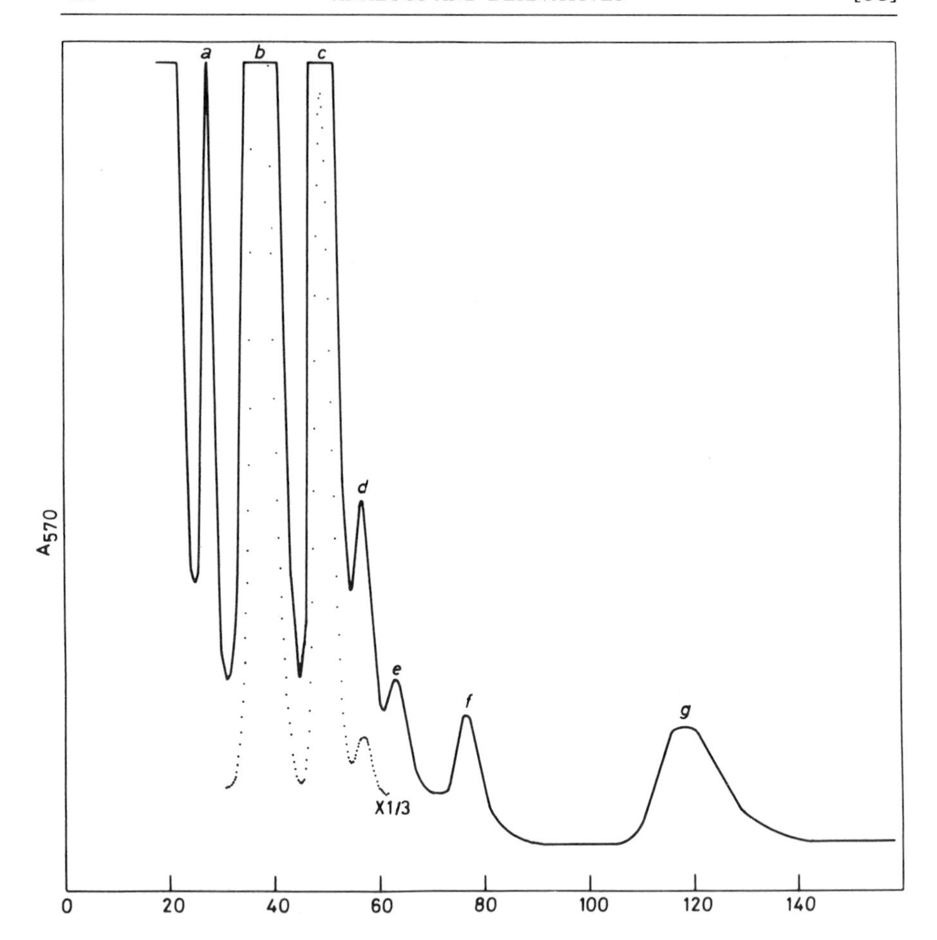

FIG. 3. Polyamines in cockroach, *Periplaneta brunnea*. From a whole homogenate of the insect (12 g), polyamines were extracted and concentrated in 2 ml as described for Fig. 2. Compound a = putrescine, b = spermidine, c = thermine, d = spermine and/or thermospermine, e = unknown, f = caldopentamine, g = unknown. Details of the study will be published elsewhere.

The novel polyamines described in this chapter were also found in a variety of organisms, such as mesophilic bacteria,[15,16] algae,[17,18]

[15] T. A. Smith, *Phytochemistry* **16,** 278 (1977).

[16] G. H. Tait, *Biochem. Soc. Trans.* **7,** 199 (1979).

[17] I. Rolle, H.-E. Hobucher, H. Kneifel, B. Paschold, W. Riepe, and C. J. Soeder, *Anal. Biochem.* **77,** 103 (1977).

[18] V. R. Villanueva, R. C. Adlakha, and R. Calvayrac, *Phytochemistry* **19,** 787 (1980).

euglena,[19] sea animals,[20] shrimps,[20,21] a higher plant,[22] a newt,[23] and halophiles.[24] Organisms belonging in Anthropoda, especially, contain novel polyamines.[20] Polyamine composition of an Antarctic shrimp, *Euphausia superba,* is shown in Fig. 2. The psychrophilic shrimp has norspermidine, spermidine, and thermine as major polyamines. Figure 3 shows the polyamine analysis of a whole homogenate of cockroach.[14] Norspermidine, thermine, thermospermine and/or spermine, and caldopentamine are found in the insect arthropod. It is noteworthy that the insect contains unidentified, strongly basic substances, i.e., compounds e and g in Fig. 3.

Recently, a new tetraamine, 1,13-diamino-5,9-diazatridecan, was detected in the seed of sword bean.[25] The plant seed contained spermidine and spermine as major polyamines and *sym*-homospermidine as a minor polyamine in addition to the novel tetraamine.

[19] H. Kneifel, F. Schuber, A. Aleksijevic, and J. Grove, *Biochem. Biophys. Res. Commun.* **85,** 42 (1978).
[20] V. Zappia, R. Porta, M. Cartenì-Farina, M. DeRosa, and A. Gambacorta, *FEBS Lett.* **94,** 161 (1978).
[21] L. W. Stillway and T. Walle, *Biochem. Biophys. Res. Commun.* **77,** 1103 (1977).
[22] R. Kuttan, A. N. Radhakrishnan, T. Spande, and B. Witkop, *Biochemistry* **10,** 361 (1971).
[23] K. Hamana and S. Matsuzaki, *FEBS Lett.* **99,** 325 (1979).
[24] S. Yamamoto, S. Shinoda, and M. Makita, *Biochem. Biophys. Res. Commun.* **87,** 1102 (1979).
[25] S. Fujihara, T. Nakashima, and Y. Kurogochi, *Biochem. Biophys. Res. Commun.* **107,** 403 (1982).

[69] Analogs of Spermine and Spermidine

By MERVYN ISRAEL and EDWARD J. MODEST

Continued interest in the biochemistry and pharmacology of the naturally occurring polyamines spermine and spermidine has resulted in the need for analogs for use as tools for biological investigation. Such compounds are particularly useful for the evaluation of structure–activity relationships. In addition, some agents, for example *N,N'*-bis(3-aminopropyl)nonane-1,9-diamine, exhibit significant intrinsic growth-inhibitory and antitumor activity.[1] This chapter is concerned with synthetic procedures for the preparation of linear aliphatic triamines and tetramines having terminal aminoethylamino and aminopropylamino functions. The

[1] M. Israel and E. J. Modest, *Abstr. Pap. Int. Cancer Congr., 10th, 1970* p. 682 (1970).

ISBN 0-12-181994-9

preparation of other analogs of spermine and spermidine has been described in this volume [68, 70].

Polyamine Analogs Containing 3-Aminopropylamino Terminal Groups

$NH_2(CH2)_3NH(CH_2)_xNH_2$ and $NH_2(CH_2)_3NH(CH_2)_xNH(CH_2)_3NH_2$

A number of homologs of spermidine and spermine have been prepared by mono- and symmetrical dicyanoethylation of the appropriate α,ω-alkylenediamines, followed by catalytic reduction of the nitriles.[2] The cyanoethylated derivatives were reduced under unusually mild conditions by means of a sponge nickel catalyst.[3]

The characterization of each of the homologs and their antitumor activity have been reported by Israel *et al.*[2] The effect of some of these analogs on the growth of a polyamine-deficient mutant of *Escherichia coli* and on the macromolecular composition of this strain has been reported by Jorstad *et al.*[4] These authors have also determined the elution pattern of some of these analogs from an automated ion-exchange column.

Procedures[2]

Freshly distilled acrylonitrile (bp 76–77°) should be used; care must be taken not to distill the acrylonitrile to dryness to avoid a potentially vigorous self-condensation reaction. The alkylenediamines containing 2–6 carbon atoms were redistilled prior to use. All distillations of cyanoethylated and polyamine products were carried out in a nitrogen atmosphere.

The following experimental procedures illustrate the general methods used to prepare these compounds.

Monocyanoethylation

N-2-Cyanoethylpropane-1,3-diamine (x = 3). Acrylonitrile (10.6 g, 0.2 mol) was added dropwise over a 15-min period to 1,3-propanediamine (14.8 g, 0.2 mol) with stirring and ice-bath cooling. The reaction mixture was stirred at 5° for 20 min, allowed to warm gradually to 45°, and held at that temperature for 0.5 hr, after which time it was heated in a boiling water bath for 2 hr. On vacuum distillation the mixture yielded two fractions, a lower boiling fraction (16 g), 90–103° (1.0 mm), and a higher

[2] M. Israel, J. S. Rosenfield, and E. J. Modest, *J. Med. Chem.* **7,** 710 (1964). This reference reports on homologs with x values of 2, 3, 4, 5, 6, 9, 10, and 12.

[3] Other catalysts for reduction of nitriles have been mentioned in this volume [68, 70, 71].

[4] C. M. Jorstad, J. J. Harada, and D. R. Morris, *J. Bacteriol.* **141,** 456 (1980).

boiling fraction (dicyanoethylation product, 3.5 g), 166–176° (1.0 mm). The lower boiling fraction was distilled twice to give 14.5 g of the desired monocyanoethylated product, bp 84–87° (0.7 mm).

The dihydrochloride salt was prepared by dissolving 3 g of the redistilled product in 40 ml of absolute ethanol and saturating the resulting solution with anhydrous HCl. Crystallization of the white precipitate (4.6 g) from aqueous ethanol gave 3.8 g (82% recovery) of white platelets, mp 216–219°.

N-2-Cyanoethyldodecane-1,12-diamine ($x = 12$). 1,12-Dodecanediamine (16.4 g, 0.08 mol) was melted in a water bath at 70°; to the melt was added dropwise with stirring 5.45 ml of acrylonitrile (4.36 g, 0.08 mol). After total addition, the bath temperature was raised to 100° and maintained there for 3 hr. On cooling, the reaction product hardened into a white solid. Vacuum distillation first removed a small amount of unreacted diamine, bp 102–140° (0.3–1.0 mm); above a temperature of 155° (0.5 mm) decomposition became excessive and the distillation was discontinued. The remaining material weighed 16.4 g (79%). Redistillation at 171–176° (0.3 mm) afforded the product as a low-melting white solid.

The dihydrochloride salt (mp 228–232°) was obtained as above.

Dicyanoethylation

N,N'-Bis(2-cyanoethyl)pentane-1,5-diamine ($x = 5$). Acrylonitrile (45.2 g, 0.854 mol) was added dropwise, with stirring and ice-bath cooling, into 43.6 g of 1,5-pentanediamine (0.427 mol). The cooling bath was allowed to warm gradually and then heated to boiling, and the reaction mixture was maintained at 100° for 3 hr. The flask was set for vacuum distillation, and the bath temperature was slowly increased to no higher than 130° with an internal pressure of 0.6 mm. A fraction containing unreacted diamine and a small amount of monocyanoethylated material was collected and the distillation was discontinued. The contents of the flask weighed 83.1 g, equivalent to a 93% yield of crude product. Vacuum distillation of 18.4 g of crude product gave, with excessive decomposition, 7.9 g (43% recovery) of material boiling at 196–204° (0.7 mm).

The dihydrochloride salt was prepared in the usual way (mp 217–220° decomp.).

N,N'-Bis(2-cyanoethyl)decane-1,10-diamine ($x = 10$). A flask containing 20.94 g (0.122 mol) of 1,10-diaminodecane was warmed to 80° in a water bath to melt the diamine. To the melt was added dropwise with stirring 16.2 ml of acrylonitrile (12.93 g, 0.244 mol). The temperature was maintained at 80° for 1 hr and then raised to 100° for 2 hr. On cooling, the reaction mixture solidified (33.6 g, 99%, mp 40–45°). The product was purified by repeated precipitation from benzene–petroleum ether (bp 60–90°) to give an analytical sample (mp 47–49°).

The dihydrochloride was obtained in 87% yield (mp 222–225°).

Catalytic Reduction of Cyanoethylated Compounds

Bis(3-aminopropyl)amine (x = 3). *N*-2-Cyanoethylpropane-1,3-diamine (24.6 g, 0.194 mol) was dissolved in 175 ml of absolute ethanol. The solution was cooled in an ice bath and saturated with ammonia gas at 0°. A teaspoonful (approximately 5 ml) of sponge nicket catalyst[5] was added, and the mixture was shaken overnight under hydrogen on a Parr, low-pressure hydrogenator at an initial pressure of 3.94 kg/cm^2 (56.0 psi) (theoretical hydrogen pressure drop, 2.21 kg/cm^2; actual 2.22 kg/cm^2). The catalyst was removed by suction filtration and washed twice with ethanol. The filtrate and washings were combined, and the ethanol was evaporated on a rotary evaporator at the water pump. The crude product showed the absence of nitrile absorption in the infrared. A fraction boiling at 72–84° (1.0 mm) was collected by fractional vacuum distillation (19.1 g, 75%). Redistillation afforded a fraction (15.5 g, 61%) boiling at 80–81° (1.0 mm).

The trihydrochloride salt was obtained by bubbling anhydrous hydrogen chloride into a cold alcohol solution containing 11.4 g of free base. The white precipitate (20.5 g, 98%, mp 260–270° decomp.) was recrystallized from 95% ethanol giving white prismatic plates (17.8 g; mp 270° decomp.).

N,N'-Bis(3-aminopropyl)pentane-1,5-diamine (x = 5). A solution containing 15.0 g of *N,N'*-bis(2-cyanoethyl)pentane-1,5-diamine (0.072 mol) and ca 5 ml of sponge nickel catalyst in 150 ml of absolute ethanol saturated with NH_3 was shaken under hydrogen at an initial pressure of 3.87 kg/cm^2 (55.0 psi) on a Parr hydrogenator. A decrease in hydrogen pressure of 1.62 kg/cm^2 was observed (calculated, 1.64). After removal of the catalyst, the ethanol was evaporated on a rotary evaporator. The residue, on vacuum distillation, yielded a fraction that boiled at 144–158° (0.2 mm) and hardened into a white solid on cooling (9.78 g, 63%). Redistillation afforded 7.48 g (76% recovery) of material boiling at 144–150° (0.2 mm); a sample boiling at 150° (0.2 mm) was used for analysis.

The tetrahydrochloride salt was prepared in the usual manner in 74% yield (mp 298–305° decomp.) after recrystallization from aqueous ethanol.

[5] Available under water in 32-kg pails, 50% solids, as sponge nickle hydrogenation catalyst, grade 986 (from the Davison Chemical Division, W. R. Grace and Co., Baltimore, Maryland). The nature and use of this catalyst has been reported by H. N. Schlein, M. Israel, S. Chatterjee, and E. J. Modest [*Chem. Ind. (London)* p. 418 (1964)].

When a warm solution of 320 mg of free base (1.5 mmol) in ethanol was added to a warm solution of 1.28 g (5.6 mmol) of picric acid in ethanol, a yellow crystalline picrate formed instantly (1.46 g, 92%). Several recrystallizations from water, with better than 90% recovery each time, gave a product decomposing at 227°. Analysis indicated the sample to be a tetrapicrate monohydrate.

Polymine Analogs Containing 2-Aminoethylamino Terminal Groups

$NH_2(CH_2)_2NH(CH_2)_xNH_2$ and $NH_2(CH_2)_2NH(CH_2)_xNH(CH_2)_2NH_2$

Linear aliphatic triamines and tetramines with 2-aminoethylamino terminal units have been derived from 1,3-propanediamine, 1,4-butanediamine, and 1,5-pentanediamine by reaction of the diamine with ethyleneimine under pressure in the presence of ammonium chloride catalyst.[6] Although the disubstituted products from these reactions could have one of three possible isomeric arrangements, rigid chemical proof established the desired bis-aminoethylation substitution pattern. Two related compounds, diethylenetriamine and triethylenetetramine, are commercially available.

Procedure[6]

2-Aminoethylalkylenediamine Derivatives. A mixture composed of 1 equivalent of α,ω-alkylenediamine, 2 equivalents of cold ethyleneimine, and 2 equivalents of NH_4Cl was placed in a 1-liter stainless steel Magne-Dash autoclave (Autoclave Engineers, Inc., Erie, Pennsylvania) and heated for 48 hr at 170–190° with continuous internal agitation. After cooling, the slight residual pressure was vented and the contents of the bomb were transferred to a round-bottomed flask with the aid of absolute ethanol. The major portion of ethanol was removed on a rotary evaporator at water-pump pressure. The remaining viscous liquid was distilled under high vacuum on a spinning-band apparatus. A low-boiling fraction consisting of unchanged diamine distilled first, followed successively by the monoaminoethyl product and the diaminoethyl derivative. The products were purified further by careful fractionation under high vacuum; center distillation cuts were taken for the analytical samples. Hydrochloride salts were prepared by saturating an ethanol solution of the base with dry hydrogen chloride; ether was added on occasion to precipitate the salt completely.

[6] M. Israel and E. J. Modest, *J. Med. Chem.* **14,** 1042 (1971). This reference reports on analogs with x values of 3, 4, and 5.

Structure–Activity Relationships for the Polyamine–Bovine Plasma Amine Oxidase System

Spermine and spermidine produce growth inhibition in mammalian and bacterial systems as a result of their conversion into cytotoxic derivatives by the action of an enzyme[7] [amine : oxygen oxidoreductase (deaminating)] present in the bovine serum used to supplement the media. Polyamine structural variants, including the 2-aminoethyl and 3-aminopropyl analogs summarized here, were used in cytotoxicity assays to define the substrate geometry required by the enzyme for conversion to its active form.[6,8] The requisite molecular feature resulting in cytotoxicity appears to be a primary amino group separated from a secondary amine by three methylene units, with the further restriction that the next basic center be separated from the secondary amine by three or more methylene units.[6] This structural requirement suggests the existence of a hydrophobic region adjacent to the active site of the bovine plasma amine oxidase. The synthesis and biological evaluation of the cytotoxic metabolites of spermine and other polyamines have been reported.[9]

[7] The formation of toxic products by the oxidation of spermine or spermidine by bovine plasma amine oxidase was shown by J. G. Hirsch [*J. Exp. Med.* **97,** 327, 345 (1953)], and by C. W. Tabor and S. M. Rosenthal [*J. Pharmacol. Exp. Thera.* **116,** 139 (1956)]. The products were characterized by C. W. Tabor, H. Tabor, and U. Bachrach [*J. Biol. Chem.* **239,** 2194 (1964)]. The purification of bovine plasma amine oxidase is described in this volume [54] and in this series, Vol. 17B [232].

[8] M. Israel, G. E. Foley, and E. J. Modest, *Fed. Proc., Fed. Am. Soc. Exp. Biol.* **24,** 580 (1965).

[9] M. Israel, E. C. Zoll, N. Muhammad, and E. J. Modest, *J. Med. Chem.* **16,** 1 (1973).

[70] Preparation of Thermospermine

By BRUCE GANEM and KAN CHANTRAPROMMA

Thermospermine has been isolated from the extreme thermophile *Thermus thermophilus*.[1] A short synthetic procedure (see Scheme 1) to prepare quantities of thermospermine from spermidine employs a simple protecting group strategy of general utility to polyamine chemistry.[2]

Spermidine (9.0 g, 61.9 mmol) in a 500-ml round-bottomed flask was dissolved in distilled water (225 ml), and the solution was cooled to 5°

[1] T. Oshima, *J. Biol. Chem.* **254,** 8720 (1979).

[2] K. Chantrapromma, J. S. McManis, and B. Ganem, *Tetrahedron Lett.* **21,** 2475 (1980).

ISBN 0-12-181994-9

$NH_2(CH_2)_3NH(CH_2)_4NH_2$ (1) ⟶ HN—ring—$N(CH_2)_4NH_2$ (2)

⟶ HN—ring—$N(CH_2)_4N{=}CHPh$ (3) ⟶ $CN(CH_2)_2N$—ring—$N(CH_2)_4N{=}CHPh$ (4)

⟶ $CN(CH_2)_2NH(CH_2)_3NH(CH_2)_4NH_2$ (5) ⟶ $NH_2(CH_2)_3NH(CH_2)_3NH(CH_2)_4NH_2$ (6)

thermospermine

under N_2. Aqueous formaldehyde (4.52 ml of 37% solution; 0.9 equivalents) was slowly added by dropping funnel to the cold solution; the mixture was then stirred for 1 hr at room temperature. The aqueous layer was saturated with solid NaCl and extracted four times with $CHCl_3$. The combined $CHCl_3$ extracts were dried (Na_2SO_4) and concentrated to dryness, giving the amine **2** (9.2 g, 94%) as a nearly pure waxy white solid.

To this amine (2.50 g, 15.9 mmol), in benzene (30 ml), was added benzaldehyde (freshly distilled, 1.69 g, 15.9 mmol) at room temperature; the mixture was then heated with refluxing for 3 hr using a Dean–Stark apparatus to remove water azeotropically. Solvent removal gave the crude imine **3,** which could be purified by distillation to give a pale yellow oil (3.30 g, 84%, bp 175° at 0.1 Torr).

To an ice-cold solution of **3** (2.44 g, 9.9 mmol) in absolute ethanol (15 ml) was added dropwise acrylonitrile (0.66 ml, 9.9 mmol, redistilled) with stirring under N_2. After 15 hr, more acrylonitrile (0.5 ml) was added, and stirring was continued for 9 hr. Ethanol was removed under reduced pressure to give the crude iminonitrile **4,** which could be purified by distillation to a colorless oil (2.9 g, 97%, bp 180–184° at 0.15 Torr).

To hydrolyze its benzylidene group, a solution of **4** (0.665 g, 2.23 mmol) in 2 *N* HCl–methanol (15 ml) was heated under reflux at 77–79° for 8 hr. After evaporation of the methanol, the resulting residue was redissolved in water (4 ml) and extracted with ether (3 × 15 ml) to remove benzaldehyde. The aqueous layer was alkalinized with 20% NaOH (8 ml) and extracted with $CHCl_3$. The combined $CHCl_3$ extracts were dried (Na_2SO_4) and concentrated. Distillation afforded pure aminonitrile **5** as a colorless oil (0.210 g, 48%, bp 165–170° at 0.25 Torr).

To reduce its cyano group, an ice-cold solution of aminonitrile **5** (0.150 g, 0.76 mmol) in dry methanol (15 ml) was mixed with $CoCl_2 \cdot 6H_2O$ (0.360 g, 1.52 mmol); then $NaBH_4$ was added in portions (0.286 g, 10 equivalents). After the addition was complete, stirring was continued for 3 hr at room temperature. The reaction mixture was acidified with 3 *N* HCl and stirred until the black precipitate (Co_2B) dissolved. After concentration *in vacuo,* the resulting aqueous solution was alkalinized with 30% NaOH and extracted with $CHCl_3$ to give thermospermine **6.** The crude natural product was purified by distillation to a colorless oil (0.105 g, 68%, bp 105–108° at 0.25 Torr).

To prepare its hydrochloride, thermospermine (0.060 g) was dissolved in 6 *N* HCl (3 ml), and the solution was evaporated to dryness. The residue was redissolved in H_2O (3 drops) and triturated with 1 : 1 ethanol : methanol to give a precipitate after refrigeration. Filtration and washing of the precipitate with ethanol furnished pure thermospermine hydrochloride (0.065 g, mp > 260°).

[71] Synthesis of Putreanine and of Spermic Acid

By HERBERT TABOR and CELIA WHITE TABOR

Putreanine[1]

$$HOOCCH_2CH_2NH(CH_2)_4NH_2$$

Putreanine

Putreanine was first isolated from brain by Kakimoto *et al.*[2] Subsequently this compound was also detected in other mammalian tissues, but at much lower levels.[3] Putreanine has also been shown to be a metabolic

[1] Carboxyethylputrescine; *N*-(4-aminobutyl)-3-aminopropionic acid.

[2] Y. Kakimoto, T. Nakajima, A. Kuman, Y. Matsuoka, N. Imaoka, I. Sano, and A. Kanazawa, *J. Biol. Chem.* **244,** 6003 (1969).

[3] T. Shiba, H. Nyzote, T. Kaneko, T. Nakajima, and Y. Kakimoto, *Biochim. Biophys. Acta* **244,** 523 (1971).

METHODS IN ENZYMOLOGY, VOL. 94

ISBN 0-12-181994-9

product of spermidine in animals,[4] presumably via the intermediate formation of the corresponding aldehyde.[5]

Two methods are available for the synthesis of putreanine.

Method A

$$CH_3CONH(CH_2)_4NH_2 \xrightarrow{CH_2{=}CHCN} CH_3CONH(CH_2)_4NHCH_2CH_2CN \xrightarrow{HCl} NH_2(CH_2)_4NHCH_2CH_2COOH$$

Synthesis of $CH_3CONH(CH_2)_4NHCH_2CH_2CN \cdot HCl$.[5,6] Monoacetylputrescine hydrochloride (53.1 g, 320 mmol) was dissolved in 400 ml of absolute ethanol and treated with 66.5 ml of 4.8 *N* NaOH and 22.8 ml (340 mmol) of technical acrylonitrile. The mixture was stirred at room temperature for 18 hr and then heated on the steam bath for 1 hr. After cooling, the pH was adjusted to approximately pH 3 with 6 *N* HCl, and the solution was evaporated to dryness under vacuum. The residue was then extracted three times with 400-ml portions of hot absolute ethanol. The filtrates were concentrated and cooled to $-10°$, and the crystals were collected; yield, 45 g (0.21 mol); mp 140–142° (capillary).

Synthesis of $NH_2(CH_2)_4NHCH_2CH_2COOH \cdot 2HCl$.[5] The above nitrile was hydrolyzed by refluxing 10 g in 200 ml of 6 *N* HCl overnight. After cooling, the residue was evaporated to dryness under vacuum.

Method B

Shiba *et al.*[7] synthesized putreanine by coupling 4-phthalimino-1-bromobutane and ethyl β-alaninate. The reaction mixture was hydrolyzed, then purified by chromatography on Dowex 50-X8. Putreanine was crystallized as the monosulfate from alcohol–water; yield 31%.

Spermic Acid[8]

$$HOOCCH_2CH_2NH(CH_2)_4NHCH_2CH_2COOH$$

Spermic acid has been isolated from nervous tissue.[9]

$$NH_2(CH_2)_4NH_2 + 2CH_2{=}CHCN \rightarrow CNCH_2CH_2NH(CH_2)_4NHCH_2CH_2CN \xrightarrow{HCl} HOOCCH_2CH_2NH(CH_2)_4NHCH_2CH_2COOH$$

[4] N. Seiler, F. N. Bolkenius, and O. M. Rennert, *Med. Biol.* **59,** 334 (1981).

[5] C. W. Tabor, H. Tabor, and U. Bachrach, *J. Biol. Chem.* **239,** 2194 (1964).

[6] E. L. Jackson, *J. Org. Chem.* **21,** 1374 (1956). Monoacetylputrescine hydrochloride was prepared as described in this series, Vol. 17B [256].

[7] T. Shiba, I. Kubota, and T. Kaneko, *Tetrahedron* **26,** 4307 (1970). This article contains considerable data on the characteristics of putreanine.

[8] *N,N*¹-Di(carboxyethyl)putrescine

[9] N. Imaoka and Y. Matsuoka, *J. Neurochem.* **22,** 859 (1974).

Synthesis of $CNCH_2CH_2NH(CH_2)_4NHCH_2CH_2CN \cdot 2HCl$.[5] This was prepared from 1,4-diaminobutane and acrylonitrile by a modification of the methods of Schultz[10] and of Jackson and Rosenthal.[11] Technical acrylonitrile (51 ml; 770 mmol), used without further purification, was added over a 2-min period to a mixture of 30 ml (about 300 mmol) of technical 1,4-diaminobutane (Aldrich) and 300 ml of 97–98% ethanol at room temperature. After 72 hr at room temperature, the solution was cooled to 0°, and cold 6 *N* HCl (approximately 100 ml) was added until the pH was approximately 1–2. A heavy white precipitate formed immediately; after several hours at 0°, the precipitate was collected by filtration and washed with cold 95% ethanol.

The precipitate was then dissolved in 150 ml of hot H_2O and recrystallized by the addition of 500 ml of hot absolute ethanol; on cooling at −10°, crystals of the desired nitrile formed. The crystals were collected by filtration, and dried under vacuum; yield, 45 g (170 mmol); mp 225–227° (capillary; decomposition). Additional material could be obtained from the mother liquors after evaporation. These samples were then fractionally recrystallized from ethanol–water.

Synthesis of $HOOCCH_2CH_2NH(CH_2)_4NHCH_2CH_2COOH \cdot 2HCl$. The above dinitrile, 29 g (110 mmol), and 6 *N* HCl, 150 ml, were refluxed for 6 hr. The mixture was then evaporated to dryness in a flash evaporator. This material was then purified by fractional crystallization from ethanol–water; mp 221–222° (capillary, decomposition).

[10] H. P. Schultz, *J. Am. Chem. Soc.* **70,** 2666 (1948).
[11] E. L. Jackson and S. M. Rosenthal, *J. Org. Chem.* **25,** 1055 (1960).

[72] Synthesis of Acetylated Derivatives of Polyamines

By HERBERT TABOR and CELIA WHITE TABOR

Monoacetyl-1,3-diaminopropane Monohydrochloride[1,2]

Monoacetyl-1,3-diaminopropane monohydrochloride is used as a starting material for the synthesis of *N*-monoacetylspermidine (see below).

[1] The synthesis of monoacetylputrescine hydrochloride was described in this series, Vol. 17B [256].
[2] Monoacetylcadaverine hydrochloride can also be prepared by a comparable procedure. We found, however, that the protocol listed did not completely remove the unacetylated cadaverine in the 2-propanol step. Therefore, the extract was chromatographed on Dowex 50-X2, and eluted with 0.4 *N* HCl. (The column size for a synthesis from 24.9 g of

ISBN 0-12-181994-9

1,3-Diaminopropane (86.1 g, 1.16 mol) is cooled in ice and added to 200 ml of glacial acetic acid with magnetic stirring and cooling. Then 95 ml of acetic anhydride (1 mol) are added dropwise with magnetic stirring over a period of 1 hr; during this time the temperature is maintained at 50–73°. The mixture is stored for 3 days at room temperature and then evaporated to dryness *in vacuo*. The residual oil is dissolved in 250 ml of 5 *N* HCl, and the solution is again evaporated to dryness. The residue is refluxed with 300 ml of 2-propanol, and filtered to remove diaminopropane dihydrochloride. Monoacetyl-1,3-diaminopropane monohydrochloride crystallizes upon cooling. Additional crystalline material can be obtained upon concentration of the mother liquor. The crystals can be recrystallized from a large volume of hot 2-propanol. The yield is 60.6 g (34%).

N^1-Monoacetylspermidine[3]

N^1-Acetylspermidine is synthesized by treatment of *N*-monoacetyl-1,3-diaminopropane with 4-bromobutyronitrile, followed by catalytic reduction.[4] To 1.45 g of *N*-acetyl-1,3-diaminopropane hydrochloride are added 45 ml of absolute ethanol and 1.5 g of K_2CO_3. A solution of 1.47 g of 4-bromobutyronitrile in 20 ml of absolute ethanol is added dropwise with magnetic stirring. After 1 hr at room temperature, the mixture is refluxed for 12 hr. After the solution is cooled, 100 ml of absolute ethanol, 0.7 ml of concentrated H_2SO_4, and 300 mg of PtO_2 are added; hydrogenation is carried out for $2\frac{1}{2}$ hr at room temperature at atmospheric pressure. The catalyst is removed by filtration; the solution is diluted with 900 ml of water and passed through a Dowex 50-X2 column (H^+ form; 2 × 20 cm). The column is washed with water and 250 ml of 0.5 *N* HCl. The acetylspermidine is then eluted with 500 ml of 1 *N* HCl; the solution is evaporated to dryness *in vacuo*, and N^1-monoacetylspermidine dihydrochloride[5] is crystallized from 2-propanol (yield 590 mg).

cadaverine was 5 × 13.5 cm). The fractions with monoacetylcadaverine were identified by chromatography on paper with a solvent containing 1-butanol, 2; acetic acid, 1; pyridine, 0.5; water, 1 (R_f for cadaverine is 0.43; R_f for monoacetylcadaverine is 0.71). These fractions were dried, then recrystallized from alcohol–ether.

[3] We have used the following numbering system as discussed in this series, Vol. 17B [256]:

$$\overset{1}{N}H_2\overset{2}{C}H_2\overset{3}{C}H_2\overset{4}{C}H_2\overset{5}{N}H\overset{6}{C}H_2\overset{7}{C}H_2\overset{8}{C}H_2CH_2NH_2.$$

[4] H. Tabor and C. W. Tabor, *J. Biol. Chem.* **250,** 2648 (1975). This method of synthesis of N^1-monoacetylspermidine is superior to the less specific procedure that we described in this series, Vol. 17B [256]. The latter procedure sometimes results in mixtures of the N^1- and the N^8-monoacetyl derivatives.

[5] A method for the unambiguous synthesis of N^8-monoacetylspermidine dihydrochloride was presented in this series, Vol. 17B [256].

Diacetyl-1,6-Diaminohexane[6]

1,6-Diaminohexane (320 ml) is cooled and mixed with 500 ml of cooled glacial acetic acid. Then 720 ml of acetic anhydride are added dropwise with stirring. The mixture is stored overnight at room temperature and finally heated for 2 hr at 60°. Water (500 ml) is added, and the mixture is evaporated to dryness on a steam bath to a volume of less than 500 ml. After cooling, crystals of diacetyl-1,6-diaminohexane appear; they are collected by filtration. Additional crystals can be obtained from the mother liquor (yield 260 g). The crystals can be recrystallized from hot water or hot acetone (mp 125–126°).

A similar method can be used to prepare diacetylputrescine.

[6] Diacetyl-1,6-diaminohexane and diacetyl-1,4-diaminobutane are often referred to as hexamethylene bisacetamide and tetramethylene bisacetamide, respectively. These compounds are of special interest, as they are capable of inducing hemoglobin formation in Friend erythroleukemia cells[7] and of inducing differentiation of neuroblastoma cells.[8] They also induce procollagen formation in malignant mesenchymal cells.[9]

[7] R. C. Reuben, R. C. Wife, R. Breslow, R. A. Rifkind, and P. A. Marks, *Proc. Natl. Acad. Sci. U.S.A.* **73,** 862 (1976).

[8] C. Palfrey, Y. Kimhi, U. Z. Littauer, R. C. Reuben, and P. A. Marks, *Biochem. Biophys. Res. Commun.* **76,** 937 (1977).

[9] A. S. Rabson, R. Stern, T. S. Tralka, J. Costa, and J. Wilczek, *Proc. Natl. Acad. Sci. U.S.A.* **74,** 5060 (1977).

[73] α-Putrescinylthymine

By A. M. B. KROPINSKI, K. L. MALTMAN, and R. A. J. WARREN

The hypermodified pyrimidine α-putrescinylthymine replaces half the thymine residues in the DNA of bacteriophage φW-14.[1,2]

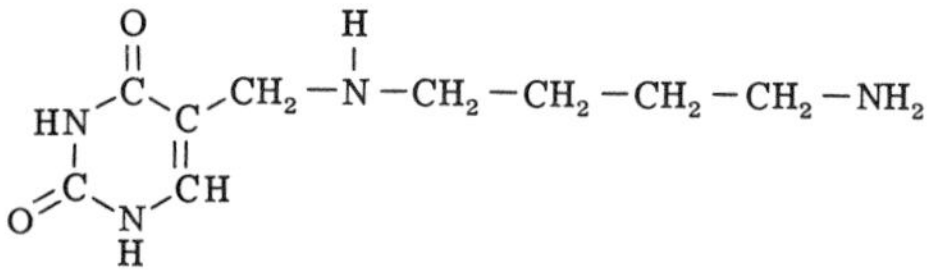

```
        O          H
        ||         |
     ,C\ ,CH2 —N —CH2 —CH2 —CH2 —CH2 —NH2
HN      C
 |      ||
 ,C\  ,CH
O     N
      H
```

[1] A. M. B. Kropinski and R. A. J. Warren, *J. Gen. Virol.* **6,** 85 (1970).

[2] A. M. B. Kropinski, R. J. Bose, and R. A. J. Warren, *Biochemistry* **12,** 151 (1973).

METHODS IN ENZYMOLOGY, VOL. 94

ISBN 0-12-181994-9

This is the only reported example of the covalent attachment of a polyamine to DNA. However, the hypermodified pyrimidine α-glutaminylthymine[3] occurs in the DNA of bacteriophage SP10.[4]

α-Putrescinylthymine is formed from hydroxymethyluracil at the polynucleotide level.[5] The presence of putrescinylthymine markedly affects the properties of φW-14 DNA.[2] The observed properties are consistent with the attachment of putrescinyl groups to the DNA.[2]

Procedures

Isolation of Putrescinylthymine from φW-14 DNA[6]

Preparation of Phage. Pseudomonas acidovorans ATCC 9355 is grown at 30° to 3×10^8 cells ml^{-1} (100 Klett units, green filter) in the following medium (grams per liter): Difco casamino acids, 3.3; Difco yeast extract, 1.7; mannitol, 1.7; tryptophan, 0.05; Tris-HCl, 8.0; pH adjusted to 7.2. Phage (ATCC 9355-B1) is added at a multiplicity of infection of 1, and incubation is continued until lysis.

Purification of Phage. Cells and cellular debris are removed by passage of the lysate at 1 liter min^{-1} through a cooled Sharples centrifuge. All subsequent steps are performed at 4°. The phage is precipitated with 7% (w/v) polyethylene glycol 6000.[7] After 2 days, the supernatant is siphoned off and the precipitate is collected by centrifugation at 6000 *g* for 10 min. The pellet is resuspended in phage buffer (0.05 *M* Tris-HCl, 0.1 *M* Na_3 citrate, 0.005 *M* NaCl, pH 8.1) and centrifuged at 6000 *g* for 10 min. The supernatant is centrifuged at 28,000 *g* for 90 min, and the pellet is resuspended in phage buffer.

Isolation of DNA. Sodium dodecyl sulfate (SDS) is added to 2% (w/v), and the suspension is heated at 45° for 15 min. If the DNA is to be used only for the preparation of putrescinylthymine, exhaustive deproteinization is unnecessary; Pronase digestion (50 μg ml^{-1}, 3 hr, 37°) is sufficient. After digestion, the crude DNA is precipitated with 2–3 volumes of ethanol, washed twice with 95% ethanol and once with acetone, and air dried.

Hydrolysis of DNA. DNA is dissolved (50–60 mg ml^{-1}) in 6 *M* HCl, then hydrolyzed under reduced pressure at 100° for 90 min. The hydrolyzate is filtered through Whatman No. 1 paper to removed precipitated

[3] M. Dosmart and H. Witmer, *Curr. Microbiol.* **2,** 361 (1979).

[4] C. B. Thorne, *J. Bacteriol.* **83,** 106 (1962).

[5] J. Neuhard, K. L. Maltman, and R. A. J. Warren, *J. Virol.* **34,** 347 (1980).

[6] Reprinted with permission from Ref. 2. Copyright 1973 American Chemical Society.

[7] K. R. Yamamoto, B. M. Alberts, R. Benzinger, L. LaWhorne, and G. Treiber, *Virology* **40,** 734 (1970).

material that interferes with subsequent purification steps. The filters are washed with 1 *M* HCl, and the combined filtrate and washings are evaporated to dryness in a rotary evaporator. Water is added to the residue, and the evaporation is repeated twice more. The final residue is dissolved in a small volume of water, and the pH of the solution is adjusted to 6.8 with 0.01 *M* NH_4OH.

Isolation of α-Putrescinylthymine. The neutralized hydrolyzate is applied to a column of CM-50 Sephadex (2.5 × 100 cm; NH_4^+ form). The column is washed with 0.005 *M* NH_4OH. Only some 15% of the UV-absorbing material applied is retained on the column. The retained material is eluted in a broad asymmetric band with 0.1 *M* HCl. The appropriate fractions (see Fig. 1) are pooled and taken to dryness. The residue is dissolved in a small volume of 0.01 *M* HCl and applied to a column of Sephadex G-10 (2.5 × 100 cm) preequilibrated with this acid. The column is developed with 0.01 *M* HCl (flow rate 0.5 ml min^{-1}) and 3-ml fractions are collected. Ultraviolet-absorbing material is eluted in two bands, the

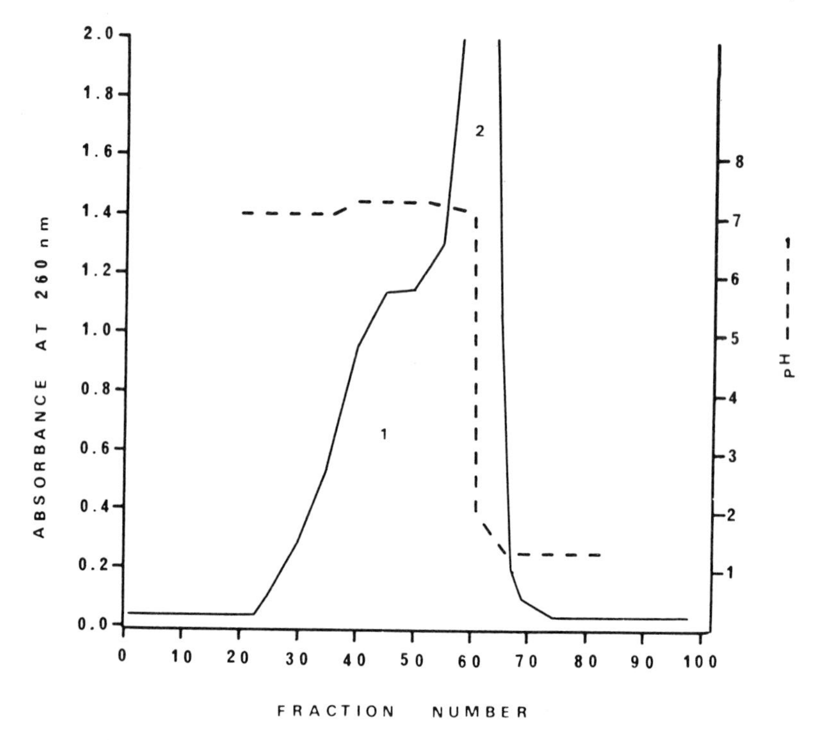

FIG. 1. Elution of bases obtained from ϕW-14 DNA and adsorbed to CM-Sephadex. Band 2 contains partially purified α-putrescinylthymine (putThy). Reprinted with permission from Ref. 2. Copyright 1973 American Chemical Society.

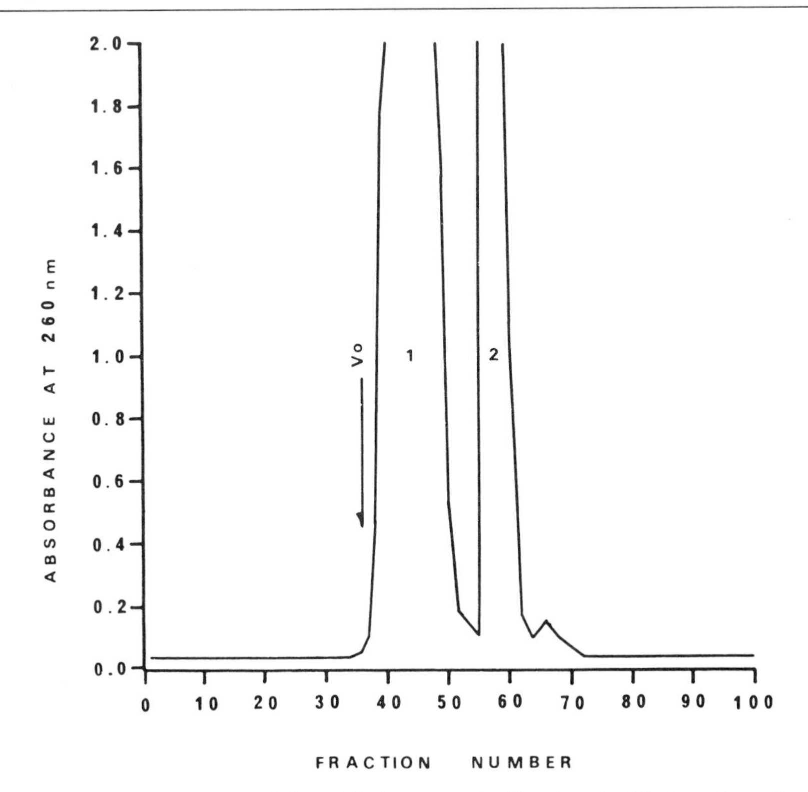

FIG. 2. Purification of partially purified α-putrescinylthymine (putThy) on Sephadex G-10. Band 1 is putThy. Reprinted with permission from Ref. 2. Copyright 1973 American Chemical Society.

first of which is putrescinylthymine (Fig. 2). Appropriate fractions are pooled and taken to dryness. The residue is triturated several times with ice-cold methanol to yield a white powder. The yield is about 40 mg (as the dihydrochloride) per gram of DNA, which is a recovery of about 45%. The material can be recrystallized from methanol–1 M HCl. It is converted to the free base by applying a solution of the dihydrochloride to a column of Dowex 50 (NH_4^+) and eluting the free base with 2 M NH_4OH.

Chemical Synthesis of α-Putrescinylthymine[6]

Bromomethyluracil[8] is synthesized from hydroxymethyluracil with the following modification: when the reaction is complete, the solution is cooled to room temperature and six volumes of anhydrous ether are

[8] J. A. Carbon, *J. Org. Chem.* **25,** 1731 (1960).

added. The precipitate is allowed to settle out in the cold. The supernatant is discarded, and the precipitate is washed several times with anhydrous ether. The finely granular product (97% yield) is dried *in vacuo* over KOH pellets.

Bromomethyluracil (1.4 g, 6.8 mmol) is stirred slowly into putrescine free base (3 ml, 30 mmol). The mixture is kept at room temperature for 30 min, then acidified with HCl (5 ml, 60 mmol). After the addition of 5 ml of water, the insoluble material is removed by centrifugation and discarded. The supernatant is applied to a column of BioGel P2 (2.5 × 100 cm) and the column developed with 0.01 *M* HCl. Fractions of 3 ml are collected. Those containing UV-absorbing material are pooled and taken to dryness. The residue is dissolved in 0.01 *M* HCl and passed through the BioGel P2 column again. This elution–concentration cycle is repeated several times until the putrescinylthymine is free of putrescine. The final product is taken to dryness, then triturated with methanol as before.

Isolation of α-Putrescinyldeoxythymidine from ϕW-14 DNA

The DNA is deproteinized by phenol extraction. Residual phenol is removed by washing with ether, and the residual ether is driven off by a stream of nitrogen. The DNA is precipitated with ethanol and dried as before, then transferred to a polypropylene tube. The DNA is dissolved (about 5 mg ml^{-1}) in 50% HF, and the solution is incubated at 37° for 85 min.[9] The HF is removed by lyophilization, with NaOH beads in the trap. This hydrolysis procedure yields the purine bases and the pyrimidine deoxynucleosides, although up to 10% of the deoxycytidine may be converted to cytosine. The material is dissolved in water (about 200 μl for each milliliter of HF solution). Any insoluble material is removed by centrifugation. α-Putrescinyldeoxythymidine is isolated by preparative thin-layer chromatography on cellulose using 2-propanol–H_2O (70 : 30, v/v) as the solvent. The α-putrescinyldeoxythymidine barely leaves the origin; all other products move well up the sheet.

Digestion of ϕW-14 DNA to Mononucleotides[10]

A putrescinylthymine deoxynucleotide is not released by digestion of ϕW-14 DNA with DNase I and snake venom phosphodiesterase. However, the DNA is hydrolyzed quantitatively to mononucleotides as described below.[11]

[9] M. S. Walker and M. Mandel, *J. Virol.* **25,** 500 (1978).

[10] Reprinted with permission from Ref. 11. Copyright 1981 American Chemical Society.

[11] K. L. Maltman, J. Neuhard, and R. A. J. Warren, *Biochemistry* **20,** 3586 (1981).

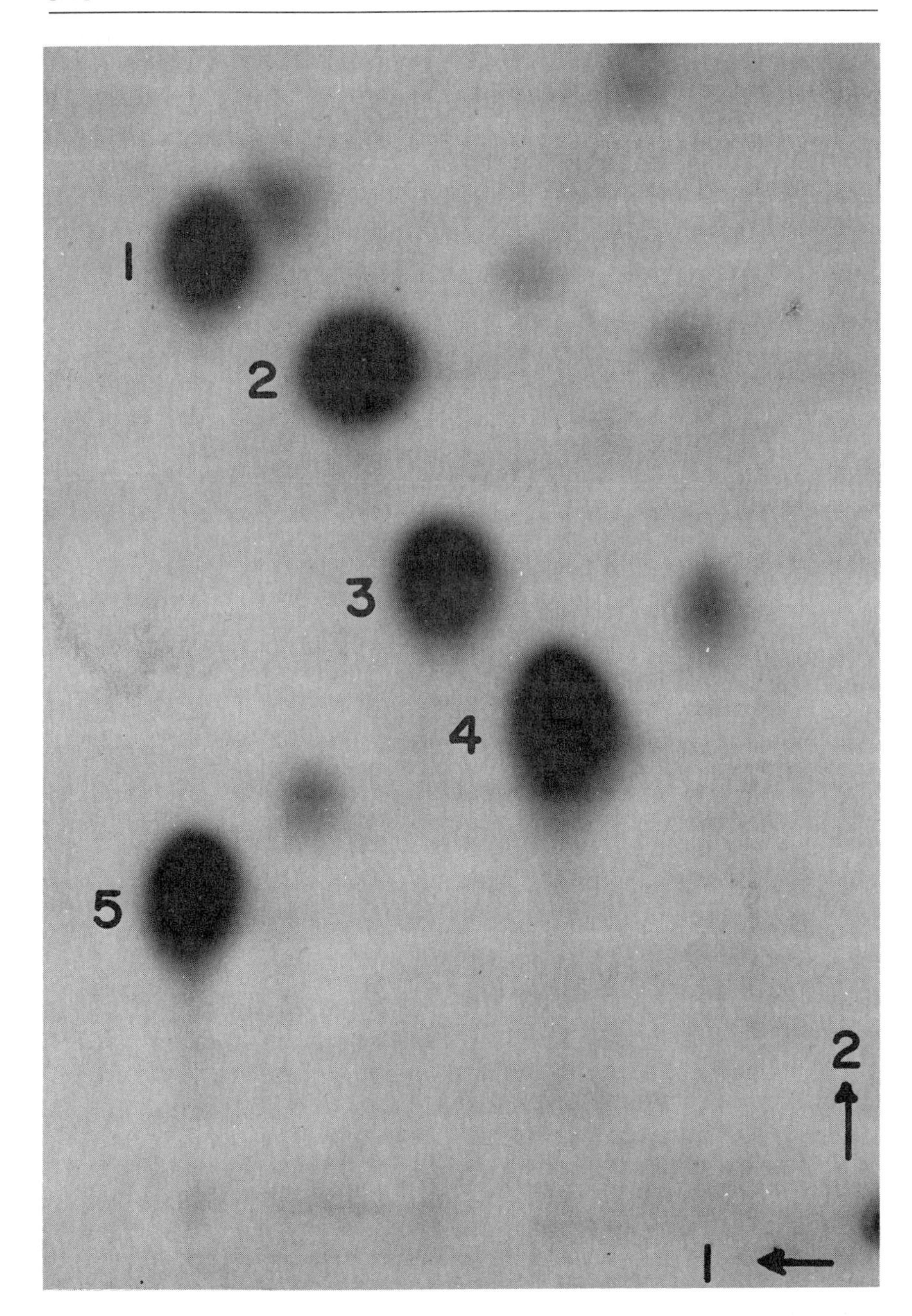

FIG. 3. Thin-layer chromatographic separation of nucleotides obtained from ϕW-14 DNA. 1, α-putrescinyldeoxythymidine; 2, dCMP; 3, dTMP; 4, dGMP; 5, dAMP.

The DNA is deproteinized with phenol. After determining the concentration spectrophotometrically, it is precipitated with ethanol and washed and dried as above. It is dissolved in the minimal volume of sterile deionized distilled water, then boiled for 5 min to inactivate residual nucleases and to ensure denaturation of the DNA. NH_4 acetate (pH 5.0) and $ZnSO_4$ are added to give final concentrations of 50 and 0.1 m*M*, respectively. After the addition of 10 units of S_1 nuclease per microgram of DNA, the solution is incubated for 4 hr at 55°, then lyophilized. The residue is suspended in deionized distilled water and lyophilized again. The residue is suspended in the minimal volume of deionized distilled water, and NH_4HCO_3 (pH 8.4) and $MgCl_2$ are added to final concentrations of 100 and 15 m*M*, respectively. After the addition of 20 μg of snake venom phosphodiesterase per milliliter, the mixture is incubated for 2 hr at 37°, then lyophilized. The residue is suspended in the minimal volume of deionized distilled water.

The mononucleotides are separated by two-dimensional thin-layer chromatography on cellulose. Samples of the final digestion mixture are applied to the sheets (20 × 20 cm). The sheets are washed twice by ascending development with 95% ethanol, run at right angles to the direction of the first solvent, with drying in between. This removes the salts in the sample. The sheet is developed in the first dimension with isobutyric acid–water–concentrated NH_4OH (66 : 20 : 1, v/v/v), and in the second dimension with saturated $(NH_4)SO_4$–1 *M* sodium acetate–2-propanol (80 : 12 : 2, v/v/v). A typical separation of ^{32}P-labeled nucleotides is shown in Fig. 3.

Comments

The isolation of the various putrescinylthymine-containing compounds is facilitated by their marked polarity and positive charges. The primary amino group of the putrescinyl side chain reacts readily with ninhydrin[2]; together with the UV absorbancy of the pyrimidine ring, this facilitates the detection and isolation of these compounds.

[74] Preparation and Purification of *N*-Carbamoylputrescine and [*ureido*-^{14}C]-*N*-Carbamoylputrescine

By K. S. SRIVENUGOPAL and P. R. ADIGA

N-Carbamoylputrescine was prepared in bulk quantities by carbamoylation of putrescine with KCNO and purified by ion-exchange chromatography on Dowex 50.[1] When [*ureido*-^{14}C]-*N*-carbamoylputrescine was required, $K^{14}CNO$ was employed, and the product was separated from unreacted reactants, and the *N*, *N'*-dicarbamoyl derivative by paper chromatography.

Large-Scale Preparation and Purification of *N*-Carbamoylputrescine

The procedure described here employs Dowex 50 (H^+) resin to separate the monocarbamoyl derivative of putrescine from residual putrescine and from dicarbamoylputrescine.

Carbamoylation Reaction. Putrescine (5.0 g, free base) dissolved in 10 ml of distilled water is treated with 5.2 ml of concentrated HCl at 0° to neutralize one of the primary amino groups. Solid KCNO (3.5 g) is then added, and the solution is heated at 100° for 40 min. The solution is then decolorized with activated charcoal and allowed to stand overnight at room temperature. The less soluble dicarbamoylputrescine crystallizes out to a large extent and is separated by decantation.

Dowex 50 Column Chromatography. Eighty milliliters of freshly regenerated Dowex Ag-50W-X12 (200–400 mesh, H^+ form) is suspended in 10 m*M* sodium citrate buffer (pH 5.0) and titrated, with stirring, with 5 *N* NaOH to pH 5.0. The resin suspended in the above buffer is packed into a column (2.2 × 30 cm). The carbamoylation reaction mixture containing *N*-carbamoylputrescine, residual putrescine, and dicarbamoylputrescine is adjusted to pH 5.0 and applied to the Dowex (Na^+) column at a slow rate. The column is then washed with sodium citrate buffer (pH 5.0) until the effluent is negative to Ehrlich's reagent. After washing the column with two bed volumes of deionized water, the resin is transferred to a 1-liter beaker; the resin is then stirred at 4° in 500 ml of 10 m*M* sodium citrate buffer (pH 4.0) to elute the *N*-carbamoylputrescine. This process

[1] Earlier procedures for the isolation of *N*-carbamoylputrescine include *(a)* paper chromatographic method [T. A. Smith and J. L. Garraway, *Phytochemistry* **3**, 23 (1964)] and *(b)* chromatography on Amberlite XE-64 [N. Akamatsu, M. Ogushi, Y. Yajima, and M. Ohino, *J. Bacteriol.* **133**, 409 (1978)].

ISBN 0-12-181994-9

(pH 4.0 buffer wash) is repeated twice to complete the elution. The pooled clear supernatant is concentrated to approximately 10 ml. The resultant citrate salt of *N*-carbamoylputrescine is passed through a column (1.5 × 10 cm) of Dowex 1 (Cl^-) and converted to the hydrochloride form.[2] The product is crystallized from water to yield needle-shaped crystals. The yield is around 1.5 g, mp 180°, and the product has an $R_f = 0.37$ on paper chromatograms in butanol–acetic acid–water (4 : 1 : 1, v/v/v).[3]

Chemical Synthesis and Purification of [*ureido*-^{14}C]-*N*-Carbamoylputrescine

$K^{14}CNO$ is prepared by fusion of [^{14}C]urea and K_2CO_3 or can be obtained commercially.

Preparation of $K^{14}CNO$. A mixture of 1 mCi of [^{14}C]urea (2 mmol) and 128 mg of K_2CO_3 (1.2 mmol) is ground into a fine powder and transferred to a small Pyrex test tube.[4] Fusion of these reactants is initially carried out at 130° and then completed with a free flame.[5] The product, $K^{14}CNO$, was allowed to cool to room temperature.

Synthesis and Purification of [ureido-^{14}C]-N-Carbamoylputrescine. The above sample containing $K^{14}CNO$ is mixed with 0.05 ml of putrescine (50 μmol) and 0.05 ml of 1 *N* HCl in a total volume of 2.0 ml, heated for 30 min at 100°, and left overnight at room temperature.[6] Aliquots (200 μl) are streaked on Whatman No. 3 filter paper and subjected to descending paper chromatography using butanol–acetic acid–water (4 : 1 : 1, v/v/v) as the solvent system for 12–15 hr. Strips (2 cm) from the point of sample application are cut out from the dried chromatographs and sprayed with ninhydrin and Ehrlich's reagents separately to locate the amine compounds. *N*-Carbamoylputrescine ($R_f = 0.37$), well separated from *N,N'*-dicarbamoylputrescine ($R_f = 0.6$), and free putrescine ($R_f = 0.2$) can be easily located this way. [^{14}C]-*N*-carbamoylputrescine is eluted, then dried *in vacuo*. The purified product has a low specific activity (6–8 mCi/mmol).[7]

[2] The conversion of the free base to hydrochloride form by addition of HCl was often found to result in inactivation of the product.

[3] T. A. Smith and J. L. Garraway, *Phytochemistry* **3,** 23 (1964).

[4] L. H. Smith, *J. Am. Chem. Soc.* **77,** 6691 (1955).

[5] D. Lloyd-Williams and A. R. Ronzio, *J. Am. Chem. Soc.* **74,** 2407 (1952).

[6] A CM-Sephadex chromatography step described earlier by us[7] can be employed to separate the unreacted putrescine and the carbamoyl derivatives in a larger-scale preparation of [*ureido*-^{14}C]-*N*-carbamoylputrescine. The exclusion of both carbamoyl derivatives of putrescine and retention of putrescine on CM-Sephadex can eliminate the interference by putrescine in a subsequent paper chromatographic separation.

[7] K. S. Srivenugopal and P. R. Adiga, *Anal. Biochem.* **104,** 440 (1980).

[75] Hydroxyputrescine: 1,4-Diaminobutan-2-ol[1]

By JIRO TOBARI and T. T. TCHEN

$$NH_2CH_2\underset{\displaystyle OH}{\underset{|}{C}}HCH_2CH_2NH_2$$

2-Hydroxyputrescine

Occurrence and General Properties

Hydroxyputrescine was first identified in a *Pseudomonas* species (ATCC 33971)[2,3] and has since been detected in other *Pseudomonas* strains,[4–6] in a few other bacteria,[4] and in a plant.[7] When the *Pseudomonas* strain of Kim[8] is grown on glucose, ammonium sulfate, and other inorganic salts, hydroxyputrescine is actually the major polyamine, present at 9.4 μmol per gram of cells (wet weight) as compared to putrescine at 4.6 μmol per gram of cells and undetectable amounts of spermidine, spermine, diaminopropane, or cadaverine. It is metabolically more stable than putrescine. This is particularly evident when the cells are subjected to nitrogen starvation, leading to a fairly rapid decline of putrescine content (50% after 1 day) and no drop of hydroxyputrescine content after 1 day.[8] Its physiological function is unknown. When tested in *in vitro* protein synthesizing systems, hydroxyputrescine behaves exactly like putrescine and very differently from spermidine or spermine (unpublished observation).

Preparation and Identification[9]

Reflux 25 g of 1,4-dibromobutan-2-ol with 30 ml of concentrated ammonia for 6 hr. Cool, add 20 ml of ethanol, and reflux for another 3 hr.

[1] Original work was supported by U.S.P.H.S. Grant AM-3724.

[2] C. L. Rosano and C. Hurwitz, *Biochem. Biophys. Res. Commun.* **37,** 677 (1969).

[3] J. Tobari and T. T. Tchen, *J. Biol. Chem.* **246,** 1262 (1971).

[4] J. Tobari and T. T. Tchen, unpublished.

[5] H. Tabor, H. Fujisawa, and C. W. Tabor, unpublished.

[6] E. Karrer, R. J. Bose, and R. A. J. Warren, *J. Bacteriol.* **114,** 1365 (1973).

[7] A. Stoessl, R. Rohringer, and D. J. Samborski, *Tetrahedron Lett.* **33,** 2807 (1969).

[8] K. Kim, *J. Bacteriol.* **91,** 193 (1966).

[9] Another method of preparation, based on the reduction of 1,4-diaminobutanone, has been described by L. Macholan [*Collect. Czech. Chem. Commun.* **30,** 2074 (1965)] and by R. Kullnig, C. L. Rosano, and C. Hurwitz [*Biochem. Biophys. Res. Commun.* **39,** 1145 (1970)].

ISBN 0-12-181994-9

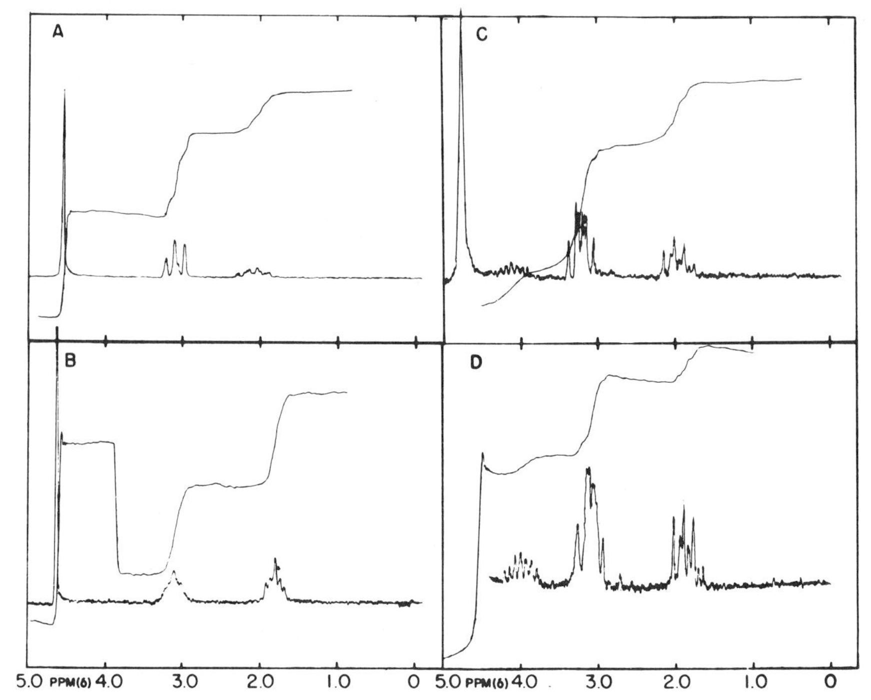

FIG. 1. Nuclear magnetic resonance spectra of the dihydrochlorides of hydroxyputrescine and related compounds in D_2O. (A) 1,3-Diaminopropane; (B) putrescine; (C) synthetic racemic hydroxyputrescine; and (D) cellular hydroxyputrescine. The ratios of carbon-bound hydrogens in these compounds are β-CH_2 : α-CH_2 = 1:2 in diaminopropane; β-CH_2 : α-CH_2 = 1:1 in putrescine; and β-CH_2 : α-CH_2 :—CHOH = 2:4:1 in hydroxyputrescine.

Evaporate to dryness. Redissolve in water and neutralize with concentrated HCl. Apply the solution to a Dowex 50 (H^+) column (column volume 500 ml) and wash with 1 liter of water. Elute with 1.5 *N* HCl, evaporate to dryness, and recrystallize from ethanol–water. The yield is 2 g of hydroxyputrescine dihydrochloride.

Biosynthetic (+)-Hydroxyputrescine

Stir 100 g of wet cell paste (grown to early stationary phase on glucose 0.1%, ammonium sulfate 0.1%, and salts) with 100 ml of 10% trichloroacetic acid. Centrifuge and collect the supernatant. Extract four times with ether to remove trichloroacetic acid. Neutralize with KOH and apply to a Dowex 50W-X4 (H^+) column (125 ml column volume). Elute with an exponential HCl gradient (300 ml of water in mixing flask and 500 ml of 2.5 *N* HCl in reservoir) and collect 3-ml fractions. Diamines are eluted between fractions 40 and 60. Dry 0.3–1.0 ml aliquots. Add 0.1 ml of dinitrofluorobenzene solution (dissolve 0.65 ml of dinitrofluorobenzene in 50 ml of acetone, store in the cold; add 10 ml immediately before use to 100 ml of 0.066 *M* sodium tetraborate) and incubate for 10 min at 65°. After cooling, add 1 ml of dioxane-HCl (100 : 1, v/v). Read absorbance at 370 nm.[10] The column eluates contain two peaks of material that react with dinitrofluorobenzene, the first one being hydroxyputrescine and the second putrescine. [If acetylputrescine were present, it would elute similarly to hydroxyputrescine.[3,11,12]] Recrystallize hydroxyputrescine dihydrochloride as described for synthetic hydroxyputrescine.

Identification by Silica Gel Thin-Layer Chromatography

The solvent system is methanol–concentrated NH_3 (7 : 3 v/v). R_f values are spermine, 0.02; spermidine, 0.06; putrescine, 0.12; cadaverine, 0.15; 1,3-diaminopropane, 0.2; hydroxyputrescine, 0.23; and monoacetylputrescine, 0.6.

Definitive Identification by Nuclear Magnetic Resonance

The simplest definitive identification is by nuclear magnetic resonance spectroscopy (Fig. 1), which shows the characteristic ratios of carbon-bound hydrogens in hydroxyputrescine: $\beta - CH_2 : \alpha - CH_2 : CHOH = 2:4:1$.

Identification by Automated Cation-Exchange Chromatography

See this volume [4].

[10] D. T. Dubin, *J. Biol. Chem.* **235,** 783 (1960).
[11] D. T. Dubin and S. M. Rosenthal, *J. Biol. Chem.* **235,** 776 (1960).
[12] H. Tabor, S. M. Rosenthal, and C. W. Tabor, *J. Biol. Chem.* **233,** 907 (1958).

[76] Glutathionylspermidine[1]

By HERBERT TABOR and CELIA WHITE TABOR

$$\underset{\displaystyle NH_2}{COOHCH}CH_2CH_2CONH\underset{\underset{\displaystyle SH}{\displaystyle CH_2}}{CH}CONHCH_2CONH(CH_2)_3NH(CH_2)_4NH_2$$

Glutathionylspermidine

Escherichia coli contains 3–4 μmol of spermidine per gram wet weight, when grown in a minimal medium. In logarithmically growing cells this amine is present in the unsubstituted form. During stationary phase, however, all the spermidine is converted to glutathionylspermidine. Under these conditions, glutathionylspermidine accounts for a large part of the intracellular glutathione.[1]

Assay

Glutathionylspermidine was assayed on the amino acid analyzer with the elution system described as Method II in this volume [4]; the preliminary low salt elution was omitted. This assay permitted the determination of both the reduced (monosulfide) and the disulfide forms of glutathionylspermidine.[2,3] Elution patterns for each form are given in Fig. 1.

In a typical assay for the content of glutathionylspermidine in *E. coli,* the cells from 25 ml of culture were collected on a Millipore filter and extracted rapidly with 2.5 ml of cold 10% trichloroacetic acid. After centrifugation, 0.01 volume of 1 *N* HCl was added, and the trichloroacetic acid was removed by ether extraction. The added HCl ensured an acid pH at the end of the extraction. If the assay was not carried out immediately, the solution was stored at −20° in a stoppered tube. Prior to assay, portions of the extract (0.15 ml) were treated with dithioerythritol (or dithiothreitol) as previously described.[4] The value for reduced glu-

[1] H. Tabor and C. W. Tabor, *J. Biol. Chem.* **250,** 2648 (1975). See also this series, Vol. 17B [253], for the preparation of an enzyme from *Escherichia coli* that carries out the synthesis of glutathionylspermidine from glutathione and spermidine.

[2] H. Tabor, C. W. Tabor, and F. Irreverre, *Anal. Biochem.* **55,** 457 (1973).

[3] We have also assayed glutathionylspermidine by chromatography of the *S*-(*N*-ethylsuccinimido) and the *S*-carboxamidomethyl derivatives.[2]

[4] To prepare reduced glutathionylspermidine, the sample (20 nmol) was incubated in 1 ml of 0.005 *M* potassium phosphate, pH 7.2, 0.0005 *M* dithioerythritol for 1–2 hr at 25°. The pH was then adjusted to 3.0 with 1 *N* HCl; 0.15 ml of the concentrated Beckman sodium citrate buffer, pH 3.25, was added, and water was added to a final volume of 1.5 ml.[2]

ISBN 0-12-181994-9

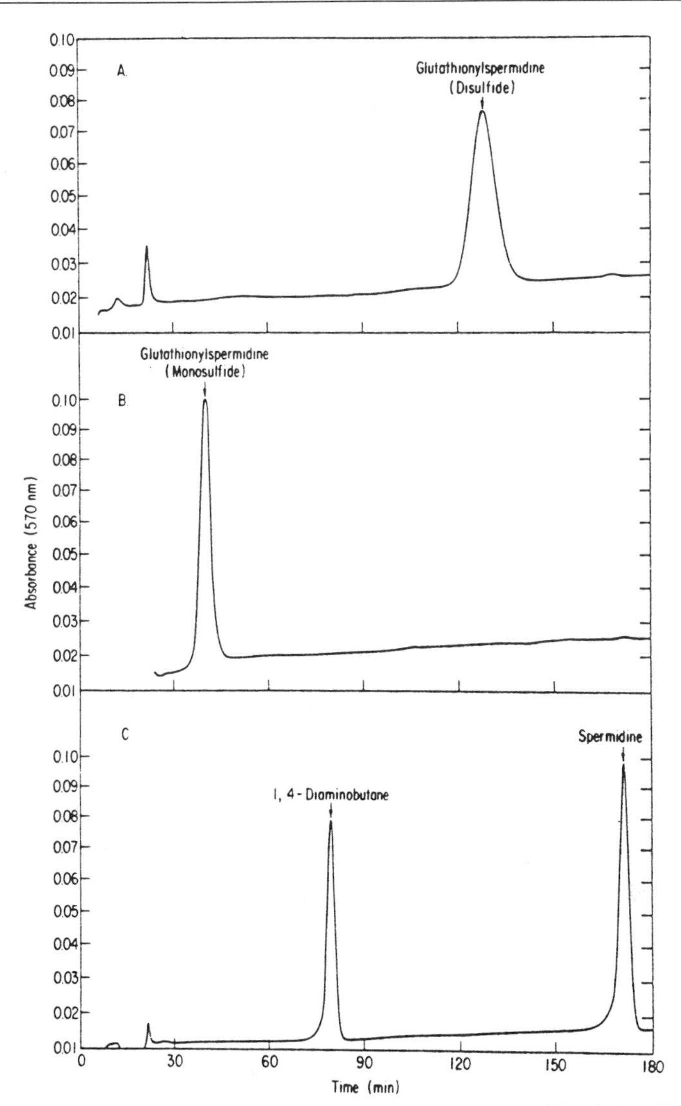

FIG. 1. Determination of reduced glutathionylspermidine and oxidized glutathionylspermidine by automated liquid cation-exchange chromatography. The color yield of reduced glutathionylspermidine is about 80% of that of spermidine; the color yield of oxidized glutathionylspermidine is about 110% of that of spermidine.

tathionylspermidine thus obtained represented the total glutathionylspermidine. If the dithioerythritol was omitted, the assay could be used to determine the proportion of glutathionylspermidine present in the reduced and disulfide forms. However, in the absence of dithioerythritol, any mixed disulfides between glutathionylspermidine and other sulfhydryl compounds would migrate elsewhere and would not be included in this calculation.

Glutathionylspermidine, as well as spermidine, adsorbs easily to glass. Therefore, solutions of glutathionylspermidine were usually made in 0.01 *N* HCl or 0.2 *N* acetic acid, and plastic containers were used wherever possible.

Isolation of Glutathionylspermidine

Growth of Cells and Preparation of Extract. Even though logarithmically growing *E. coli* had no glutathionylspermidine, >85% of the spermidine in the organism was converted to this derivative during stationary phase, particularly if the culture was anaerobic during this period. In order to obtain the maximal conversion, sufficient glucose had to be present to permit the pH to fall, usually to pH 5.5–6.0.

A 320-liter culture of *E. coli* B was grown at 37° with aeration in a minimal medium[5] containing 0.5% glucose. At the end of logarithmic phase, 640 g of glucose were added, and the temperature was maintained at 37°, but the aeration was discontinued. After 2 hr the cells (yield, 1980 g) were collected by centrifugation and stored at −20°. For each isolation, 25-g portions were thawed and extracted with 4 volumes of 5% trichloroacetic acid.

To facilitate isolation and characterization of glutathionylspermidine, isotopically labeled material was prepared. The conditions used were the same as above, except that the culture volume was small (50–1000 ml). ^{35}S-labeled cells were grown in 50 ml of medium containing 0.4 mCi of $^{35}SO_4^{2-}$ and 35 μmol of $^{32}SO_4^{2-}$. Alternatively, ^{14}C or ^{3}H labeling was carried out by adding [^{14}C]spermidine or [^{3}H]spermidine when the cell density was 7×10^8 cells/ml; 3 aliquots of 3–15 μCi of the desired isotope were added per liter of culture at 10- to 15-min intervals. When the culture reached stationary phase, the incubation was continued as described above. The labeled cells were collected by Millipore filtration and extracted with 5% trichloroacetic acid.

The labeled extracts were then added to the trichloroacetic acid extract from 25 g of unlabeled cells; the trichloroacetic acid was removed either by extraction with ether (three times, each with 2 volumes of diethyl ether) or by passage through a column of Amberlite CG-45 (column dimensions: 3.1 cm diameter, 4.2 cm height, for 110 ml of a 5% trichloroacetic acid extract).

[5] H. J. Vogel and D. M. Bonner, *J. Biol. Chem.* **218,** 97 (1956). See also this series, Vol. 17A [1], p. 5.

Chromatography of Glutathionylspermidine on Amberlite CG-50. The extract was treated with 0.01 *M* dithiothreitol, adjusted to pH 7 with 0.1 *N* NH_4OH, and stored in a stoppered vessel for 24 hr at 0°. If this reduction step was omitted, the chromatography was poor, presumably because of the formation of mixed disulfides of glutathionylspermidine with glutathione and other sulfhydryl compounds during the isolation procedure. The reduced extract (110 ml) was passed through a column of Amberlite CG-50 pyridine.[6]

The column was washed with 150 ml of 0.001 *M* dithiothreitol and 1000 ml of water, in order to remove most of the sulfhydryl compounds other than glutathionylspermidine. The column was then washed with 1000 ml of 1% pyridine over a 48-hr period. Under these conditions glutathionylspermidine was converted to the disulfide form, presumably by dissolved oxygen. This step was introduced since the disulfide form of glutathionylspermidine was adsorbed more tightly to the resin than the reduced form, and hence was purified more readily from contaminants.

The column was then washed successively with: (*a*) 750 ml of H_2O; (*b*) 1000 ml of 0.2 *N* acetic acid; (*c*) 1000 ml of 0.4 *N* acetic acid; (*d*) 1000 ml of 0.5 *N* acetic acid; and (*e*) 2000 ml of 1 *N* acetic acid. Most of the glutathionylspermidine was eluted by 1 *N* acetic acid. This fraction was evaporated to dryness *in vacuo* and stored in 5–10 ml of 0.2 *N* acetic acid. The yield of glutathionylspermidine disulfide was 13.2 μmol from 25 g wet weight of packed *E. coli.*

[6] Amberlite CG-50 (XE-64) carboxylate resin was suspended in 5 volumes of 10% pyridine, packed in a column (2 cm diameter; 24 cm height), and washed with 200 ml of 1% pyridine just before use. All chromatographic procedures were at 20–25°.

[77] Isolation and Assay of 2,3-Dihydroxybenzoyl Derivatives of Polyamines: The Siderophores Agrobactin and Parabactin from *Agrobacterium tumefaciens* and *Paracoccus denitrificans*

By J. B. NEILANDS

At the present time two siderophores, defined as microbial ferric ion transport compounds,[1,2] are known that may be classified chemically as catechol derivatives of a threonylspermidine tertiary amide. Agrobactin

[1] J. B. Neilands, *Annu. Rev. Biochem.* **50,** 715 (1981).

[2] J. B. Neilands, *Annu. Rev. Nutr.* **1,** 27 (1981).

ISBN 0-12-181994-9

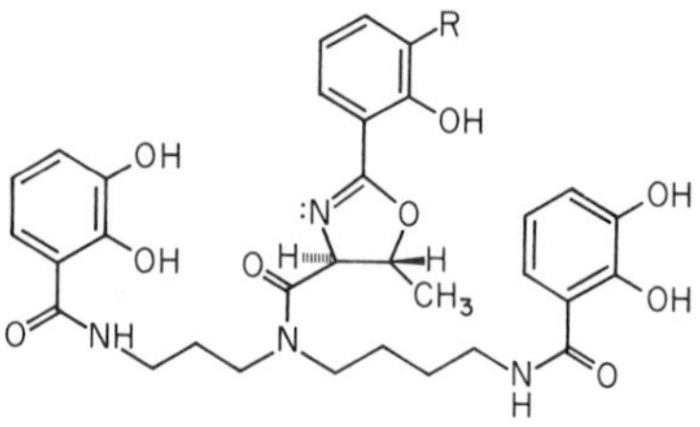

(I)
Parabactin, R = H
Agrobactin, R = OH

(I, R = OH), *N*-[3-(2,3-dihydroxybenzamido)propyl]-*N*-[4-(2,3-dihydroxybenzamido)butyl]-2-(2,3-dihydroxyphenyl)-*trans*-5-methyloxazoline-4-carboxamide, was isolated from low-iron cultures of the crown gall organism, *Agrobacterium tumefaciens,* and the proposed structure[3] was confirmed by high-resolution nuclear magnetic resonance (NMR) spectroscopy[4] and X-ray crystallography.[5] Parabactin (I, R = H), *N*-[3-(2,3-dihydroxybenzamido)propyl]-*N*-[4-(2,3-dihydroxybenzamido)butyl]-2-(2-hydroxyphenyl)-*trans*-5-methyloxazoline-4-carboxamide, was isolated by Tait[6] from *Paracoccus denitrificans,* named Compound III, and assigned the structure I, R = H, except with the threonyl residue not in an oxazoline ring. The subsequent finding[7] of this ring in the *P. denitrificans* product prompted the change of name from Compound III to parabactin. The opened forms of the two siderophores are designated agrobactin A and parabactin A.

Agrobactin

Isolation[3]

An overnight culture of *A. tumefaciens* B6[8] in Luria broth is transferred at 1% inoculum into 50 ml of neutral Tris medium[3] modified to

[3] S. A. Ong, T. Peterson, and J. B. Neilands, *J. Biol. Chem.* **254,** 1860 (1979).
[4] T. Peterson, K.-E. Falk, S. A. Leong, M. P. Klein, and J. B. Neilands, *J. Am. Chem. Soc.* **102,** 7715 (1980).
[5] D. L. Eng-Wilmot and D. van der Helm, *J. Am. Chem. Soc.* **102,** 7719 (1980).
[6] G. H. Tait, *Biochem. J.* **146,** 191 (1975).
[7] T. Peterson and J. B. Neilands, *Tetrahedron Lett.* **50,** 4805 (1979).
[8] A large number of other strains of *A. tumefaciens* produce agrobactin [S. A. Leong and J. B. Neilands, *Arch. Biochem. Biophys.* **218,** 351 (1982)]. Strain B6 will be made available by the author.

contain 0.1 μM $FeSO_4$, 1.0 μM $MnSO_4$, and 0.4% glucose, the latter added after sterilization. After 24 hr of growth at 30°, a 10-liter batch of Tris medium is inoculated with 20 ml of the passage culture. After 2 days of growth with vigorous aeration, the cells are separated by centrifugation and the agrobactin is extracted into three 1-liter portions of ethyl acetate. The combined solvent extracts are evaporated to about 100 ml and washed successively with 0.1 M citrate buffer (pH 5.5) and water. The extract is then dried overnight over anhydrous $MgSO_4$, filtered, and concentrated to a few milliliters; the agrobactin is crystallized by slow, dropwise addition of *n*-hexane. Brownish oxidation products may be removed by chromatography on silicic acid, as described by Tait.[6] The yield is ca 75 mg.

Properties

Agrobactin is a white, crystalline solid, mp 108–112° with decomposition. It is soluble in the lower alcohols, ether, ethyl acetate, tetrahydrofuran, and dioxane. It is sparingly soluble in water, benzene, and hexane. The absorption spectrum in ethanol shows a broad peak characteristic of 2,3-dihydroxybenzoyl compounds with $\varepsilon_{316\ nm} = 9.6 \times 10^3\ M^{-1}\ cm^{-1}$ and a sharper peak deeper in the ultraviolet with $\varepsilon_{252\ nm} = 28.3 \times 10^3\ M^{-1}\ cm^{-1}$. There is an even more intense peak centered at about 218 nm. The R_f on thin-layer silica gel with 4 : 1 chloroform–methanol is 0.64. Agrobactin is detected by its blue fluorescence in the ultraviolet, by ferric chloride spray, or by I_2 vapor reactions.

The 2-oxazoline ring is stabilized by electronic effects emanating from the *o*-hydroxy substituent in the phenyl group and is relatively resistant to hydrolysis.[4] Cis-trans isomerization around the tertiary amide bond gives rise to duplicate NMR signals.[4]

The wine-colored ferric complex[9] of agrobactin contains iron bound in an apparent Λ, cis configuration to the two distal catecholate and central *o*-hydroxyphenyl oxazoline functions. The stability constant with ferric ion is comparable to that of enterobactin.

Assay

Solutions of agrobactin in ethanol can be standardized by use of the ε_{mM} of 9.6 at 316 nm. This absorption band is nonspecific and occurs in all 2,3-dihydroxybenzoyl compounds.

[9] J. B. Neilands, T. Peterson, and S. A. Leong, *in* "Inorganic Chemistry in Biology and Medicine" (A. E. Martell, ed.), p. 263. Am. Chem. Soc., Washington, D.C., 1980.

A more specific, semiquantitative assay[3] for agrobactin is based on its reversal of iron starvation in *A. tumefaciens* invoked by the presence of ethylenediaminedi(*o*-hydroxyphenylacetic acid) (EDDA). Luria broth agar is prepared to contain 1 mg of purified[10] EDDA per milliliter and allowed to stand at 4° for 24 hr. The molten agar is seeded with 1000 colony-forming units per 25-ml batch and poured into petri dishes. Filter paper disks (6 mm) impregnated with 10 μl of ethanolic solutions of agrobactin are placed on the agar surface, and the plates are incubated at 30° for 2–3 days. A 10-μl aliquot of 25 μM agrobactin gives a halo of colonies about 30 mm in diameter. The diameter of the halo varies monotonically with the quantity of siderophore applied, although the assay contains too many variables to be strictly quantitative.

Parabactin

Isolation[6,7]

Paracoccus denitrificans NCIB 8944 is grown in the low-iron medium described by Tait,[6] minus added molybdenum. Inoculum is prepared in rich medium containing, per 100 ml: 0.4 g of peptone, 0.2 g of yeast extract, 1.0 g of K_2HPO_4, 2.0 g of KNO_3, and 1.0 g of glucose (sterilized separately). After overnight growth the organism is transferred at 0.4% inoculum into 10 liters of minimal medium containing, per liter: succinic acid, 5.9 g; NaOH, 4.0 g, K_2HPO_4, 4.9 g; $MgSO_4 \cdot 7H_2O$, 0.2 g; and NH_4Cl, 1.6 g. After neutralization and autoclaving, a sterile solution of ferric citrate is added to a concentration of 20 μM. After 2 days the cells are separated by centrifugation, and the parabactin is extracted and isolated from the neutralized cell-free supernatant as described above for agrobactin. The yield is ca 50–100 mg.

Properties

The general properties of parabactin are similar to those of agrobactin, with notable exceptions attendant upon replacement of a 2,3-dihydroxybenzoyl group by a salicyloyl group. The R_f[7] on silica plates in 4 : 1 chloroform–methanol is greater than that found for agrobactin, agrobactin A, or parabactin A. The absorption band in the near ultraviolet is shifted to the blue and peaks at 309 nm (ε = 10,300). The two other bands in the ultraviolet have the general character of those seen in agrobactin. The pK_a of the oxazolium nitrogen is 2.3, as in agrobactin.

[10] H. J. Rogers, *Infect. Immun.* **7,** 445 (1973).

Assay

All catechol compounds give a pink color with the nitrite–molybdate reagent of Arnow.[11] The ε_{mM} at 510 nm for 2,3-dihydroxybenzoic acid is about 8.8.

A bioassay has not been reported for parabactin, but one could presumably be devised based on relief of EDDA inhibition of the source organism. There is some evidence that agrobactin A and parabactin A form Δ, cis coordination isomers with ferric ion,[9] which may account for their activity in *Escherichia coli* RW193, an organism defective in the synthesis but competent in the transport of ferric enterobactin, which yields the Δ, cis ferric coordination isomer.

Note

Although siderophores appear to be formed generally by phytopathogenic microorganisms,[12] production of agrobactin in *A. tumefaciens* is not correlated with virulence,[13] as seems to be the case for several microorganisms attacking mammalian species.[14] Agrobactin may hence be involved in iron assimilation *ex planta*. Agrobactin is effective in removing iron from heart cells in culture and may be of potential use in chelation therapy for iron overload.[15] A siderophore structurally related to agrobactin occurs in *Vibrio cholerae*.[16] Agrobactin and parabactin have been prepared by chemical synthesis.[17]

[11] L. E. Arnow, *J. Biol. Chem.* **118,** 531 (1937).
[12] S. A. Leong and J. B. Neilands, *Arch. Biochem. Biophys.* **218,** 351 (1982).
[13] S. A. Leong and J. B. Neilands, *J. Bacteriol.* **147,** 482 (1981).
[14] E. D. Weinberg, *Microbiol. Rev.* **42,** 45 (1978).
[15] B. R. Byers, *Microbiology* in press (1983).
[16] G. Grifiths, S. M. Payne, and J. B. Neilands, unpublished.
[17] R. Bergeron, personal communication.

[78] Edeine A, Edeine B, and Guanidospermidine

By Z. KURYLO-BOROWSKA and J. HEANEY-KIERAS

The antibiotics edeine A and edeine B are linear oligopeptides conjugated with polyamines: spermidine (edeine A) and guanidospermidine (edeine B).[1–3] The structures of edeine A and edeine B are shown in Fig. 1.

[1] T. P. Hettinger, Z. Kurylo-Borowska, and L. C. Craig, *Ann. N.Y. Acad. Sci.* **170,** 1002 (1970).

ISBN 0-2-181994-9

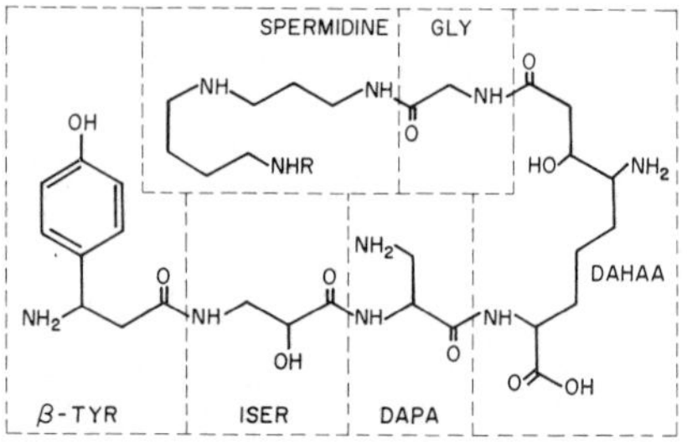

EDEINE A, R = H
B, R = C(=NH)NH_2

FIG. 1. Structure of edeine A and B; GLY, glycine; DAPA, 2,3-diaminopropionic acid; DAHAA, 2,6-diamino-7-hydroxyazelaic acid; ISER, isoserine; β-TYR, β-tyrosine.

Edeines are produced by *Bacillus brevis* Vm4 during late log phase of growth.[2,4] The biosynthesis of edeines occurs through a nonribosomal process involving activation of amino acids and transthiolation via 4′-phosphopantotheine.[5,6] Biosynthesis of the peptidyl fragment of edeines proceeds bidirectionally from the 2,6-diamino-7-hydroxyazelaic acid residue, which throughout synthesis remains esterified through its C-1 carboxyl group to a non-pantotheine-containing protein in the edeine synthetase complex.[7,8] The *in vivo* synthesis of edeine B precedes edeine A[9]; the *in vitro* syntheses are simultaneous, and equal amounts of edeines A and B are produced.[6] Edeines A and B, which are bound to the complex of edeine-synthesizing enzymes (edeine synthetase), can be released from the complex by treatment with mild alkali or methoxamine; the freed antibiotics are biologically active.[7,9] Extracellular edeines are free and biologically active.

Edeines have a broad spectrum of antimicrobial activity[4,10] and little or

[2] Z. Kurylo-Borowska and E. L. Tatum, *Biochim. Biophys. Acta* **133,** 206 (1966).

[3] T. P. Hettinger and L. C. Craig, *Biochemistry* **9,** 1224 (1970).

[4] Z. Kurylo-Borowska, *Bull. Inst. Mar. Med. Gdansk* **10,** 151 (1959).

[5] Z. Kurylo-Borowska and E. L. Tatum, "Progress in Microbial and Anticancer Chemotherapy," pp. 1123–1127. Univ. of Tokyo Press, Tokyo, 1970.

[6] Z. Kurylo-Borowska and J. Sedkowska, *Biochim. Biophys. Acta* **351,** 42 (1974).

[7] Z. Kurylo-Borowska and J. Heaney-Kieras, *Proc. Int. Congr. Biochem., 10th, 1976* Abstract 04-3-351, p. 181 (1976).

[8] Z. Kurylo-Borowska and J. Heaney-Kieras, *in* "Peptide Antibiotics-Biosynthesis and Functions" (H. Kleinkauf and H. von Döhren, eds.), p. 315. de Gruyter, Berlin, 1981.

[9] Z. Kurylo-Borowska and W. Szer, *Biochim. Biophys. Acta* **418,** 63 (1972).

[10] Z. Kurylo-Borowska, *in* "Antibiotics" (J. W. Corcoran and F. E. Hahn, eds.), Vol. 3, pp. 129–140. Springer, New York, 1974.

no cytotoxicity against some normal or neoplastic cells in culture.[11,12] In prokaryotes, edeine A inhibits the DNA polymerase II-dependent replication of DNA[13] and protein synthesis by binding to the initiation codon of the 30 S (in eukaryotes, 40 S) ribosomes.[14] While edeines can enter both prokaryotic and eukaryotic cells, these cells do not modify the antibiotics.[11,15]

Guanidospermidine is produced during late log phase of growth by *Bacillus brevis* Vm4. It can be isolated from the growth medium as a free polyamine and as a constituent of edeine B.[16]

Edeine A and Edeine B

I. Preparation and Isolation

A. *Isolation from Culture Medium of Bacillus brevis Vm4*

Reagents

K_2HPO_4, 1.0 *M*
NH_4OH, 0.1 *M* and 1.0 *M*
Culture medium of *B. brevis* Vm4

Procedure. Bacillus brevis Vm4 is grown in Bactopeptone-yeast extract medium at 30° with shaking; at late log phase the culture medium (4 liters) (pH 7.8) is collected by low-speed centrifugation.[2] This supernatant is shaken with 200 ml of AG 50-X4 (H^+ form, 50–100 mesh; Bio-Rad Laboratories) for 15 min. The resin is allowed to settle, and the supernatant is decanted and discarded. The resin is washed with water, adjusted to pH 7 with 1.0 *M* K_2HPO_4, washed three times with 20 volumes of water, resuspended in water, poured into a column, and washed with 3 volumes of 0.1 *M* NH_4OH. Then, 1.0 *M* NH_4OH is used to elute the edeines from the resin; the edeines are located in fractions of pH 10.5–11.5. These fractions are pooled, concentrated under reduced pressure,

[11] Z. Kurylo-Borowska and J. Heaney-Kieras, *Exp. Cell Res.* **124**, 371 (1979).
[12] Z. Kurylo-Borowska and J. Heaney-Kieras, *Fed. Proc., Fed. Am. Soc. Exp. Biol.* **35**, 1573 (1976).
[13] Z. Kurylo-Borowska and W. Szer, *Biochim. Biophys. Acta* **287**, 236 (1972).
[14] W. Szer and Z. Kurylo-Borowska, *in* "Molecular Mechanism of Antibiotics Action on Protein Biosynthesis and Membranes" (E. Muñoz, F. Garcia-Ferrandiz, and D. Vazquez, eds.), pp. 57–74. Elsevier, Amsterdam, 1972.
[15] J. Georgiades, Z. Kurylo-Borowska, and L. Dmochowski, *75th Annu. Meet. Am. Soc. Microbiol.* Abstract K107, p. 164 (1975).
[16] T. P. Hettinger, Z. Kurylo-Borowska, and L. C. Craig, *Biochemistry* **7**, 4153 (1968).

and neutralized with glacial acetic acid. The concentrate is subjected to Sephadex G-50 (Pharmacia Fine Chemicals) column chromatography (2.8 × 90 cm) in water (flow rate 80 ml/hr), which separates the accompanying yellow pigment from the edeines. The antibiotic fractions have a partition coefficient (K_{av}) of 0.59 (530–850 ml effluent on this column), and the position of edeine B precedes slightly that of edeine A. This procedure yields approximately 0.20–0.25 g of edeines A and B from 4 liters of culture medium containing 32 g wet weight of cells.

B. Synthesis in Vitro

Reagents

Tris-HCl buffer (Sigma Chemical Co.), 1.0 *M*, pH 7.6 and pH 7.9
Morpholinopropanesulfonic acid (Sigma Chemical Co.), 0.1 *M*, adjusted to pH 7.2
EDTA, 2.5 m*M*
Lysozyme (Sigma Chemical Co.), 10 mg/ml
DNAase (Sigma Chemical Co.), 10 mg/ml
Magnesium acetate, 1.0 *M*
KCl, 1.0 *M*
Dithiothreitol (Sigma Chemical Co.), 0.1 *M*
ATP (Sigma Chemical Co.), 0.1 *M*
Phosphoenolpyruvate kinase (Sigma Chemical Co.), 10 mg/ml
β-Tyrosine, 5 m*M*
2,3-Diaminopropionic acid, 5 m*M*
2,6-Diamino-7-hydroxyazelaic acid, 5 m*M*
Isoserine, 5 m*M*
[U-^{14}C]Glycine
[U-^{14}C]Spermidine
Solution of edeine synthesizing enzymes, crude, 5–10 mg/ml (protein)

Procedure. From a 10-liter culture at late log phase of growth, *B. brevis* Vm4 cells are collected by centrifugation. After washing three times with 600 ml (each time) of cold 0.1 *M* Tris-HCl buffer (pH 7.6) in 2 m*M* dithiothreitol (buffer A), the pellet (about 120 g wet weight) is suspended in 60 ml of buffer (pH 7.2) consisting of 0.1 *M* morpholinopropanesulfonic acid and 2.5 m*M* EDTA. The cells are lysed by incubating the suspension with 36 mg of lysozyme and 1.2 mg of DNAase at 30° for 20 min and then at 0° for 20 min. The lysate of cells is centrifuged for 90 min at 30,000 rpm at 4°. All further operations are conducted at 4°. To the supernatant fraction solid ammonium sulfate is added to 30% saturation.

After 1 hr the precipitate is removed by centrifugation and the concentration of ammonium sulfate of the supernatant fraction is adjusted to 55%. After 1 hr the precipitate is collected by centrifugation. The precipitate containing the crude enzymes is dissolved in 5 ml of 1.0 *M* Tris-HCl (pH 7.6), and dialyzed for 5 hr against 2 liters of buffer A, changing the buffer twice. The purified enzymes are prepared by subsequent fractionation on DEAE-cellulose chromatography and Sephadex G-200 column chromatography, as described previously.[17]

The reaction mixture consists of enzymes (100 μg of protein), 80 μl of 1.0 *M* Tris-HCl buffer (pH 7.9), 20 μl of 0.1 *M* dithiothreitol, 10 μl of 1.0 *M* KCl, 10 μl of 1.0 *M* magnesium acetate, 5 μl of 0.2 *M* phosphoenolpyruvate, 5 μl (10 mg/ml) of phosphoenolpyruvate kinase, 25 μl of 0.1 *M* ATP, 10 μl of each 0.1 *M* edeine constituent amino acid, including [U-^{14}C]glycine (25,000 cpm/μmol), and 10 μl of 0.1 *M* [U-^{14}C]spermidine (25,000 cpm/μmol), in a total volume of 1.0 ml. After incubation at 35° for 30 min, samples are acidified to pH 5.5 with 1.0 *M* acetic acid. The precipitate is removed by centrifugation. The resulting supernatant solution is neutralized with 1 *N* NaOH and adsorbed to AG 50-X4 (H^+ form) on a column (2 × 1 cm). The resin is washed successively with 20 ml of 0.5 *M* ammonium formate buffer (pH 7.0), 20 ml of water, and 5 ml of 1.0 *N* NH_4OH. The fraction eluted by the NH_4OH is concentrated under vacuum. The radioactive products may be separated by the chromatographic procedures described in Section II,D.

II. Separation of Edeine A from Edeine B

A. Partial Separation by Phenol Extraction

Reagents

Phenol (redistilled), 88%
CH_3COOH, 5 *N*
NaOH, 1 *N*
NH_4OH, 1 *N*
Ethyl ether
Mixture of edeines A and B

Procedure. On a relatively large-scale purification of edeines (0.5 g or more), a partial separation of the antibiotics can be achieved conveniently with phenol extraction, so that edeine A and a mixture of edeines A and B are obtained.

[17] Z. Kurylo-Borowska, this series, Vol. 43, p. 129.

The residue (0.5 g) obtained from the Sephadex G-50 column fractionation (see Section I,A) is dissolved in 120 ml of 88% phenol and shaken with an equal volume of 5 *N* acetic acid for 1 hr. The aqueous phase contains mostly edeine A, whereas the phenol phase contains a mixture of edeines A and B. The yield of edeine A is about 0.08 g, and in the mixture of edeines A and B, 0.1 g of edeine A and 0.22 g of edeine B. To remove traces of phenol from the aqueous phase, it is extracted three times with 1 volume of ethyl ether, concentrated under reduced pressure, and neutralized with 1 *N* NH_4OH. The phenol phase is adjusted to pH 7.5 with 1 *N* NaOH, extracted with 3 volumes of ethyl ether followed by three times with 1 volume of water; the clear aqueous phase of this extraction is then concentrated as described above. Further separations of the antibiotics are performed by methods described in Section II,D.

B. Separation by Carboxymethyl Cellulose Chromatography

Reagents

CH_3COOH, 0.03 *M*
Ammonium acetate, 0.25 *M*
Mixture of edeines A and B

Procedure. Carboxymethyl cellulose (Whatman Ltd.) is equilibrated with 0.1 *M* ammonium acetate buffer, pH 4.5, and packed into a column (2.5 × 25 cm). The edeines (100–200 mg) are applied to the column, followed by a gradient consisting of 400 ml of 0.03 *M* acetic acid and 400 ml of 0.25 *M* ammonium acetate. Edeine B is eluted at pH 4.5 and edeine A at pH 5.6; these may be desalted after adsorption to AG 50-X4 (H^+ form) as described in Section I,A. From 100 mg of a mixture of edeines A and B, 30 mg of edeine A and 30 mg edeine B are recovered.

C. Separation by Countercurrent Distribution

Reagents

Phenol (redistilled), 88%
Ammonium acetate, 0.15 *M*
CH_3COOH, 0.3 *M*
Ethyl ether
Mixture of edeines A and B

Procedure. A complete separation of small amounts of the edeines (50–100 mg) can be achieved using countercurrent distribution.[16] Edeines A and B (100 mg) from the Sephadex G-50 column chromatographic fractionation are dissolved in 30 ml of a system consisting of 1 : 1 88% phenol in water and 0.15 *M* ammonium acetate–0.3 *M* acetic acid, and distributed

TABLE I
SEPARATION OF EDEINE A FROM EDEINE B

Procedure	Edeine A	Edeine B
Paper chromatography (R_f)		
1-Butanol : acetic acid : H_2O (12 : 3 : 5)	0.04	0.11
1-Butanol : acetic acid : pyridine : H_2O (6 : 3 : 2 : 3)	0.11	0.14
2-Propanol : NH_4OH : H_2O (4 : 1 : 1)	0.19	0.11
Ethanol : NH_4OH : H_2O (60 : 35 : 5)	0.25	0.15
Thin-layer chromatography (silica gel) R_f		
1-butanol : acetic acid : pyridine : H_2O (6 : 3 : 2 : 3)	0.12	0.15
High-voltage electrophoresis (centimeters toward cathode)		
Paper: 0.2 *M* sodium citrate, pH 3.5 (65 min, 40 V/cm, 0°)	0.80	0.85
Cellulose acetate: pyridine : acetic acid : H_2O (100 : 4 : 896) pH 6.4 (65 min, 40 V/cm, 0°)	1.27	1.24

over the first 6 tubes of a 500-tube countercurrent distribution apparatus. The volume of these 6 tubes is brought to 10 ml using equal volumes of the phenol phase and the buffer phase. The remaining tubes (494) of the apparatus are filled with 5 ml of the phenol phase per tube. Buffer phase (5 ml) is added to the first tube after each transfer, and the distribution is carried out to 600 transfers. The lower phases are analyzed for antibiotic activity after the phenol has been extracted with ethyl ether (Section IV,B). Edeine A is located in tubes 240–290 and edeine B in tubes 120–170. The yield from 100 mg is 30 mg of edeine A and 30 mg of edeine B.

D. Separation by Chromatographic and Electrophoretic Methods

The separation of the edeines can be accomplished by chromatography on paper (ascending, Whatman No. 1; Whatman Ltd.), silica gel, and by electrophoresis on paper (Whatman No. 3 MM), cellulose acetate (Eastman Chemicals Co.).[16] The separation of these antibiotics under various conditions is summarized in Table I.

III. Identification of Edeine A and Edeine B

Edeines are amorphous and colorless. Edeines A and B are soluble in water and insoluble in many organic solvents. At pH 7.5–8.0 edeines A and B are equally soluble in phenol, but at pH 4.0–4.5 edeine B is more soluble than edeine A in phenol. The molecular weight of edeine A is 730, and of edeine B, 810.

A. *Spectrophotometric Methods*[18]

Aqueous solutions of edeines A and B have an absorption spectrum analogous to β-tyrosine with a maximum at 272 nm when measured at neutral pH. Their molar extinction at 272 nm is 1310 (liter mol^{-1} cm^{-1}).

B. *Colorimetric Reactions*[16]

Reagents

Ninhydrin reagent: ninhydrin, 0.25%, in 0.5 *M* CH_3COOH in 1-butanol

Pauly reagents: a1, $NaNO_2$, 5% in water; a2, sulfanilic acid, 10% in concentrated HCl; b, Na_2CO_3 (anhydrous), 10% in water

Sakaguchi reagents: a, oxime, 0.1% in acetone; b, bromine liquid, 0.3 ml in 100 ml 0.5 *N* NaOH

Procedure. Edeine A develops a light purple color and edeine B a blue color when allowed to react with ninhydrin reagent, then heated for 1 min at 100°. In the Pauly reaction,[19] edeines A and B react with diazotized sulfanilic acid, and a yellow color (~400 nm) results that is due to the coupling of the reagent with the β-tyrosine moieties. Equal volumes of reagents a1 and a2 are mixed and kept for 5 min at 15°; then 2 volumes of reagent b are carefully added. The chromatogram (dried at 100° for 10–15 min) is dipped through this solution and laid on a glass. The yellow color develops within a few minutes. In the Sakaguchi reaction,[20] edeines A and B react with 8-hydroxyquinoline and then with alkaline hypobromite. Edeine B develops a reddish-brown color, edeine A develops a blue color. The Sakaguchi reaction is positive only for the free-base form of edeine A. The chromatogram (dried as described above) is dipped through reagent a and left until the acetone has evaporated. It is then dipped through reagent b. The color develops immediately.

IV. Biological Assay

A. *Spectrum of Biological Activity*

Edeines A and B have a broad spectrum of *in vivo* activity against gram-positive (except species of *Neisseria*) and gram-negative bacteria,

[18] J. Geatano, Z. Kurylo-Borowska, and L. C. Craig, *Biochemistry* **5,** 2153 (1966).

[19] I. Smith, "Chromatographic and Electrophoretic Techniques," Vol. 1, p. 218. Wiley-Interscience, New York, 1960.

[20] I. Smith, "Chromatographic and Electrophoretic Techniques," Vol. 1, p. 97. Wiley-Interscience, New York, 1960.

mycoplasma,[21] fungi, and yeasts.[10] The edeines are stable as dry powders at 4° for a year and in solution for several days at 37°.

B. Determination of Antibiotic Activity[16]

Biological assay uses the agar plate diffusion method with *Bacillus subtilis* as the standard. Incubation on plates containing 20 μg of edeines A or B in 5 μl for 24 hr at 25° gives an inhibition zone of 20 mm (error = 0.5 mm $\pm$ 15%). A 10-fold increase in antibiotic concentration increases the inhibition zone by 7 mm.

Guanidospermidine

I. Preparation and Isolation

A. Chemical Synthesis[16]

Reagents

Agmatine-2HCl (1-amino-4-guanidinobutane hydrochloride)
Acrylonitrile
Methanol
NaOH, 5 *N*
HCl, concentrated
Ethyl alcohol (absolute)
Platinic oxide

Procedure. Agmatine-2HCl (1.4 mmol; 284 mg) is suspended in 1.5 ml of methanol. Then 5 *N* NaOH (0.28 ml) is added with constant stirring, followed by 1.5 μmol (100 μl) of acrylonitrile. The mixture is kept for 2 hr at room temperature and then refluxed on a steam bath for 1 hr. The pH of the mixture is adjusted to pH 3 with concentrated HCl, and the solvent is removed by vacuum distillation. The residue is extracted with 9 ml of hot absolute ethanol. The extract is evaporated to dryness *in vacuo,* then dissolved in 3 ml of absolute ethanol. To hydrogenate the substance, 200 μl of concentrated HCl and 50 mg of platinic oxide are added, and the reaction is carried out at a pressure of 2.5 atm for 3 hr at 30°. Water (10 ml) is then added, the catalyst is removed by filtration, and the acidic solution is evaporated. The resulting residue is dissolved in 5 ml of water and purified by ion-exchange chromatography on a column containing AG 50-X2 (H^+ form, 200–400 mesh; 2.2 $\times$ 20 cm) that has been equilibrated with 3 *N* HCl. The same solvent is used to elute guanidospermidine at a

[21] J. Borysiewicz, *Appl. Microbiol.* **14,** 1049 (1966).

volume of 340–460 ml. After evaporation the resulting guanidospermidine-3HCl is a glassy, colorless substance. The yield is 42% based on agmatine-2HCl.

B. *Purification from Culture Medium of B. brevis Vm4*

A culture of *B. brevis* Vm4 (4 liters) grown to late log phase is adjusted to pH 5, kept for 1 hr at 0°, centrifuged at 4000 *g* for 15 min at 0°, and the sediment is discarded. The supernatant is applied to a column containing AG 50-X2 (H^+ form; 2.2 × 24 cm), equilibrated with 1 *N* HCl, and adjusted to a flow rate of 26 ml/hr. The resin is washed with 650 ml of 1 *N* HCl, and then the polyamines are eluted with 3 *N* HCl. Spermidine is eluted between 225 and 260 ml and guanidospermidine between 450 and 480 ml. The yield is 0.5 g of guanidospermidine per 4 liters of culture.

C. *Isolation from the Acid Hydrolysis of Edeine B*[16]

Edeine B (0.1 g) is dissolved in 3.6 ml of 6 *N* HCl and kept at 105° for 24 hr *in vacuo* in a sealed tube. The hydrolyzate is evaporated to dryness several times with the addition of water. The residue is dissolved in 5 ml of water and applied to a column containing AG 50-X4 as described in Section I,A; guanidospermidine is eluted by 1 *N* NH_4OH at pH 10.5–11.5. The yield from 0.1 g of edeine B of guanidospermidine-HCl is 9.5–10.5 mg.

TABLE II
SEPARATION OF GUANIDOSPERMIDINE FROM SPERMIDINE

Procedure	Spermidine	Guanidospermidine
Paper chromatography (R_f)		
1-Butanol : acetic acid : pyridine : H_2O (6 : 3 : 2 : 3)	0.27	0.33
Thin-layer chromatography (silica gel) R_f		
Ethanol : NH_4OH : H_2O	0	0.13
High-voltage paper electrophoresis (centimeters toward cathode)		
Pyridine : acetic acid : H_2O (10 : 100 : 890) pH 3.5 (1.5 hr, 45 V/cm, 4°)	47.8	41.5
Separation on ion-exchange resin AG 50-X2 (H^+ form) (milliliters of effluent)[a]	225–260	460–480

[a] See Section I,B for methods.

II. Identification[16]

Guanidospermidine is a strongly basic compound of amorphous glassy appearance. It is water soluble and yields a crystalline picrate (mp 204–205° with decomposition) and chloroplatinate (mp 212–213° with decomposition). Guanidospermidine gives a blue color in the ninhydrin reaction and a reddish-brown color in the Sakaguchi reaction (see Section III,B for details of these methods). Guanidospermidine can be purified by chromatography on paper, on silica gel, or by electrophoresis. The separations from spermidine that can be achieved by these methods are summarized in Table II. Guanidospermidine is stable to acid (6 *N* HCl, 105°, 24 hr) and is degraded to spermidine in alkali (saturated $Ba(OH)_2$, 100°, 2 hr). Dinitrophenylation of Edeine B forms a mono-DNP guanidospermidine with the DNP group located on the secondary amine function; this structure is indicated by the positive ninhydrin and Sakaguchi reactions of the DNP-guanidospermidine isolated from acid hydrolyzates. This DNP derivative has an absorption maximum in 1 *N* acetic acid of 387 nm.

[79] Synthesis of N^1-(γ-Glutamyl)spermidine, N^8-(γ-Glutamyl)spermidine, N^1,N^8-Bis(γ-glutamyl)spermidine, N^1-(γ-Glutamyl)spermine, N^1,N^{12}-Bis(γ-glutamyl)spermine, and N^1,N^4-Bis(γ-glutamyl)putrescine[1]

By J. E. Folk

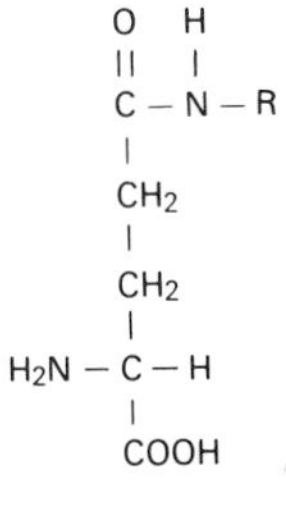

```
      O  H
      ||  |
      C – N – R
      |
      CH2
      |
      CH2
      |
H2N – C – H
      |
      COOH
```

WHERE R =

— $(CH_2)_3NH(CH_2)_4—NH_2$, N^1—(γ-glutamyl)spermidine

— $(CH_2)_4NH(CH_2)_3—NH_2$, N^8—(γ-glutamyl)spermidine

— $(CH_2)_3NH(CH_2)_4NH(CH_2)_3—NH_2$, N^1—(γ-glutamyl)spermine

[1] The nomenclature used here for the amine derivatives follows that recommended in H. Tabor, C. W. Tabor, and L. de Meis, this series, Vol. 17B [256].

ISBN 0-12-181994-9

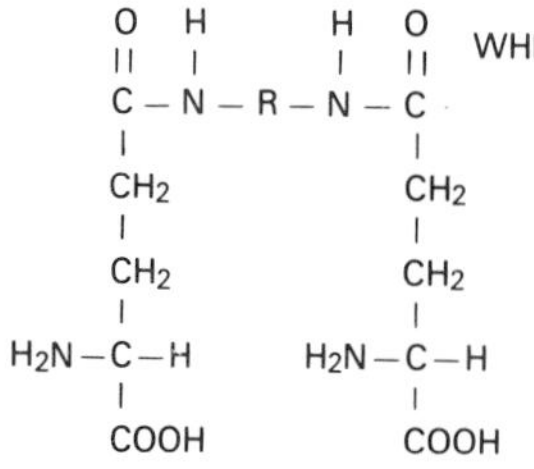

$$HOOC-CH(NH_2)-CH_2-CH_2-C(=O)-NH-R-NH-C(=O)-CH_2-CH_2-CH(NH_2)-COOH$$

WHERE R =

$-(CH_2)_4-$, N^1,N^4–bis(γ-glutamyl)putrescine

$-(CH_2)_4NH(CH_2)_3-$, N^1, N^8–bis(γ-glutamyl)spermidine

$-(CH_2)_3NH(CH_2)_4NH(CH_2)_3-$, N^1, N^{12}–bis(γ-glutamyl)spermine

Synthesis Methods[2]

Principles. Each of the compounds, with the exception of bis(γ-glutamyl)putrescine, is prepared by a combination of enzymic and chemical steps. The general procedure is summarized in the following equations, in which Z = *N*-benzyloxycarbonyl and R = $(CH_2)_3NH(CH_2)_4$, $(CH_2)_4NH(CH_2)_3$, or $(CH_2)_3$ $NH(CH_2)_4NH(CH_2)_3$.

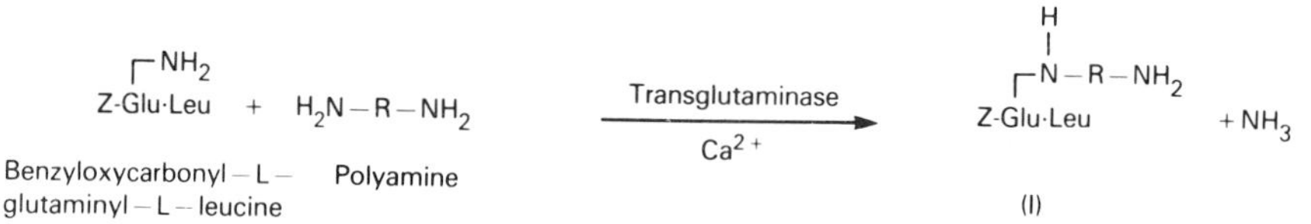

$$\text{Z-Glu(NH}_2\text{)·Leu} + H_2N-R-NH_2 \xrightarrow[Ca^{2+}]{\text{Transglutaminase}} \text{Z-Glu(NH-R-NH}_2\text{)·Leu} + NH_3 \quad (1)$$

Z-Glu(NH₂)·Leu = Benzyloxycarbonyl–L–glutaminyl–L–leucine; $H_2N-R-NH_2$ = Polyamine; product = (I)

$$\text{Z-Glu(NH}_2\text{)·Leu} + \text{Z-Glu(NH-R-NH}_2\text{)·Leu (I)} \xrightarrow[Ca^{2+}]{\text{Transglutaminase}} \text{Z-Glu·Leu(NH-R-NH)Z-Glu·Leu (II)} + NH_3 \quad (2)$$

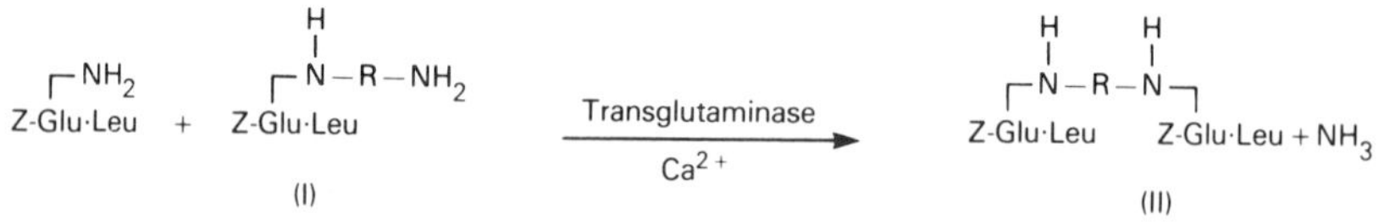

$$\text{Z-Glu(NH-R-NH}_2\text{)·Leu (I) or Z-Glu·Leu(NH-R-NH)Z-Glu·Leu (II)} \xrightarrow{\text{Carboxypeptidase A}} \text{Z-Glu(NH-R-NH}_2\text{) (III)} + \text{Leu or Z-Glu(NH-R-NH)Z-Glu (IV)} + \text{Leu} \quad (3)$$

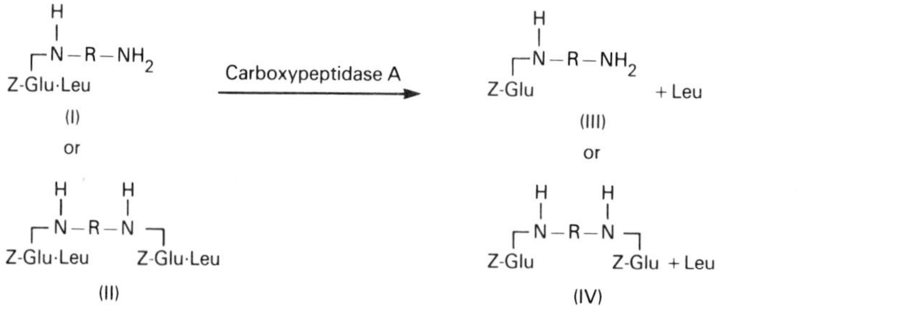

[2] J. E. Folk, M. H. Park, S. I. Chung, J. Schrode, E. P. Lester, and H. L. Cooper, *J. Biol. Chem.* **255**, 3695 (1980).

$$\underset{\text{(III)}}{\text{Z-Glu}-\text{N(H)}-\text{R}-\text{NH}_2} \text{ or } \underset{\text{(IV)}}{\text{Z-Glu}-\text{N(H)}-\text{R}-\text{N(H)}-\text{Z-Glu}} \xrightarrow{\text{HBr in AcOH}} \underset{\text{(V)}}{\text{Glu}-\text{N(H)}-\text{R}-\text{NH}_2} \text{ or } \underset{\text{(VI)}}{\text{Glu}-\text{N(H)}-\text{R}-\text{N(H)}-\text{Glu}} \quad (4)$$

Guinea pig liver transglutaminase catalyzes a Ca^{2+}-dependent acyl transfer reaction in which the γ-carboxamide group of a glutamine residue in a benzyloxycarbonyl-(Z)-glutaminyl peptide may function as an acyl donor and the primary amino groups in a variety of amines can serve as acyl acceptors.[3,4] With diamines and polyamines, either one or both of the primary amino groups in each compound may participate with the subsequent formation of mono(γ-glutamyl)amine derivatives (I) or bis(γ-glutamyl)amine derivatives (II) [Eqs. (1) and (2), respectively].[5] The amounts of mono and of bis derivatives formed vary with the time of reaction. However, the final yield of each is a function of the molar ratio of amine to Z-glutaminyl peptide in the starting reaction mixture. The carboxyl-terminal L-leucine is removed from the isolated derivatives by the use of the enzyme carboxypeptidase A (Eq. 3). Finally, the free (γ-glutamyl)amines (V and VI) are obtained through chemical deblocking with HBr in acetic acid [Eq. (4)]. Bis(γ-glutamyl)putrescine is prepared by a conventional chemical procedure.

Reagents for Preparation of (γ-Glutamyl)polyamines and Their Derivatives

Stock buffer solution: 1 *M* Tris-HCl (pH 7.5) containing 0.3 *M* NaCl, 0.5 *M* $CaCl_2$, and 10 m*M* ethylenediaminetetraacetic acid

Z-L-Glutaminyl-L-leucine, 50 m*M*. The Z-dipeptide[5] is dissolved in 1 equivalent of 0.1 *M* NaOH, the pH of the solution is brought to 7.5, and the volume is adjusted with water.

Spermidine · 3HCl, 200 m*M*, in water

Spermine · 4HCl, 200 m*M*, in water

Purified guinea pig liver transglutaminase,[6] 7–14 mg/ml

[3] J. E. Folk and P. W. Cole, *J. Biol. Chem.* **240,** 2951 (1965).

[4] J. E. Folk and S. I. Chung, *Adv. Enzymol.* **38,** 109 (1973).

[5] J. Schrode and J. E. Folk, *J. Biol. Chem.* **253,** 4837 (1978).

[6] J. M. Connellan, S. I. Chung, N. K. Whetzel, L. M. Bradley, and J. E. Folk, *J. Biol. Chem.* **245,** 1093 (1971); also see this series, Vol. 17A [127] for an earlier preparative procedure.

RETENTION TIMES FOR γ-MONO- AND BISPOLYAMINE DERIVATIVES OF Z-L-α-GLUTAMYL-L-LEUCINE AND Z-L-GLUTAMIC ACID[a]

	Polyamine			
	Spermidine		Spermine	
Derivative Number	Mobile phase[b]	Retention time (min)	Mobile phase[b]	Retention time (min)
I	60:40	4.6, 5.1[c]	80:20	6.8
II	50:50	6.9	65:35	5.6
III	55:45	3.0, 3.4[c]	55:45	7.0
IV	40:60	5.4	45:55	6.4

[a] From Folk *et al.*[2] The reference times are only approximate and vary from column to column and with the age of an individual column.

[b] Methanol–10 m*M* NH_4HCO_3 (v : v).

[c] Retention times for the N^1,N^8-spermidine derivatives. The identities are not known, and the derivatives are collected as mixtures.

Bovine carboxypeptidase A (3× crystallized) 10 mg/ml in 2 *M* NH_4HCO_3

Anhydrous hydrogen bromide, 30% in glacial acetic acid

High-Performance Liquid Chromatographic Procedures for Separation of (γ-Glutamyl)polyamine Derivatives. Analytical and preparative separations are carried out with the use of a Waters Associates instrument equipped with a Model U6K injector. Monitoring is conveniently performed at 210 or 254 nm. The column used is a 3.9 mm × 30 cm μ-Bondapak C_{18}, and the isocratic mobile phases are composed of various ratios of methanol and 10 m*M* NH_4HCO_3. All runs are made at 2.5 ml/min. The mobile phases used and the approximate retention times are given in the table.

Procedures for Preparation of (γ-Glutamyl)polyamines. Reaction mixtures are prepared by adding together 0.05 ml of stock buffer solution, 0.2 ml of Z-L-glutaminyl-L-leucine, 0.025 ml or 0.20 ml of spermidine or spermine and water to make 0.475 ml. A 0.025-ml portion of transglutaminase is added to start the reaction. A blank is prepared by substituting water for the polyamines. At appropriate times after the start of the reactions, commencing at about 10 min, 5-μl portions of reaction mixtures and blanks are examined by high-performance liquid chromatography. When the reactions are essentially complete (at between 30 min and 4 hr) as indicated by no further changes in the chromatographic patterns, separations are performed on approximately 0.1-ml portions of reaction mixtures. Collections of portions of effluents containing the products are made using the analytical runs as guides. In some cases chromatography of the larger samples results in elution of individual products in what appear to be multiple overlapping unsymmetrical peaks. This elution feature encountered with preparative samples may occur as a consequence of column overloading. However, there is little difficulty in obtaining pure products because all the products, with the exception of the N^1- and N^8-(γ-glutamyl)spermidine derivatives, are well separated by the mobile systems listed in the table. No attempt is made to separate the mono(γ-glutamyl)spermidine derivatives. They are collected together and carried through the remaining steps as a mixture. The reaction mixtures prepared with the lower concentrations of spermidine or spermine (10 m*M*) provide high yields of bis derivatives (II); those with the higher concentrations (80 m*M*) give high yields of mono derivatives (I).

The effluent fractions containing the individual products (I and II) are combined, and the solvents are removed under a stream of air at 35–40°. Each of the residual products is dissolved in 0.19 ml of water and treated with 0.01 ml of carboxypeptidase A. The reaction is usually complete within 15 min, as evidenced by analytical high-performance liquid chromatography. The products (III and IV) are purified by chromatography as described for the preceding step.

Again, the effluent fractions containing the individual products are combined and the solvents are removed as described above. Each of the residual products is further dried over P_2O_5 under vacuum. A 0.1-ml portion of HBr in acetic acid is added to each. After 60 min the excess reagent is removed by placing the samples under vacuum in a desiccator over KOH and P_2O_5. The dried samples (V and VI) are dissolved in water and stored frozen at −20°.

Procedure for Preparation of Bis(γ-glutamyl)putrescine

Bis(α-benzyl-Z-L-glutamyl)putrescine. To a solution of 3.7 g of α-benzyl-Z-L-glutamate and 1.4 ml of triethylamine in 10 ml of dimethylformamide at 0° is added dropwise 1.3 ml of isobutyl chlorocarbonate. After the mixture has stirred for 20 min at this temperature, a solution of 0.45 g of putrescine (free base) in 5 ml of dimethylformamide is added. Stirring is continued overnight at room temperature. The solvent is removed at 50° under high vacuum and is replaced by 100 ml of hot ethyl acetate. The ethyl acetate layer is washed in the usual manner with 1 *N* HCl, 1 *N* $NaHCO_3$, and water and dried over Na_2SO_4. The solutions are kept warm during the washing and drying steps in order to prevent precipitation of product. After removal of the ethyl acetate under vacuum, the crystalline

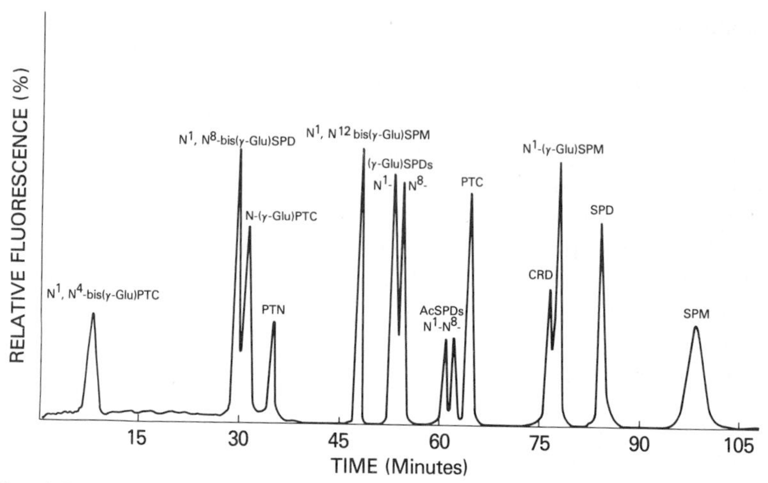

FIG. 1. Ion-exchange chromatographic separation of polyamines and polyamine derivatives. The separations are carried out on a Dionex D-400 analyzer using a column (0.4 × 8 cm) of DC-6A resin (Dionex) at buffer and reagent flow rates of 43.5 and 21.0 ml/hr, respectively. Mixing of reagent with column effluent and fluorometric detection are as previously described [J. R. Benson and P. E. Hare, *Proc. Natl. Acad. Sci. U.S.A.* **72,** 619 (1975)]. The column is maintained at 66°, and the elution program is as follows: sodium citrate buffer (0.2 *N* Na^+), pH 3.31, 12 min; sodium citrate buffer (0.2 *N* Na^+), pH 4.31, 19 min; sodium citrate buffer (0.6 *N* Na^+), pH 5.80, 17 min; sodium citrate buffer (1.5 *N* Na^+), pH 5.55, 25 min; sodium citrate buffer (3.0 *N* Na^+), pH 5.55, 30 min. All buffers are 0.2 *N* in citrate and contain 0.5% thiodiglycol, 0.1% Brij, and 0.1% phenol. Glu, glutamyl; SPD, spermidine; SPM, spermine; PTC, putrescine; PTN, putreanine; Ac, acetyl; CDR, cadaverine. The basic amino acids and ammonia elute between 28 and 35 min.

residue is recrystallized from ethanol–ether to yield 2.6 g (65%) of product, mp 166°: $C_{44}H_{50}N_4O_{10}$ (794.9); calculated: C 66.5, H 6.3, N 6.9; found: C 66.7, H 6.4, N 7.0.

Bis(γ-glutamyl)putrescine. To 0.8 g of the benzyloxycarbonyl ester in 30 ml of methanol is added 10 ml of water, a few drops of acetic acid, and palladium black catalyst. The compound is hydrogenated in the usual manner. The product (0.28 g, 80%) is obtained by crystallization from water–ethanol and is dried at 75° overnight under high vacuum, mp 230° (decomp.): $C_{14}H_{26}N_4O_6 \cdot 0.5H_2O$ (355.4); calculated: C 47.3, H 7.7, N 15.8; found: C 47.2, H 7.7, N 15.3.

Chromatography[2]

Each of the individual γ-glutamylamines, for which the preparative methods are given here, chromatographs as a single component in the ion-exchange chromatographic system described in Fig. 1. Ion-exchange chromatography of acid hydrolyzates prepared from these γ-glutamylamines shows the expected ratios of glutamic acid and amines.

The N^1- and N^8-(γ-glutamyl)spermidines, which are obtained as a mixture, are partially resolved in the ion-exchange system. They are shown to elute in the order given in the figure by collection of fractions directly from the column and analysis by the following procedure.[5] The *N*-(γ-glutamyl)spermidine isomer in each fraction is derivatized by treatment with 5-dimethylaminonaphthalene-1-sulfonyl chloride. After removal of the glutamyl group by acid hydrolysis, the free amino group of the partially derivatized polyamine is acetylated and the site of acetylation is determined by thin-layer chromatographic comparison with standard monoacetyl di(5-dimethylnaphthalene-1-sulfonyl)spermidines.

Substrates for γ-Glutamylamine Cyclotransferase[7]

Each of the γ-glutamylpolyamines, as well as bis(γ-glutamyl)putrescine, serves as a substrate for γ-glutamylamine cyclotransferase. The products of enzymic action on the mono-γ-glutamylpolyamines are pyroglutamic acid and the free polyamines. With the bis(γ-glutamyl)-amines, there is a transient formation of mono-γ-glutamylamines. The enzyme is useful in verifying the identity of γ-glutamylamines after their separation by ion-exchange chromatography.[8]

[7] M. L. Fink, S. I. Chung, and J. E. Folk, *Proc. Natl. Acad. Sci. U.S.A.* **77,** 4564 (1980); see also Ref. 8.

[8] This volume [61].

[80] Chromatographic Identification of Hypusine[N^1-(4-amino-2-hydroxybutyl)lysine] and Deoxyhypusine [N^1-(4-aminobutyl)lysine][1]

By MYUNG HEE PARK, HERBERT L. COOPER, and J. E. FOLK

Hypusine	Deoxyhypusine
NH_2	NH_2
CH_2	CH_2
CH_2	CH_2
CHOH	CH_2
CH_2	CH_2
NH	NH
CH_2	CH_2
CH_2	CH_2
CH_2	CH_2
CH_2	CH_2
$CHNH_2$	$CHNH_2$
COOH	COOH

The unusual amino acid hypusine was discovered in extracts of bovine brain by Nakajima and co-workers.[2] These investigators determined its chemical structure[2] and provided evidence that it is a constituent amino acid of animal protein.[3] We have reported that hypusine occurs in human lymphocytes[4] and in Chinese hamster ovary cells[5] predominantly in one major protein, M_r 18,000, with a relatively acidic p*I*. Furthermore, in all other eukaryotic cells examined this amino acid occurs in the same or a very similar protein.[6] The 4-amino-2-hydroxybutyl moiety of hypusine derives in part directly from the butylamine portion of the polyamine

[1] The procedures given here are taken from Şhiba *et al.*[2] and Park *et al.*[4,5]

[2] T. Shiba, H. Mizote, T. Kaneko, T. Nakajima, Y. Kakimoto, and I. Sano, *Biochim. Biophys. Acta* **244,** 523 (1971).

[3] N. Imaoka and T. Nakajima, *Biochim. Biophys. Acta* **320,** 97 (1973).

[4] M. H. Park, H. L. Cooper, and J. E. Folk, *Proc. Natl. Acad. Sci. U.S.A.* **78,** 2869 (1981).

[5] M. H. Park, H. L. Cooper, and J. E. Folk, *J. Biol. Chem.* **257,** 7217 (1982).

[6] H. L. Cooper, M. H. Park, and J. E. Folk, *Cell* **29,** 791 (1982).

ISBN 0-12-181994-9

spermidine[4] and its amino acid precursor is protein-bound lysine.[5] Biosynthesis proceeds by way of a site-specific protein modification reaction in which the lysyl residue is first converted to the transitory intermediate deoxyhypusine. This, in turn, is hydroxylated to form hypusine.[5] Although the precursor protein is actively synthesized by resting lymphocytes and a pool of this protein is maintained in these cells, the post-translational events leading to hypusine production in this protein occur only after initiation of cell growth.[6]

In this chapter we describe an ion-exchange chromatographic method for separation and determination of hypusine and deoxyhypusine. Verification of the identities of these amino acids is made by means of their 2,4-dinitrophenyl (DNP) derivatives, for which the thin-layer chromatographic properties are given. In addition, hypusine, by virtue of its vicinal amino and alcohol groups, is subject to oxidative cleavage with periodate.[2] The production of β-alanine and lysine as end products of oxidation provides further evidence for the identity of this unusual protein constituent. Reference is made to methods for the isolation of hypusine from the nonprotein and the protein fractions of bovine brain and to a chemical procedure for synthesis of deoxyhypusine.

Hypusine and deoxyhypusine are more basic amino acids than is arginine. Consequently, they are not detected by the ion-exchange systems normally employed for determination of the usual component amino acids of proteins. The procedure described here for ion-exchange chromatography of hypusine and deoxyhypusine makes use of a small column of cation-exchange resin and elution buffers of relatively high ionic strengths. DNP derivatives may be prepared using effluent fractions collected directly from the ion-exchange column. Satisfactory oxidative cleavage of hypusine, however, requires isolation of this amino acid free of chromatographic reagents.

Procedures

Ion-Exchange Chromatography.[7] Separations are conducted with the use of a Dionex D-400 analyzer equipped with a column (0.4 × 8 cm) of DC-6A resin (Dionex). The column is maintained at 66°, and buffer and reagent flow rates are 43.5 and 21.0 ml/hr, respectively. The elution program is as follows: sodium citrate buffer (0.6 N Na^+), pH 5.80, 17 min; sodium citrate buffer (1.5 N Na^+), pH 5.55, 25 min; sodium citrate buffer (3.0 N Na^+), pH 5.55, 30 min. All buffers are 0.2 N in citrate and contain

[7] The ion-exchange system described here for separation and determination of hypusine and deoxyhypusine is used routinely in our laboratory for polyamine analyses. It is basically that described by Dionex for this purpose.

TABLE I
ELUTION TIMES OF HYPUSINE, DEOXYHYPUSINE, POLYAMINES, AND SOME RELATED COMPOUNDS[8]

Compound	Elution time (sec)
Hypusine	1638
N^1-(γ-Glutamyl)spermidine	1694
N^8-(γ-Glutamyl)spermidine	1764
Deoxyhypusine	1764
N^1-Acetylspermidine	2000
N^8-Acetylspermidine	2080
Putrescine	2295
Spermidine	3320
Spermine	3880

0.5% thiodiglycol, 0.1% Brij, and 0.1% phenol. No column regeneration is needed; the column is equilibrated by passing the first buffer for 15 min. Mixing of reagent with column effluent and fluorometric detection are as described.[7] Collection of effluent for external analyses is made directly from the column.

The elution times of hypusine and deoxyhypusine are listed in Table I,[8] together with those of putrescine, the polyamines, and several derivatives of these oligoamines. All of the usual constituent amino acids of proteins, including lysine and arginine, are eluted within the first 1000 sec. Hypusine is eluted just ahead of, and overlapping, N^1-(γ-glutamyl)spermidine.[9] Deoxyhypusine chromatographs together with N^8-(γ-glutamyl)-spermidine.[11]

DNP Derivatives of Hypusine and Deoxyhypusine.[4,5] Reaction of these amino acids with 2,4-dinitrofluorobenzene is carried out in a single-phase

[8] For the elution times of a number of these compounds and several additional compounds in a similar ion-exchange chromatographic system see this volume [79]. The elution positions of hypusine and deoxyhypusine relative to N^1- and N^8-(γ-glutamyl)spermidines are the same in the two systems.

[9] For this reason the ^{3}H-labeled hypusine in proteolytic digests of the protein fraction of human lymphocytes that had been grown in the presence of [^{3}H]putrescine was initially thought to be N^1-(γ-glutamyl)spermidine.[10] It was subsequently identified as hypusine[4] following the observation that, unlike the γ-glutamylspermidines (see this volume [79]), it is stable to the conditions used for acid hydrolysis of proteins.[10]

[10] J. E. Folk, M. H. Park, S. I. Chung, J. Schrode, E. P. Lester, and H. L. Cooper *J. Biol. Chem.* **255,** 3695 (1980).

[11] Deoxyhypusine, like hypusine, is stable to the conditions used for acid hydrolysis of proteins.[5] Thus, these two unusual amino acids can easily be distinguished from the γ-glutamylspermidines.[9]

TABLE II
R_f VALUES FOR DNP DERIVATIVES OF HYPUSINE, DEOXYHYPUSINE, AND THE PRODUCTS OF OXIDATION OF HYPUSINE

Compound	R_f in solvent[a]			
	1	2	3	4
Hypusine	0.25	0.16	0.22	0
Deoxyhypusine	0.33	0.37	0.44	—
Lysine	0.39	0.44	0.58	0.08
β-Alanine	0.65	0.70	0.85	0.5
Hypusine, after oxidation	0.39, 0.65	0.44, 0.70	0.58, 0.85	0.08, 0.5

[a] Thin-layer chromatography was carried out on silica gel G (Merck) in: 1, chloroform–methanol–acetic acid (95 : 5 : 1); 2, chloroform–benzyl alcohol–acetic acid (70 : 30 : 3); 3, chloroform–*tert*-amyl alcohol–acetic acid (70 : 30 : 3); 4, benzene–pyridine–acetic acid (80 : 20 : 2).

system composed of ethanol and aqueous $NaHCO_3$ according to a published procedure.[12] In Table II are listed the R_f values for the DNP derivatives of hypusine, deoxyhypusine, and some related compounds.

Oxidative Cleavage of Hypusine.[2,4] Conversion of hypusine to β-alanine and lysine is accomplished by oxidation with HIO_4 and $KMnO_4$. Although, oxidation can be carried out on hypusine isolated by the ion-exchange procedure described, it is first necessary to obtain the amino acid free of chromatographic buffer components. The following procedure has been satisfactory in our hands. The fractions containing hypusine are combined and taken to dryness under vacuum. The residue is dissolved in a small volume of water (0.2 ml), and the solution is adjusted to pH 7–7.5 with NaOH and then made 1 *M* in NaOH by the addition of 10 *M* NaOH. A 5-μl portion of benzoyl chloride is added. After vigorous stirring, the solution is allowed to stand for 30 min, after which time a second 5-μl portion of benzoyl chloride is added with stirring. After an additional 30 min, the mixture is made 1 *M* in HCl, and the benzoylated hypusine is separated from buffer salts by extraction into ether (three extractions with 1-ml portions). The combined ether extracts are taken to dryness, and to the residue is added 0.5 ml of 6 *M* HCl. After hydrolysis for 18 hr at 105° in a sealed tube, the acid is removed under vacuum and the residue is dissolved in 0.2 ml of water. Benzoic acid is removed by extraction with ether and the free amino acid in the aqueous layer is treated with 2 mg of HIO_4. After 1 hr small portions of dilute $KMnO_4$

[12] F. J. Lucas, T. B. Shaw, and S. G. Smith, *Anal. Biochem.* **6**, 335 (1963).

solution are added until the purple color persists. Dinitrophenylation[12] may be performed directly on the oxidation mixture. The R_f values of the DNP derivatives of the amino acids formed as a result of this oxidation procedure are given in Table II and correspond to those of the DNP derivatives of lysine and β-alanine.[13]

Isolation of Hypusine and Chemical Synthesis of Deoxyhypusine. Hypusine was isolated by Shiba and co-workers from the trichloroacetic acid extracts of whole bovine brain by means of a series of ion-exchange chromatographic steps. It was obtained in crystalline form as the dihydrochloride salt, mp 234–238° (decomp.). This amino acid is more conveniently isolated in small quantity from an acid hydrolyzate of the protein fraction of bovine brain by the use of a similar series of ion-exchange steps.[3] The R_f values of hypusine in several paper chromatographic systems are given, as are those of its 5-dimethylaminonaphthalene-1-sulfonyl derivative in thin-layer chromatography.[3]

Deoxyhypusine was prepared by Shiba and co-workers through the reaction of 1-bromo-4-phthaliminobutane with N^{α}-benzyloxycarbonyl-L-lysine in slightly alkaline solution. The blocking groups were removed by acid hydrolysis, and the free amino acid was partially purified by chromatography on the NH_4^+ form of Dowex 50 ion-exchange resin. Final purification was achieved through its crystallization as the *p*-hydroxyazobenzene-*p*′-sulfonic acid salt.[14]

[13] Direct fluorometric detection of products of hypusine oxidation by the use of the ion-exchange system described here has proved to be unsatisfactory because some component of the oxidation mixture precludes visualization by fluorometry. The ion-exchange method has been employed successfully, however, for detection of radiolabeled products of hypusine oxidation.[4,5]

[14] Shiba and co-workers[2] provided analytical data as evidence of an anhydrous trisulfonate salt of deoxyhypusine and reported mp 173–188° (decomp.). We obtain deoxyhypusine as the trisulfonate dihydrate after drying for 24 hr under high vacuum at 100°, mp 160–180° (decomp.), $\gamma(^2H_2O)$; 1.3–1.9 (10 H), 2.9–3.1 (6 H), 3.6–3.8 (1 H) $C_{10}H_{23}N_3O_2 \cdot 3C_{12}H_{10}N_2O_4S \cdot 2H_2O$ (1088.2). Calculated: C 50.77, H 5.28, N 11.59, S 8.84; found: C 50.50, H 5.33, N 11.31, S 8.95. The amino acid displays the following R_f values on thin-layer chromatography on silica gel G (Merck): 0.04, 1-butanol–acetic acid–H_2O (4 : 1 : 1); 0.08, chloroform–methanol–17% NH_4OH (2 : 2 : 1).

[81] Isolation of S-Adenosyl-3-thiopropylamine

By SHOSUKE ITO

S-Adenosyl-3-thiopropylamine may be structurally regarded as a product of decarboxylation of S-adenosylhomocysteine or of demethylation (transmethylation) of S-adenosyl-3-methylthiopropylamine (decarboxylated S-adenosylmethionine). The compound has been found only in the eye of the sea catfish (*Arius felis* L.).[1] Chemical synthesis of the compound was first described by Jamieson[2] and was successfully repeated in other laboratories.[3-5] The compound may also be prepared from adenosine and 3-mercaptopropylamine by the action of a condensing enzyme from rat liver.[6]

Isolation Procedure

The following procedure allows the isolation not only of S-adenosyl-3-thiopropylamine, but also of S-adenosyl-3-methylthiopropylamine.[7] Since these compounds are unstable in either strongly acidic or alkaline solution, and the eyes of the catfish contain other basic compounds,[1,7] a chromatographic procedure was adopted that permitted the separation of these basic compounds at near neutral pH.

Extraction. Forty eyes (wet weight 14.8 g) of the sea catfish (*Arius felis* L.) were homogenized in 60 ml of 0.5 *M* H_2SO_4 using a Waring blender and centrifuged at 10,000 *g* for 10 min. The precipitate was again homogenized in 30 ml of 0.5 *M* H_2SO_4 and centrifuged. The combined extract was brought to pH 4 by adding a freshly prepared suspension of 0.5 *M* $BaCO_3$[8] with stirring. The resulting precipitate of $BaSO_4$ was filtered off and washed with water. The filtrate and washing were combined and again adjusted to pH 3 with 0.5 *M* H_2SO_4. The mixture was kept overnight at 2° to ensure the precipitation of $BaSO_4$ and filtered.[9] The

[1] S. Ito and J. A. C. Nicol, *Biochem. J.* **153,** 567 (1976).

[2] G. A. Jamieson, *J. Org. Chem.* **28,** 2397 (1963).

[3] R. T. Borchard and Y. S. Wu, *J. Med. Chem.* **17,** 862 (1974).

[4] K. Samejima, Y. Nakazawa, and I. Matsunaga, *Chem. Pharm. Bull.* **26,** 1480 (1978).

[5] See also this volume [8] for the separation of other adenosyl sulfur compounds.

[6] J. A. Duerre, *Arch. Biochem. Biophys.* **96,** 70 (1962).

[7] S. Ito and J. A. C. Nicol, *Proc. R. Soc. London, Ser. B* **190,** 33 (1975).

[8] Prepared by adding pieces of Dry Ice to a solution of 0.5 *M* $Ba(OH)_2$ until neutral. A commercial preparation of $BaCO_3$ could not serve the purpose.

[9] Although tedious, these steps were necessary for complete removal of $BaSO_4$ and other insoluble materials that were found to interfere with the CM-Sephadex chromatography.

ISBN 0-12-181994-9

filtrate was passed through a column (1.9 × 3.0 cm) of Dowex 1-X8 resin (100–200 mesh; acetate form) in order to remove residual H_2SO_4 and other acidic compounds. The column was washed with 20 ml of water. The effluent and washing were combined and concentrated to about 20 ml in a rotary evaporator at 40°. The insoluble material was removed by centrifugation at 10,000 *g* for 10 min, and the supernatant was evaporated to dryness at 40°. The residue was taken up in 3 ml of 0.5 *M* ammonium formate, and the solution was kept overnight at 2° and cleared of precipitate by centrifugation.[8]

CM-Sephadex Chromatography. The clear supernatant was applied to a column (1.05 × 40 cm) of CM-Sephadex C-25, equilibrated with 0.5 *M* ammonium formate, pH 6.5. The column was eluted with the same buffer at a flow rate of 24 ml/hr; fractions of 3.8 ml were collected, and their absorbances at 260 nm were measured. Fractions 16–21 showing spectral features of adenosine were combined and evaporated to dryness at 40°, and the remaining ammonium formate was removed by sublimation under high vacuum at 50°. The residue was dissolved in 2 ml of 1 *M* formic acid and applied to a column (1.0 × 27 cm) of CM-Sephadex C-25, equilibrated with 1 *M* formic acid. The column was eluted with the same solvent; fractions of 5.5 ml were collected and monitored by their absorbances at 260 nm. Fractions 19–24 containing the adenosyl compound were evaporated to near-dryness at 40°, transferred to a small centrifuge tube, and evaporated with a stream of nitrogen. The residue was crystallized by adding 0.1 ml of 0.5 *M* H_2SO_4 and 2 ml of ethanol. The crystals were collected by centrifugation and washed with ethanol. *S*-Adenosyl-3-thiopropylamine bisulfate was obtained as needles in a yield of 1.6 mg (3.6 μmol).

To isolate *S*-adenosyl-3-methylthiopropylamine, fractions 44–55 of the first CM-Sephadex chromatography were further purified by the second chromatography on CM-Sephadex equilibrated with 1 *M* formic acid, as described above. The yield of *S*-adenosyl-3-methylthiopropylamine was 9.6 μmol, as determined by ultraviolet spectrophotometry.

Properties

The chromatographic and electrophoretic properties of *S*-adenosyl-3-thiopropylamine were compared with those of related *S*-adenosyl compounds.[1] The compound exhibits ultraviolet absorption spectra almost identical with those of adenosine. The infrared spectrum has been published.[1] The compound is stable in alkaline solution, but rather unstable in acidic solution; heating in 1 *N* HCl at 100° for 10 min resulted in complete decomposition yielding adenine and *S*-ribosyl-3-thiopropylamine.[1,6]

Author Index

Numbers in parentheses are footnote reference numbers and indicate that an author's work is referred to although his name is not cited in the text.

A

B

C

E

F

G

H

I

J

K

N

O

P

T

U

V

W

Y

Z

Subject Index

A

B

C

D

F

G

H

I

L

M

P

Q

R

S

T

U

V

Y